The Delight of Thinking

The Delight of Thinking

*The Life of Tatiana Afanassjewa
and Paul Ehrenfest*

MARGRIET VAN DER HEIJDEN
Eindhoven University of Technology, Eindhoven, Netherlands (the)

Translated by

BRENDAN MONAGHAN

OXFORD
UNIVERSITY PRESS

Great Clarendon Street, Oxford, OX2 6DP,
United Kingdom

Oxford University Press is a department of the University of Oxford.
It furthers the University's objective of excellence in research, scholarship,
and education by publishing worldwide. Oxford is a registered trade mark of
Oxford University Press in the UK and in certain other countries.

Copyright © 2021 by Margriet van der Heijden

Originally published in 2021 by Uitgeverij Prometheus, Amsterdam.

First published in English in 2025 by Oxford University Press.

The publisher gratefully acknowledges the support of the Dutch Foundation for Literature.

Nederlands letterenfonds
dutch foundation
for literature

© Margriet van der Heijden 2021

The moral rights of the authors have been asserted.

All rights reserved. No part of this publication may be reproduced, stored in a retrieval system, transmitted, used for text and data mining, or used for training artificial intelligence, in any form or by any means, without the prior permission in writing of Oxford University Press, or as expressly permitted by law, by licence or under terms agreed with the appropriate reprographics rights organization. Enquiries concerning reproduction outside the scope of the above should be sent to the Rights Department, Oxford University Press, at the address above.

You must not circulate this work in any other form
and you must impose this same condition on any acquirer.

Links to third party websites are provided by Oxford in good faith and
for information only. Oxford disclaims any responsibility for the materials
contained in any third party website referenced in this work.

Published in the United States of America by Oxford University Press
198 Madison Avenue, New York, NY 10016, United States of America

British Library Cataloguing in Publication Data

Data available

Library of Congress Control Number: 2025946177

ISBN 9780198927099

DOI: 10.1093/9780198927112.001.0001

The manufacturer's authorized representative in the EU for product safety is
Oxford University Press España S.A. of Parque Empresarial San Fernando de Henares,
Avenida de Castilla, 2 – 28830 Madrid (www.oup.es/en or product.safety@oup.com).
OUP España S.A. also acts as importer into Spain of products made by the manufacturer.

Contents

Author's Note   vii

PART 1

1. You shall see the ocean later: Growing up with older brothers   3
2. Not moonlight but rather sunlight: Well brought up in St Petersburg   13
3. Boltzmann and Columbus: Voyages of discovery in physics   21
4. Sailing the Sahara: Raised and constrained in St Petersburg   27
5. Between gods and humans: The Mecca of science   34
6. Don't ever smoke: A shared passion kindles love   41
7. Gleaning knowledge: Leiden canals, Russian novels and Göttingen mathematics   47
8. Aversion and love: Seeking a sense of self   52
9. Ill at ease in Vienna: Study, discussions and strolling with a pram   56
10. First publications: Unemployed but working   66
11. Happy in Göttingen: Travelling, thinking, writing   74
12. Dangerous experiments: A Russian beard and feeling at home   83
13. A sombre honorary secretary: Arduous success in Russia   92
14. Rest, cleanliness and regularity: A daughter and a spa   100
15. It could have been so wonderful here: Inaccessible German-speaking universities   109
16. A big, dear boy: Meeting Einstein   118
17. Coincidence and new opportunities: Farewell to St Petersburg   126

PART 2

18. The man in the empty sphere: A flying start in Leiden   139
19. Water to the left, water to the right and water in between: Cosmopolitans in a provincial town   145
20. War and friends: Leiden 1914   154

21. Oasis in an ugly world: Work, war and Witte Rozenstraat	162
22. The red professor and the Russian princess: Happiness in wartime	172
23. Ups and downs: Bankruptcy, peace and another son	181
24. Physics curator: Bohr as Rembrandt and Einstein as Holbein	190
25. Physics at the highest level: Just not for women	202
26. Einstein, Wassik and Russians: Concerns far and wide	211
27. Second rate and second class: Capitalism and socialism, men and women	222
28. Two worlds: American girls and 'male logic'	230
29. Warm people and thermodynamics: Paul's network and Tanya's science	240
30. Physics curator 2.0: Passionately interested in people's fate	250

PART 3

31. Falling behind: Mathematics in Moscow and mathematics as a blight	261
32. Escapism: Radios and travelling	271
33. Love is no longer enough: Nelly, Russia and the Nazis	283
34. Broken: A hopeless deed in a hopeless world	294

PART 4

35. Loss after loss: If only we could just be eyes	303
36. Finding joy in less: Writing and thinking to the end	311
Epilogue	321
Acknowledgements	329
Notes	331
Bibliography	375
List of abbreviations	393
Index	394

Author's Note

A more than sizeable archive, two eventful and active lives: trying to extract their essence requires making choices that are obvious, but not always. Sometimes over almost finicky matters. Like, do I keep the old classification of the Ehrenfest archive at Rijksmuseum Boerhaave, or do I use the new one that was introduced after I had been working on this project for a while? The new one, of course; the main characters in my book would also agree: one should remain as scientifically up-to-date as possible.

This brings me straight to choices related to the protagonists. It is customary to refer to the protagonist of a biography by his or her surname. Using the first name can easily sound too familiar and suggests that the biographer has taken too little distance from his or her subject. Admittedly, it also took me a while to get used to seeing the first names Paul and Tanya in the same sentences as surnames like Ioffe, Burgers or Einstein. But it was a deliberate choice. An important reason was that I regularly write about the relationship between these two people and draw on their correspondence, which would make the use of their surnames somewhat absurd. Sentences like 'Ehrenfest wrote to Ehrenfest' could also be confusing for the reader.

An additional issue is that of women's surnames. Before marrying Paul Ehrenfest, Tanya's surname was Afanasyeva—spelled here according to the customary English-language transcription of her Russian name. After she married, she initially used the name Ehrenfest, including in their joint and her own publications. But in Leiden, she used Ehrenfest as a surname in her personal life and Ehrenfest-Afanassjewa for her publications (and in her passport)—with her name now thus spelled according to the German transliteration. And sometimes, as in her pamphlet on mathematics didactics, this was misspelled Afanaassjewa. This illustrates why the contributions of women, and of scholars from languages other than those dominant in science, are often difficult to trace. And this again demanded choices: using first names in the main body of text circumvents this problem; in the endnotes and bibliography, I follow her own use of names.

Hans Driessen, a Dutch Slavist with a long and impressive track record, helped me with the Dutch version of this book by deciphering the names of Russian scholars in letters and translating numerous (mostly handwritten) Russian texts—in letters, diaries, manuscripts and reports—for me. For the English translation, we chose to translate the Dutch translation into English. And although the original source texts of the remaining quotations are shown in the endnotes, it also took it too far to show the original Cyrillic in these notes. So here the reader has to rely on collective expertise, not in the least also of Brendan Monaghan who rendered the Dutch into English.

Then something else about the many languages spoken by the Ehrenfest family and in their correspondence. Paul Ehrenfest could play wonderfully with language; it is clear from the letters and also from anecdotes about him. Even in Russian, which he learned fluently (though certainly not flawlessly) in St Petersburg, he played on words and made puns. But when he wrote, he almost invariably resorted to German. His Dutch students, such as Rutgers and Goudsmit, received long letters from him in German, and although Lorentz always wrote to him in Dutch, Paul usually replied in his mother tongue.

Paul also wrote to Tanya mostly in German, with the exception of short notes on cards and short passages in Russian in letters. Tanya wrote to Paul in German at first, and later increasingly in Russian, especially when she spent long and frequent periods in the Soviet Union. In addition, manuscripts from the 1930s show that she had a very good command of French. She could read English, but when she spoke it, according to family tradition, no one could understand her. Here Paul cut a better figure. As for the Ehrenfest children, they could even switch from Dutch to German to Russian in a single letter with ease. Galinka also learned to speak English very well during her long journey through the United States, and Pawlik also mastered French during his years of study in Paris. As mentioned above, the passages used in the book are reproduced in the original language in the endnotes where necessary.

Finally, another choice: in how much detail should the scientific work be discussed? Apart from the contributions to the debates on the didactics of mathematics, Ehrenfest and Afanassjewa's work consisted of subtle contributions and side notes to complex and often subtle problems in physics—specifically in statistical mechanics, thermodynamics, special relativity, quantum theory and quantum mechanics. Such work can't be well summarised without becoming technical or specialised, while a longer and more comprehensive explanation that does justice to all the subtleties would have turned this book into a textbook. I have chosen to summarise the scientific theories and the relevant difficulties in broad outline, and then situate their work in them. Non-physicists or non-mathematicians who find this overwhelming can probably skip the abstract sections. Enthusiasts will find references to articles and books in the text and in the endnotes that provide more background.

PART 1

1
You shall see the ocean later
Growing up with older brothers

Emil swung his legs over the windowsill. Arthur and Hugo climbed through the window after him, and finally Otto also stepped carefully onto the zinc roof. Beneath their feet the dark storeroom smelt of soap, and the stacked sugar loaves, sacks of coffee beans, tins of petroleum and limewater eggs were stored in. All these wares were sold in Sigmund and Johanna Ehrenfest-Jellinek's grocery, below the sitting room where Paul stood on his own at the window.

He saw how his brothers clambered over the slanting storehouse roof. Emil, who was already twenty and worked in the shop, was on a break. Arthur, the eldest at twenty-two, had come home late in the afternoon from the *Technische Hochschule* (now the Vienna University of Technology) in central Vienna. Hugo was fifteen and had had Latin and history lessons at his *Gymnasium* (grammar school), and thirteen-year-old Otto had spent the whole day at the *Realschule*, a secondary school focused on technical subjects and science. At the peak of the roof, they could just sit in the evening sun.

Without Paul: he was only five and too young to climb out the window. 'What do you see?' he asked, standing next to the windowsill. With Minna, his Czech nanny, he had watered the fuchsias and begonias in the hall that morning. Minna had darned socks, and they had been for a walk. First down 'their' Himberger Strasse, past the adjacent Weniger butchery, then farther northwards and back through side streets, where new houses were still being built in some places.[1]

The Viennese district Favoriten, where Paul and his family lived, was expanding at a meteoric pace. When Paul's newly wed parents moved there from Moravia in 1860, it had approximately 20,000 inhabitants; half a century later, there were eight times as many people. Most were labourers from Moravia, Bohemia and Italian-speaking parts of the empire, who toiled for long hours at the soap works, the tar factory and the tile factory.[2] In Paul's parents' shop, their families bought their everyday necessities—flour, peas, rice, lard, potatoes, plum jam, soap, brushes. For a reasonable price, as Paul's parents bought the products—never more than two kinds of each article—in bulk directly from producers.

Their 'supermarket formula' was a success. Even middle-class families from more central parts of Vienna sometimes sent servants to Favoriten to buy cheap supplies. The horse tram from the centre stopped almost in front of the shop.

That afternoon Paul and Minna had heard the clattering hooves of the horses drawing the wooden tram car along the rails.

'What do you see?' he asked again. He heard his brothers laughing and talking. Their world was so much bigger than his, with the house, the shop and Minna, so full of things Paul knew nothing of.

Emil looked affably at him. 'Do you know what you see here on the roof?'

'The sky?' Paul asked. 'The courtyard?'

'That too', Emil said, 'but in the distance, really far away, you can see the ocean'.

The ocean? With waves as far as you can see? Paul had occasionally seen photos of it in magazines. At the other side of this gigantic mass of water were other countries, he had been told.

'Isn't the ocean too far away?' he asked, looking tentatively at his eldest brother.

'It's far away', Arthur replied, 'but we can see it. I think I even see froth on the waves. It looks like a storm'.

'I think so too', Hugo agreed. 'Look at the white horses'.

'And a ship', Otto exclaimed. 'I can see its masts'.

Paul hesitated and craned his neck. 'No', Emil said. 'You're too low. You have to look over that high roof over there'.

Paul looked attentively at his brothers; all four were staring at the ocean. The silent room behind him smelled of floor polish. Vaguely he heard a hubbub in the shop. He could have climbed onto the flat roof of his neighbour Weniger's storehouse, which was also in the courtyard of their semi-detached house. But his brothers insisted that even there he wouldn't see anything. He looked below. The zinc roof had taken on a whitish hue. The paint on the windowsill was peeling. If Paul were to sit on it, his feet would dangle high above the zinc, and then he would have to jump, which he didn't dare. 'Can't I come up for just a moment?'

'No, you know it's too dangerous', Emil said. 'Perhaps after the summer when you've grown some more. Honest, you'll get to see the ocean'.

In bed Paul thought about the storm raging across the ocean while the sun warmed the roof in Vienna. It was only much later that he realised that his brothers had pulled the wool over his eyes.

Paul was a latecomer in the house on Himberger Strasse. He certainly didn't understand everything he was told, but he soaked up odours, colours and sounds like a sponge. The domestic sphere was a world in itself. Not only his brother Emil helped in the shop; there were also four servants (Aloys, Martin, Moritz and David) and two shop-girls (Milly and Emma), while above the shop a cook and a maidservant ran the household. All these people, including Minna, slept in the cramped house at night, Paul in his parents' bedroom.

Early in the morning when the others started work, he sat with Minna, who crocheted or mended clothes. He told her stories he made up and listened to Czech songs she sang. At quiet times of the day, he was allowed to help downstairs in the shop. He felt important when he gave customers things—a bar of soap, a brush, a bag of coffee beans his father had weighed. Even better was when he sat on his father's lap in the shop window. This gave him a direct view of the bottom of the glass display case, which was filled with merchandise. When people on the street outside walked past the shop window, he saw their legs transmute in the reflective glass into a second pair walking in tandem. The best was when a little dog trotted by, with a total of eight legs. But mostly it was too busy for this. When the factory workers were paid on Saturday evening, a long queue formed in front of the shop. Then his brothers lent a hand wrapping articles. Everyone laughed and talked at once, long rows of numbers were scribbled on strips of paper and his mother dealt with the payments in her glass booth. She knew who to give extra credit as well as almost everyone's name.

When it was quieter, Paul liked to listen into the conversations in her booth, messing about with the stamp his mother used to put the date under bills and admiring the telephone which was connected to the home upstairs. But on such busy evenings, he had to keep his head down, especially when it was late, and his mother was tired. Decades later he still vividly remembered how one evening he had lingered too long by the door of the booth, with his fingers clasped around the doorpost, while his mother added up long lists of numbers. She had warned him several times already. 'Oh, it hurt so much', Paul wrote years later. 'I'd never thought she'd actually close the door while my finger was still on the doorpost'.[3]

His father was far more protective. Paul often sat with Sigmund while he ate the lunch especially prepared for his sensitive stomach. Then his father gave him a piece of shortbread or a vanilla or jam puff, which he kept in the closet for Paul. As the week went by, these Sunday biscuits tasted more and more like mothballs, but 'how nice it was to have this privilege over my brothers'.

The ultimate was when his father entrusted Paul with a job in the family business. When a sugar delivery arrived, for example, he was allowed to count the sugar loaves that were packaged in deep-grey and blue cardboard. Long beforehand he waited in the shop window for the sugar cart, and he went outside with his father and brothers before the waggon had stopped. While the horses panted with fodder bags around their necks, Paul's brothers and the shop staff formed a winding chain, and the sugar loaves flew from the cart, through the shop, to the storehouse. These oblongs were about a metre long, and their rounded ends were immediately struck off, to be pulverised later—'the air smelled so sweet'.[4]

Paul, on a chair next to the shop door, drew a pencil line on a sheet of paper for every loaf. He was proud when all the sugar was in the storehouse and his count— almost two hundred or even three hundred—equalled his father's tally at the cart. Then they took a stack of white tissue paper upstairs. Johanna always removed them from inside the grey cardboard around the loaves, smoothed them out and cut them into straight pieces. The paper was so fibrous that you could only write

6   THE DELIGHT OF THINKING

Figure 1.1 Photograph of Paul Ehrenfest with the text 'Paul in 1889, aged nine'.—*Photograph Ehrenfest Family Archive.*

on it with a soft pencil, even then only with fluid motions, but Paul liked drawing and 'writing' on it. This brought him closer to his brothers, with their homework and textbooks.

Paul wanted to follow in his four older brothers' footsteps (see Figure 1.1). He was shy, but also proud when, as a six-year-old, he entered the classroom at the *Volksschule* on Keplerplatz, a straw hat on his dark head of hair. His mother had taught him how to read, write and do simple arithmetic, so he was allowed to go straight into the second year. The children burst into laughter when he walked to his desk: he hadn't taken off his straw hat! Even aged fifty, he could feel how his cheeks flushed. Falling flat on your face wasn't something a sensitive and ambitious child like Paul, who strove for straight As and wanted to be as brilliant as his brother Arthur, took lightly.[5]

Arthur was the eldest of the four brothers. He made his own crystal models out of cardboard and paint and built a camera. From university practicals he had taken carbon–zinc batteries and electrical doorbells to connect to the batteries. Sometimes he used a lens to project the oil lamp on the wall or played around with a telegraph key, an electromagnetic device used by a telegraph operator to encode and transmit text messages in Morse code. Awed, Paul would look on while Arthur described such phenomena at the dining-room table with formulae that were even more mysterious than Morse code.

Arthur and his brothers taught Paul how to draw floor plans—which he threw himself into with such dedication that he neglected his domestic chores and was beaten. This didn't stop him from diving into something else his brothers told him about. By his own account, it happened so naturally that he couldn't say when he was first introduced to the energy principle or for example how an electric clock worked any more than when he learnt the words 'bird' or 'house'. It was as though he 'had always known' them, he wrote. And when Hugo, who was learning the classics at his Gymnasium, declared that Socrates was the greatest of all wise men, Paul posed just one question: 'Wiser than Arthur?'[6]

This inevitably resulted in practical jokes.

'What's that?' Paul asked Arthur one day when he came home tired and put a device in the sitting-room cabinet.

It was a small, enamelled black box with a handle, which Paul wanted to hold. 'Is it a little telephone?' he asked.

'You're quick off the mark', Arthur replied, taking the object back out.

'Then where's the cable?' Paul asked pedantically. Earlier, Arthur had explained to him how their mother could call from the shop to the bedroom upstairs or vice versa. Thus, Paul knew that sound, transformed into an electric signal, was carried through a long cable connecting the two devices. This enabled the sound to cross large distances without fading or being dissipated by walls.

But this new device was different, Arthur explained. It could transport the sound of a voice even without a cable. 'That's the amazing thing about it. This invention is a wireless telephone'.

'You're joking', Paul said uncertainly.

'Certainly not', Arthur replied. 'You know what, I'll show you. I'll call Emil and give him a message. Then you can go and ask him if he got it'.

Arthur held the device to his face, pulled the handle and called Emil downstairs in the shop. 'Emil, are you there? Were you snoozing? Listen, I'm going to give you a message for Paul. He'll ask you in a bit whether you heard it. Alright? Can you hear me clearly? Tell him, without making any mistakes: ...'

Paul was already scampering off. Through the two rooms to the winding stone staircase, across the courtyard, through the storehouse and the shop to Emil, who much to his astonishment repeated exactly what Arthur had said. It was years before he realised that Arthur had had plenty of time to call Emil on the 'ordinary' telephone while Paul scurried down the stairs on his little legs.

Johanna and Sigmund Ehrenfest had worked their way up after moving from their native village Loschitz, present-day Loštice, to Vienna in 1860. This village, situated in the rolling countryside of Moravia, with babbling brooks and with a square white synagogue, offered a poor weaver like Sigmund few prospects. In Vienna, they escaped from the strictures of this small and probably somewhat oppressive rural community. They had started their grocery in the capital. Not among the Jewish families in Leopoldstadt or Alsergrund, where large Jewish communities lived and Yiddish could be heard everywhere, but in Favoriten, where almost everyone was Roman Catholic (Goldhammer 1927, 10).[7] Not wishing to stand out too much from their customers they had assimilated, and they had greatly increased the small dowry that Johanna had been given as a sixteen-year-old bride.

Their lives were busy in this bustling neighbourhood, in an overcrowded house and working six long days a week, from six o'clock in the morning to ten o'clock at night. But their five sons went to excellent schools. In the sitting room, they had piano lessons on the small grand piano from Ludmilla Gross, who said her father had been friends with Beethoven. What is more, they had a bookcase—filled with cheap editions, but all the same. Paul devoured *Gulliver's Travels*, *Robinson Crusoe*, *Around the World in Eighty Days*, *Grimm's Fairy Tales* and Wilhelm Busch's illustrated stories, which were published every two weeks in the *Münchener Bilderbogen*. Minna also often told him new versions of *Aladdin and the Wonderful Lamp*. The sapphires, emeralds and diamonds in the trees in the garden and in front of the windows of the Persian palace sparkled in his head.

This contrasted sharply with life at home, as Sigmund and Johanna had remained frugal. They thought Paul, born in their prosperous years, was perfectly fine with his brothers' hand-me-down pencils. No new school supplies, colour pencils or shiny pencil case for him. And the very plain meals were an ordeal for Paul. Later he wrote that he had 'detested' the food: in his memory the rice porridge and other porridges were invariably burnt, the meat was tough and the soups watery. Beans were cooked with sugar, salt, cinnamon, some herbs and garlic— Paul loathed them—and rice was cooked with an onion and then eaten with sugar. Paul often only finished his plate because Arthur, who sat next to him at the table, threatened never to explain anything to him again otherwise.

How was Johanna supposed to find the time to organise complicated meals? She depended on her cook and had abandoned Kosher dietary practices. The same went for the sabbath, the day on which the family worked overtime in the shop. Paul's only acquaintance with Jewish customs was on Sundays when the whole family went to Sophie Jellinek, his stately maternal grandmother, who had moved to Vienna after her husband's death. She lived on the Raaber-Bahn-Gasse with her son Fritz and her two much younger daughters, Mathilde and Josephine.

Her sitting room was full on these Sundays. Paul's aunt Lina usually came too, with her husband Emmanuel Friedmann and their three children. They sang and danced with Paul's brothers, played piano and put on plays. Paul, the youngest, played with a box of little clocks and watches which he could take apart and reassemble. And when he later asked why his grandmother lived by different precepts to them at home, he was told to 'shut his mouth'.

Confusingly, it was in the streets that Paul was most often conscious of being a Jewish boy. Going out on his own was forbidden for a long time. As a Jewish boy, he was too often jeered at—by children and adults—and sometimes threatened or had stones thrown at him. Even the girl next door, the butcher's daughter Bertha, with whom he sometimes furtively gawked at the butchered pigs in her father's barn or roamed the streets for a bit, called him a 'Jewish swine' when they argued. Decades later, Paul still recalled how fruit sellers tried to give him inferior apples and pears, and how they swore when he stubbornly gazed at the better produce.

Did this reinforce Paul's fits of devotion, which the rest of the family struggled to understand? Dismayed, they saw his total absorption in Old Testament stories and how after visits to their Orthodox Jewish friends, the Hoffmann family, he was deeply impressed by the songs and ceremonies. They teased him for his questions about angels, occasioned by a golden Christmas angel that he had seen dangling from a rubber band. For a while, Paul wrote later, he had even believed that 'heaven with its angels' was on the other side of the street: in the attic with the small shutters beneath the roof trusses above Mrs Schumann's flat. 'That is why I loved visiting her so much. You were close to heaven, and she always gave me delicious biscuits and pastries'.[8] As usual, it took him some time to understand how things really were: in answer to his question where God and the angels lived, someone had probably pointed mockingly to the sky above the roof.

At school, Paul's interest in angels was replaced by a fascination with the Roman Catholic faith: the exquisitely illustrated Bible, the hymns, the prayers. Even though the slowly growing Jewish community in Favoriten had a Mikva by this time, the neighbourhood was and remained predominantly Roman Catholic, as did his school. Sometimes, Paul snuck into the church when the other children had a morning service before school. 'The priest didn't seem to object, and so I just sat there'.

Almost every day, he wrote later, he also peered through the shop window of the undertaker at the ivory Christ on a dark wooden cross. 'How dearly I would have liked to have bought it!'[9] He crafted a little cross for himself from two small pieces of wood and the silver and gold decorations on a toy soldier, which he wore 'day and night', hidden 'under my shirt', because he didn't want anyone at home to see it.[10] Until his mother inevitably discovered it, one night just before he fell asleep, and confiscated it without a word.

Did she regard the incident as one of Paul's 'quirks' which so tried her patience? With all the prosperity life had brought her, Johanna must have had a sense of time slipping through her fingers, like the sugar, beans, coffee and peas for her customers. Her long workdays left no space for boy who cried from frustrated ambition, had nightmares when his imagination ran wild and suffered nosebleeds when overexcited. 'Mother's pride: Arthur; mother's favourite: Hugo', Paul later noted in brackets in his written recollections.[11]

Even when they escaped the heat of the Viennese summer in cooler and calmer Brunn am Gebirge, he felt inferior. Together with the Friedmanns (Johanna's sister's family), their friends the Koch family, servants, cooks and piano teacher Ludmilla Gross, the Ehrenfests regularly rented a 'summer house' in this village, 20 kilometres south-west of Vienna. Clothing, toiletries, mattresses, furniture and even a piano were transported ahead of time on a large cart. Paul and his mother followed the jam-packed waggon by train. The summer houses weren't up to much, Paul concluded as an adult. Usually, they had a house on the long high street: two stories, a sandy courtyard, at the back of which was an entrance to two lower, subsequently built wings. This is where 'the paying little man was tucked away' in rooms that were 'dark, damp, with small windows'.[12] Although with any luck, there was a garden with flowers and bushes, and a view of the fields.

At a rickety garden table in the bright sun and 'next to mum in all her strictness', Paul practised doing sums and writing words.[13] He also studied the company that, apart from the occasional outing or swim, sat in the garden. The mothers played cards (Paul parroted his father, calling such activities a regrettable waste of time) and listened to the children playing the piano with Ludmilla Gross—'from Chopin to operettas'. Lina Friedmann sang the praises of her three beautiful daughters, who had French governesses; Paul's mother boasted about her brilliant eldest son Arthur, who sometimes came over in the weekend; and Mrs Koch, who lived with her three sons and a daughter a few streets away in Favoriten, derived status from her husband's profession, a doctor.

Much later, Paul recalled an occasion when, singing together one evening, he suddenly lost his voice upon hearing how beautifully one of the girls sang. Specifically, how much more beautifully she sang than he did. The incident caused such an intense clash with his mother that only his father could calm them down during the weekend. Similarly, conflicts between the women, who bickered and argued, slowly resolved over the weekends. Sometimes, they wouldn't speak to each other for days. They were 'well-intentioned and helpful, but a lack of education and development meant they lacked competence in the techniques of maintaining good relations with neighbours', Paul subsequently observed.[14] Most annoying of all was the authoritarian landlord's attempts to exploit these disputes—'tyranny', Paul later concluded. So, when these weeks came to an end, it was also pleasant to return to the crowded, familiar little world on Himberger Strasse in Vienna.

But just as Paul thought he could finally secure a more significant place in this little world, when he started secondary school, so he could come home with more interesting stories and seriously participate in the family business, it fell apart. Sigmund and Johanna had had enough of the hard work. Their four eldest sons could fend for themselves: Arthur had graduated as an engineer, Hugo as a doctor, Otto as an electrical engineer and Emil worked in the shop. It was time to sell the firm, which was worth a considerable sum by then. Sigmund invested in a more tranquil business: a pawn broker on the well-heeled Kärntner Strasse in the centre of Vienna. Around the corner, at Singerstrasse 2, they rented a house.

Sigmund and Johanna, from the impoverished Moravian village Loštice, could be proud of their achievements. The Kärntner Strasse and Singerstrasse were in a chic central neighbourhood, a stone's throw from Saint Stephen's Cathedral and within the Ringstrasse boulevard, along which grand public buildings were erected in Renaissance, gothic and neoclassical styles. In the evening, smartly attired Viennese promenaded up the Kärntner Strasse to the Ringstrasse and back down to exchange news and gossip. In this neighbourhood, Austrian nobility and notables gradually intermingled with members of the new bourgeoisie. These included rich Sudeten Germans, who had reached the upper echelons of the civil service, industry or academia, as well as Jewish families who, like Sigmund and Johanna, had profited from finally being granted equal rights when the constitution was changed in 1867 but still mostly worked in trade, banking, medicine and academia, which traditionally they hadn't been excluded from (Beller 1991, 33–42).

There was also another reason for their decision. Johanna had developed breast cancer, which was almost untreatable at the time. She died in early 1892, two months after her younger sister Line, when Paul was in his second year at the Akademische Gymnasium on Beethovenplatz.[15] Was he thinking about this when, much later and a father himself, he wrote: 'All things considered, the sick are better off quietly disappearing somewhere in the south, basking in the sun, rather than squandering the rest of their lives in the north "in their family's bosom", burdening their children. ... It is absolutely unacceptable that ... the cheerful personal development of children is stifled or fundamentally damaged by illness of their parents'.[16]

Making matters worse, Arthur, Paul's hero, married Regine Egger and moved to Berlin one-and-a-half months after Johanna's death.[17] Who could he talk to about the latest inventions? Or about the books he had read or should read? Gone was the bustle of the shop. Gone were his brothers' and the servants' hubbub and jokes. Who was still interested in his reports, progress and news?

With growing concern, Sigmund watched his youngest son sink into despair. Two years after Johanna's death, he married Johanna's youngest sister Josephine, or Pepperl, with whom he travelled to elegant Baden.[18] But even jolly Pepperl, who at twenty-six was younger than Paul's two eldest brothers, couldn't cheer Paul up. Renouncing religion, he replaced the 'sham and lies' with 'thinking things through', he later wrote. He 'took great pleasure, if it arose, pointing out to everyone around me as sharply and harshly as possible all the nonsense of popular religious beliefs'.[19] A single word summed up this period for him: 'ghastly'.

His academic performance was almost as precarious as his spirits. The Akademische Gymnasium wasn't at all to his liking. The school was only for boys, and the education was based on humanist ideas. Almost half the pupils were Jewish (Beller 1991, 53). Its teachers declared that they set great store by personal development and social skills as well as aiming to cultivate highly educated, well-rounded citizens, schooled in the classics, who could effortlessly take their place in the upper social strata—in the spirit of a liberal education, known in German as the *Bildungsbürgertum*. But Paul, who had already learnt so much from Arthur and Hugo, certainly didn't think highly of all his subjects and teachers. He also struggled to deal with the anti-Semitic classical-language teachers.[20]

By then the optimism engendered by the new laws that should have granted Jews equality and freedom had ebbed. Anti-Semitism was a cornerstone of the Christian Social Party, founded in 1893. In the ensuing years its leader, Karl Lueger, made anti-Semitism almost commonplace thanks to the tone of his ultimately successful mayoral campaigns. It gave the Viennese carte blanche to insult and to discriminate against Jews: crudely and directly in working-class districts like Favoriten and subtler in more refined quarters, like around the Kärntner Strasse, and evidently in the classroom too (Wistrich 1987).

Paul read Nietzsche and questioned everything. His reports describe his behaviour as barely 'adequate', his effort was 'variable'; he thought Latin and Greek were pointless subjects; and he had to repeat his fourth year.[21] As Paul's guardian, Arthur had to do his utmost to ensure that Paul didn't throw in the towel. Years later Arthur and his brothers were still amazed by what they called Paul's impossible emotional volatility.

What gave him just enough energy to keep his other subjects up to scratch was his interest in philosophy and the extra mathematics and physics books a kind teacher had given him. The clear reasoning in mathematics and physics were a solace to him. His life seemed chaotic and not very fair in this period. On top of everything, in 1896 he also lost his father, who had wasted away from a stomach ulcer.[22]

Paul's salvation was a chance encounter on one of his sombre walks in the grand Stadtpark, near his school.

'Hey, you're about to drop that book. Look out ...'

Perhaps it was because he had seen thousands of times how his parents struck up a conversation with customers, perhaps it was innate, but despite his waves of gloom, Paul could easily break the ice when he had to. In the mellow sunlight, he recognised not just the mathematics book, but he also had a hunch who the boy was under whose arm it had been tucked. He helped him intercept the book. 'Isn't this ...?' The boy—longer than stocky Paul, with dark-blond curls, a shy expression and an elegant coat—looked nonplussed. 'Do you know it?'

Gustav Herglotz went to the k.k. Franz-Joseph Gymnasium on the other side of the park, just within the Ringstrasse. Like the Akademische Gymnasium, it was a boys' school with eight years, where about half the pupils were also Jewish. Gustav was the school's 'maths genius', and the two boys were soon engaged in conversation.[23] It was the start of a lifelong friendship between lively and unsettled Paul—from a boisterous and hardworking grocer's family—and phlegmatic and calm Gustav—the only son of a notary who died young. At the end of Paul's seventh year, he moved to Gustav's school, and even though its teachers were also authoritarian and not always pleasant, he passed his final exams with relative ease in 1899.[24] The sun shone once more.

2
Not moonlight but rather sunlight
Well brought up in St Petersburg

The Austro-Hungarian Empire was vast and powerful, but Russia was immense. Vladivostok at one end was 6400 kilometres from Moscow at the other. The capital St Petersburg, in the north-west, was a thousand kilometres from Kiev (Kyiv)[1] in the south. In this country with innumerably more fir trees than people, and snow and ice or scorching sun, cooperation was compelled with an iron fist. To retain their scarce agricultural labourers, large landowners simply declared them their property in the sixteenth and seventeenth centuries. And no one in either the countryside or large cities would be surprised to see someone being lashed by a superior.

But this seemed to become a thing of the past as the second half of the nineteenth century progressed. With a graceful signature, enlightened Tsar Alexander II restored serfs' freedom in 1861. Even Jews, readily suspected of lacking patriotism, were granted certain rights. Alexander didn't go as far as his Austro-Hungarian counterpart, but he did permit Jewish craftsmen, merchants, scholars and soldiers—albeit with some restrictions—to live outside the Pale of Settlement instituted during the reign of Catherine the Great[2]; Jewish men were no longer barred from university. And in an effort to modernise the huge country, the Tsar also deployed a new technology to link cities and create a network of connections across the country: the train. Almost 50,000 kilometres of railway lines were laid between 1866 and 1899 (Westwood 1964, Ch. 2).

In St Petersburg, Pyotr Afanasyev profited from these developments. He was a chief engineer at the famous imperial railways and worked his way up to a professorship at the St Petersburg Polytechnic Institute. With his wife Sofya Maslova, he lived a respectable life. Their salon might not have been graced by nobility, but their guests were upstanding members of the community. They so valued etiquette that seven-year-old Tatiana Alexeyevna had already made a faux pas by asking the wrappers back of the chocolates she had just presented in the salon. Not that she did do so as an avid collector of wrappings. Tanya, as she was called, didn't attach much value to the carefully selected items and the airs and graces in the salon. She simply liked to draw, which is what had drawn her to the hand-painted wrappers.[3]

Tanya was a calm, slightly timid girl with a sharp mind. On weekdays, tutors gave her arithmetic, writing and drawing lessons. And when she walked through the streets, St Petersburg must have impressed her. In summer, the piercing golden

The Delight of Thinking. Margriet van der Heijden, Oxford University Press. © Margriet van der Heijden (2021).
DOI: 10.1093/9780198927112.003.0002

Spire of the Admiralty flashed against the blue sky. In winter, the dark ice on the frozen Neva River lay at the foot of the sea-green Winter Palace, and snow muffled the tinkle of sleigh bells on troikas. She wasn't much fond of cold and disliked foggy autumnal days though, perhaps also because St Petersburg wasn't her native city. Sonya and Pyotr weren't her parents either.

Tanya came from more southerly Kiev in Ukraine. The city sat in green hills, with church domes that glowed in the sun, and the muddy waters of the River Dnieper flowed far below. Her mother grew up here in humble circumstances: Ulyan Ivanov, Tanya's maternal grandfather, was a freed serf. Or had he already bought his freedom before 1861? Had that given him the time to raise his four children in relative freedom and send them to school? Tanya didn't know much about her family history. She did know that her grandfather had run a buffet at the railway station in Kiev.[4]

It was always busy in this newly built station, completed in 1870, in old-English gothic style. Steam trains brought people from all over Russia to the city with its sugar trade, grain exchange and breweries. With what now seems a sluggish top speed of at most 50 kilometres an hour, railway travel in the empire often took people days. In Kiev, they swarmed out of the waggons: traders and farmers, soldiers and agricultural workers, fortune seekers and aristocrats—from Odessa, Kursk, Brest, Moscow and later even Poland. They wore kaftans or fur coats, hats or shawls, sirwals or costumes, embroidered skirts or long frocks. They carried cabin trunks and baskets with sausages, bread, drink and live chickens. The din of their voices filled the cavernous station hall (Hamm 1993, 34).

They had come to do business, buy or sell things, visit one of the monasteries or family, and many of them also came to stay for good. For in his munificence Alexander II might have emancipated 23 million serfs, but all these people then needed to earn a living too—and while the population rapidly swelled, there was precious little land to divide and cultivate. As a result, many destitute farmers and agricultural labourers and their families migrated to cities like Kiev. Women often ended up in subordinate positions as maids, nannies, barmaids or shop girls. Men toiled for a pittance—lugging clay and bricks in brickworks, sweating next to copper kettles in breweries or steel barrels in sugar refineries, or working their fingers to the bone in the factories where these barrels were cast (Bezemer and Janssen 2008, 138–139).

Ulyan Ivanov escaped this fate with his station buffet. He even managed to help his children climb the social ladder. His three sons became a doctor, a veterinarian and a railway official, and his daughter Yekaterina—called Katya at home—attended school too. She was still a girl when she came across Alexei Afanasyev. Had he disembarked from the train in Kiev, where she had just started helping her father in his buffet? Had she given him a coffee? A sandwich? According to her passport, she was only sixteen years old when her daughter was born, a girl with grey, slanted eyes, whom she called Tatiana—Tanya at home—and whom she took for walks in the parks of Kiev.[5]

Two years later, Alexei Afanasyev was permanently committed to a psychiatric clinic after a nervous breakdown. He had been an engineer at the waterworks and was said to have had a misunderstanding with his boss. The issue had kept him awake for nights on end, until Alexei, who loved trees and music, had suffered a collapse. At least, this is as much as Tanya knew.[6] But the only thing that is really certain is that Katya took her daughter in tow and caught the 'iron horses that devour the forests' as the trains were called (Hamm 1993, 34). To St Petersburg, where her childless brother-in-law and sister-in-law were pleased to have a little girl at home.

Just two or three years old, Tanya ended up in the Russia's 'Window to the West'. In summer, the sun light flashed in the waters of the Moyka, the Fontanka and the canals Peter the Great had copied from Dutch cities. The pistachio-green, lemon-yellow and apricot-coloured stucco of the Renaissance palaces and neoclassical townhouses gave the city Italian refinement. On the Nevsky Prospekt, shops sold jewellery, porcelain tableware and fur coats. And it wasn't peopled by priests and monks, as was so often the case in Kiev with its cathedrals and monasteries, but rather by diplomats and businessmen, and ladies in seal fur. The kiosk sold newspapers in many languages.

As in other fast-growing European cities, the rest of the population—poor and mostly illiterate—paid the price for the affluence of this internationally oriented upper class. In drab backstreets, exhausted workers lived in decrepit boarding houses. Wan mothers cooked cabbage or onion soup in rooms where often two or three families lived—the beds screened off with sheets. Long queues snaked in front of the soup kitchens (Clements 2012, 141). 'Among the gentry and the merchants Fatherland does not exist', liberal writer and educational theorist Vladimir Stoyunin[7] wrote in his diary around 1850 (Hans 1963, 130). 'Civil servants amass wealth by cheating the Treasury, landowners by exploiting the peasantry, and the merchants by cheating everybody'.

But Tanya was well cared for (see Figure 2.1). In the portrait photos her aunt had made every year, her hair gleamed, and she wore dresses with frills and bows. By then her intelligent mother Katya had to earn her own living. She trained to become a midwife and was determined to fend for herself while her daughter was brought up as a lady.

Much later, Tanya would tell her granddaughter how at school she had learnt to walk gracefully and erectly. When she was eleven, her aunt and uncle sent her to a recently opened private gymnasium for girls. The school, in a distinguished building on Ulitsa Kabinetskaya, south of the Fontanka, was the twelfth gymnasium and the third private gymnasium in the city (Margolis 2004). It was run by progressive

16 THE DELIGHT OF THINKING

Figure 2.1 Tanya aged eight or nine.—*Photograph Ehrenfest Family Archive.*

Maria Nikolayevna Stoyunina, who implemented ideas that her husband Vladimir also propagated in his lectures and speeches: no corporal punishment, much attention to individual talent and independent thinking, a fifteen-minute break after every lesson in which to jump, run, chat and recharge, and gymnastics every day. Classical languages weren't on the syllabus. 'The programme was a watered-down version of boys' gymnasiums', Tanya said later (Ehrenfest-Afanassjewa 1961, Introduction). 'But the education was how I would like to see it everywhere. We were encouraged to think for ourselves and never had any rote learning.'[8]

That Tanya spent five of her last seven years of secondary school at a more conventional gymnasium, closer to home, was because she contracted almost 'all the childhood diseases that exist', as she put it later (Ehrenfest-Afanassjewa 1961, Introduction). The education was broad-based at this Mariinsky Girls Gymnasium, named in honour of Tsarina Maria Feodorovna and closer to Tanya's home, with less tiresome travel. The pupils were given lessons in Bible study, Russian language and literature, French, German, English, pedagogy, mathematics, history, geography and the sciences. They also learnt calligraphy and were given lessons in drawing, dancing and singing.[9] And at the gymnastics hall, Tanya learnt to walk gracefully: with a broomstick between her tucked-back elbows to keep her back straight, first her heels on the ground and then her toes.[10]

The other lessons were also thorough. 'We have already covered twenty-three postulates in geometry', wrote a school friend when Tanya had a prolonged illness in the third year.[11] Tanya didn't fall behind, catching up effortlessly. She found learning easy and particularly enjoyed mathematics and drawing. She was popular with the other girls and had many friends: 'My dearest, darling, golden Tanitchka', a fellow pupil wrote.[12] It seemed fitting that the old-fashioned spelling of Tanya's surname with a *fita* (derived from the Greek theta) on her school-leaving diploma renders Afanasyeva more like 'Athanasia', which means 'immortality'. In any case, when Tanya was awarded her diploma in May 1893, she received a gold medal for exceptional attainment.[13]

But when she subsequently wanted to read mathematics and physics at university, her uncle, Pyotr Afanasyev, was displeased. He thought her health was too delicate for a university study (Ehrenfest-Afanassjewa 1961, Introduction). Or was it also because higher education for women was still tinged by a hint of rebelliousness? Traditionally, Russian higher education had sought to educate young men. These students, donning caps, green trousers and blue smocks, appeared at first sight able to freely wander around the city, develop themselves and even express criticism of their surroundings. But ultimately these sons of the upper bourgeoisie, the clergy and the nobility, were meant to become engineers and civil servants, often in the large tsarist bureaucratic apparatus, and toe the line. As did the smartest boys from more modest backgrounds whose intelligence earned them a place at university.

For Russian intellectuals who had acquired humanist ideas in Western Europe, such developments were all too slow. Already in 1840, intellectual and literary critic Vissarion Belinsky was opposed to merely drilling children in useful subjects, wanted to exclude religion from schools and valued science (Brower 1970; Hans 1963, 35–36). 'Without a knowledge of the natural sciences, there is no salvation for the modern man', Belinsky's friend, writer and thinker Alexander Herzen, had also concluded (Hans 1963, 39). And like the nineteenth-century British philosopher John Stuart Mill, for these Russians, this 'modern man' could be male or female. 'When a woman is given a good education, she can develop as well as a man in science, art and public life', brilliant professor and surgeon Nikolay Pirogov had written in the year of Tanya's birth (Hans 1963, 62–63).

Aristocratic Russian women had already enthusiastically embraced such ideas. Around 1860, almost thirty 'Sunday schools' had been established in St Petersburg. Some four hundred teachers gave unpaid lessons in the stately living rooms of affluent families to maids, shop girls, cooks and their daughters, and even factory workers. 'Every family that considered itself part of the St Petersburg intelligentsia felt obliged to support the Sunday schools', as one of these women observed in this period (Pietrow-Enker 1999, 316–317).

This gave women courage: they were even allowed to attend university lectures as non-examination students. That is, until student revolts broke out in 1862, and Alexander II banned in one stroke Sunday schools and denied women access to universities. But the genie was out of the bottle by then. Many former non-examination students decided to leave St Petersburg for Zurich and Paris, where the first universities had started accepting women. Here, they became pioneers of higher education for women: of the 203 women admitted to the University of Zurich between 1864 and 1872, an astounding 148 were Russian, and these proportions were similar in Bern and Geneva (Koblitz 2000, 10).

For Tsar Alexander II and his minister of education Dmitry Tolstoy, things had thus gone from bad to worse. In dismay they looked on as women not only obtained diplomas in Western Europe but also regularly acquired ideas about society that the Russian government considered radical. Their decree in 1873, ordering all women studying abroad to return to Russia, was ineffectual (Stites 1990, 82; Koblitz 2000, 17).

If the Tsar wanted to keep ambitious women in Russia, he would have to offer them education. Minister of Education Tolstoy had already started grudgingly down this path in 1870 by authorising the new *Vladimirskii* courses in a school building on the Vladimirski Prospekt. 'Ladies with ponytails and close-shorn hair, Garibaldi supporters and a great deal more' listened 'breathlessly' to St Petersburg professors, according to philanthropist Anna Filosofova[14], who initiated the course together with her fellow feminist aristocrats Nadezhda Stasova and Evgenia Konradi (Pietrow-Enker 1999, 323). With enormous reluctance, Minister Tolstoy then also sanctioned new medical studies for women—financed by a wealthy student on the Vladimirskii courses—which focused particularly on women's health issues. And in 1878, just before Tanya and her mother came to St Petersburg, he failed to thwart the opening of the 'woman's university' on Ulitsa Gorokhovaya in the imperial capital. It was the fourth women's higher education institution in Russia, after those in Moscow, Kazan and Kiev (Stites 1990, 81–82).

Tanya was bound to have walked past this institution on occasion, which had soon moved to Vasilyevsky Island. Originally, Peter the Great had wanted to make the island the heart of his new city, but the idea didn't work well in practice, and the canals cutting across the island overflowed so often at high tide that they were soon filled in. Instead, it became a tranquil university island. The red baroque university building on the Strelka, the spit at the tip of the island pointing at the Winter Palace, was the workplace of the likes of chemist Dmitri Mendeleev, who created the first Periodic Table. Ivan Pavlov, famous for the Pavlovian response, regularly came here too. The leafy *Linii*, or lines as the filled-in canals were called, were home to mathematicians, biologists and philologians.

The new higher education institution for women was on the tenth *Linii*. Perhaps Tanya had at times seen young ladies with long skirts and lambskin mittens entering. The institution was chiefly known as the Bestuzhev Courses, after Konstantin Bestuzhev-Ryumin, its first director. Students specialised in history and literature or physics and mathematics. Sometimes lectures were by well-known academics like chemist and composer Alexander Borodin and Mendeleev, who, as well as being a chemist, was progressive and emancipated, and even by women lecturers.[15]

Professors weren't paid much for their lessons, as government subsidies were minimal. The institution was mainly financed by tuition fees and contributions from the Association for the Support of Women's Higher Education, which had about one-thousand members and raised money through book sales, lectures, concerts and lotteries. The Association also gave scholarships to intelligent students without means (Stites 1990, 82–83). So, the *bestuzhevki* in their lambskin mittens and long skirts weren't only from the elite and upper bourgeoisie but from the lower classes too.[16] What they shared was a yearning for knowledge, something Tanya must have empathised with.

But Tanya's uncle insisted that she did a teacher training instead (Ehrenfest-Afanassjewa 1961, Introduction). Was the Bestuzhev Courses' status as a bastion of social criticism since the 1880s a factor too? When the populist-nihilist *Narodniki* blew up Alexander II's carriage in 1881, a leaden pall descended over the period of reform. Alexander III, who witnessed his father's violent death first-hand, immediately undid all changes. He increased censorship, expanded the Okhrana (secret police) and introduced new anti-Jewish laws: no more than 3 per cent of students at universities and colleges could be Jewish. Women were also put in their place. For example, feminist and philosopher Anna Filosofova was exiled for two years; and three of the four women's universities established in the late 1870s were closed by the mid-1880s. The police called them 'a veritable sewer of anarchist disease'; and hadn't a woman—Sophia Perovskaya—been involved in the plot to assassinate the Tsar's father (Stites 1990, 168)?

Only the Bestuzhev Courses in St Petersburg survived—although no new female students were allowed to register for four years (Morrissey 1998, 13). The regime imposed new management and an inspector and banned assembly—apart from lectures of course. Moreover, students were obliged either to live at home or at the institute and strictly forbidden from renting a room somewhere. Finally, they also had to show proof of means for their maintenance and their family or guardian's consent for their study (Stites 1990, 169). And this last condition was one that Tanya couldn't meet because her uncle refused to give his blessing.

Probably filled with supressed anger, Tanya plodded through the years of 'Pedagogical courses at the Girls' Gymnasium'. The courses prepared her for a job as a teacher in the first years of girls' schools. The new Russian woman desired 'not the moonlight but rather the sunlight', feminist Nadezhda Stasova had said in 1859 (Stites 1990, 34), but Tanya lacked the opportunity to shine like a young man with her talent and from her social background likely would have.

Would she have heard that the poorer bestuzhevki shared beds with two or three other students and ate stale bread and dry sausage at night, warming their hands on a cup of tea, in a 'cloak full of wind', as unlined jackets were known (Valk 1918, 10; Stites 1990, 170)? Probably she had understood that Tsar Nicolas II, who had succeeded Alexander III on his sudden death in 1894, regarded female students with at least as much suspicion as his father had. And despite their long skirts and high-necked blouses, the bestuzhevki were called street girls and strumpets by conservatives (Morrissey 1998, 157).

But Tanya still shared the ideal of this assortment of women: daughters of high-ranking military officials, Orthodox clergy and members of the civil service, impoverished nobility, merchants, farmers and Cossacks. She wanted the same as the bestuzhevki in their lambskin mittens: to acquire knowledge and contribute towards the advancement of their motherland. For *bestuzhevska* also means 'idealist'.

The only impediment was her uncle. He was adamant, even after Tanya graduated from the teacher training—again with a golden medal (Ehrenfest-Afanassjewa 1961, Introduction). She could hardly object to the man who had taken her in out of the goodness of his heart. Instead of studying, she travelled with her restless and adventurous aunt to Berchtesgaden in Bavaria.

Of course, the two women could not have known that about half a century later Hitler would turn nearby Obersalzburg into a hideous Nazi mountain, with a complex of tunnels and bunkers for his unsavoury associates Albert Speer and Hermann Göring. For them, Obersalzburg was still the charming village where composers like Clara Schumann, Johannes Brahms and Gustav Mahler dined together with well-known painters, inventors and even royalty at the famous pension Moritz. The two women saw the mountain slopes turn white with the first winter snow. After which, clad in muffs and fur hats, they boarded the train—Berchtesgaden was on the railway line between Munich and Salzburg—to celebrate Christmas in Vienna.

Naturally, Sonya and Tanya were also quite unaware of the existence of a disconsolate, seventeen-year-old teenager in the city called Paul. In Vienna, their trip took an unexpected turn when they received the message that Tanya's Uncle Pyotr, who had fallen ill shortly before, had died in St Petersburg.[17] Not long afterwards, in January 1898, Tanya enrolled in the Bestuzhev Courses halfway through the academic year. Her Aunt Sonya paid the tuition fees and stood surety for her.[18] She was going to read mathematics and physics.

3
Boltzmann and Columbus
Voyages of discovery in physics

'Everything I have: myself, my entire way of thinking and feeling ... That is what I have to offer you'. In the lecture hall on Türkenstrasse in Vienna, physics students sat with notepads on their laps. The Physics Institute building was old and dilapidated, and the floors shook when a tram went down Währinger Strasse around the corner. The cramped lecture hall only had space for benches without desks.

But the run-down setting did nothing to diminish the formidable reputation of the heavyset man at the blackboard. Professor Ludwig Boltzmann was perhaps the most famous academic in Austria-Hungary. His combed-back hair and untidy beard gave him a slightly wild appearance, but his eyes sparkled with enthusiasm behind his glasses.

Boltzmann hardly had to tell his students to listen attentively to his lectures, work hard and rack their brains. Not Paul, his friend Gustav Herglotz or any of the other students: young men in suits—often moustachioed, imparting a somewhat elderly demeanour to their faces—and the odd young woman. Nor was it paramount to him: 'First of all I ask of you what means most to me: your friendship, your love—in a word, the best you can give: yourself'.

This appeal must have dumbfounded many students. Such things weren't often said to them and certainly not at the gymnasium. But this was the rather informal Physics Institute of the University of Vienna, where Professor Ludwig Boltzmann was a touch more informal than the rest of his colleagues. When he made a mistake writing a formula on the blackboard, he would say matter-of-factly: 'Oh, how stupid of me'. He encouraged students to ask questions and held discussions about the curriculum. More than anything, Boltzmann was a compelling lecturer. 'He was so enthusiastic about everything he taught us that we left every lecture with a sense that a magnificent new world had been revealed to us', students said (Lewin Sime 1996, 13–14; Cercignani 1998, 37–38).

For Paul and Herglotz, this must have been a breath of fresh air after their strict secondary schooling. Boltzmann lectured them on Maxwell's work on electricity and magnetism as well as on thermodynamics and on statistical mechanics, the topic he was famous for.[1] His metaphors, how he used the blackboard and his clear language deeply impressed Paul. This was a scientist who wanted to understand the world profoundly and who considered debate an essential element of science and the transfer of knowledge.

Awed, Paul sometimes gazed afterwards into the coffeehouse next to the institution, where you were allowed to write with chalk on the marble tabletops. Theoretical and experimental physicists often had long discussions there—partly because the old institute building was so airless and overcrowded (Rentetzi 2007, Ch. 3).[2] But Paul observed chiefly how some people made passionate thinking and debate central to their lives.

Earlier that year, in October 1899, Paul had in fact enrolled at the Technische Hochschule in Vienna.[3] He had had long conversations about it with his brother Arthur. Of the four elder Ehrenfest brothers—Hugo the doctor, Emil the businessman, Otto the electrical engineer and Arthur the engineer—Paul followed in Arthur's footsteps. For the most part at least, as Arthur had studied mechanical engineering, and Paul chose chemistry. Had Arthur advised him to do so because chemical engineers had good employment prospects in industry?

It matters little. Students at the *Technische Hochschule* could also take courses at the University of Vienna, and chemistry students could enrol for mathematics and physics courses at both institutions. Paul did so immediately. Alongside his obligatory chemistry courses and 20 hours of chemistry practicals a week, he chose a whole range of physics and mathematics courses: classical mechanics, electricity and magnetism, hydrodynamics, differential equations and much else.

Another advantage of these extra classes was seeing Gustav Herglotz. Naturally enough, the former 'maths genius' at the *kaiserlich-königlich (k.k.) Franz-Joseph Gymnasium* had decided to read mathematics. Two other students joined them. Hans Hahn had initially read law for a year, before switching to mathematics and physics in 1899.[4] Heinrich Tietze had started the same study a year earlier in 1898.[5] They soon became friends (see Figure 3.1).

All four were from the upper middle classes. Tietze's father oversaw the geological institute of the university; Hahn's father held a senior post at the *Telegraphen-Korrespondenz-Büro*, the imperial press agency; and Herglotz's father had been a notary. Paul's father was the only one with little education, but he had been a successful businessman. The young men themselves were only interested in one thing: the sciences. When other students saw them walking in city parks or the streets of Vienna, they were invariably deep in discussion about mathematics. Even on their weekend excursions to the hills of the Wienerwalt woods mathematics monopolised conversation.

Hahn, Herglotz and Tietze soon forged ahead in the subject. Paul didn't. He wasn't as good a mathematician as his friends, even though the extra mathematics had helped him get through his boring school years. For that matter, chemistry hadn't captured his heart either, but his theoretical physics lectures captured his imagination. This was thanks to Boltzmann, who had shown that physics was a human creation and that something was really at stake in the field (Klein 1970, 35–36).

Figure 3.1 Paul Ehrenfest, Heinrich Tietze, Hans Hahn and Gustav Herglotz as students in Vienna.—*Photograph National Museum Boerhaave, Leiden.*

Boltzmann spoke freely about his struggles with the subject. 'He told us the amount of opposition he had encountered because he was convinced of the existence of atoms and how philosophers attacked him for this, without always understanding what they had against his work', Lise Meitner, Boltzmann's most brilliant student, later said (Lewin Sime, 1996 p. 13). Although this opposition was likely also caused by Boltzmann's colleagues not always understanding what he meant. Unlike his lectures, his essays and scientific articles were woolly, and his subtle ideas drowned in lengthy expositions on paper. Boltzmann 'could not understand me on account of my shortness, and his length was and is an equal stumbling block to me', Maxwell wrote (Harman 1995, no. 474).

These ideas had ancient roots. Ancient Greek philosophers such as Democritus had proposed two millennia earlier that the world consisted of indivisible building blocks, atoms. According to these 'atomists', atoms moved about in all directions through space, sometimes gathering by chance. Haphazardly, without a preconceived plan, trees and stars thus came into existence, elephants and the moon, humans and the earth—a vision of creation that didn't always find favour with other Greek philosophers and later certainly not with the Church Fathers.

But these pragmatic chemists' ideas weren't always taken seriously by all academics. Some even dismissed working with atoms as trickery. 'I don't believe they exist!' Viennese philosopher and physicist Ernst Mach famously declared during a lecture by Boltzmann on atoms in 1897. Mach was a positivist: science could only describe facts and must do so as parsimoniously as possible, without embellishments such as the existence of tiny particles nobody had ever observed directly.

Hence, not everyone believed Boltzmann when he assumed that atoms were part of reality and even started making calculations with these particles. In fact, despite the imperceptibly small size of atoms and there being too many atoms in the tiniest gas bubble to describe their behaviour individually, Boltzmann even linked the movements of atoms to the behaviour of a gas on a large scale. This association relied on statistical calculation methods, which Boltzmann used to link the average behaviour of atoms to measurable properties of a gas such as pressure, temperature and volume. In doing so, he gave a statistical underpinning to nineteenth-century thermodynamics, which hitherto had chiefly described the behaviour of gases and steam in steam machines.

His most important formula linked the behaviour of atoms to the most abstract quantity in thermodynamics: entropy. This concept is connected to the second law of thermodynamics, formulated by German scientist Rudolf Clausius. In a concise form it states that heat never flows spontaneously from a cold object (or reservoir) to a hot object (or reservoir). This phenomenon had far-reaching consequences because it implies that all heat will spread evenly throughout a system with enough time. Even the universe would eventually end up as a uniformly lukewarm bath, some commentators thought. In any case, the passage of heat indicated a direction in processes and therefore in time; it indicated both a past and a future.

For this process, too, Boltzmann constructed a statistical underpinning. In his view, gas atoms shot in all directions through space, regularly colliding with and bouncing off each other. Temperature was a measure of these movements: if the atoms moved quickly on average—with a great deal of kinetic energy—then the gas had a high temperature. If they moved more slowly, then the gas was cooler. The essence of Boltzmann's reformulation of the second law was that all these continuous collisions help to distribute the kinetic energy more or less evenly distributed over all atoms. In practice: if you let a gust of cold wind flow briefly into an otherwise enclosed space, then after a while it will be a little cooler throughout this space. Translated to liquids: if you pour some cold milk into a hot cup of tea then the tea and the milk will soon mix 'spontaneously' into a uniform and light-brown liquid.

But you will never see the reverse take place. A cold gust of wind will never spontaneously arise in an enclosed space. The uniform light-brown liquid in the cup will never suddenly disintegrate into dark tea with a dash of cold milk. This is because, as Boltzmann showed, there are many more ways for colliding atoms to throng happily around each other than for them spontaneously to form a 'hot' group here and a 'cold' group there through random collisions. Equilibrium states is what Boltzmann called these states in which this happy thronging prevails and randomness, or entropy is maximal. By doing so, Boltzmann had elegantly rewritten the second law as a tendency in nature towards randomness.

Yet not all of Boltzmann's colleagues saw his statistical reasoning as the key to the problem. Equilibrium states went against their intuition. Critics included venerable Austrian physicist Johann Loschmidt and German logician and mathematician Ernst Zermelo, who had worked as an assistant of eminent physicist Max Planck. German chemist and philosopher Wilhelm Ostwald articulated the objections most directly. His belief in the idea that energy—and hence not entropy—was the driving force in nature was so ardent that he called his house in Leipzig 'Villa Energie' (Görs et al. 2005, Introduction). He bombarded Boltzmann with what he had learnt of Newton's work. Hadn't Newton demonstrated that his laws of motion have no preference for the future or the past? In other words, precisely defined movements—the collision between billiard balls, the trajectory of a bullet—could in principle just as well be played in reverse order? If you conceive of atoms as miniscule billiard balls, then an inverse should exist for every collision, Ostwald argued, and it must therefore be possible for the film of a gas or a cup of tea with milk to be played in reverse as well.

Boltzmann was vulnerable. This was because he was born on the short night between the end of carnival and the start of the strict Lenten season, he liked to joke. Throughout his life, he lurched between vivacious enthusiasm and lethargic depression. Between a love of bustling Vienna and a longing for tranquillity at the more formal German research institutes. Between the fresh vigour of new beginnings and subsequent disappointment, such as the reaction of colleagues like Ostwald. These often-vehement reactions made his life tempestuous. This didn't change after his marriage in Graz in 1876 to Henriette von Aigentler, a progressive mathematics and physics teacher at a girls' school (Cercignani 1998, 11–12). Boltzmann went back to Vienna with her, returned to the University of Graz, then went to Munich in 1890 and back to Vienna in 1896. And just after Paul finished his first year of studies in 1900, Boltzmann left for Leipzig, where he earned twice as much and was farther removed from his critical Viennese colleague Mach (Uffink 2017).

Yet Boltzmann had already imparted something to Paul: the idea that there really was something at stake in physics. Physical theories weren't only useful because they gave rise to steam trains and telephones. They weren't merely adequate descriptions of lifeless nature. Boltzmann promulgated the idea that physics revolved around the search for 'the ultimate truth'. This quest became Paul's new lodestar.

Years later, Paul would copy a text in which Boltzmann, when asked and 'without hesitation', identified the explorer Columbus as the 'happiest person' of all time. 'Not that other discoveries weren't equally worthy as his; the German Gutenberg's to name just one'. But happiness is also sensual, Boltzmann held. Columbus felt the wind in his hair and heard the roll of the waves as America loomed in front of him. 'I can't go ashore in America without a twinge of jealousy or perhaps feeling some kind of excitement at sharing in his joy'.[6] (Boltzmann 1905, 409)

It concurred with how Paul, thanks to Boltzmann, thought physics should be: not just a field in which you studied and derived formulae, but one in which you talk, discuss and search for clear images that bring formulae to life. A lively field too, in which you sail unknown seas to discover what lies beyond the horizon. Preparing himself for this, he threw himself into vector calculus, kinetic gas theory, physics practicals as well as other mathematics and physics courses in the ensuing semesters.

By then his 'inseparable friends' had already left Vienna to complete their studies at universities in other cities, as young men of means did. Hahn was in Strasburg, Herglotz in Munich and Tietze was performing military service, after which he went to Munich. A year later Paul, too, decided to seek greener pastures. Very little still bound him to Vienna after he had finished his courses. His brothers had their careers and families. Pepperl had married Emmanuel Friedmann, the widower of her likewise deceased sister Lina.[7] And it was little fun without his friends.

So it was that in autumn 1901, he travelled to Göttingen, a quiet little town in the Province of Hanover, in the valley where the Weende River rises and the Leine River flows between the rolling hills. Herglotz planned to visit later, and it was on Hahn's wish list because sleepy Göttingen, with just 30,000 inhabitants, was a Mecca for mathematics and physics.

4
Sailing the Sahara
Raised and constrained in St Petersburg

St Petersburg didn't have a coffee culture. A good conversation started with tea. Over a pot of tea, you could talk for hours. It was something the students at the Bestuzhev Courses often did. 'The education wasn't modern', Tanya said later, referring particularly to the sciences (Ehrenfest-Afanassjewa 1961, Intro). The syllabus covered a broad array of subjects: on top of mathematics and physics, science students were taught biology and chemistry, mineralogy and crystallography, as well as physical geography (Vakhromeeva 2018). Of course they did tests, and they concluded each year with an examination, but the latest developments in all these disciplines were not always discussed (Ehrenfest-Afanassjewa 1961, Intro).

Physics was taught by then well-known Orest Khvolson, one of several professors at St Petersburg University who also lectured at the Bestuzhev Courses. His five-volume *Physics Course* was the leading series of physics textbooks and was translated into German, Spanish and French. He was versatile, handsome, and Tanya was impressed by him, as her friend Kitty Miloradovich noticed. 'Is Orest Danilovich still your favourite teacher?' she wrote teasingly, once she had become a teacher in Izmail, 250 kilometres southwest of Odessa.[1]

The mathematics teachers also had a good name. Dmitry Selivanov, for example, had been a student of famous German mathematicians Weierstrass and Kronecker. But Tanya was most fond of Vera Iosifovna Schiff. A member of the first class to graduate from the Bestuzhev Courses in 1882, she had proceeded to study in Paris before returning to the Bestuzhev institute as a teacher. She not only gave mathematics courses, such as geometry and calculus but also published scientific articles in Russian mathematics journals.[2]

With dark curls and a pince-nez, Schiff was one of the first Russian women to dedicate her life to mathematics, and she conveyed her passion for the field to at least some of her students. Kitty, for instance, in a letter to Tanya in early 1900, gleefully recounted how she had met a charming officer and solved a fourth-degree mathematical equation with him during the Christmas holidays.[3]

Tanya was still studying that year. The institute seemed to flourish. To the right of the institute, a new wing with a laboratory and dormitories had recently been opened; a left wing with a library and an auditorium was under construction. Tanya herself had much studying and reading to do, and she had many friends, who asked her for advice and practical help or just wanted to talk. But it was a year of complications too.

Predictably, the gap between the impoverished, predominantly illiterate masses and the small, often erudite elite widened even further under the conservative and authoritarian rule of Tsar Nicholas II, while institutions giving the populace a voice remained non-existent. Some students had encountered the adverse effects of this social inequity in their childhood; others only experienced it upon entering university. For many, it led to inner conflicts: they saw the unjust division of the country, but at the same time, their education destined them to become part of the elite or at least to hold a position in the vast tsarist civil service.

While the Tsar tightened his grip, some of the students radicalised, also at the Bestuzhev Courses, where the secret police extended its tentacles even farther. In 1896, when Tanya was still travelling in Bavaria with her Aunt Sonya, the Okhrana arrested bestuzhevka Mariia Vetrova for anti-tsarist publishing activities. Shortly thereafter, in early February 1897, Vetrova was said to have used remnants of lamp oil to set herself on fire in a dark dungeon in the infamous Peter and Paul Fortress; her fellow bestuzhevki could only speculate about the exact circumstances of Vetrova's death (Morrissey 1998, 178–179).

Dismayed, the women had sat through the official commemoration at the institute. A day later, they congregated on the other side of the Neva, in the square enveloped by two immense colonnades facing Kazan Cathedral. Since commemoration of a suicide wasn't permitted inside a church, they stayed outside—the chanting of their hymn 'Eternal Memory' rolling over the square. But even if students themselves viewed their gathering as a commemoration and not a protest, police and Cossacks with whips intervened. Over 1200 students got police summonses, and hundreds of students, among whom were about 160 bestuzhevki, were arrested and jailed for three days in the infamous Kresty Prison (Morrissey 1998, 29–30).

The anger of male students, two years later, appeared to have a less significant cause. After a year of incidents with drunken students, the rector of the university, Vasily Sergeyevich, threatened to come down hard on disruptions during the annual celebration of the founding of the university. His admonition outraged the students, who felt that the rector had seized upon a few indiscretions to tar all of them with the same brush. During the celebration, they subjected him to quiet but forceful catcalls, and the rector kept his word: afterwards, mounted Cossacks lashed the students in their winter coats in every street and alley (Kassow 1989, 91; Morrissey 1998, 45–46).

Several days later, a majority of students concluded that striking was their only viable option to uphold 'rights and laws against arbitrariness and violence', and to emphasise that 'science, enlightenment and universities cannot thrive where arbitrariness holds sway'. They also made clear that they intended to strike in a controlled and dignified way (Kassow 1989, 94–95, 103; Morrissey 1998, 47–48).

Should the bestuzhevki join them? The question swept through the corridors, flew over the shallow wooden stairs, up the stairwells and through the lecture halls of the Bestuzhev Courses. On the evening of 13 February, the women held a ballot in one of the lecture halls, with the door locked. Unlike the men, the women had divided loyalties: they wanted to fight for a more democratic Russia and the rule of law, as Vetrova had done. But they also wanted to defend women's right to education. The Bestuzhev Courses was the sole remaining institution of higher education for women in the entire country. What if the Tsar closed it in reprisal for the strikes (Morrissey 1998, 81)?

Most men had little time for such discussions. Either they didn't see the problem or prioritised the people's liberation: freeing women would follow later. A few weeks later, a group of male students reproached their female counterparts who chose not to strike, branding them selfish and obedient in a 'poetic' proclamation. 'Let she who is gifted with pitiful conscience /Not carry her share /Let the well-behaved children study (modestly) /In happiness, under the rod'. (Morrissey 1998, 83–84). One of the 'well-behaved children' was Tanya. She was surprised how the young first-year students at the ballot took the lead and called for action. Even more surprising to her was that the majority silently joined them. A day later she wrote a long appeal to her fellow students.[4]

She couldn't of course set aside the fact that the decision to boycott classes was made following a vote. 'The vote yesterday was a reminder of how things are done in states where matters are decided by majority vote' was her opening. But: 'I believe many of you are guided by that example, so it is necessary to examine why things are done this way and not differently'.

Tanya discussed the idea of such votes, which were so foreign to Russia. She, too, seemed to have only a general understanding of the concept, offering a basic explanation. 'These people live on the same piece of land where, for thousands of material and spiritual reasons, they don't want to and cannot leave. ... When deciding on an issue, such as the choice of a president, when the votes are divided between two factions, people understand that if neither side concedes, they will remain indefinitely in conflict and the state will degenerate into chaos and disorder more awful than having the worst president. ... For this reason they agree in advance that one of the two factions will submit to the other, and it is clear why the minority submits to the majority: so that as few people as possible suffer or are dissatisfied'. This ultimately amounted to: 'One submits to the majority only to maintain some semblance of order; this is the basis of the majority's power'.

Then Tanya came to the point she wanted to make: 'Is something of this kind happening to us?' In other words, should the bestuzhevki who didn't agree with the boycott hold their tongues to prevent chaos at the institute? No, Tanya argued, because there was little likelihood that the free exchange of ideas would result in total chaos, whereas, conversely, a boycott entailed a grave risk of permanent closure of the courses. This is why opponents should indeed speak out, she believed.

'Our reluctance to boycott the courses isn't an expression of indifference ... it is an act to prevent the loss of the courses'. Precisely this conviction had led her to write the epistle. 'What moral argument could compel someone, deeply convinced of the disastrous nature of an enterprise, to permit its execution?' Phrased differently: 'You must defer to the majority, but not to the degree that everything you wish to belong to falls apart'.

In the ensuing days, Tanya and a group of like-minded bestuzhevki collected signatures against the boycott. They approached students who hadn't been at the meeting or perhaps hadn't dared to say anything.[5] And they were appalled when they learned that the strike committee had disseminated their list with names and signatures in pamphlets and on placards. Tanya would keep drafts of the letters she wrote in protest—littered with deletions and corrections—for the rest of her life.

Didn't the majority of students object to lists of strike breakers' names 'being hung up and circulated to other institutions?' Did the majority concur 'with those who cast them out of the fellowship and heap insults upon them?'[6] No? Then the lists should be destroyed, Tanya wrote. She even dubbed the strike committee a 'comrades' court' that, she suspected, acted on their own authority.

But it was futile. The lecture halls remained empty; no books were strewn on the shelves beneath the writing desks; the conical flasks and pipettes lay still in the lab; the corridors were silent and even some of the professors were absent. They feigned illness or cited other reasons for their inability to work; Vera Schiff was among them (Morrissey 1998, 81; Valk 1918, 34).

Authorities at the institute and civil servants couldn't do much. The students stuck to their guns, even when forced to report individually to the authorities. The more high-handed authorities were, the more students joined the strike. But not Tanya. Was she perhaps also a conservative who was interested only in 'pure science', separated from society, and who wanted to avoid political discussions in lecture halls? This assertion is countered by the fact that she enjoyed reading popular writer Gorky, whose books accused 'petty bourgeois academics' of blindness to the lives of Russians at the bottom of the social ladder. Moreover, her Aunt Sonya had raised her on the ideas of Tolstoy, who mightn't have been very feminist but was antiauthoritarian. And wasn't Tanya's mother a working woman? Most likely Tanya did indeed simply value women's rights to higher education and feminism more than the student protests against police abuses and autocracy and for respect of their basic rights.

From a more practical perspective, it is perhaps not so surprising that she wouldn't want the long-awaited opportunity to pursue her studies jeopardised by what she considered a frivolous protest. But what seemed most important was that Tanya subordinated the slogans of the more revolutionary students to feminism and women's rights to higher education, rather than the other way around, as many students did.

More than anything else, the strikes and their consequences ultimately highlighted the rigidity of the hierarchical system, in which all forms of inequality—whether of sex, rank or class—were entrenched (Kassow 1989, 94–100). It was nearly inevitable that this resulted in many bestuzhevki radicalising further and playing prominent roles in the revolutions, first in 1905 and then in 1917. Just as it is easy to imagine that others lost all faith in mass movements and preferred guidance from their own, personal ethics (Morrissey 1998, 97). 'Certain phrases are often repeated among your enraptured and enthusiastic words: it is necessary to develop *lichnost*' [character and ethics] in oneself, the striving towards the freedom of opinion, toward full equality. But we consider these words to be beautiful only when they leave the realm of theory and enter that of practice', women guided by personal ethics wrote of their more radical sisters during new unrests in 1904 (Morrissey 1998, 86). And: 'It is easier to be "radical" than to be yourself'. Such words must have resonated with Tanya, who later in life would strive to put personal responsibility and critical thinking above politics or personal gain.

In 1900, Tanya's fear that the Bestuzhev Courses would close didn't materialise, although unrest at the institute persisted for some time. Twenty-one students were suspended and, after another round of strikes and action in March, almost 300 bestuzhevki were sent away for the rest of the semester. The institutional authorities were replaced, as was the inspector, after having refused to name activist students. In the partially empty classrooms, lessons continued in a subdued atmosphere, while bureaucratic meddling undermined what had been a high academic standard (Morrissey 1998, 82–83; Valk 1918, 37). It took doggedness and perseverance to complete the study.

Tanya was well-endowed with both. She didn't allow herself to be distracted. Not by strikes, not by unrest which continued to flare up and not by a lack of prospects. What was the point of educating women if they weren't allowed to use their knowledge and talent afterwards, lamented founders of the Bestuzhev Courses when access to jobs beyond teaching assistant at a college or teacher at a girls' gymnasium was once again denied. It was like giving people a long education at a nautical college only to send them to the Sahara. But Tanya soldiered on. She saw the suspended bestuzhevki return after the summer, she saw the promises of reform come to nothing, and she graduated with a gold medal in 1901 (Ehrenfest-Afanassjewa 1961, Inleiding).[7]

The following summer, Tanya went to her uncle's dacha in Kukarka village (present-day Sovetsk), which was little more than a wooden church and a collection of wooden and brick houses scattered along the steep green banks of the Vyatka, Pizhma and Nemda rivers in central Russia. It smelled of hay and honey. In the workshops lining the streets, carpenters crafted the sturdy sledges Kukarka was renowned for, while potters, basket weavers and lacemakers plied their trades. Perhaps an eleven-year-old boy, Vyacheslav Skryabin, hung around the houses. Later

32 THE DELIGHT OF THINKING

Figure 4.1 Tanya in Kukarka in 1901.—*Photograph Ehrenfest Family Archive.*

he would rise to prominence as a Bolshevik and then Stalin's henchman, better known as Molotov—a name that lives on in the infamous Molotov cocktail (Watson 2005). But in 1901 it was still peaceful in Kukarka, and Tanya could calmly prepare for the new year (see Figure 4.1).

The postman brought her a letter from Vera Schiff, who was spending summer in Germany.[8] Tanya's mathematics teacher wanted to know 'how your examinations at the gymnasium went. I am confident that they were a resounding success. Any shortcomings that may have arisen shall have been discernible only to your extremely strict critical eye'. As her teacher had suspected, Tanya had passed the examination for a teaching position for the lowest classes at a private gymnasium in St Petersburg with flying colours. She had also secured a place as a teaching assistant at the Bestuzhev Courses. This gave her a place to sleep at the institute as well as a modest income—a nice start.

And perhaps things would get even better. Schiff continued: 'I was much pleased to read in the newspaper that the Tsar has decreed that women must be admitted to the University of Helsingfors [Helsinki] on the same terms as men. Now we must simply wait and see whether women shall be admitted to Russian universities. I do not see any logical reason for women not to attend university with men. ... Here in Germany, for example, society, and especially the students, are not in the least inclined to support university education for women, yet women are admitted to universities without any issues. And then [in Russia] they claim that education at our girls' gymnasiums is insufficient for pursuing a university education, while there are only two girls' gymnasiums in Germany, one in Karlsruhe and one in Stuttgart, and girls here typically attend the *Höhere Mädchenschule* [a German secondary school for girls], which has a far lower level than our Mariinskaya gymnasiums, they don't even know what the word algebra means, yet despite this, anyone [here in Germany] who wants to do so can sit for final examinations'.

Vera Schiff had a point. Even conservative Austria had admitted girls to university since 1897. In Vienna, Paul's mentor Boltzmann was one of the advocates for admitting women to the physics department. But most western and central European countries had a dearth of secondary schools that prepared girls for university, whereas Russia had founded the famous Mariinskaya gymnasiums decades earlier, named after Tsarina Maria Fyodorovna.[9] But perhaps more important to Tanya was that the letter brought something else to her attention: the opportunities in Germany for well-educated Russian women. People like herself.

5
Between gods and humans
The Mecca of science

In 1901 and 1902, most people weren't used to the new century yet; they still had to assimilate the changes of the previous century that had brought coal and steam, trains, factories. Even though people in the countryside still rose and went to bed with the sun, their clock towers were now synchronised to other clocks, farther west or east, to prevent railway timetables from being thrown into disarray. And a growing number of whistling, hissing and thudding locomotives brought increasing numbers of people to distant countries where the sun rose or set hours earlier or later—countries where people wore different clothing, where they ate different food, read other books and had another legal system. It allowed people to make new contacts and exchange ideas more easily. Thanks to trains, post arrived faster, and a flood of new manufactured wares inundated ever-larger areas. In the cities, department stores appeared, and gas lighting meant that life simply continued after sunset. The changes were such that they overstimulated some people; the latest condition in these years was 'neurasthenia'.[1]

In Vienna, Paul learned first from his brothers and then at university how much these innovations were based on mathematical and physical principles. The driving rods that propelled steam engines transmitted forces according to the old Newtonian laws. Thermodynamics quantified the energy required for glowing coals to set these rods in motion through steam. James Clerk Maxwell had demonstrated that electromagnetic radiation also transferred energy: in the form of light, heat or, almost imperceptibly, as invisible radiation. And in 1896, one year before Tanya went to the Bestuzhev Courses, Alexander Popov had even used these electromagnetic waves on the University of St Petersburg campus to send messages from one building to another.[2] Genuine wireless communication, unlike the joke Arthur had once played on Paul.

Taking all these discoveries together, you might be tempted to think that mathematicians and physicists had tamed nature. That they could train steam and steel like ponies and horses; that this would enable them to smoothly transform a jumbled world into an orderly entity; that they could use railway timetables, artificial lighting, paved roads and rising stock prices to turn an anarchic jungle, where everyone did their own thing, into a tidy garden. And that just a few well-hidden rough edges would be left over here or there.

So, when Paul took the train to Göttingen, he didn't expect to encounter untamed phenomena. Mounting evidence even existed for the hypothetical atoms that were fundamental to the work Paul's mentor Boltzmann. In the years leading up to the turn of the century, French physicist Jean Perrin and his British colleague J.J. Thomson discovered cathode rays, for instance. These rays were emitted by the material at one end (the negative pole) of a vacuum tube after applying a large difference in voltage, causing a fluorescent screen near the other end of the tube (the positive pole) to softly light up. It soon became clear that this glow was caused by the impact of negatively charged particles that were accelerated towards the positive pole, and that were for example also at the heart of Dutchman Hendrik Lorentz' electron theory. And even though their role was still uncertain, at the very least these so-called 'electrons' made clear that materials were comprised of smaller components. Likewise, X-rays emitted by materials or radiation emitted spontaneously by uranium salts indicated internal activity and structure in materials.[3]

Paul wanted to attend twenty-seven-year-old Johannes Stark's lectures on such phenomena. This young professor already had forty articles to his name, mostly about cathode rays, a subject in vogue.[4] But after his arrival, Paul also enrolled in no less than seven other courses, including electromagnetism with Max Abraham and mechanics with famous mathematician Felix Klein.[5] Soon he also made a new friend: Walter Ritz, son of a well-known landscape painter who grew up in a large house in the mountains around Sion in Switzerland.

Paul looked up to Ritz, who arrived in Göttingen at about the same time to write his dissertation under Woldemar Voigt, director of the centre for mathematical physics. Ritz was composed, purposeful, had a sarcastic sense of humour and, like Herglotz, could tackle complex mathematical formulas with great skill (Hsu and Zhang 2001, 572–573). Once again, Paul had found a role model, but it was also simply pleasant to have a connection with someone in Göttingen.

The city was tiny compared to Vienna, but walking past the half-timbered houses, leaning genteelly with age, it was easy to feel intimidated. Göttingen had no clattering horse trams, broad avenues, towering government buildings. You could cross the centre by foot in 15 minutes; the central town square was no bigger than the squares in Favoriten, and its few churches were unimposing. But it exuded scholarship: it was as though some eminent professor had lived in almost every house. The Brothers Grimm, of course, were professors of literature who had used the babbling brooks, murky ponds, swaying fields and rustling woods around the town as decor in their fairy tales. Or bricklayer's son Carl Gauss, prince of mathematicians, who was so attached to Göttingen that after his study, he stayed at the local Georg-August University for the rest of his life because 'no matter what advantages large cities may have in other pleasures, nowhere else is the pursuit of prizing nature's great secrets from her received with warmer enthusiasm'.[6]

It goes without saying that Paul visited the Göttingen Observatory, the astronomical observatory just outside the town walls of which Gauss had been the first director. Doubtless he also heard about Gauss' sophisticated calculations—when still a student—on the orbits of planets and planetoids. Or about the precision instrument Gauss developed to determine the geomagnetic field. And particularly about all the mathematical theorems Gauss discovered and proved starting in his secondary-school days (Dunnington 1955; Bühler 1981).

It was almost surprising that anyone in Göttingen dared to venture into mathematics under Gauss' intimidating shadow. Yet number theorist Johan Dirichlet had already charted new directions in the field during Gauss' lifetime. And Gauss' student Bernhard Riemann's work would be crucial for Einstein's relativity theory in the twentieth century. Their former homes in the narrow streets of Göttingen, like that of many other learned men, were easy to spot. And this long, venerable tradition augmented the standing of professors who, like Paul, strolled through the town at the dawn of the twentieth century. The streets were imbued with reverence.

Fortunately, there was more than just stagnant erudition, Paul observed. In the late nineteenth century, Felix Klein had brought fresh vitality to Göttingen's august but by then rather torpid mathematics. Klein was a versatile mathematician and a superior teacher who prepared his classes down to the smallest detail and had thought about every drawing on the blackboard. 'He was like a prince who wanted to show his admirers his splendid realm, leading them along endlessly winding paths through an obviously impenetrable landscape, pausing on every hill to look back at the path they had travelled', according to his former student and future Nobel Prize laureate Max Born, who incidentally was quite vexed by all the winding paths (Thiele 2011, 196).[7]

Klein was also an organiser. The technical ingenuity of the Chicago World's Fair in 1893 had deeply impressed him, and he was determined to create linkages between mathematics and industry in Germany too. He founded the 'Göttingen association for the advancement of applied physics and mathematics', where professors and industrialists deliberated over cooperation, and he tried to develop new branches of research (Tobies 2021, 439–450). Paul's eldest brother Arthur would have loved it.

Paul was more inclined towards 'the search for truth', discovering the essence of nature. Another aspect of Klein's work appealed more to him. Klein wanted to make mathematics teaching more *anschaulich* (intuitively visualisable) and more expressive, freeing formulas from paper. A literal illustration of this effort was the collection of mathematical models in the display cases next to the lecture halls on the third floor of the stately auditorium of the Georg-August University. These models, often delicate and made from materials like enamel, gypsum, glass, paper, plywood or wire, and featuring sensual curves or angularly symmetries, seemed to bring to life the mathematical formulas, rendering functions, curves and objects visible and tangible.[8]

But Paul's attention was primarily drawn to the adjacent reading room established by Klein. In the distinguished hall on the ground floor of the building, the black-and-white chequered floor tiles shone beneath ornate ceilings; farther up in the building broad floorboards creaked. The bookcases at the back of the reading room held several thousand books, next to binders containing scientific publications. They couldn't be loaned and had to be read at one of the long reading tables in the room (Rowe 1989, 202). A solemn silence pervaded the space. Sometimes the rustle of pages being turned or the movement of a neighbour's pencil tip softly scraping over paper could be heard. And yet in the reading room you felt part of a group. A group of people who were different from others: in some senses more intelligent, insightful and original; in other aspects perhaps unworldly or eccentric. They were in their element, and Paul often spent hours here (see Figure 5.1).

Perhaps most rewarding were Klein's discussions of the latest mathematics and physics publications in the library during weekly mathematics seminars with a

Figure 5.1 Paul in Göttingen in 1901.—*Photograph Ehrenfest Family Archive.*

small club of students and teachers. He often gave an introductory lecture himself, but students were also permitted to introduce articles. In any case, they could listen, think along with the group and participate in discussions. Klein wanted to give them an idea of the scientific enterprise, sharpen their thoughts in exchanges and let them partake in new developments not covered in the lectures (Tobies 2021, 394–395).

Naturally, as is only fitting between 'gods and mortals', a suitable distance was maintained between students and supremely learned Klein, who had a cast of Jupiter's head in front of his study door at home (Thiele 2011, 203).[9] At the same time Paul felt, thanks to his sessions with Klein, strengthened in his plan to devote himself to theoretical physics. He wasn't going to be discouraged by everything he didn't yet know. 'Believe me and learn now', he wrote in a black leather notebook in his third semester in Göttingen, in 1902. 'Try to make progress, write down when you don't understand something, when you discover shortcomings and continue; otherwise, you might end up getting stuck on a subject that could easily consume your entire life'.[10]

Hungarian mathematician Farkas Bolyai, a friend of Gauss, had once impressed this adage on his mathematical son, Janos. Paul often walked past the house in Kurze Strasse, where Bolyai had lived for a few years when elderly, and soon adopted this paternal advice as his guiding principle. For the rest of his time as a student—and for many years afterwards—he scribbled comments in a growing pile of notebooks about lectures, textbooks, articles and, most of all, a great many questions.

But in other ways shaping his future wasn't so simple. Paul took classes with mathematicians such as Klein, Otto Blumenthal and Zermelo, and physicists such as Voigt, Abraham and Karl Schwarzschild. He learned how physicists tried to tame gravity, planetary orbits, atoms, electricity and other physical phenomena with mathematical formulas, occasionally borrowing these formulas from existing mathematics and at other times developing their own. He grappled with all this knowledge as much as he could, but what would come next?

Physicists with a comprehensive understanding of the entire field became a thing of the past. Experiments had grown more complex, as had the mathematics needed to interpret results. Theoretical physics was becoming an increasingly independent discipline. Mathematicians, too, specialised more and more and shifted their focus. They placed less emphasis on arithmetic, puzzle solving and the technical aspects of the field and more on abstraction, on concepts and ideas and on structures in logical reasoning and proofs.

These developments were intertwined. Klein's colleague, mathematician David Hilbert, brought them together concretely when, in 1900, he compiled a list of the twenty-three most important mathematical problems of the twentieth century (Reid 1996a). Based on a few basic assumptions, or 'axioms', Hilbert still wanted to use logical reasoning to construct a sturdy edifice that would establish the concepts and notions in each mathematical field as well as the relationships between them. He had thus observed with disapproval the latest discoveries in physics—such as cathode rays, X-rays and electrons—and how physicists attempted to encapsulate them in formulas. To his mind, this was stopgap mathematics—too intuitive, too off-the-cuff, and the kind of mathematics from which physicists often drew overly far-reaching conclusions. Hilbert believed this could, and must, be done better and more rigorously, and he decided to incorporate the entire field of physics as problem number six on his list.

It certainly wasn't a call to engineers and physicists like Klein's to use mathematics as a tool. Hilbert believed that nature itself was expressed in the language of mathematics. He was convinced that physics would be more consistent, clear and convincing if physicists, from the outset, constructed their theories according to the rules of mathematical language rather than piecing together formulas ad hoc. And his opinion carried weight. Hilbert's list, which he presented to colleagues in Paris, definitively established his name and put Göttingen on the map as a leading centre of mathematics alongside Berlin and Paris. It also immediately became evident that Klein's decision to persuade Hilbert to move from Königsberg to the town in 1895 had been a masterstroke.

Paul soon noticed that students sang the praises of Hilbert's lectures. Hilbert's teaching style was entirely different from that of 'the Great Felix', as Klein was sometimes ironically called (Tobies 2019, 5). Hilbert loosely prepared his lectures, got muddled sometimes, lost his way in a proof or had to seek his assistant's advice. But this spontaneity was his goal: Hilbert didn't want to show the result of thinking but the process. He didn't give elaborate tours of the mathematical landscape like Klein but, like a mountain guide, solely sketched the shortest and surest route to the soaring peak: a result (Reid 1986, 103–104; Thiele 2011, 96). 'You could feel his intellectual muscles working'[11] (Reid 1996b, 23).

It made Hilbert unconventional. His status allowed him to disregard the self-imposed punctiliousness and conceited formality of the provincial town, simply cycling to the university in a short-sleeved shirt. Or go on his bicycle to his latest girlfriend, perhaps with a bouquet from his own garden. And if he felt the slightest draught during a concert or play, he would casually flip up the fur collar of one of the ladies next to him in the audience (Reid 1986, 131). But Hilbert was strict and selective when choosing his '*Wunderkinder*', the students allowed to join long walks filled with mathematical discussions (Reid 1986, 52). Much to Paul's disappointment, despite attending his lectures, he wasn't included in this circle. Nor could he become part of the small group of students taking their doctoral degrees under Hilbert, often focusing on topics in physics.

Fortunately, he had Ritz, who delved into current issues in physics on electrons, loved complex mathematics and was eager to share his insights with Paul. Things improved even more when Hahn and Herglotz arrived in Göttingen in autumn.

They continued their tradition of strolling in the hills and engaging in endless conversations in coffeehouses, even though Hahn and Herglotz had surpassed of Paul by then. Herglotz had already taken his doctoral degree in astronomy at Munich and was going to do his *Habilitation* under Klein. This German degree—a kind of second doctoral degree—would allow him to become a professor. Hahn, who had taken his doctoral degree in mathematics in Vienna, had received a scholarship to continue research on the subject in Göttingen. However, Paul hadn't stood still either: he had discovered something that melancholy Herglotz and more philosophically inclined Hahn didn't yet know.

6
Don't ever smoke
A shared passion kindles love

Unsurprisingly, Tanya was conspicuous when she first appeared in the lecture halls in Göttingen in 1902. At Georg-August University, she stood out for being a woman. Her straight short-cut hair and blouses with tie-like bows did little to obscure her presence in the almost-exclusively male lecture halls. Outside, her appearance only made her more noticeable: pinned-up hair and embroidered clothing with rounded lines were more conventional for women in Göttingen. But her exotic air made sense when one heard her clear voice: Tanya's German was perfectly good, but her accent suggested Russian provenance.

This wasn't necessarily an advantage. Schiff had rightly observed the paucity of female German students. The presence of Polish and Russian women in lecture halls, often with better prior education, irritated many Germans—even more so as lecture halls became increasingly overcrowded and universities more overburdened. Many blamed foreign students, especially women. In 1899, German students in Halle demanded the exclusion of foreign women from medical school. And when the University of Leipzig withdrew its recognition of Russian girls' gymnasium diplomas in 1901, the medical faculty in Göttingen followed suit (Neville Bonner 1992, 117–118).

But not the mathematicians in Göttingen. Almost ten years earlier, in 1893, after a visit to the United States, Klein had opened the mathematics lectures in Göttingen to women. That same year, Americans Mary Winston and Margaret Malby (who would go on to do a doctorate in physics under Walther Nernst) and Englishwoman Grace Chisholm attended lectures as 'auditors', with the support of Klein, who had also helped with the necessary paperwork. A few years later, the Education Ministry in Berlin granted Grace Chisholm special dispensation to complete her doctorate in Göttingen under Klein (Jones 2000).

Such experiences convinced Klein that women should be admitted to university as students on an equal footing with men—and that there was no need for separate women's colleges. 'I would simply like to point out that during this semester, for instance, no fewer than six women have participated in our higher mathematics courses and practicals and have continually proven themselves to be equal to their male classmates in every respect', he said in an 1897 survey on the subject. And he added that 'the nature of the situation is that, for the time being, these women have been exclusively foreigners: two Americans, an Englishwoman and three Russians, but certainly no one would wish to assert that these foreign nations possess some inherent and specific talent that we lack, and thus that, with suitable preparation, our German women should not be able to accomplish the same thing'. Klein thus shared Schiff's view that German secondary education didn't adequately prepare girls for university studies. And he did much to change this and to ensure that women were welcome not only at university but also in other academic bodies such as learned societies (Tobies 2021, 412–418).

Others emulated his example. For instance, Hilbert followed in Klein's footsteps by supervising a total of six female doctoral students (out of a total of 69 doctoral students). Among them were the Russians Lyubov Zapolskaya and Nadezhda Gernet, who had taken their doctorates in Göttingen in 1901 and 1902 (Tobies 2022). In other words, when Tanya arrived in Göttingen with Aunt Sonya, a small group of women had already begun to pave the way.

Probably, the most famous female mathematician of all to complete her studies in Germany was already dead. Sofya Kovalevskaya, another Russian, had been the first (active) woman professor of mathematics in the world.[1] According to—often romanticised—tales of her life, as a child she had been intrigued by the strange symbols on her bedroom walls: not having wallpaper, her parents had pasted a mathematics lecture dictation instead. Years later, she read mathematics under Carl Weierstrass in Berlin, 'illicitly' attending his lectures. Although she was barred from the ceremony in Göttingen in 1874, her dissertation was awarded the highest honour, *summa cum laude*.

Yet regardless of her colleagues' appreciation of her work, Kovalevskaya failed to secure a position in Germany and could teach only at primary schools in Russia. She started writing novels, her talent dissipated somewhat, her love life was turbulent and Weierstrass had to use 'all his powers of persuasion to secure her a professorship in Sweden. Kovalevskaya did important work there until her untimely death (Koblitz 2000, Ch. 5).

Despite, or perhaps also because of, these examples Tanya felt intimidated when she first entered the lecture halls and especially the reading room in Göttingen. The bookcases contained so many tomes. Where should she begin? Which authors were important? In which books would she find basic knowledge; and which articles described interesting developments?

Assisting at the Bestuzhev Courses and teaching lower gymnasium classes had made her realise the previous year how much more there was to study than she had yet done. It had inspired her decision to go to Göttingen, with Zapolskaya and Gernet as examples. Moreover, she hoped to learn more about physics in Göttingen. But most of all she hoped the study and research in Göttingen would be more modern.[2]

This last was of course so. Klein's attempts to connect mathematics and physics to industry had produced an electrifying and enthusiastic atmosphere. Johannes Stark conducted extensive experiments on radiation. Hermann Simon headed an institute for applications of electricity; Emil Wiechert was establishing a geophysics institute; Ludwig Prandtl's work, starting in 1904, would result in the renowned aerodynamics institute; and Carl Runge would oversee the new department of applied mathematics from 1904 onwards.

All of this was still out of Tanya's reach. At the long table in the silent reading room, high above the botanical garden where gardeners raked autumn leaves, she staked out her own little territory, burying herself in her books and spreading out her papers around her. She felt less exposed than in the lecture halls to the curious gaze of students wondering who this woman was. After a year of working at the gymnasium and the Bestuzhev Courses, she had to readjust to life as a student. Moreover, here and there, critical voices from Halle and Leipzig could be heard, and living among male students was new to her anyway.

Things only became easier when one of these students acted. Paul, direct and curious, had inquired about the slender Russian who had suddenly appeared at Klein and Hilbert's lectures. He asked why she never attended the mathematics students' club weekly meetings and was annoyed by the answer: women were only welcome during festivities. It took time and energy to convince the other students that Tanya ought to be invited. 'But he succeeded', Tanya wrote later. 'Afterwards, he checked with Russian students, whom he had waited with to see whether I would accept the invitation. Well, I was happy with any offer giving me the opportunity to better understand life in Germany. And I greatly desired scientific contact with other mathematicians and physicists'.[3]

Tanya was settling in. Sometimes she was invited to tea or dinner at professors' homes, and the fact that her uncle had been a mathematics professor in St Petersburg opened many doors. She often invited other students to Kirchstrasse, where she rented a flat with her Aunt Sonya, who continued to chaperone her niece.[4] Herglotz, Hahn and Paul enjoyed visiting. Sonya loved a lively atmosphere, served tea, dished out biscuits, chivvied the cook in the kitchen and encouraged Tanya's friends to play the grand piano in the living room. It was always light and sunny there, Herglotz later wrote in a letter.

And then there was Tanya. With her delicate figure and youthful face, it wasn't very noticeable that she was older than most other students; she was turning twenty-seven that winter semester. Increasingly she was seen with Paul. He was thickset, with bristly hair, sparkling eyes and agile hands; she was slight, calm, with an amused smile and slanted grey eyes. While she was surprised by his casual colloquialisms and Viennese expressions, he listened enraptured to her rolling Rs. And she would watch amusedly how he wanted to know all the ins and outs during lectures, while he enjoyed it when she engaged in logical reasoning.

Their conversations, which had started at a long table at the restaurant on Theaterplatz, were soon about more than just mathematics and physics.[5] In the hall in front of the reading room, on walks through the long and narrow botanical garden, in the coffee house on the town-hall square and at Tanya's home, they talked about music, literature, Vienna and St Petersburg, friends, professors and what to do after their studies. They discovered a shared love of playing the piano and dislike of timber-framed houses.[6]

Paul had grown up with brothers and at boys' schools; Tanya with women and at girls' schools. Both were unaccustomed to the opposite sex. To some extent, they were both alone: Paul was parentless, while Tanya couldn't remember her father and saw her mother only sporadically. Just as Arthur, Paul's guardian, had expressed misgivings about the dreams his youngest brother was pursuing, Tanya's uncle Pyotr had responded dismissively to her dreams of her future.

Did they also see things in this light when they walked around Göttingen? Insecure Paul could show Tanya around the streets and squares that had been unfamiliar to him just a year before; he could explain to her what he had already studied. Tanya walked next to a man who listened to her logical reasoning, involved her in discussions and genuinely took her seriously, not half-heartedly because of fashionable feminist 'ideals' or simply not at all (see Figure 6.1).

But Paul's support didn't mean that the rest of the world treated Tanya as an equal. Many, and probably most, people still regarded women such as Zapolskaya, Gernet and Chisholm, as well as Tanya, primarily as exceptions confirming the rule: clever and tenacious but in their isolation not really threatening to the status quo. What this status quo was like becomes clear from the letter Göttingen professor Ludwig Prandtl wrote around this time to his fiancée Gertrud Föppl: 'Unfortunately, we men can but rarely give our wives enough in return for the devoted love they give us. A wife dedicates herself entirely to the household for the sake of her husband—while she must share him with his professional interests'.[7] His view was shared by Föppl's mother, who subsequently wrote back, assuring the professor that she had raised her daughter 'to pursue the profound happiness that comes from faithfully fulfilling duties at home'.[8]

It left Tanya and the few other women at university in an ambiguous position. They knew that other women did *Herr Professor*'s housework, raised his children and sometimes—in the case of Käthe Hilbert, for example—even typed his publications and lecture notes while he stood weightily in a lecture hall. Doubtless, they sensed that the professors were aware that they knew this. They were in a no-man's-land between the sexes.

It also meant that Paul and Tanya—who were soon inseparable—had few role models. Paul of course knew that his Viennese mentor Boltzmann had married a qualified mathematics and physics teacher. Perhaps he knew even that Boltzmann had encouraged his fiancée Henriette von Aigentler to appeal when the university had refused to allow her to attend lectures as a non-examination student because the presence of women would have 'endangered the institution'. But like so many of his contemporaries, Boltzmann was emancipated only to a certain degree. He

Figure 6.1 Paul, Tanya and Walter Ritz in Göttingen.—*Photograph Ehrenfest Family Archive.*

hadn't envisioned an independent career for von Aigentler. 'Love cannot endure, I think, if one's wife cannot show understanding and enthusiasm for her husband's efforts and is merely a housekeeper instead of a companion in his struggles', he explained in a letter during their engagement (Lewin-Sime 1996, 14). This role as an intellectually capable companion—but no more than that—is exactly what von Aigentler was given.

Did Paul and Tanya imagine something similar for themselves? Tanya came from another tradition. She had the example of the Bestuzhevki, Gernet, Zapolskaya, and Schiff with her independent publications. On the other hand, she had her Aunt Sonya, a widow who leant on her perseverant niece and who enjoyed life in high circles with their own conventions. Tanya certainly wasn't as progressive or even radical as some bestuzhevki. She was tight-lipped about the idea that women should be economically independent, perhaps partly because she never had to worry about money. And she accepted that her aunt never left her side.

As early as in 1843, influential Russian writer and philosopher Herzen described marriage as a worthless institution in his diary: 'In the future there will be no marriage, the wife will be freed from slavery; and what sort of a word is "wife" anyway? Woman is so humiliated that, like an animal, she is called by the name of her master. Free relations between the sexes, the public education of children and the organisation of property, morality, conscience, public opinion, and not the police—all this will define the details of family relationships' (Stites 1990, 22). Others also made critical comments. For example, one of the characters in a mid-nineteenth-century novel by influential writer Nikolay Chernyshevsky says to his bride: 'You should look on me ... as a ... comrade who is striving toward one goal together with you' (Koblitz 2000, 67). But nothing indicates that Tanya thought marriage was outmoded.

Tanya had other concerns. Together with Paul, she wanted to contemplate the things that truly mattered to them, and she wanted to do all that thinking and working together for the rest of her life. 'Don't ever smoke', she wrote in Paul's notebook on 12 December 1902.[9] Evidently, she hoped to have a long life with him.

7
Gleaning knowledge
Leiden canals, Russian novels and Göttingen mathematics

'As a reminder of the pleasant preparation', Ritz wrote on the copy of his dissertation he gifted Tanya in spring 1903.[1] Ritz obtained his doctorate well ahead of either Paul or Tanya—if she would be able to find a doctoral position in the first place. Attaining a doctorate was par for the course for Paul. Aged twenty-three, he had done more than enough courses in his four semesters in Vienna and three in Göttingen to graduate and take the next step. His friends Hahn and Herglotz had already obtained their doctorates and were making a furore in Göttingen with their mathematical work. Ritz gave 'dear Paul' some extra encouragement in his copy of Ritz' doctorate: 'To redoing this as soon as possible and better!'[2]

But how? For his oral defence of his work, Ritz had received the highest praise, a *summa cum laude*. How could Paul surpass this? First, he had to come up with a subject and find a supervisor. Klein in Göttingen had no time. Hilbert surrounded himself with his own '*Wunderkinder*'. Paul wasn't keen on the applied or more technical-mathematical physics that other professors in Göttingen specialised in. And he was at a complete loss for a subject. So, a proposal by Ritz came at exactly the right moment: in March he planned to go to the Netherlands. This was where Hendrik Lorentz was based; whose work on a refined electron theory had won him the Nobel Prize in 1902. Paul's Swiss friend, who had his own ideas about electrons, hoped to meet him. Naturally, Paul wanted to join him (Klein 1970, 45–46).

A few weeks later, the duo moved into a small flat on Apothekersdijk in Leiden, a city of about 50,000 inhabitants, surrounded by marshy polders and traversed by canals and waterways. The waters of the Nieuwe Rijn lapped in front of their doorstep, and a ten-minute walk away, on the Steenschuur canal, was the Physics Laboratory. The building was a maze of winding corridors and laboratories where physicist Heike Kamerlingh Onnes was creating the coldest spot on earth by liquefying helium.

But Paul and Walter Ritz had another goal when they crossed the bridge at Galgewater and walked down the cobblestone streets along Leiden's Physics Laboratory. Or when they walked along the Nieuwe Rijn, past the old Burcht van Leiden castle and over the quays of the Fish Market and the Butter Market to the Lab. They had come for the lectures of Lorentz, a man renowned not only for his Nobel Prize but also for his ability to articulate the latest developments in physics with utmost clarity.

Church bells rang, polished windows reflected the clouds and servant girls swept the doorsteps. Milkmen and postmen still used 'dogcarts' to negotiate the steep bridges, Paul noted in his travel diary. And in the shabby alleys behind the canals, small mirrors often protruded next to windows: *spionnetjes*, 'little spies', mirrors which functioned as spyholes. In even dingier backstreets lived workers from the factories and beer breweries, and 'street urchins aged six or seven smoked' cigarettes on the sly.

Over the canals, snippets of 'piano music [occasionally wafted] through the open windows'. The quiet façades exuded an understated gentility not unlike that of the scholars and dignitaries living behind them. People in Leiden also lived in a class-based society, albeit a veiled one. Poverty was hidden in the alleys, and wealth wasn't conspicuous. Only the colourful 'shiploads of flowers' which sailed in and around the city in springtime were a joy everyone shared.[3] Cognoscenti knew something else also rose above the city's mix of genteel repose and impoverished drabness: physics research was blossoming in Leiden.

It hadn't always been so. When Peter the Great visited the Netherlands in the late seventeenth and early eighteenth centuries, the country was powerful, with wealth accumulated from its colonies in the Orient and the West. Science flourished in this period. In Amsterdam, Peter the Great attended anatomist Frederik Ruysch's dissection lessons; in Leiden, he visited physician Herman Boerhaave and peered down microbiologist Antonie van Leeuwenhoek's microscope at the blood circulation in the tail of an eel (Driessen 1996, Dekkers 2024, 124–125). He took these new discoveries and knowledge with him to his newly founded city, Tanya's beloved St Petersburg.

But in the early nineteenth century, the 'black-garbed' professors of Leiden, with 'genteelly puckered' faces and speaking in 'measured tones', chiefly rested on their laurels. They sat on passably interesting committees and replicated spectacular experiments from abroad at gatherings of societies pursuing 'the cultivation of a useful and agreeable society' by 'promoting knowledge in experimental philosophy and other sciences'. None of which achieved much scientific progress (Maas 2001).

Only when the looms, spinning frames, steamships, steam trains and steam engines of the Industrial Revolution reached the Netherlands did Dutch science regain momentum. To keep abreast of all the new developments, Minister of Education Johan Thorbecke introduced the *Hogere Burgerschool* (HBS; Higher Civic School) in 1863. Thorbecke, considered the founder of modern Dutch parliamentary democracy, thanks to his revision of the Dutch Constitution in 1848, modernised the Dutch education system. The HBS wasn't an elite school where sons of affluent burghers learnt Latin and Greek to prepare for legal or medical studies, but a school focused on the sciences, like

Realschulen in Germany and Austria-Hungary and *realnoye uchilichye* in Russia. Intelligent sons of less wealthy or powerful families immediately seized the opportunity, such as Lorentz, son of a horticulturalist and a small-scale landlord in Arnhem. For Johannes van der Waals, son of a carpenter, this new type of school was a stepping stone on his ascent from the MULO, a level academic lower, to headmaster of an HBS and then to a physicist whose equation of state for gases—which, like Boltzmann, he based on the existence of atoms— earned him a Nobel Prize (Baneke and Maas 2018; Kox and Schatz, 2021, Ch. 5).

A second law on higher education in 1877 gave scholars like him an extra boost. For the first time Dutch law defined a university as an institution that not only provided education but also facilitated scientific research. This scientific dynamism gave rise to new businesses, such as the light-bulb factory founded by the Philips brothers (Maas 2001, 21).

Paul and his friend Walter Ritz were oblivious to all of this. Paul was disappointed to discover that Van der Waals, whose work was so closely related to Boltzmann's, was based in Amsterdam and was difficult to approach. Kamerlingh Onnes' work was too experimental for him. But like Ritz, he was impressed by Lorentz's meticulous lectures, and Ritz managed to arrange an appointment with the professor.

'Unfortunately, I cannot be at home tomorrow afternoon. Therefore, I would appreciate if you and Mr Ehrenfest would come at half-past nine in the morning', Lorentz had written on a card in German on 19 April.[4] 'Home' was a house with a modest stepped gable, small-paned windows and a low doorstep on Hooigracht, just fifteen minutes' walk from Apothekersdijk. It was large by Dutch standards, with a magnificent living room, but probably looked modest rather than imposing to Viennese or other foreign eyes. As did Lorentz, who awaited them in his office, seven stairs up on the corridor.

At his desk and surrounded by bookcases, Lorentz had worked for years in complete tranquillity, going his own way. After studying, he stayed in Arnhem for a while, where he taught at an HBS and wrote his dissertation in the evenings. But just three years after his doctorate, he was appointed professor in Leiden in 1878. His marriage to Aletta Kaiser, the youngest daughter of the director of the Rijksmuseum in Amsterdam, soon elevated him to higher social circles (Kox and Schatz 2021, Ch. 3). There he moved with the same combination of affability and inscrutability as he displayed when greeting Paul and Ritz and pulling up two chairs to his desk.

In physics, Lorentz was a celebrity, even though the Nobel Prize hadn't yet gained its later prestige in the few years since it first had been awarded. Like Boltzmann and Van der Waals, he was working out the building blocks of matter. Not atoms in his case, but components that carried an electric charge and that he initially called 'light-ions'.

This work stemmed from Lorentz' research into light, which he began in his dissertation. Frenchman Augustin Fresnel had been one of his heroes in this regard. At the start of the nineteenth century, Fresnel had opposed the idea—from Newton himself—that sources of light were fountains of shooting 'light particles'. Light is transmitted like waves on the sea, Fresnel demonstrated with elegant experiments. Lorentz later rediscovered these waves in the work of Scottish physicist James Clerk Maxwell, who expressed all the newly discovered electrical and magnetic phenomena in mathematical formulas in 1861 and 1862. In these formulas, light rays were rendered as 'electromagnetic waves' in which an alternating electric field and an alternating magnetic field generated each other in one continuous motion, like an endless dance. Lorentz was one of the first people to fathom and defend Maxwell's theory.

But these light waves also exhibited unexplained characteristics. How did they propagate? Did they travel through ether? What did they look like? How was it possible that light sometimes refracted at the boundary of two materials? How could it break up into colours? Or be reflected? Put in another way: why does a straw in a glass of water 'bend' when you look at it from the side? Why do clouds float in the windowpanes of canal houses? How does a prism or rain disperse white light into a spectrum of colours? Lorentz puzzled over these questions in the quiet seclusion of his study. Imperturbably, he identified three main elements in nature. First, ether, which carried light and heat radiation. Second, light and heat radiation, both forms of electromagnetic radiation. And third, 'ponderable' matter: oxygen, carbon, iron—everything you can weigh. As the link between light and matter, he then proposed electrically charged particles found in ponderable matter: 'light-ions' (Kox and Schatz 2021, Ch. 4).

Lorentz used Maxwell's insight that an electric current can produce electromagnetic waves, such as light, and, conversely, electromagnetic waves can generate a small electric current in a wire. This current consisted of moving electric charges, Lorentz hypothesised, and these charges, were evidently carried by components of matter, the 'light-ions'. Indeed, Lorentz contended that passing light waves interact with these ions and vice versa. The light caused the ions to sway like buoys at sea, and in turn, the swaying electrons slightly altered the direction of the light waves, causing refraction, as in the case of the bent straw. Or they threw back some light: the reflection in canal water.

Furthermore, he realised that these ions were slower than fast-moving light because they had mass. As such, they always lagged in this interaction; more noticeably in short waves of blue light and less so in longer waves of red light. This elucidated how white light can, at times, disperse into the spectrum of colours observed behind a prism or in front of a curtain of raindrops at the conclusion of an interaction.

What subsequently attracted most attention was an extension of this work in which he investigated the interaction between charged particles and magnetic fields. In 1892, Lorentz demonstrated how freely moving electrons are deflected by a strong magnetic field and their trajectory is curved to a greater or lesser degree (Lorentz 1892, Lorentz 1895).[5] Three years later, he also calculated the effect of a magnetic field on the interplay between passing light waves and ions in matter to explain experiments by Pieter Zeeman, his friend and former assistant. Zeeman had placed a piece of asbestos, soaked in a solution of table salt (sodium chloride), in a flame and then placed this arrangement between the poles of a magnet, and saw a sharply defined spectral line (orange) of hot, glowing sodium split into several (yellowish) components.[6] The so-called Zeeman effect and Lorentz's explanation of it led to a small flurry of similar experiments abroad and eventually to a joint Nobel Prize for the two men in 1902.[7]

Lorentz didn't emphasise how follow-up measurements by Zeeman showed the ratio of charge to mass of the light-ions to be very small—much smaller than that of known chemical ions (Kox 1997). This result fell into place in 1897, when J.J. Thomson conducted a meticulously designed experiment with 'his' cathode rays at the University of Cambridge, confirming this charge to mass ratio with high precision. In other words, the cathode rays were made up of small particles, electrons, and these appeared to be the same constituents of matter that Lorentz had described in what was by then called his electron theory.[8]

In the hour in Lorentz' study, Ritz and Lorentz discussed the latter's theory and the interactions between light and matter.[9] Ritz also expressed some reservations, as he had his own ideas about the nature of light, ether and electrons. Paul mostly listened, and when they left the quiet house and walked back along the canals, he decided to devote as much energy as possible to the remaining lectures: up to the third week of May, his black leather notebook contains even more lecture notes and questions than he had written previously. Would he use all this knowledge he had gleaned and these questions he had noted in Bolyai's style to come up finally with a subject for his dissertation?

It was mainly level-headed Tanya who, from a distance, helped him also to look beyond the bounds of physics. 'You must buy a piano by the beginning of the summer semester at the very latest—and not for it to gather dust; whoever has ears, let them hear', she had instructed him in his notebook before his departure for the Netherlands.[10] 'Be good!' For his trip, she had compiled a list of books for him to read as well.[11] 'If possible, quickly read Leo Tolstoy: *War and Peace, Anna Karenina, Family Happiness, The Coffee-House of Surat, Sevastopol Sketches*'. When not preoccupied with electron theory, Paul immersed himself in Tolstoy's Russia.

8

Aversion and love

Seeking a sense of self

By this time, the chestnuts and lime trees in Göttingen were blossoming, and mathematics and physics students discovered formulas that they hadn't seen before. 'There was a great difference between what the professors in St Petersburg had taught us and what was discussed in Göttingen' (Klein, Hilbert, Minkowski), Tanya later said (Ehrenfest-Afanassjewa 1961, Introduction).[1] 'People came here who had completed their official studies in other places in the world. They were mostly younger than me and knew more'.

All the agitation during her studies at the Bestuzhev Courses certainly hadn't had a positive effect on what were already broadly oriented courses. 'Yet I had the courage in the years 1902–1904 to give presentations twice in Klein, Hilbert and Minkowski's seminars', she added.[2] 'Conversations with Paul and Walter Ritz, who I was frequently with, contributed much to my further orientation'.

Like Paul, Tanya was also enthusiastic about the ideas of Klein, who wanted to put more emphasis on *Anschaulichkeit* in mathematics and physics and in the didactics of these disciplines—that is, pay more attention to visualising the concepts and clarifying the relations between them. They were both impressed by Hilbert too. In his lectures on geometry, they learnt his way of thinking about axioms and, using logical reasoning, constructing a mathematical structure from them. But Hilbert didn't only appreciate strict logical thinking. Like Klein, he also recognised the important role played by *Anschaulichkeit*. Later he wrote that there are two tendencies in mathematics: 'The tendency towards abstraction seeks to crystallise the logical relations inherent in the maze of material that is being studied, and to correlate the material in a systematic and orderly manner. On the other hand, the tendency toward *Anschaulichkeit* fosters a more immediate grasp of the objects one studies, a live rapport with them, so to speak, which stresses the concrete meaning of their relations' (Hilbert and Cohn-Vossen 1952, iii). This last tendency plays an important role in geometry, he continued. If people were taught geometry in a lively, clear and visual way, they would understand that mathematics is far more than glorified arithmetic. They could see the beauty of mathematics. Tanya took these considerations to heart.

Paul was still brooding over a subject for his dissertation. From his time in Leiden, he had taken away three things. In restrained and internationally oriented Lorentz, he had encountered a different kind of scientist to domineering Klein, extravagant Hilbert or changeable Boltzmann. It had been good to see that such different styles could coexist side by side.

He had also finally seen the sea, which promised more vistas beyond the horizon. Chugging past the tulip fields, he and Ritz had regularly taken the steam tram from Leiden to Katwijk, where fishermen's wives walked in the wind with golden head brooches, black shawls draped over their shoulders and fluttering skirts, while children in white shirts jostled in the sand and green-grey waves crashed onto the beach.[3]

But most valuable of all seemed to be his notebook, and especially his notes on Lorentz's lectures about German physicist Max Planck's latest work. Planck's research was also on the interaction between light and matter but on a different aspect to Lorentz. His subject was 'black bodies': idealised physical bodies that—unlike all other objects—absorb all light and thermal radiation from their surroundings and then re-emit a characteristic mixture of such radiation, without heating up or cooling down. The point wasn't so much that they were black, but that they were in equilibrium with their surroundings, not gaining or losing any net energy and therefore always staying at the same temperature. In fact, to a good approximation, real objects at a fixed temperature, such as the white-yellow sun, could also be described as black bodies.

The light emitted by such objects was a mixture of colours around the dominant colour, and the temperature of the object determined which colour predominated. If an object was as hot as a glowing log in a fireplace, then red predominated. If the object was even hotter, such as the sun, then yellow was prevalent. In stars hotter than the sun, blue prevailed. In cooler objects, invisible infrared light predominated. And the shape of the light spectrum—the mixture of colours organised by frequency or by wavelength—was always the same as Planck's eminent colleague Wilhelm Wien had demonstrated at the end of the nineteenth century.

Wien couldn't completely explain this shape. For blue and ultraviolet—the high frequencies and short wavelengths—the formulas derived by Englishmen Lord Rayleigh and Sir James Jeans' description worked better. But Wien did demonstrate that the spectrum simply shifts a little from red to yellow to bluish white as the temperature of a black body increases. And Planck was intent on finding out why this was so.

Paul was enthusiastic when he noticed that reputedly conservative Planck, who in the past had often objected to Boltzmann's use of statistical methods, was now inspired by those same methods. Boltzmann had demonstrated that the temperature of a gas is a measure of the average kinetic energy—velocity—of the gas atoms shooting around, most of them with energies close to the average energy. Planck wondered whether the temperature of a black body would somewhat similarly correlate with the energy—and thus the dominant colour—of the light rays in the emitted.

In his calculations, Boltzmann had intricately divided gas atoms into small groups, each with their own specific average kinetic energy. Just as intricately, Planck divided the energy black bodies and light rays exchanged.[4] His calculations started to work when he described a black body as a collection of oscillators, each absorbing or emitting precisely determined parcels of light energy, depending on the frequency of their vibration. It was a complex calculation, and there hardly seemed to be physical grounds for the energy parcels, but amazingly, all this computation produced a description of the spectra of black bodies that was almost the same as Wien's for high frequencies and Rayleigh and Jeans' for low frequencies, describing the entire spectrum for the first time. Planck appeared to have found a way to a deeper explanation of the interaction between light and material.

Yet despite his Boltzmannian background, Paul couldn't distil a subject for his dissertation from this. He hadn't found a supervisor in Leiden either. Nor had he managed to shake off the unease he had felt since before his trip to Leiden. Under no circumstances did he want to end up in a stolid job at a technical company, like his brothers Arthur and Otto, but he didn't envision a career in mathematics either. Theoretical physics is what appealed to him, and after long conversations with Ritz and Tanya, he decided to return to Vienna.[5] There, Paul renewed contact with Boltzmann, who had also returned to the Austro-Hungarian capital and who soon submitted a small article on a minor aspect of Van der Waals' work on behalf of Paul to the Imperial Academy of Sciences (Ehrenfest 1903). It was Paul's first publication and a leg-up towards writing his dissertation in Vienna.

But when the universities in Vienna closed for the summer vacations, Paul fled the city once more. On the West-Frisian island of Schiermonnikoog in the Netherlands—no more than a few square kilometres of sand, bathed by the sea and swept by the wind—he felt 'more at home among the fishermen ... than with my countrymen in Vienna'.[6] Rested, he then started a trip around Italy.[7] Like many privileged young men who had undertaken such a Grand Tour before him, he marvelled at the Roman statues, renaissance paintings, rococo palaces and much more.

'Lord God, if only I had seen a tiny morsel of this as a child', he wrote dumbfounded in his travel diary after visiting the Pantheon in Rome.[8] He noted the names of the buildings and the painters: 'Titian, Da Vinci, Michelangelo, Botticelli'. Italy was a treasure trove, like the Persian garden in the tales of Aladdin. Everywhere there was other art, new cityscapes and landscapes where the mellow light brushed over rocks or olive groves. Paul, once the boy who had counted sugar loaves in a poor, working-class neighbourhood in Vienna, felt overwhelmed.

In Florence, which gently bustled beneath the evening sky, the 'string of pearls of the lanterns down the Arno' caught his eye. 'One ought to take one's children to Florence for their upbringing'.[9] In Venice, he hadn't chosen modest accommodation: 'the view of the Grand Canal from the hotel!' He ended his trip at Lake Como.

'Tatjana!!' he wrote alongside a pencil sketch, and he listed all the colours: golden treetops, violet hills, russet rocks ...[10]

On the way, he also learnt a thing or two about himself. He was well aware that he easily made contact and dared to approach people with enthusiasm. Much later, his brother explicitly wrote in a letter: 'You have that rare ability of smelling unusual personalities & to quickly approach them'.[11] But just as often people wearied him. Constantly, he was overtaken by an 'indescribable aversion' towards people with whom he 'must be together only for a short time'.[12] On the shores of Lake Como— 'Waves on the water'—he was inundated with gloom. 'Get to work, you!'—'Anxiety' he noted.[13] 'Isn't this feeling of nervous aversion that often overtakes me simply another form of the same psychological condition that used to manifest as Sunday boredom, etc.?' With thick lines, he had drawn a square around the question that evidently preoccupied him the most: 'How to heal?'[14]

'You poor, poor *Kindjutsch* [tender heart]', Tanya replied on one of the postcards she sent Paul, with her pet name for him. 'You write such sad letters, but I hope you shall soon feel brighter'.[15] Her consistent refusal to be daunted by anything helped Paul to regain his footing. In autumn 1903, he finally found a subject for his dissertation and approached Boltzmann. While Tanya completed her third and fourth semesters in Göttingen, Paul wrote his dissertation in Vienna and on the Dalmatian coast. He used late German physicist Heinrich Hertz's method to describe the movement of rigid bodies in an incompressible fluid, and he wrote quickly (Ehrenfest 1904). Perhaps 'only love' can overcome this aversion; he had scribbled next to his remark about his aversion towards his fellow humans.[16]

9
Ill at ease in Vienna
Study, discussions and strolling with a pram

In Vienna, you could go to the opera. In Vienna, you could have endless discussions and study in coffeehouses. Without a job, you could even do so with an inheritance, like Paul, who had received a sizeable sum from his parents, and like Tanya and her Aunt Sonya, who weren't without means thanks to Uncle Pyotr's prudent investments in Russian railway stocks. That Paul and Tanya married on 21 December 1904—just days before Western-Christian Christmas—was coincidental. Perhaps Paul remembered how a tin Christmas angel had enchanted him long ago. If Tanya valued any Christian festival, it was Easter. Be more importantly, they had formally abandoned church calendars and religious practices because marrying in Austro-Hungary obliged them—as a Russian Orthodox woman and a Jewish man—to renounce their faith.

They had quarrelled about it on postcards sent between Vienna and St Petersburg. '1) You still don't know me entirely if you believe I would convert to another religion after giving up the religion of my youth', Tanya had written resolutely on a Christmas card. On the front, a Christ-child cheerfully stretched his arms towards his mother. '2) I find it absurd that even for a moment you presume I would leave the church while you stay Jewish!!!'[1]

Clear language and a completely different tone from previous cards. After which Tanya's 'whole world was reduced to a single point' because she hadn't received any post from 'dear, bad and my only *Pauljutsch* [little Paul]' for five days.[2] On these, she wrote that she hoped 'soon finally to be together again' and asked Paul whether he had considered his 'university affairs and entire career' adequately in his plan to renounce his religion.[3]

Had he come up with this plan first? Had she then become annoyed that he backtracked and thought she alone should renounce her faith? Did she think they should be consistent, and both abandon their religions? Had it then annoyed her that he baulked at the decision?

As a teenager, Paul had vehemently opposed any form of religion and had wanted to do everything a little differently to his more conservative brothers. At the same time, he had felt badly let down when shortly after his mother's death his eldest brother had married, moved to Berlin and converted to Lutheranism. Evidently, things were more complicated for Paul than for Tanya, who could summarise her considerations on a card: 'No denomination = not irreligious'.[4]

Paul's recollections of the togetherness of his large Jewish family were perhaps deeper-rooted than he realised or cared to admit.

Harmony was restored on postcards from St Petersburg to Vienna a little later. They featured a couple standing upright in a storm at sea; on the other, Aladdin floated through the air on a carpet. 'My boundless love, do you have an inkling how certain my future seems to me?' Tanya wrote on them. 'What else could happen? I am utterly unable to worry about anything anymore. I am just impatiently waiting until we are finally together again. *Kindjutsch* [tender heart], look at me. Yes?!'[5] The matter seemed completely settled when Tanya visited the Russian consul a day after their wedding. Usually, Russians who had renounced their Russian Orthodox faith abroad weren't allowed back into the country, but Tanya managed to convince the consul—'a kind man'—that she and Paul 'were really harmless to the Russian government despite our being areligious'.[6]

For the time being, they were staying in Vienna anyway. Their address was on the simple wedding card they sent to friends and family: Radeckgasse 1.[7] At the far end of this street was the railway line, with working class Favoriten, where Paul and his brothers were raised. The brand new, modern residential block with their flat was situated at the other end, where the Radeckgasse ended in green and pedestrianised Alois-Drasche-Park. In the bright rooms high above the park, they had placed a grand piano and hung a vista of St Petersburg, and in the study, portraits of scholars (see Figure 9.1). The dining area had a round table and the sitting area a sofa covered with rugs. There was a room for Aunt Sonya, and they had found a housemaid who also cooked.[8]

Tanya could sense that Paul's elder brothers looked at them with disapproval. The carefree manner she and Paul placed science at the centre of their lives and assumed everything else would work out was unfamiliar to them. The brothers had wisely invested their inheritances: Emil had become a successful businessman, with his own Villa Ehrenfest in the well-heeled Hietzing district. Otto had a flourishing electro-technical company; Arthur was a widely respected engineer; and Hugo had used the money to build a career as a gynaecologist in the United States. They were pleased with their social status and glad their wives didn't have to toil as hard as their mother had. But their little brother wore a Tolstoy shirt, followed Tolstoy in refraining from smoking and drinking and made no effort to look for paid employment. He preferred discussions at home or at Viennese coffeehouses with physicist friends or going to lectures with Tanya. His brothers doubted all this studying and thinking would get him anywhere.[9]

Meanwhile, Tanya was having difficulty settling into Vienna. At the Bestuzhev Courses, despite everything, she had been surrounded by women who, like herself, demanded the right to education. And although she had been the odd one out in Göttingen, she had still been one of the students. But her status in Vienna was unclear. There were almost no foreign students and few women students, let alone married women students. It made her an outsider when attending lectures

Figure 9.1 Tanya and her aunt Sonya in the study in the new flat in Vienna.—*Photograph Ehrenfest Family Archive.*

as a non-examination student, even though she had received the official seal of approval.[10] And at home, she struggled with her role as spouse: once when Otto and his wife Olga visited while both her aunt and cook were out, Tanya wasn't able to offer them dinner. She was tired and had never learned how to cook: 'I really couldn't cope' she later wrote to Paul, 'I'm still ashamed.'[11]

Paul knew how much Tanya pined for her native country. After their honeymoon, with snow, sleighs and friends in Bavaria, he made three promises in their red journal. 'I promise to move to Russia with you and to live there', he wrote, 'as soon as it is reasonably possible and if you still want to.'[12] Paul would also try to 'learn to read, write and hopefully speak as soon as possible.'[13] Russian is of course what he meant, the language in which Tanya felt at home. Moreover, he also wrote a third promise: 'to have our children baptised and raised in the religion of your youth; so that, unlike me, they can find a homeland in it!'[14] Did he think that any children they had would feel more at home in the world if they belonged to a religious community? Did he want to please Tanya? All she had promised was 'to energetically overcome every sadness—never to shy away from your help.'[15]

Both of them were also well aware that St Petersburg wasn't the most attractive destination at the time. Trade was flourishing, and an abundance of new Art Nouveau buildings and neoclassical palaces were being erected. But the gap between rich and poor, illiterate and learned, powerful and powerless was immense. In early 1905, a month after their wedding, the square in front of the Winter Palace filled with demonstrators. This time it wasn't students who sparked the protests but workers, who toiled long hours in dangerous factories and lived in appalling conditions. The procession was led by charismatic priest Georgy Gapon, who wanted to petition the Tsar. 'Sire, ... we have no strength left. The awful moment has come when death is better than the prolongation of our unendurable tortures'.[16] Were they aware that the Tsar was elsewhere, 'playing dominos' in Tsarskoye Selo?[17] His guardsmen, perhaps partly in panic, fired upon the crowd, and the country was rocked by anger, strikes, unrest, violence, executions, attacks and mourning.

Only in April, when passions in St Petersburg had subsided somewhat and the spring sun had loosened winter's grip, did Tanya introduce Paul to her friends and 'her' city, with its canals and palaces, wharfs and the Neva, with the shopping galleries and gold-clad church domes. They travelled to Petergof, a small town southwest of St Petersburg on the Gulf of Finland, where Peter the Great built Peterhof, a series of palaces including his country retreat Mon Plaisir and the imposing main building of the Peterhof Palace, in gardens filled with fountains, like a Russian Versailles. Here Paul also finally met Tanya's mother Yekaterina—Katya for short—who worked in an orphanage not far from the lush gardens.

Had Tanya already realised that she was pregnant by then? At the end of June, back in Vienna and more than four-months pregnant, she contracted mumps. 'My husband has shown himself to be the best-possible, compassionate nurse', she scribbled next to Paul's concerned notes about her bouts of fever.[18] 'I thank God that he so quickly averted this great danger to us', Paul subsequently wrote with relief.[19] To help her recover, Paul and Aunt Sonya took 'poor, weak *Jutscherl*' to Weesen on Lake Walen in Switzerland.

Were Paul and Tanya insouciant, living in such a nice flat, travelling and holding onto their dream of an academic life without an income? St Petersburg was on edge, but Vienna held no prospect of a university position for either Tanya or Paul. Boltzmann was reserved in his judgement of Paul's rather hastily written dissertation. In search for a topic, Paul had seized upon a remark Boltzmann made during a seminar in Vienna in 1903 about Hertz' novel perspective on the very classical topic of mechanics. He decided to follow Boltzmann's suggestion on this occasion to try and extend Hertz' work, which was formulated in a general and overarching way, to the more specific case of fluid mechanics. But Boltzmann described his dissertation merely as 'diligently and cleverly worked out'—hardly the key that would grant Paul direct access to the academic world, which his friends Herglotz, Tietze, Hahn and Ritz seemed to have entered with such ease.[20] Gaining entry to this world seemed even less likely for Tanya, who belonged to the half of humanity, which according to most was simply inapt for an academic career, leaving exceptional cases aside. But with her railway stocks and bonds, with his inheritance and thanks to Aunt Sonya, they could get by for a while.

It wasn't easy for anyone to enter academia. A dissertation was just the first step. After that, young doctors had to demonstrate that they could survive in the academic world without a supervisor. This required writing a habilitation, entitling them to give lessons and set exams. The next step was to become a *Privatdozent* (the equivalent of an unpaid senior lecturer), associate their name with a university and make a living by being paid by students. And if they were lucky, they could then finally earn a real salary—and standing—by being appointed a professor.

Paul had merely reached the first hurdle in this course: writing a successful habilitation. He had to produce a subject linked to current research that was manageable while also appealing to the university committee tasked with evaluating the work. No mean feat, all the more so because, in his student years, he hadn't really connected with a mentor who could have offered him a helping hand (see Figure 9.2). What Paul was exceptionally good at was explaining existing work. It earned him his friends' appreciation and esteem as well as Tanya's admiration. But his ability to explain professors' theories sometimes better than they themselves, or thinking he had this ability, wasn't a quality that necessarily made him popular with them.

Figure 9.2 Paul in the study in the new flat in Vienna.—*Photograph Ehrenfest Family Archive.*

Strong-mindedness wasn't necessarily smiled upon in the academic world anyway, even when accompanied by an exceptionally original mind. Albert Einstein, one year Paul's senior and living just over 700 kilometres to the west in the Swiss capital Bern, was also left out in the cold—as Paul would find out seven years later. By then, Einstein and Paul would also discover how much else they shared in common. Both came from secular-Jewish middle-class families and were shaped by their Jewish background—probably far more than either man yet imagined. They had both been to a Catholic primary school and had developed an aversion to the authoritarian school system, especially at secondary school. They both liked making music, Paul on the piano and Einstein on the violin. And both had a close group of friends with whom they discussed physics.

Perhaps surprisingly, Tanya and Mileva Marić, Einstein's wife, had much in common too. Like Tanya, Marić grew up in a well-heeled milieu with an Orthodox background, Serbian Orthodox in her case. She too was some four years older than her partner and she too had excelled in high school, in her case a boys' school where she was admitted on an exceptional basis. In Zurich, Marić was the only woman in her year at Eidgenössische Technische Hochschule (ETH), as was Tanya in Göttingen. And like Tanya and Paul, she and Einstein met during their studies and had children soon afterwards. Hans Albert, their son, celebrated his first birthday while Tanya was pregnant.

Finally, Einstein and Marić also almost certainly planned to do things differently—less conformist. In his letter to Marić, 'my little witch',[21] Einstein made her promise that they would never become bores. And they planned to share their lives and work, much as Tanya and Paul did.[22] In both cases, this was more easily said than done.

Marić's academic performance deteriorated after she met Einstein. Had she focused more on Einstein than her studies? Had she felt insecure as the only woman at lectures? She twice failed a final examination in 1901 and 1902, and she permanently abandoned plans to develop her thesis into a dissertation after giving birth, in 1902 and in seclusion, to her daughter Lieserl, who, it is believed, was put up for adoption shortly afterwards (Isaacson 2007, 93–94). And even if to Tanya this may perhaps still have seemed a bit uncertain in 1905, Marić must already have been painfully aware that motherhood posed an additional obstacle to a career.

That is not to say that everything went smoothly for Einstein and Paul. Einstein, like Paul, couldn't get even the simplest of jobs at a university. Their loathing of deferential behaviour certainly didn't help. When he applied for a position with one of his former professors at ETH, Einstein wrote blithely that he seldom had attended the man's calculus lectures.[23] And although Paul would have appreciated this antiauthoritarian disposition if he had known about it, the professor in question certainly didn't (Isaacson 2007, 74–75). It was thus good fortune when Einstein's friend Marcel Grossmann found him a job as a clerk at the

Swiss Patent Office, the more so since Einstein and Marić couldn't afford to spend their time thinking and discussing without a regular income, like Paul and Tanya.

But the year 1905 also revealed a key difference between uncertain Paul and Einstein, who was brimming with ideas and didn't doubt which path he had to take in physics. Indeed, Einstein had a clear vision of the direction of physics. While Paul hesitated and wavered and Tanya was still boldly expanding her knowledge, Einstein had already started to change the course of the field by looking at problems from entirely new perspectives.

In fact, while Paul and Tanya visited St Petersburg, struggled through Tanya's bouts of fever and walked in the mountains around Lake Walen, Einstein did so no less than four times. He knew Viennese philosopher Mach's work, even admiring his idea that physical theories should only describe observed facts, without any extra frills or assumptions. Yet in his work, measured data turned out never to be the starting point. Instead, Einstein built his work on one or more assumptions, axioms, you could say. And even though he insisted that the resulting theoretical formulas subsequently described and predicted measurement results, colleagues were still startled by the boldness with which he chose these axioms.

Still modest in this respect, was his paper for *Annalen der Physik*, in which he elaborated on the central question of his dissertation: how can you demonstrate that atoms (and molecules) exist and that they move around haphazardly in a gas or liquid? Einstein sought the answer in relatively large particles that sometimes float among billions of minuscule liquid molecules, such as fat particles in milk or pollen in a pond. These particles are nudged first one way and then another—randomly, depending on the chance movements of liquid molecules. The beauty was that his formulas precisely described a frequently observed phenomenon: such large fatty proteins or pollen grains indeed go on a random, meandering 'drunken walk'.

Far bolder was the article he submitted to *Annalen der Physik* shortly before and in which he shifted Planck's perspective. Two years earlier, in Leiden during Lorentz's lectures, Paul had studied Planck's model for black bodies, in which light and matter exchange energy through 'oscillators' or 'resonators', a spring-like mechanism that absorbs or releases energy solely in small chunks of a specific size. The size of these chunks was related to the frequency at which the 'resonators' vibrated. More precisely, it was a multiple of an 'energy parcel' that was defined as the product of the resonator frequency and a constant, which came to be known as the Planck constant. But, Einstein suggested, what if light rays themselves were composed of such energy parcels, which were determined by the frequency of the light that then set the corresponding resonator in motion?

This was very different from Paul's approach. He was still working on a short paper that Boltzmann was to present to the Vienna Academy of Sciences in November 1905. Based on his knowledge of Boltzmann's work and what he had learned from Lorentz in Leiden, he carefully compared Planck's statistical methods with Boltzmann's. He showed that Planck's definition of 'entropy' for blackbody radiation differed slightly from Boltzmann's for atoms in a gas. And he went on to discuss Planck's hypothesis that 'the energies of the different colours are composed of energy particles'. This was a misinterpretation of Planck's work, in which the 'resonators' absorbed or emitted parcels of energy of a well-defined size. But Paul didn't seem to take energy particles too seriously anyway. 'The hypothesis, that in its present form apparently is only meant formally, shall then require further reduction', he wrote, interpreting it as an irregularity that needed to be straightened out (Ehrenfest 1905; Klein 1970, 232–234).[24] Paul certainly didn't take the bold step Einstein did, who interpreted 'energy particles' as a true physical phenomenon.

Einstein essentially proposed that different colours of light, infrared rays or, for example, X-rays each carry energy quanta of a specific size—large quanta for high frequency ultraviolet light for example and smaller quanta for low-frequency infrared light. From this perspective, 'when a ray of light propagates from a point, the energy is not continuously distributed over a larger and larger space but consists of a finite number of quanta of energy located at points of space, which move without dividing, and can only be absorbed and emitted as a whole', he wrote (Einstein 1905b). In short, light waves must then also have a particulate character. At the end of his article, Einstein even presented various experimental results that could be explained by such a particulate character. The most well-known was German physicist Philipp Lenard's work, demonstrating that ultraviolet light—in Einstein's interpretation corresponding to energy quanta of a rather large size—could cause the ejection of electrons from a metal.

But neither did Einstein completely return to the old Newtonian idea that light consists of particles. It was undeniable by then that light also had a wavelike nature. Einstein noted that this duality would cause a great deal of puzzlement and in doing so, displayed his ability to look farther ahead than many others. The duality—the particulate and wavelike character—would form the basis of what would develop into quantum mechanics in the ensuing decades.[25]

While Paul, unaware of all of this, was still fathoming Planck's work, Einstein sent yet another new article to *Annalen der Physik* six weeks later. It was a 31-page exposition in which he questioned the nature of time and space as well as the almost intuitive assumption that all events on Earth—and in the firmament—take place against a fixed backdrop in which time is kept by an imaginary cosmic clock.

Again, this started with light, but now Einstein focused on another aspect of light, related to Maxwell's treatment of light as consisting of electromagnetic waves (now perhaps also to be considered as particles). One of the consequences of Maxwell's work was that light must always travel at the same constant velocity. And this velocity was determined by the ether, Maxwell and for example also Lorentz suspected, a postulated medium which carried light waves.

It was a concept with puzzling consequences. If one compared the ether to a dead, calm sea and travelling light waves to a dolphin swimming at constant speed, such consequences became clear. Someone on shore would see the dolphin pass by at a certain constant speed, but someone on a ship moving towards or away from the dolphin would measure a different swimming speed relative to the ship. Why then, according to Maxwell's theory, and indeed confirmed by the measurements of the American physicists Albert Michelson and Edward Morley, should all observers always measure the same speed of light (dolphin), regardless of their own relative speed (ship and shore)?

Lorentz seemed to have salvaged the situation between 1892 and 1904, both for the ether and the results of Michelson and Morley's experiment. He assumed the ether was stationary and rigid and that it exerted a small pressure on objects moving through it—much like how you feel the wind in your hair when travelling in a convertible or riding a bicycle on a calm day. The pressure of this ether wind, Lorentz subsequently explained, would slightly contract objects. Indeed, Lorentz's elegant formulas describing this contraction precisely nullified the disruptive effects of the constant velocity of light.

The consequent need to also adjust time in his formulas had no further physical significance as far as Lorentz could see. More importantly, his formulas did something else too: they could unambiguously translate measurements of all electromagnetic phenomena described by Maxwell—closely related to light—from one system (like the ship) to another (like the quay). Lorentz had thus saved not only the ether but Galileo Galilei's relativity principle as well, which states that physical phenomena must obey the same rules for observers in all systems moving at constant velocity relative to each other.

But Einstein had upended Lorentz's work and, with a firm swing, jettisoning ether entirely. Ether was like the static decor in a firmament where a universal clock kept time and synchronism was something absolute. But why should such decor exist? Why couldn't time and space be 'elastic' concepts, depending on an observer's frame of reference? Much has been written about the role of Einstein's job at the patent office, where so many patents seemed to be related to clocks and light, and which made him walk past the big station clock in Bern every morning. What is certain is that he retained only the principle of relativity and the constant speed of light and then reinterpreted Lorentz's formulas in such a way that clocks in systems moving relative to each other no longer had to run at the same rate and read the same time. And it wasn't the objects that shrank but space itself.

A few months later in a new publication, Einstein pointed to another consequence of his 'special relativity theory'. If a train were travelling nearly at the speed of light, then a clock inside this train, seen from the platform, would barely tick: time would almost freeze as the train approached the speed of light. What would happen if coal and steam kept providing energy to the drive rods while the train was barely moving forward relative to this nearly stationary clock? It would be as if the train were getting effectively heavier and heavier, as if energy were being converted into mass. In other words: $E = mc^2$, Einstein wrote.

He had changed the face of nature for ever with a series of clear principles: with atoms, light with a particulate character, time and space as relative concepts, and energy and mass as two sides of the same coin.

But Paul and Tanya were unaware of any of this. They found rest that summer at 'Frau Pferniger's' in Weesen, strolled along Lake Walen, took afternoon tea at Hotel Swert and gazed at the 'starry sky above the treetops and the lake' in the evenings, after dinner. And despite the admonishments of Aunt Sonya, who travelled in their wake, they took the post bus up to Obstalden and walked back down on a steep path. The crickets sung, the grasshoppers hopped, the view was panoramic, and Paul supported Tanya 'very meticulously', he noted.[26]

They visited Ritz in Sion, listened to his '1001 theories' and hiked with him to waterfalls and around Vex, 'under the evening sky above the alpine meadows' with a view of Dent Blanche, causing Aunt Sonya constant anxiety. They read Freud and Gorki, Boltzmann, Gibbs and Mach. They enjoyed the company of new acquaintances and friends, such as French physicist Edmond Bauer, with whom Paul discussed Planck's theory he had been lectured on in Leiden. Aunt Sonya must have been relieved when they returned to Vienna before hail and snow started falling on the lower-altitude paths and mountain roads.[27]

They continued their lives there much as Einstein and Marić did in a relatively similar manner: outside the official academic circuit but with many informal contacts, with extensive discussions about physics among themselves and a deep desire to understand. And at the end of October, they welcomed a healthy and magnificent daughter, who made their lives more complicated and busier. Only after two months did Paul have time to describe this in short sentences in the red notebook: 'Already God has granted us a *Jutscherl*', and what a girl!! She was born on 28/15 October 1905 at ¾ 6 in the morning. Her name is Tatiana'. He added in full and in Cyrillic script: 'Tatiana Pavlovna'.[28]

A Russian nanny helped with the care of Jutscherl', as he called her with a distinguishing dash that could be borrowed from mathematics and mean 'the derivative of'—in this case, the derivative of Paul's great Jutscherl', Tanya. 'The nappies—camomile tea—Never minds' and so on 'almost flew through the air', according to Paul. 'And when Tanya's mother came at the end of the fourth week, Jutscherl' screamed for a second regime. And Jutscherl' will have reason to be grateful to Mama for her endless work in these weeks. But whether Jutscherl' liked it, none of them know!'[29]

10
First publications
Unemployed but working

Sometimes Tanya put her daughter on the grand piano. Then she played for the little girl, who, wrapped in knotted white sheets, gazed at her (see Figure 10.1). She looked like a doll when Tanya put her in the wicker basket of her pram with large, spoked wheels and carefully pushed her through the streets of Vienna. At home, Paul took affectionate pictures of Tanya nursing. Aunt Sonya delightedly lifted her granddaughter into the air.

'The entire day Tanitchka was very sweet and so cheerful, she slept well, she had a bath and finally went to bed at eight o'clock. In the evening Aunt asked: 'Where is papa?' and she immediately looked towards your room—she was on the balcony at the time', Tanya wrote in mid-March to Paul, who was visiting Ritz in Switzerland.[1]

But it wasn't actually a Viennese idyll. Now Tanya and her aunt weren't there as tourists, as they had been years earlier, they discovered other aspects of Vienna. The city liked to present itself to tourists as a convivial place, with coffeehouses, concerts, cakes and entertainment. Connoisseurs also knew Vienna as the city of the *Bildung*, the Austro-German tradition of self-cultivation, with an intellectual and artistic avant-garde, among whom composers such as Arnold Schönberg and Gustav Mahler, writers such as Arthur Schnitzler and scholars such as Sigmund Freud. But in the end—as anyone who spent more time there knew—the old Catholic aristocracy called the shots in the Austro-Hungarian capital. And it prized loyalty to authority and reliability—important qualities for the military and civil service—above brilliance of mind or creativity.

This wasn't at all to Tanya's liking, who—with a little help from fate—had just broken free of her authoritarian uncle's grasp. But once noticed, it was as difficult to ignore as the condescending anti-Semitism that permeated all layers of society. Vienna gave Jewish subjects of the Austro-Hungarian Empire opportunities but also belittled them. Employment in government, diplomacy and the military was still out of bounds for large groups of Jews from Galicia and Moravia who tried to improve their lot through hard work and study (Pauleys 1991, 45).[2] The fact that they sought refuge in the thriving business and banking world as well as journalism and indeed constituted a large part of the cultural avant-garde, roused envy too. It was as if both success at the top and poverty in Jewish ghettos were used to justify more or less overt anti-Semitic expressions.

In his youth, Paul had already seen how this vile tendency could be stirred up by politicising it, as anti-Semitic mayor Lueger and his Christian Social Party did. They claimed Jewish workers had 'dangerous socialist ideas' and took away jobs, while accusing the liberal 'Jewish' elite of prejudicing ordinary Austro-Hungarians with its policies. In short, 'it was impossible, especially for a Jew in the public eye, to ignore the fact that he was a Jew, for the others did not', writer Arthur Schnitzler observed (Beller 1991, 205). 'Vienna oppresses me, perhaps more than is good [for me]', wrote Freud, who like Paul was the son of a Jewish merchant from Moravia (Gay 1988, 9). And the same happened to Paul too. 'As soon as a Jew talks of "we" in another sense [than that of race], I always see how the rest of the company, that he embraced so naively with "we", hastily begins "crossing themselves"', he wrote many years later.[3] 'I don't see how a Jew can use that word with conviction'.

As a result, both Paul and Tanya felt out of place in Vienna. Tanya as a free-thinking Russian with 'deviant' ideas about work and life; Paul as a Jewish man without a job. Things didn't improve when they gained only limited access to the world where they did feel at home: academia. Certainly, Tanya still regularly took the tram through the stately boulevards of central Vienna and the untidy streets of Alsergrund to the decrepit lecture hall at Türkenstrasse, where she listened to

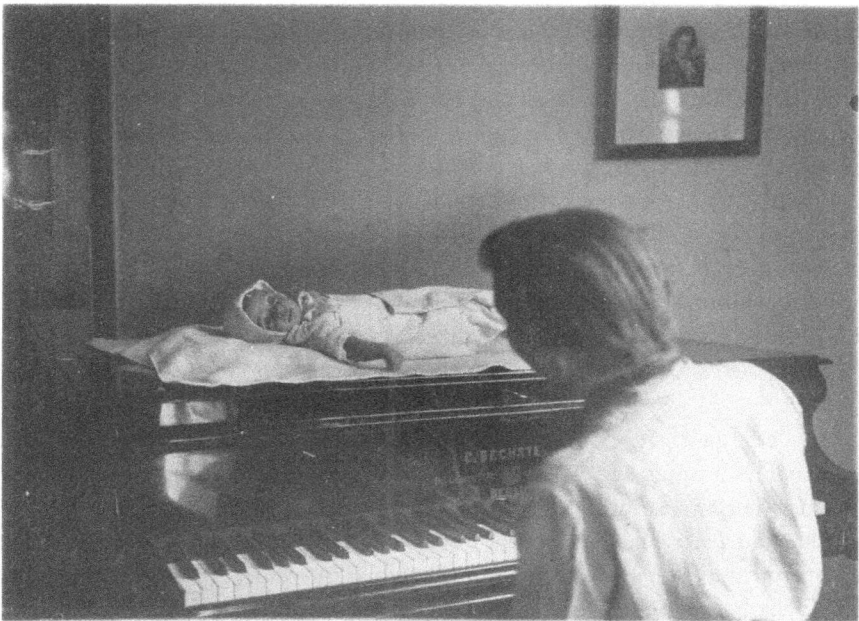

Figure 10.1 Tanitchka on the grand piano in the apartment in Vienna, with Tanya in the foreground.—*Photograph Ehrenfest Family Archive.*

Viennese professors as a non-examination student. But in a city where people still were adjusting to the idea of women studying, a student who was a young mother stood out as even more unusual. On top of that, Tanya's Bestuzhev diploma was a cryptic document of uncertain value here.

Paul couldn't find his feet at Türkenstrasse either. Its relatively small group of physicists, led by Franz Exner, focused primarily on experimentation and had fallen under the spell of the new phenomenon 'radioactivity'. Paul's interest was far more theoretical, but Vienna's great theoretical physicist, his tutor Boltzmann, had been admitted to a clinic for depression in spring 1906 (Cercignani 1998, 245). Paul and Tanya had been lucky to have the chance to attend a last series of Boltzmann's lectures in autumn and winter 1905.[4]

What Paul enjoyed most were the discussions with his old friends Tietze and Hahn as well as a new Viennese talent: Lise Meitner. She had passed her final oral exams with Boltzmann and Exner *summa cum laude* at the end of 1905, and a version of her dissertation had been published in the proceedings of the Viennese Academy of Sciences shortly afterwards. Meitner had been forced to take a different path to Paul. As a girl, studying seemed out of reach for her. Until the twentieth century, Vienna had only one gymnasium for girls, and even this one gymnasium had a crucial limitation: pupils graduated without sitting a university matriculation examination, as they were not allowed to go to university anyway. It was Meitner's good fortune that this situation changed in 1897. After long discussions about the effects of higher education for women on natality, children's welfare and women's mental health, the faculties of the arts and of physics started admitting women that year. As long as they had a diploma of course, and a transitional arrangement was established to enable women to pass their university entrance examination (Tichy 1993).

In 1901, Meitner was one of the few women to pass this examination after a two-year crash course. Together with Boltzmann's daughter Henriette and others, she did her exams in the intimidating Akademische Gymnasium building where Paul had been so miserable. At university, under Exner's and especially Boltzmann's wing, she had proved herself more than worthy of *summa cum laude* (Lewin Sime 1996, 18). Yet, she too found herself without a position or any prospects in early spring 1906.

It at least gave Paul and *Fräulein Doktor* Meitner the opportunity to review Boltzmann's lectures and works during 'conversations and cheerful skirmishes' in 'coffeehouses and milk halls' and in 'Stadtpark and Rathauspark', with Tanya in 1905 and also while Tanya spent her days with little Tanitchka in 1906.[5] Paul benefited from Meitner's insights from Boltzmann's lectures on topics, which he himself had studied elsewhere. For Meitner it was interesting to hear what Paul had learned in Leiden and Göttingen.[6] But they didn't really become friends. Meitner, raised in a warm and reasonably well-heeled Jewish household in the Leopoldstadt district, quite possibly felt distant to Paul whose path to university had been far less arduous than hers. Foremost, she seemed to find him too intense and was sometimes 'disturbed' by his questioning that she described as an 'inclination to put [sic] questions about altogether personal things'.[7]

Paul and Tanya soon ended up mainly working together. Mostly, this contented Paul: 'After turning 25, once I was married, I no longer struggled with a gloomy disposition', he wrote years later.[8] In their light-filled, music-filled home, Tanya didn't let him wallow in despair. Moreover, their talents and skills were complementary. Paul had an endless urge to reduce things to their core, which he then tried to express in clear formulations and metaphors. Tanya was trained to reason rigorously in Russia and further developed her analytical and mathematical skills in Göttingen.

At the large oak desk in the study, they considered the so heavily criticised part of Boltzmann's work: the idea that processes in nature are unidirectional and move towards a state of equilibrium. For a gas or a liquid, Boltzmann linked this idea to the collision of atoms or molecules moving around. In such successive collisions, the particles stayed intact, but their kinetic energy—or velocity—was more or less evenly distributed across all the particles. This meant they all eventually moved with greater or lesser variations around the same average velocity, he thought. But how could this process, which always moved towards a state of equilibrium, be irreversible while the collisions could just as easily play in reverse, like the balls on a billiard table?

Paul and Tanya's model clarified Boltzmann's conception. Later Paul received most of the credit. For example, the explanation in an earlier Ehrenfest biography begins by stating that 'Ehrenfest described his model in two papers written jointly with his wife, Tatiana, who had worked with him in exploring its various subtleties'. But in accordance with what was then still the common notion that women served primarily as sounding boards or secretaries, it exclusively refers to Paul afterwards: 'Here is Ehrenfest's own description'; 'Ehrenfest's model captures'; 'But now Ehrenfest could raise the old paradox in an especially sharp form'; 'Ehrenfest showed how'; 'Ehrenfest also pointed out that' and so on (Klein 1970, 115–117).

Yet, the statistical model Paul and Tanya used to address the problem likely was largely Tanya's conception—probability theory and probability calculus were among her favourite subjects at the Bestuzhev Courses. And who knows, perhaps Meitner also played a part because Paul had worked with her on Boltzmann's formulas for these equilibrium states.[9]

In any case, the recipe was simple. Take two large urns. Take one hundred balls numbered 1 to 100 and take one hundred lots, also numbered 1 to 100. Place ten balls in one urn and 90 in the other and start drawing lots. Each time a lot is drawn, move the ball with the corresponding number to the other urn. What happens?

Initially, it is more likely the ball drawn by lot is in the full urn. This probability gradually decreases until the distribution is approximately equal. If you return the lots and keep playing this game, the distribution will eventually fluctuate around an even fifty-fifty split, sometimes with more balls in one urn and sometimes more in the other. The balls have reached a state of equilibrium, like the atoms of a gas or a liquid, all of which move at more or less the same velocity. But if you draw lots indefinitely, there is a chance that an incredibly rare sequence of lots occurs, resulting in 90 balls ending up in one urn and ten in the other. Such an extraordinary situation wouldn't last long: the distribution would soon revert to approximately equal. Yet, it does demonstrate that the equilibrium state of such a system is—at least on paper—never completely and fully irreversible, meaning there wasn't a paradox.

The publication of this model in autumn 1906 must greatly have pleased Tanya and Aunt Sonya. Her uncle had relegated her to a place in the 'moonlight', but now Tanya's name basked in the sunlight, printed in the *Mathematisch-Naturwissenschaftliche* Blätter (Ehrenfest and Ehrenfest 1906b). Even though students mainly read the journal, it was still an accomplishment. And their joy must have been even greater since a second joint article, focused on the work of Josiah Gibbs—who had also laid the foundations of thermodynamics and the phenomenon of entropy with an elegant and mathematically vastly different approach to Boltzmann's—had already been published earlier that year. This time with Tanya as the first author and in the *Sitzungsberichte der Kaiserliche Akademie der Wissenschaften in Wien*, perhaps even more prestigious (Ehrenfest and Ehrenfest 1906a)![10]

In the meantime, Paul continued to ruminate on a metaphor that could capture the story of the urns in a few simple images. The metaphor he finally came up with would make him famous among physicists. Paul replaced the two urns with two dogs and the balls with fleas, who without any particular preference could jump from one dog to another. Everyone immediately sees that fleas initially infesting just one dog will quickly spread evenly between the two canines. And after thinking about it for a bit, everyone can appreciate that 'reversible events' such as flea jumps back and forth thus will indeed inevitably almost always lead to the same equilibrium state, resulting in both dogs being covered with more or less equal amounts of fleas.

The metaphor illustrated Paul's strength. He didn't produce new measurement results, like Meitner, who was exceptionally good at experimentation. He wasn't as brilliant as Einstein, who turned physics on its head. Paul analysed other people's insights and ideas and tried to apply what Boltzmann impressed upon his students: you must express in words what can be expressed in words and use as few formulas as possible. Modesty about what knowledge can achieve and capturing the essence of a theory in striking and unpretentious images was what it was about for Boltzmann: 'no theory can be objective, actually coinciding with nature, but rather ... each theory is only a mental picture of phenomena ... From this it follows that it cannot be our task to find an absolutely correct theory but rather a picture

Figure 10.2 Paul with Tanitchka, one-and-a-half, at Walensee, Switzerland, summer vacation 1907.—*Photograph Ehrenfest Family Archive.*

that is as simple and as possible and which represents phenomena as accurately as possible (Boltzmann 1974, 90–91)'.[11] And although Paul wasn't as esteemed by Boltzmann as he had hoped, he was still his model pupil in this regard.

The whole previous year, as fallen leaves blew through the streets of Vienna, horses steamed in the autumnal air, the first snow powdered the city's parks and as little Tanitchka was breastfed by her mother and pampered by her grandmother, Paul had tried to apply part of this approach to Planck's work as well. He had combined his notes from Leiden and Vienna and, step by step, compared Planck's calculations on 'black bodies' with Boltzmann's work.

He had already expressed his feeling that this comparison was flawed during a conversation with French physicist Bauer in the summer of 1905 in Weesen (Bauer 1963). It had gnawed at him in the Viennese coffeehouses and in his flat high above Alois-Draschepark, but only during another visit to Weesen in early summer 1906 did he finally commit his perspective on the matter to paper. Spring was in the air, snow still capped the mountaintops and T' had come along this time—she enjoyed outings in her pram, staring with her large blue eyes at Lake Walen (see Figure 10.2).[12]

Meanwhile her father wrote about the colours in the light spectrum of these 'black bodies' and about Planck's mathematical model for this. He wrote that Planck's calculations were so complex and extensive that, at first, they appeared to obstruct any clear understanding. Therefore, he wanted to approach their meaning in another way: since Planck had applied Boltzmann's statistical methods for describing the behaviour of atoms in a gas to the case of light waves being absorbed and emitted by a black body, Paul thought he would learn most by comparing Planck and Boltzmann's work in more detail than he had done in his short note in 1905.

Boltzmann had shown that atoms shooting around in a gas or liquid eventually had roughly the same average kinetic energy and velocity everywhere due to collisions between these atoms, resulting in a state of equilibrium. Planck had described how a black body was also in a state of equilibrium and how it continuously and uniformly emitted a characteristic mixture of the incident light in all directions. Simply put, Planck imagined covering the surface of a black body with imaginary 'springs', or oscillators, which were compressed when they absorbed a light wave and recoiled when they emitted this light wave in all directions.

But unlike Boltzmann's colliding atoms, in this process the light waves retained their energy, which corresponded not to a velocity but to a colour in this case. In other words, the spectrum could become more disordered because the resonators emitted light in all directions (as a 'diffusely reflecting mirror', Paul wrote), but it didn't tend towards light waves all having the same average, relatively low, energy. It didn't get 'blacker' over time, to use Paul's word.[13]

This also implied that something else must underpin the spectrum, giving it its specific shape and keeping it in this shape. These, Paul argued, were indeed the energy quanta introduced by Planck. He even cautiously speculated that electrons, which were thought to be responsible for the interaction between light and matter and thus to play the role of 'oscillators' in nature, might have specific characteristics that resulted in this selective absorption and emission of light colours.

But Paul preferred to end his conclusions with something else. He reasoned that Planck's 'energy atoms' prevented a few high-energy ultraviolet light waves from suddenly carrying away all available energy, leaving no energy for other colours. Unless a black body was inconceivably hot and energetic, there didn't seem to be any chance that such a massive amount of energy would suddenly coalesce into a single 'energy atom'. This explained why Planck's formulas accurately described the spectrum for these colours, while British scientists Rayleigh and Jeans' more classical descriptions incorrectly predicted that ultraviolet rays would 'empty' the spectrum by carrying away all the energy, he concluded, in very cautious wording.

Later, Paul would further elaborate on this phenomenon and call it the 'ultraviolet catastrophe'. The term stuck, as did the dogs with fleas. But in 1906, he just noted 'Weesen on Lake Walen, 28 June' beneath the somewhat coy text published in the *Physikalische Zeitschrift* (Ehrenfest 1906d). Without realising it, he once again shared something with Einstein, at least to a certain extent. Paul too had realised that Planck's energy quanta revealed something fundamental about matter and light and were more than just mathematical frippery. At the same time, it was clear he lacked Einstein's calibre: Einstein used recent work as a starting point to depart in entirely new directions, while Paul cautiously added annotations.

What did apply to both men was that their articles were at odds with the views of the prominent physicists at the beginning of the twentieth century. Leading physicists such as Lorentz, Arnold Sommerfeld or Max Abraham hoped to reduce all forces to electromagnetic fields and even describe mass as a product of electromagnetism. 'The world would then consist exclusively of positive and negative electrons and of the electromagnetic fields generated in space by them', Max Abraham, for example, had written (Klein 1970, 146–147).

It was no longer in vogue to describe the cosmos in Newtonian terms as a splendid mechanism where one mass set another into motion or to compare it to a mechanical clockwork. Force fields were the rage. This also made Einstein's idea that light was particulate feel a little old-fashioned. It was perhaps too reminiscent of 'outdated' Newton, who had ushered in mechanics and ideas of precise timepieces. Something similar probably also went for Paul's comments on electrons. Altogether, the independent young physicists, unpaid or working at night, struggled to find their place in academia.

Einstein was promoted from third-class to second-class clerk in 1906, and he briefly considered a career in education (Pais 1982, 184). Meitner decided in the summer of 1906 to give lessons at a girls' school and conduct her experiments at night (Lewin Sime 1996, 19).[14] And Paul and Tanya? Their work was a mere exercise compared to Einstein's work and the well-conceived trials Meitner conducted under Exner. And Paul's quantum article not only lacked Einstein's wide overview, daring and unerring intuition but also clear mathematical foundations. As a result, their prospects of a warm reception in another university city seemed even slimmer, while theoretical physics in Vienna had also stalled, not in the least because of Boltzmann's admission to a clinic.

What should they do? Paul's parents had amassed a fortune, but it was divided between five sons, and Paul's share was dwindling rapidly. Could they go to Switzerland? Paul once more sounded out Ritz during a visit to the country in March 1906. 'I keep thinking it's you when a door opens or I hear someone walking; I think you might come in at any moment', Tanya wrote to him shortly after his departure. 'God be with you'.[15] But when Paul did return safely, their hopes for a position in Bern or Zürich evaporated.

Their dream to keep thinking, writing and learning on the other hand was still very much alive. In May 1906, they gave up their flat in Vienna and arranged to return to Göttingen in September, after their summer vacation in Weesen.[16] No matter how uncertain their future there, and even though they had mocked Göttingen's half-timbered houses, its mathematics and theoretical physics were vibrant. And in the end, it was the sciences that gave them the pleasure, camaraderie and clear thinking that made them feel at home.

11

Happy in Göttingen

Travelling, thinking, writing

Paul and Tanya had just furnished their home at Wilhelm-Weber-Strasse 44 in Göttingen.[1] The flat was in a freestanding house just outside the city walls, across the street from Hilbert's house. If you stuck your neck out the window, you could see his garden, with a six-metre-long blackboard under a small roof next to the flowerbeds so that Hilbert could think and do mathematics outside even when it was drizzling (James 2002, 250). If instead you looked inside, you would see the sunny rooms where Tanya and Paul lived. But the news that reached them at the end of summer 1906 via Vienna cast a shadow over things. Boltzmann, who had been holidaying at Duino, near Trieste, in modern-day Italy, had taken his own life.[2]

Meitner, who also knew Boltzmann's wife and children well, later attributed this to Boltzmann's unstable temperament and his hypersensitivity. He 'may have been wounded by many things a more robust person would have hardly noticed', she said (Lewin Sime 1996, 15). Others, such as Boltzmann's friend and academic adversary Ostwald, thought the quest for truth had exhausted Boltzmann and spoke of 'sacrifices' for science. Paul made no mention of Boltzmann's death in his brief outline of the autumn months in the red family journal. In the obituary he wrote in October for the German universities' mathematics and physics journal, *Mathematisch-Naturwissenschaftliche Blätter*, he only discussed Boltzmann's work (Ehrenfest 1906f).

Was this because Paul never became part of Boltzmann's circle? Rather shocked, fellow students heard Boltzmann exclaim in 1905, after a barrage of questions from Paul, that he was 'not a lemon to be squeezed dry (Klein 1970, 48)'. Boltzmann had not been particularly enthusiastic about Paul's dissertation either. Perhaps not entirely without reason because loyal but honest Tanya also conceded that it wasn't 'to tell the truth, much good, but was good enough to earn a PhD', as she later wrote.[3]

Had Paul secretly hoped Boltzmann would discern more intelligence behind his hastily prepared dissertation? 'I'm diving into mathematics as much as possible to perhaps prove that foolish B. wrong', he wrote in early 1906 on the first page of what by then was his seventh notebook.[4] Did B. stand for Boltzmann and did 'foolish' refer to his judgement?

In any case, Paul could never again show his supervisor how thoroughly he had studied his work and mathematical formulas. Conversely, Boltzmann could no longer write Paul letters of recommendation—even if he had started producing better work. It was a poor start to his time in Göttingen.

On a brighter note, Herglotz lived nearby. Paul's melancholic friend often came over to listen to music in the evenings, and Paul regularly visited Herglotz too, who lived in a flat and had a garden with chickens.[5] Often Paul brought T' along, a precocious girl with a magnificent baby face, who, aged one-and-a-half, had uttered her first complete sentence in spring 1907: 'Listen, mama, music'.[6] (See Figure 11.1) Her mother took Tanitchka with her to other acquaintances in Göttingen. Proudly, Tanya recounted how Tanitchka had enchanted even mathematician Hermann Minkowski, a scholar who, according to others, felt more at ease with children than with his rather solemn students and colleagues in Göttingen. In the family journal, Tanya described how she had visited Minkowski's family with a 'very sweet' Tanitchka, 'but when Minkowski came in, she started running after him. We wanted to leave but Minkowski started dancing in a circle with his wife and two daughters to please Tanitchka'.[7]

Did Tanya make friends with the Minkowskis because of their shared Russian background? Minkowski was born in what is now Lithuania but as a boy had moved with his family to Germany to escape the anti-Jewish laws. He had studied at Königsberg, and before Hilbert persuaded him to move to Göttingen in 1902, he had given lectures to Einstein as a professor in Zürich. A few years later, Einstein's special relativity theory astounded him. At university, Einstein struck him as 'an idler'. And 'he didn't care about mathematics at all'.[8] Minkowski certainly did, and he started elaborating Einstein's theory in a much more mathematical way in 1906.

Perhaps the visits even helped spark Paul and Tanya's interest in Einstein's work too. In a brief article he sent to the journal *Annalen der Physik* in early 1907, Paul joined a related and lively discussion about Walter Kaufmann's measurements (Ehrenfest 1907b). With ingenious experiments, this German experimentalist had measured that it takes more and more energy to make increasingly fast electrons go even faster (Kaufmann 1906). Was it, as Abraham hypothesised, because moving electrons produced an electromagnetic field that opposed them? Or was it, as Einstein thought, because more and more energy was converted into mass and the electrons effectively gained weight as they went faster and faster? Or did it precede in a completely different way, as Frenchman Paul Langevin proposed?[9]

In an initial article addressing these discussions, Paul had ignored Einstein's work (Ehrenfest 1906b). But in this 1907 article, he shifted the discussion towards Einstein's work and especially towards Lorentz's formulas that Einstein had used in his work. Following Kaufmann, Paul proposed that an electron moving with a high, constant velocity contracted slightly in the direction of motion and as a result

76 THE DELIGHT OF THINKING

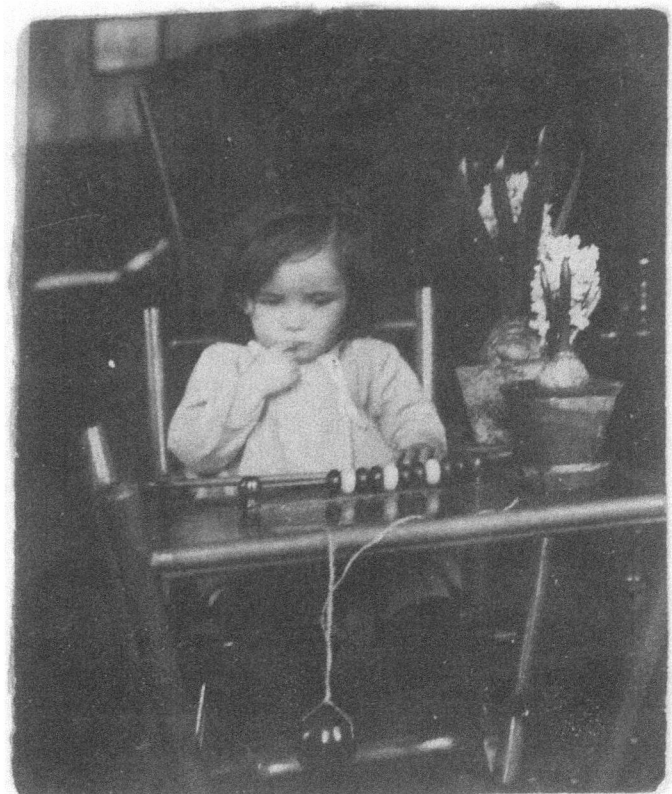

Figure 11.1 Tanitchka in their apartment in Göttingen.—*Photograph Ehrenfest Family Archive.*

became more disc shaped. But such a disc would wobble in the electric field it generated, whereas a stationary electron wouldn't. And this distinction in behaviour went against the principle of relativity on which all physics rested, including Einstein's theory: any law of nature must always give the same results, whether in a speeding train or on a platform, whether for an electron whizzing by at high velocity or for a stationary electron. Didn't this suggest that Lorentz's formulas and 'Einstein's system' must be adjusted so that nature could behave the same everywhere, Paul wondered.

Einstein, who had previously ignored Kaufmann's work, made the effort to respond at this point. His work wasn't, he wrote in a response in the same journal, about certain 'systems' or the distortion of electrons but dealt with 'a heuristic principle which ... contains only statements about rigid rods, clocks, and light signals (Einstein 1907)'.[10] In other words, the theory makes it possible to transform clocks and measuring rods from one reference frame to another and doesn't concern itself with (the distortion of) objects. And even though Paul might have misunderstood him, in any case Einstein had taken Paul's article seriously and he had seen that Paul was part of the 'modest group' of physicists who took his ideas seriously (Seth 2004; Janssen and Mecklenburg 2006).

These ideas subsequently gained clarity when Minkowski successfully reformulated them in a superb mathematical framework at the end of 1907. 'From now onwards space by itself and time by itself will recede completely to become mere shadows and only a type of union of the two will still stand independently on its own', he concluded (Minkowski 2012, 39). With elegant mathematics, he had seamlessly fused relative space and time into a four-dimensional spacetime. Objects or people travel as it were through this four-dimensional world, Minkowski showed, tracing a path through spacetime, which he called a 'worldline'. Minkowski also clearly defined the leeway they have. It is bounded by light, which draws two cones from every point in spacetime. Within these cones lie the past and future, while all spacetime outside the cones is inaccessible, he explained in a lecture to the Göttingen Mathematical Society.[11] And even though Einstein initially complained that he no longer understood his own theory, thanks to this 'superfluous learnedness', he subsequently came to appreciate Minkowski's approach using mathematical vectors (Pais 1982, 152).

Conversely, Minkowski's clear and concise presentation made other physicists value Einstein's theory more, even though not all of them were convinced of it yet. Sommerfeld in Munich, for example, continued to complain that Einstein's reasoning was too abstract-conceptual and couldn't resist connecting this to his Jewish background. 'No matter how brilliant they are, there is still something unhealthy in this irreproducible, non-intuitive dogmatism. An Englishman would have found it difficult to produce such a theory. Perhaps there is something here that ... corresponds to the abstract-conceptual style of the Semite', he wrote at the end of 1907 to Lorentz (Kox 2008, doc. 165).[12]

Paul and Tanya were still engrossed in the nineteenth-century physics of Boltzmann's disorder and equilibria. After Paul delivered a lecture on their urn model with balls at the Mathematical Society in early November 1906, they gave a more detailed description of the model in an article published in 1907 in *Physikalische Zeitschrift* (Ehrenfest and Ehrenfest 1907). Even before the article was printed, Felix Klein wrote a letter to the 'Dear Doctor'. The lecture had given him 'much pleasure', he wrote.[13] An unexpected twist followed: 'Would you, possibly with your wife, be prepared to write the article about statistical mechanics for volume four of the *Encyclopaedia*?'

Klein's ambitious *Encyclopaedia of Mathematical Sciences* aimed to provide an overview of the entirety of modern mathematics and its applications, including those in physics (Tobies 2021, 425–433). He had strongly encouraged Boltzmann to write about his statistical mechanics for the fourth volume of the extensive series, which dealt with mechanics—opening with the 'Foundations of Mechanics', discussing all sorts of applications of classical mechanics, and ending with Boltzmann's novel ideas on the 'Mechanics of systems comprised of very numerous discrete particles'. But Boltzmann was dead. It must have surprised Paul and Tanya too that Klein had approached them instead: a physicist still lacking a habilitation and his mathematically trained wife, who didn't even have a doctorate. 'In any case, I would like to request that you and your wife come around to my home, perhaps tomorrow (Sunday evening) at six o'clock, so we can discuss the whole matter'.[14]

Had Klein seen how well Paul could explain things? Had he noticed how calm and analytical Tanya kept Paul on track? Naturally, Paul and Tanya would still have to get to grips with many of the finer points of Boltzmann's work to present it in a balanced and clear manner. Yet evidently, Klein had enough confidence in their capacity to have them contribute to the same encyclopaedia as Hilbert in Göttingen, Lorentz in Leiden, formidable Sommerfeld in Munich, Tanya's former Bestuzhev lecturer Selivanov in St Petersburg, and Paul's promising friend Hans Hahn in Vienna. It was the start of a lengthy process in which Paul and Tanya methodically trawled through Boltzmann's work and through articles in which his colleagues had responded to this work.

This was far from easy. In hindsight, once ideas have been developed completely, it seems as if the course to this endpoint always existed. But for a full understanding, Paul and Tanya had to retrace the wandering path taken by Boltzmann in his verbose writings. 'Paul had a quite special talent for quickly having an overall view of every article, while I read very slowly and usually first sat with my own reflections', Tanya wrote much later.[15] 'Without him I would never have been able to begin'. Then in parentheses: 'Still I found one or several contributions which he alone, perhaps, wouldn't have found: these are the logical distinctions between the two objections, the reversibility and recurrence paradoxes'.

The reversibility paradox, articulated by Boltzmann's critics Loschmidt and Ostwald, was elucidated by Paul and Tanya with their urn model, which Paul subsequently simplified to the flea model: when two dogs lie down next to each other, one covered in fleas and the other with hardly any fleas in its coat, their owners will soon find about as many fleas on each dog, even though an individual flea jumps as easily one way as the other (reversibility). Only if the owners had all the time of the world, would they perhaps have the chance of observing the extremely rare occurrence of a random sequence of flea jumps resulting in one dog briefly carrying many more fleas than the other.

However, this would give rise to a new paradox, the so-called recurrence paradox, Planck's assistant Zermelo claimed.[16] Zermelo agreed that the distribution of fleas over the two dogs will eventually revert—for a brief time—to the initial state if you wait long enough. In other words, if people, fleas and dogs had eternal life and endless patience, then dog owners would see the flea distribution thrown out of kilter as often as it quickly reverted to an even fifty-fifty distribution. But why then, he asked, did Boltzmann's so-called H-theorem, and the ensuing Second Law of Thermodynamics state that every system inevitably heads towards equilibrium—in this case, the equal distribution of fleas?

With a simple probability calculation, Paul and Tanya illustrated why Boltzmann's theorem consistently turns out to be correct—at least for short-lived humans in a finite world. There are simply more ways (in this case, series of flea jumps) to return to equilibrium (in this case, fifty-fifty) from an imbalance than to end up in an even more imbalanced situation. In other words, the tendency towards a state of equilibrium is the greatest—for fleas as well as for randomly moving atoms and molecules in Boltzmann's statistical mechanics. Paul and Tanya also made clear that the confusion about the paradoxes stemmed from Boltzmann interchangeably using different concepts of probability (for a much more in-depth analysis of Boltzmann's subtle work, see Darrigol (2018)). All in all, it would take them five years to piece together Boltzmann's ideas and criticism of them.[17]

But in 1907, they were above all glad to finally count. They were contented, making new friends and meeting old acquaintances. Gernet, the bestuzhevka who had taken her doctoral degree under Hilbert in 1902, stayed in Göttingen for some time, and Tanya's mathematics teacher, Vera Schiff, passed by as well. Paul even 'became reconciled' with Abraham, whose work he had been critical about. And in June, they took the train with Aunt Sonya via Paris to the white beaches of Berck-sur-Mer, a northern French fishing village turned into a health resort.[18] That thalassotherapy was devised here, and the Rothschild family had built a magnificent paediatric hospital and a sumptuous villa, catered to Aunt Sonya's desire for society life and accommodated her fretting about her niece's health, while Paul could hardly refuse such a trip.

All in all, their life wasn't bad, but it was also clear that it wasn't sustainable. There was no prospect of a paid position in Göttingen. Was it time to go to St Petersburg, where Tanya could work? 'My boundless love', she wrote to Paul in autumn 1904, 'do you have any idea how absolutely certain my future seems to me? I've crossed a threshold in the meantime, probably when I received the dispensation from college [the Bestuzhev Courses] because there were actually only two possibilities: to come to you either now or later—didn't you know that?!'[19] She had negotiated this dispensation well: even in 1907, three years later, Tanya could still return to her job.

Was this the right moment to do so? It would provide them with some income. Tanya would also be able to study on the side at the 'normal' and previously exclusive male university in St Petersburg that began admitting women in 1906. And it would enable her to finally obtain a degree more widely recognised abroad—and in Russia—than her Bestuzhev diploma. An optimist could even have argued that Russia was on the up: the economy had been flourishing since the 1890s and student numbers were skyrocketing (Kassow 1989, 16). A parliament was established after the Russian Revolution of 1905 and although this State Duma had limited powers, it was a start, as was suffrage for male farmers and workers.[20] Moreover, the enthusiasm for science and technology was evident in the expanding railway network, modern ports and large new factories. Russia was booming, westerners visiting the country noted (Schlögel 2002, Ch. 2).

Whether Paul with his theoretical interests, western European education and Jewish background would benefit from any of this was another matter. Universities were granted more autonomy after the 1905 revolution, allowing them to determine more easily, whom they employed. Many universities, including in St Petersburg, exploited this policy relaxation to admit women and scrap restrictions on Jewish student numbers. But it was uncertain whether this policy would endure in the long run. In 1906, the Tsar disregarded the advice of his new Minister of Education, Count Ivan Tolstoy, to abolish official quotas for Jewish students (Kassow 1989, 271–278).

Tolstoy's successor, Pyotr von Kaufmann, likewise failed to implement this proposal as policy. Even strict tsarist Prime Minister, Pyotr Stolypin, appointed in late 1906, couldn't persuade the Tsar to grant his Jewish subjects equal civil rights (McMeekin 2017, 52–55). The time wasn't ripe, according to the Tsar, perhaps influenced by the same ministers who had warned Tolstoy that 'there is far more antisemitism in Russia than you think' (Kassow 1989, 298). Indeed, anti-Semitism brutally manifested itself to the world during the Kishinev pogrom in 1903, and again during subsequent pogroms in October 1905 in cities including Odessa, Sevastopol, Minsk, Bialystok, Tiflis, and attacks on universities and students in Odessa, Kazan, Kiev, Kharkov and Moscow.

Even so, Paul and Tanya didn't seem overly pessimistic in 1907. Did they think their secularity solved the issue? That in the new Russia, and certainly St Petersburg, things weren't as bad as they appeared? Paul's greatest objection to the city seemed to be that its physicists were reputedly rather old fashioned. But unless he secured a position in the more modern physics circles of the German-speaking world, this objection carried less weight.

Still, it was not without reason that in June and July Paul made last-minute trips to German university cities such as Tübingen and Munich, where he hoped to earn his habilitation. In the 'shuddering railway carriage' on his way to Tübingen, which he shared with a 'young man, the pharmacist type', and not just a little but 'terribly reeking of smoke', he reassured Tanya in his scrawl.[21] 'You needn't fear that I'll commit myself to Munich and that your St Petersburg dream won't come to pass. I just want to see if I can possibly find accommodation in Munich'. Accommodation to habilitate remotely? From St Petersburg and while visiting Munich from time to time? Tanya must have been well aware that Paul hadn't given up his dream of an academic career at a German, Austrian or Swiss university.

Perhaps she also sensed the trip wouldn't be a success. Paul hadn't expected much good from Tübingen, even though the work he and his 'wife' had done on Boltzmann's H-theorem and entropy had been well received by the local mathematics society.[22] He had pinned most of his hopes on his acquaintance Gustav Hertz. They were going to travel together from Tübingen to Munich, where Hertz studied, so that Hertz could bring Paul up to speed on the Bavarian capital on the way.

Things went as expected. Upon arrival, Tübingen proved to be as dismal as Paul had imagined. Over a meal in 'a dimly lit dining hall' with a 'sleepy waiter', he had a 'slowly meandering conversation' with Gustav Hertz and his local host, Hans Happel. As he and Happel subsequently climbed the hill on which the city was built—over 'stairs, around corners, past lanterns'—, the 'cloisters, churches and storehouses' seemed 'haphazardly thrown together'. There wasn't a soul about. 'Everything strangely lifeless', Paul scribbled on a card to Tanya at 10:30 pm.[23] 'If everything is as it seems to me now and as it appears through Hertz and Happel, then I would rather be dead than ever have to live like this or in such a manner'. A few days later and after long conversations with Hertz, the possibility of a better life in Munich seemed out of the question too. 'It might well be entirely impossible for me to habilitate in Munich', Paul wrote to Tanya in the train as 'antisemitism' and the 'requirement to have conducted experimental work' irrevocably impeded him.[24] 'These are reprehensible matters'.

It seemed to leave 'only St Petersburg and Austria'—with even there probably at most 'a job outside the university. ... Without having seen too much, I have discovered, almost right in front of me on this trip to Tübingen, the boundaries confining me, while I had only just been under the illusion that the entire world lay open before me. ... In Tübingen, I realised how demanding I am', he wrote despondently.[25] And: 'I'm very, very attached to you, Tanya!'

A few weeks later, after another futile trip via Hanover and Heidelberg, Paul collected their last furniture and things from storage in Vienna. It was hot, he wrote to Tanya.[26] Did she fear he would have second thoughts in his native city? 'Dearest, I had such a bad headache yesterday, and I didn't feel at all well ... and it seemed to me that you should come and take my head in your hands', she wrote.[27] He replied that he had read 'a drama by Ibsen – *The League of Youth*' without much pleasure.[28] She told him 'T' had fallen and had a swollen lip. In a long missive, he complained that his brother Otto's 'very sweet' daughter was 'kept ignorant', and his intelligent cousin Bertha Friedmann's son was curtailed in a 'bleak environment', and you constantly come across such 'unwholesome milieux' in Vienna.[29]

Having started the letter in Russian—written phonetically and in Cyrillic block letters—he ended it that way too. Paul had kept his promise to learn to read and write Tanya's language, although a few German words slipped in. 'Sweetheart, my sweetheart. I'm very close to you, take our baby on your lap, give her a kiss from me and tell her that her daddy never feels at home anywhere unless he is with Tanitchka and her mama', he wrote to Tanya from Vienna, again with a smattering of German words in the Russian sentences.[30]

For Tanya it was probably more important that he had written Vienna was definitively off the table. 'It does me good seeing everyone again, but if I ever doubted the correctness of the decision to go to your homeland ... now that my "fatherland" is right in front of me, I'm suddenly very sure of it'.[31]

12
Dangerous experiments
A Russian beard and feeling at home

Paul often sounded confident but was never sure of himself. Shortly after they arrived in St Petersburg with the piano, cabinets, Aunt Sonya, Tanitchka and boxes full of books, he was already questioning the whole project again. His new situation was 'unstable' and 'risky', he wrote to his Austrian friend Heinrich Tietze. 'The main argument remains of course that I don't have a real Heimat anywhere— whereas Tanya really belongs here. We'll see. At worst, one to two years and a corresponding amount of money will get wasted'. The whole thing, according to him, was primarily an 'interesting (though dangerous) experiment to me' (Huijnen and Kox 2003, 187).

The mathematicians and physicists in St Petersburg felt the same way. All kinds of people from many different places lived in St Petersburg, including a large German-speaking community. But in science, the custom was that Russian academics went to Berlin, Munich, Göttingen and Paris, not the reverse. Grandees such as mathematician Andrei Markov worked in St Petersburg, and of course, world-famous chemist Mendeleev, who had died in early 1907, but the heart of scientific research was in the West.

Even the civil service in St Petersburg was at a loss. Paul's 'secularity' caused particular bafflement. 'Yes, don't you understand? In which cemetery should we bury you if you die here?' they asked him every time he extended his residence permit (Frenkel 1977, Ch. 3).[1] Yet almost right away Paul also seemed to feel at ease. He grew a Russian beard, bought an astrakhan hat and made a great many new acquaintances.

'It's almost grotesque', he wrote much later to a friend about the country.[2] 'Everyone here treats me like a very close and trusted member of the family. Not only my friends and family. No. Also children in the street, workers and their wives, among whom I found myself during a festive gathering. And yet from my accent they can immediately tell that I am a foreigner'.

Despite the puzzlement, Paul was indeed welcome in Russia. His potential eccentricities were less noticeable. What some had previously considered his 'impertinent' questions could now easily be attributed to his foreign background, and no one was surprised by the Viennese and idiosyncratic peculiarities of his German. Neither was his passion for physics out of place: in academic circles in St Petersburg, it was quite normal for a conversation to move almost directly on to radio waves or higher mathematics. Furthermore, young scholars yearned for modernisation. The plans Minister of Education Tolstoy had presented one-and-a-half years earlier—more autonomy for the universities, more opportunities for young researchers and free admittance for everyone regardless of sex, religion or nationality—were still stalled. And his successors, together with more conservative professors, had reverted to repressive policy. Consequently, a modern curriculum and better-paid jobs for young scholars hadn't materialised nearly as much as hoped.

Disappointment with the stifling atmosphere also prevailed at the Physics Institute, led by Ivan Borgman. A grave character, Borgman made a name for himself in 1897 by demonstrating X-rays could induce crystals and minerals to glow green, red or blue (thermoluminescence). He had given the Tsar physics lessons and represented science in the Duma, but the inadequate laboratories and stagnating plans seem to have discouraged him. 'It would already be nice if we could simply manage to duplicate the measurements that have been done in other countries', he said to young experimental physicist Abram Ioffe, when he returned from Munich, having earned a doctorate *summa cum laude* and buzzing with plans for further experiments with X-rays (Klein 1970, 85). He called Ioffe a 'know-all'. Little wonder that the most creative Russian physicist of the time, experimental physicist Pyotr Lebedev had departed for more progressive Moscow (Frenkel 1977, Ch. 3).

Likewise, the theoretical side of physics lacked vibrancy in St Petersburg. In Western Europe, theoretical physics was developing into a fully fledged branch of physics, with its own professorships and lecture courses. But in the Russian capital, theoretical formulas were still considered the domain of mathematicians. Those wanting to pursue an academic career in physics after obtaining their doctorate were even obliged not only to obtain a magister's degree (comparable to the German habilitation) in physics but also one in mathematics. Moreover, to young physicists' annoyance, updating these old-fashioned mathematical magister's exams to reflect recent developments in physics was out of the question (Frenkel 1977, Ch. 3).

This context meant that Paul suddenly transformed from an enthusiastic outsider in Göttingen into a leading light in St Petersburg. He had studied under Boltzmann in Vienna ('an extremely thorough education'), spent years in Göttingen (knowing 'all the great mathematicians and physicists' and surrounded by 'very good students') and had also spent time with Lorentz in Leiden: 'He was probably the best educated theoretical physicist of the younger generation in the world', Russian physicist Paul Epstein later noted about these years (Epstein 1965, 69–70). Furthermore, he was well read: he knew the works of Tolstoy and Marx, he was interested in Russia and had a Russian wife who had studied.

For Tanya, it was chiefly a homecoming. She immersed herself in her mother tongue. With Tanitchka, she regularly stayed with her mother at Petergof. She rekindled old friendships and enjoyed metropolitan life. It was so much livelier than Göttingen, even on tranquil Vasilyevsky Island, where they rented a flat.³ Unlike before her move to Göttingen, there were electric trams on the other side of the water on Nevsky Prospekt and new apartment blocks were under construction throughout the city. But her friends' effusiveness was still the same. There was only one minor blemish to her 'Petersburg dream', as Paul called it.

At the beginning of September, she had wanted to resume her part-time position at the Bestuzhev Courses, but mathematics instructor Boris Mikhailovich Koyalovich had enlisted her only for his practicals.⁴ 'Because N[adezhda] N[ikolaevna] Gernet has resigned her position, we have a vacancy at the Course for two hours of practicals with students in my subject (differential and integral calculus). I would like to inquire if you would be willing to undertake these practicals', 'your sincerely dedicated Koyalovich' had written to her.⁵ But in the ensuing months, Tanya increasingly sensed that she had been assigned these two hours merely out of 'courtesy'. She was supposed to be Koyalovich's assistant until her former position as supervisor of the mathematics laboratories could be formalised again in a 'senate meeting' with the dean. However, this still hadn't occurred by the final meeting in 1907.

'As you yourself stated in a conversation with me at the beginning of the semester when you invited me to the Courses and reconfirmed ... I was entitled to an appointment ... due to my prior experience in leading the mathematics practicals', she wrote to Koyalovich at the close of the year. She removed the strident sentence that she initially drafted, 'That this did not happen for an entire semester signifies contempt for my legal rights and suggests to me that my work at the Courses is not valued'. Nevertheless, she resigned: 'no matter how close this activity is to my heart, I am compelled to resign from my duties as supervisor of the work in your subject'.⁶

By this point, Tanya had also spoken to Gernet, whom she knew from the Bestuzhev Courses and met again in Göttingen. When Tanya arrived at the town in 1902, Gernet had just finished her doctorate under Felix Klein. Shortly afterwards Gernet returned to St Petersburg, taking over Tanya's duties at the Bestuzhev Courses and leading practicals on number theory and analytical geometry together with Schiff (Valk 1918, 124). After a sojourn, which included Göttingen in 1907, she now wanted to resume this work, she told Tanya. 'From the conversation I had with you yesterday, it is clear to me that any reduction in work would present you with a serious financial problem', Tanya summarised their conversation in a letter to her the next day.⁷ And she reiterated 'very explicitly' her decision 'not to resume the duties the Senate entrusted to you after my departure'.

In the first days of January, Tanya repeated this decision in a letter to Boris Mikhailovich, who had written that he was 'very surprised' by the whole affair and was 'absolutely unable to recall anything with which I might have offended you or infringed upon your rights, since I have never treated you with anything other than the greatest kindness'.⁸ If others reproached him for her departure, Koyalovich should simply repeat her motive, she replied.⁹ 'I see that my presence at the Courses is not at all necessary, as there is no need to hire a new person for two hours of practicals. ... Of course, I could have realised this sooner, but I felt this most strongly when I was at the Courses'. And: 'The Courses will always remain dear to me, and I shall always be happy to help with anything I can, but not to the extent that I displace others from their positions for the pleasure teaching gives me. Please convey my warmest regards to the highly esteemed Vera Semyonovna [his wife] and do not be annoyed with me'.

It would seem Koyalovich wasn't annoyed, as at the end of the second semester in April he asked if he should discuss Tanya's 'appointment as supervisor of the practicals' at the Senate meeting anyway. 'Let me hope that if I have ever wronged you in any way or if there have been other frictions, this has now been erased by time and no longer poses an obstacle to discussing your appointment. I wish you a happy Easter and sincerely hope you shall spend the holiday in good health'.¹⁰

Were Russian teachers always this polite? Did Tanya's membership in the esteemed Afanasyev family play a role? In any case, the affair had been disappointing, especially because her work at the private girls' gymnasium was not warmly received either (Ehrenfest-Afanassjewa 1961, Introduction). Inspired by what she had learnt in Göttingen from Hilbert and Klein about pedagogic teaching, Tanya wanted to reorganise the geometry lessons. Pupils would better retain knowledge if they could visualise what the formulas described and what mathematicians aimed to prove with them, she argued to the school administration. This was at odds with the prevailing approach, with every new theorem derived from previous ones, all in the tradition of great Greek mathematician Euclid's geometry writings.

Everything started with five axioms in plane geometry: five 'truths' which you could understand without additional proof. A straight line can be drawn from any point to any other point. Any line segment—a line of a particular length—can be extended infinitely in a straight line. Any line segment can also be the radius of a circle, with one end of the segment at the centre of the circle. Right angles always remain right angles, regardless of rotation, reflection or translation. And the fifth axiom states that exactly one line passes through a point outside a given infinite line without intersecting it. Put differently: two parallel lines on a flat plane don't intersect anywhere. From these axioms, secondary-school pupils learnt to deduce many other theorems. The sum of the three angles of a triangle is one-hundred-and-eighty degrees, for example. Or the Pythagorean theorem.

In everyday plane geometry, these axioms were absolutely valid, but there were also caveats to this approach. The fifth axiom had already been the subject of much debate among mathematicians. Early in the nineteenth century, Hungarian mathematician Janos Bolyai, Russian Nikolai Lobachevsky and—in private—their German colleague Gauss had realised that other kinds of spaces were also conceivable. They developed a self-consistent geometric system that discarded the fifth axiom. This non-Euclidean geometry described relationships between lines, points and planes on curved surfaces, such as the surface of a saddle (hyperbolic), of the globe (spherical) or of an egg (elliptic; equivalent to spherical). These relationships were different to Euclidean geometry: for example, the sum of the three angles in a triangle on such surfaces wasn't 180 degrees.

But Tanya's most important objection to existing teaching methods was pedagogic. Euclid produced his system only after observing space, identifying concepts such as points, lines and angles, and thinking about their relationships. Euclid's true genius lay in his subsequent formulation of the five axioms and development of a self-consistent system in which the other theorems about space fell into place. But how many pupils made the connection to the space around them when struggling on a slate or in a notebook through this jungle of theorems, formula after formula and proof after proof? How much meaning did the underlying concepts still have for them? What understanding did they gain from all this theory?

Tanya wanted to encourage pupils first to explore space themselves, as Euclid had done. Besides, didn't Klein and Hilbert also emphasise that *Anschaulichkeit*— that is visualising concepts and making the relations between concepts clear— was crucial in geometry? Before immersing pupils in cool reasoning, Tanya first wanted to familiarise them through play with concepts such as planes, points, lines and angles. By expressing the concepts and discovering the patterns themselves, pupils could better understand what geometry was about and what the theorems meant. They would gain more insight into the relationships between concepts and theorems, and instead of merely reproducing theorems, they would have to think for themselves.[11]

To encourage her thirteen-year-old gymnasium students to do so, Tanya devised a set of practical assignments. She arranged them into themes such as symmetry, the relationship between straight lines and planes, parallel lines and so on. She also tried to make them easy to perform. Tanya wanted pupils to cut symmetrical figures out of paper or look for lines of symmetry in building façades as if they were in a 'geometry laboratory'. She wanted them to discover trihedral angles in the classroom, such as the one between two walls at right angles and a ceiling. She wanted them to consider how much information was needed to determine where a treasure was buried or, more abstractly, the position of a dot on a sheet of paper. And what if this dot was on a globe?[12]

While she continued to develop these assignments and a friendly colleague piloted them on the class of thirteen-year-olds in the school year 1907–1908, the school administration looked on disapprovingly. Every time the lessons were discussed at school, a teacher of the higher grades 'vehemently protested that this was not an ordered course', while some of the parents also grumbled. 'She speaks of a sphere, while the pupils haven't even learned the definition of a straight line!' (Ehrenfest-Afanassjewa 1961, 8)'[13] It sometimes seemed as if Tanya's work here, like at the Bestuzhev Courses, was only tolerated out of courtesy.

Did this contribute to her feeling tired, having respiratory issues and even staying with her mother at Petergof for a while? It was perhaps a relief that she could pack all their household goods and leave St Petersburg in mid-spring. As was the custom of the St Petersburg elite, Paul, Tanya, Tanitchka, Baba Sonja, the nanny Xenia and a cook spent the summer on the Baltic coast. They rented a house in Kannuka, in present-day Estonia, where the 'firs [grew] right behind the beach, on mossy ground with mushrooms, emitting a glorious scent ...' as Paul wrote later (Klein 1970, 93). This was better than the sickening vapour of poverty drifting in the streets of St Petersburg during the summer. In Kannuka, Tanya's sensitive lungs enjoyed the fresh air; all she needed to do was protect herself from the sun. She tied wide-brimmed hats around her head with scarves.

Paul strolled casually with Tanitchka on the path along the coast, often to the steep stairs at Klint, a cliff on the road to Sillamyagi (now Sillamäe). Sometimes they stopped at the little church to listen to the organ playing. And occasionally they walked all the way to Sillamyagi, a village that hosted even better-heeled guests than Kannuka. In summer, people including physiologist Pavlov walked through the sunny streets, and painters such as Ilya Repin and his student Konstantin Somov were drawn there. When Tanya rested and studied at home, Xenia tidied and the cook Anna prepared the evening meal, Tanitchka sometimes walked along the coast with her baba Katya, who visited regularly. They ate raspberries from the garden. With Anna Föhringer, Ioffe, Yakov Tamarkin, Karl Baumgart and other guests they continued discussions from St Petersburg. White nights melded the languid days together.[14]

Paul and Tanya had plenty of time to introduce Tanitchka to the world. The girl had her own wall with drawings on the veranda of the wooden house. Just as Paul's elder brothers had spoon-fed him the natural sciences in his youth, Paul and Tanya familiarised Tanitchka with mathematical and physical phenomena. She was only two-and-a-half but very perceptive, for example, showing her mother 'how flies flew—she made a zigzag motion with her hand—and how mosquitoes flew—she made a smooth circular motion with her hand'. They were Tanitchka's 'own observations', Tanya added, although the girl must have become interested in some phenomena on account of hearing her parents talk about them. 'Once I showed her a rainbow and told her what the phenomenon was called', Tanya wrote for instance. 'Some time later, every morning when she lay in her bed, on which the sun shone directly, Pavel [Paul] used a glass prism to show her the spectrum. Tanitchka was delighted; she held her small, pale hands under the rays, which were painted in all sorts of colours. Every morning, she called for her father and asked him to show her the spectrum. When a rainbow appeared in the sky again, I pointed it out to Tanitchka and asked what it was called and what it resembled, and T' replied that it was a rainbow and looked like a spectrum'.[15]

Nearly a year later, Tanitchka would also bring up Tanya's favourite subject: probability. 'You're probably coming' means: 'I'm not sure, perhaps you'll come and perhaps you won't', the little girl observed. She produced this 'definition of probability', once again, 'completely independently and of her own initiative', Tanya wrote proudly. And: 'This is evidently the conclusion of numerous observations. In any case, she loves explaining the meaning of words from time to time.'[16]

Not only for Tanitchka was the world new and full of discoveries. Her mother had to learn how to secure a place in academia and how to give 'thinking' a central place in her life. Her father had ventured into completely unfamiliar territory in Russia. He was surprised by the difficulty of the magister's examinations he had to pass to advance in Russian physics: one for physics and one for mathematics.[17]

It was good fortune that Paul could work on them with his new friend Abram Ioffe, who was in the same boat. They first met in 1905 in Munich, where Ioffe was researching the structure of crystals with X-rays for his doctorate at Wilhelm Röntgen's laboratory. Ioffe was pleased to see Paul again in St Petersburg in 1907, if only because he was familiar with and understood his research. Likewise, Paul was also pleased with Ioffe, who kept up to date with the latest developments in physics. In winter and early spring 1908, he travelled every week by horse tram to Ioffe's home in the northern district Lesnoy, and often they worked and talked for so many hours that Paul stayed the night. 'Every day Paul wrote me letters, six to twelve pages long, laying out his thoughts and calculations', Ioffe later wrote (Ioffe 1967, 43).

Others also noticed his ardent enthusiasm—the directors of the institute Khvolson and Borgman as well as younger Karl Baumgart, for example, who invited Paul to attend the Russian Society of Physics and Chemistry lectures. Impressed by his probing questions and broad knowledge, in mid-1908 they asked him to become editor of the *Russian Society of Physics and Chemistry Journal* (Frenkel 1977, Ch. 3). Soon Paul's contributions and reviews regularly appeared in its pages (Tsipenyuk 1973).[18]

Did this also increase his hopes for a real job? Did he have an eye on a position at the Saint Petersburg Polytechnic Institute, where Ioffe, Khvolson and Borgman were based? In any case, after returning from Kannuka, Paul and Tanya moved to Aptekarsky Island, which was home to the institute. With Tanitchka, Xenia and a new cook, they moved into a flat on the first floor of a two-story house on Ulitsa Lopukhinskaya.[19]

The district was farther from the centre than Vasilyevsky Island and even quieter (see Figure 12.1). It was named after the old Main Pharmacy (Aptekarsky) and its large botanical garden that Peter the Great established to grow medicinal plants and herbs. In later periods, rich Petersburgians built dachas around this garden for the weekend and afterwards apartment blocks were built between them, and houses with notable residents, such as Prime Minister Stolypin (whose house was blown up in 1906).[20] Not far from Paul and Tanya's flat was also the laboratory for experimental medicine, where Ivan Pavlov had discovered his famous Pavlovian response. They could hear the dogs barking, and, of course, in no time Paul struck up a conversation with Pavlov on the street.

Figure 12.1 Paul and Tanya in 1912 in Leiden, with their second daughter Galinka, in front of the Leiden Observatory. The building was inspired by the Pulkovo Observatory near St Petersburg, making this the place that most reminded them St Petersburg and St Petersburg winters (few of their photographs of St Petersburg have survived).—*Photograph Ehrenfest Family Archive.*

Yet it was Tanya who first found a job. She had given up her position at the gymnasium, although not because of the criticism of her mathematics laboratory. On the contrary, in the new school year, the previously so sceptical mathematics teacher of the higher grades had said that no pupils had progressed so easily with systematic geometry as the group from the experimental laboratory. But the school was far from Aptekarsky Island, and Tanya lacked the energy to travel back and forth (Ehrenfest-Afanassjewa 1961, 8, 12).

Instead, she started work at the Pedagogic Museum of the Military Academy, a typically Russian institution focused on conscripts. Recruits, since 1874 conscripted to serve six years in the army, came from all corners of the vast empire and all levels of society (Bezemer and Jansen 2008, 54–56). Even poor peasant boys could rise to the rank of officer, or sometimes even general, in this army of over a million soldiers. At least in principle, as peasant soldiers' starting point was often wanting. Less than half could read or write Russian or perform relatively simple calculations and to ameliorate this situation, the Pedagogic Museum had for decades given recruits reading and writing lessons and educational lectures. This was of course primarily in the army leadership's self-interest as well. Some degree of literacy was crucial for reading maps, working with new communication systems or even learning to operate a three-line rifle or field cannon. Moreover, speaking, reading and writing in the same language were meant to foster a sense of unity within the army (McMeekin 2017, 54–56).

But instead of offering them remedial education, naturally it would be even better if recruits could already read, write and do arithmetic from the start. Which is why the museum organised increasing numbers of lectures for teachers and others wanting to improve and expand primary and secondary education.[21] It led to lively debates in the museum's mathematics department, inspired by the educational reforms in mathematics in Germany and France.[22] Tanya's proposal to discuss Hilbert's ideas on the foundations of geometry in spring 1908 thus went down well. 'Your lecture can, of course, expect much interest from the mathematics department. I gratefully accept your kind offer to give the lecture next year', the director of the institute, general and mathematician Zakhar Maksheyev, wrote on 1 April 1908, ending courteously: 'Please accept my sentiments of high esteem'.[23]

Subsequent lectures, such as 'Intuition and Logic in Geometry Education', were also so successful that they were printed and distributed, and the general asked Tanya to become a member of 'his' mathematics work group.[24] And while bellicosity and military discipline conflicted with her ideas influenced by Tolstoy and Gorky, Tanya didn't let this opportunity pass by to spread her ideas about mathematics education and to make more people think. Thus, an unexpected collaboration proliferated between Tanya and an institute of the Military Academy.

Something else was at least as good. At Aunt Sonya's insistence, from the autumn onwards, Paul and Tanya organised *kruzhoki*, discussion groups, on mathematics and physics, usually at their home. Aunt Sonya thought Paul in particular shouldn't languish in isolation but should engage in discussions, make contacts, laugh and flourish. So, they put a blackboard in their study. Every two weeks, they spent an afternoon there with friends and colleagues in a semi-circle discussing the latest publications in mathematics and sometimes physics too (Frenkel 1977, Ch. 3).

Over the months, little Tanitchka saw a stream of visitors at their house. Ioffe of course, as well as the inseparable and fervent mathematics students Yakov Tamarkin and Alexander Friedmann—one stout and solid, the other slender with a bowler hat and walking stick—fat uncle and thin uncle, as the girl called them.[25] In addition, there were about ten recently graduated physicists, including Karl Baumgart, Alexander Dobiash, Leonid Isakov and Dmitry Rozhestvensky, as well as a number of university students, among them Yuri Krutkov, Victor Bursian and Georgi Weichardt, plus some students from the Bestuzhev Courses, such as Valentina Doynikova, who had a brief relationship with Friedmann, and Anna Föhringer, who was in a relationship with Ioffe at the time (Tropp et al. 1993, 34; Frenkel 1977, Ch. 3). The sessions were animated and cheerful, more exuberant than in aloof Vienna and solemn Göttingen. St Petersburg started feeling like home.

13
A sombre honorary secretary
Arduous success in Russia

'Well, what questions have you got?' Paul invariably asked when kruzhok participants entered. Valentina Doynikova recalled later how she once teased him. 'Why, Pavel Sigizmundovich, can't we come and visit you unless we have questions?' Paul would respond with genuine surprise: 'What do you mean, unless you have questions? If you study physics, you can't not have questions!' (Tropp et al. 1993, 35) Paul led the discussions, she remembered. For him two questions were paramount: what is the essence of a physical problem or a physical theory, and how do you present a problem? And even if you couldn't completely understand such a theory or problem, it was still good to note it down, mentally or preferably in a notebook, so you could discuss it later.

Ioffe sometimes poked fun at Paul's zeal for rules, single-mindedness and impatience. 'Dear Ehrenfest', he wrote on a card to him in January 1909. 'In response to your request I inform you that 1) your lecture shall take place on . . . . 2) the next physics kruzhok shall take place on . . . . 3) unfortunately, it is not possible for me to provide you with accommodation on the night of Friday to Saturday. Warm greetings from all of us!' And below on the left, he scribbled: 'Well, what do you think of my quick response?!'[1]

For Paul, the kruzhoks were more important than for Ioffe. He lacked the security of a stable, paid position and couldn't take the company of colleagues for granted. He stayed on course thanks solely to his immense drive to mean something in physics. Conversations with friends were about physics. Discussions at home often centred on physics. Physics was prevalent even on 'Tanitchka's wall' in the kitchen. Not only were there children's drawings on it, but also geometric figures such as a cube, a sphere, a tetrahedron and a dodecahedron. And a shelf stood all kinds of physics gadgets, such as a homemade electric coil which generated a magnetic field when the power was turned on, causing an iron ball to spin (see Figurer 13.1).[2]

At the same time, it was difficult to maintain such enthusiasm without the clear structure of a job. So, it was fortunate that, thanks to Ioffe, Paul secured one in 1909, even if it was a temporary appointment. As a result of a budget surplus at the Polytechnic Institute, he could give senior students mathematics lectures. 'And such classes!' Ioffe said later. Paul tackled mathematical differential equations and

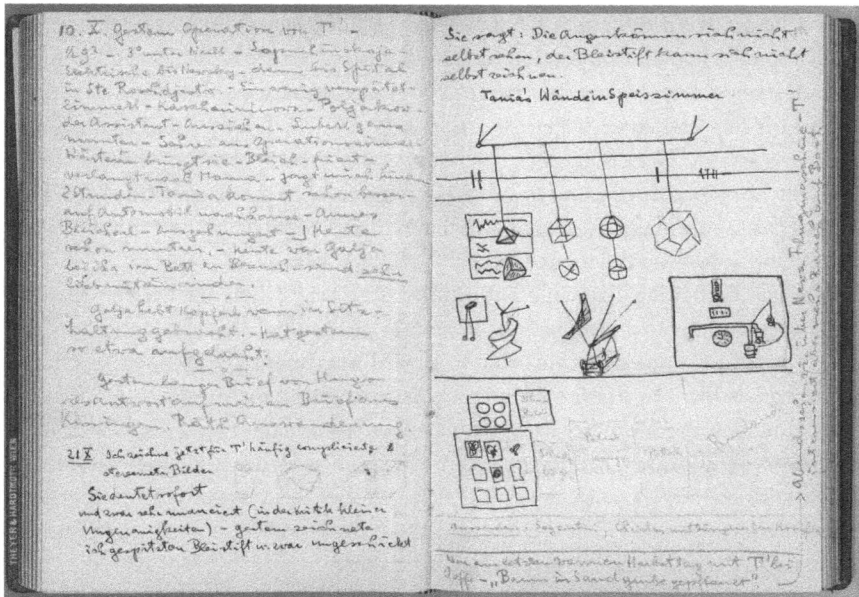

Figure 13.1 A drawing by Paul of 'Tanitchka's wall' in the kitchen of their flat in St Petersburg, in the red family diary.—*Ehrenfest Family Archive.*

other topics using lively examples from physics and with infectious enthusiasm. He enjoyed teaching, and the students enjoyed him (Ioffe 1967, 44).

Yet the job gave him only limited satisfaction in the end. Mathematics simply wasn't his chosen field. He avoided difficult calculations when possible and had an aversion to meticulous proofs of things that seemed understandable at first glance. He enjoyed mocking mathematicians who insisted on such careful proofs, dubbing them 'gendarmes' or 'epsilonians'—after the auxiliary entity epsilon which they frequently deployed in such proofs (Frenkel 1977, Ch. 3). When Paul wrote two short articles on mathematical functions for the German *Mathematisch-Naturwissenschaftliche Blätter* in 1909, even using an epsilon himself, this stemmed mainly from the lectures he gave and not from a shift in his interests (Ehrenfest 1909b; Ehrenfest 1909c).

The fact was that Paul had lost his heart to the concepts and principles in physics that revealed a profound truth about nature. He wanted to participate in 'expeditions' as Boltzmann had done and take them to the boundaries of natural science or even beyond. And he knew he needed strong credentials to join such expeditions, on which leading physicists from Western and Central Europe usually took the lead. Paul had to show a larger audience than just his students and visitors to the kruzhoks what he was worth. He needed to publish in international scientific journals.

But on what? His 1905 and 1906 articles on Planck's work contributed, according to some, to Planck's subsequent reformulation of his ideas from a slightly more focused perspective (Navarro and Pérez 2004). But they brought Paul as little exposure as his first article in St Petersburg in 1908, in which he pointed out a small error in versatile Nobel laureate Lord Rayleigh's calculations on stellar aberration (the apparent displacement of stars) and the velocity of light from 1892. 'It is terribly easy to make mistakes!' the magnanimous patrician conceded to Paul subsequently, settling the matter (Klein 1970, 143-144).[3]

Even Paul's Russian-language publication on the Le Chatelier-Braun principle failed to give him the recognition he sought (Ehrenfest 1909d). Had Ioffe, the experimentalist, suggested this topic to him? Experimentally inclined physicists and chemists liked to use the principle when designing measurement configurations. It helped them predict how disturbances would affect systems, which were initially in equilibrium. According to French chemist Henri le Chatelier, German physicist Karl Braun, and Dutch Nobel Prize winner Jacobus van 't Hoff, such systems inherently resisted disturbances and had more possibilities to do so effectively as the number of determining variables increased.

The electric coil with an electric wire on the shelf near Tanitchka's wall was an example of this. When a magnetic field was applied to the coil, it induced an electric current in the wire, which generated a magnetic field in the opposite direction, weakening the initially applied field as much as possible. Le Chatelier also gave examples from thermodynamics, such as gases destabilised by heating or cooling. He demonstrated that such disturbances are counteracted more strongly when volume *and* pressure are also allowed to vary. Such examples were just up Paul and Tanya's street, seamlessly complementing their ongoing reflections, amidst other responsibilities, on their review article on Boltzmann's work for Klein's *Encyclopaedia*.

In their discussions, they soon come across counter examples to Le Chatelier's loosely formulated principle. 'This time, Tanya and I are certainly victorious over Uncle Experimentalist', Paul wrote triumphantly to Ioffe in 1909, who had previously defended the principle (Klein 1970, 158). While Paul once more summarised the counter examples in thermodynamics in his article, Ioffe, in Munich, sent a response in the same style to 'nephew Ehrenfest'. Paul had, in Ioffe's words, 'destroyed the last refuge of teleology—the Le Chatelier principle. It withstood an incessant stream of waffle without issue but collapsed under analysis. Don't you feel sorry for him?'[4]

Probably not, but Paul's publication hadn't necessarily given him the 'right credentials' either. For pragmatic chemists and physicists, when used smartly, the principle was still a handy guide. For theoreticians seeking to fathom nature, the principle wasn't all that fundamental. There was even irritation, as Ioffe indicated, at philosophising about it, as if every system had some sort of equilibrium as a preconceived goal. What mainly hindered Paul, however, was publishing his ideas in Russian—incidentally, without Tanya. It didn't gain him the attention he most desired, that of progressive, German-speaking physicists.

Only a subsequent article, touching on Einstein's special theory of relativity and published in the *Zeitschrift für Physik*, brought him more success (Ehrenfest 1909e). Previously, Paul had engaged in the discussion about, among other things, what the theory predicted for the behaviour of fast-moving electrons. Einstein had pointed out that his special theory of relativity 'only contains assertions about rigid bodies, clocks, and light signals' (Einstein 1907; CPAE 2, 268). Subsequently, a subtle debate arose about these rigid bodies, involving prominent physicists such as Planck. They wondered what happened to rigid bodies travelling close to the speed of light. Paul, too, demonstrated that the 'contraction' of bodies solely in the direction of motion was problematic for the internal proportions of rigid bodies.

The example he chose was a rotating, non-deformable disc, like the records he enjoyed listening to at home with Tanya and Tanitchka. In a thought experiment, he imagined the disc rotating at nearly the speed of light. The radius of the disc would remain the same from any perspective, as it was perpendicular to the direction of rotation. But this didn't apply to the edge of the disc. To an observer beside the disc, each segment of the edge would appear contracted, like a measuring rod whizzing past. And since more measuring rods would fit around the circumference in this manner, the circumference would appear longer.

Paul had found a wonderful example to illustrate that the special relativity theory didn't tell the full story. The circumference (C) of the disc was no longer equal to what an ancient formula had dictated for centuries: twice the radius (R) multiplied by the number pi (π). Yet the crux of the relativity principle was that physical laws have the same effect in all systems moving at a constant speed relative to each other and that underlying mathematical rules, such as $C = 2\pi R$, are also equally valid in such systems. Didn't this hold for accelerating systems, such as a rotating disc, where the speed continually increased in one direction and decreased in the other? Although Einstein had grasped the problem by this point, and the discussion about rigid bodies had been ongoing for some time, the clear simplicity of Paul's example was compelling. No less than three prominent physicists—Max Abraham, Max Planck and Max Born—subsequently employed the 'Ehrenfest paradox' in the discussion about rigid bodies.[5] It was exactly what he needed!

Was 1909 the propitious year in which the 'dangerous experiment' of moving to St Petersburg started paying off? Paul had secured a job and written a remarkable publication, and Khvolson and Borgman showed their appreciation when, for the physics magister's examination in April, they chose only subjects such as thermodynamics, which Paul was well-versed in. With this boost and his sharp mind, it should have been easy for Paul to overcome the final hurdle to an academic career in Russia: the mathematics magister's exam.

Or wasn't he actually so keen on a career in Russia? That summer in Kannuka, Paul struggled to focus on the mathematics material for the examination. The enthusiasm with which he had embraced Russia, the Polytechnic Institute and everything that went with it was quickly overtaken by his critical aversion. He ranted about the stuffy old guard at the physics institutes. He grumbled about the lack of organisation in Russia.[6] And he complained so often about the 'Magister *Chineserei* (hocus pocus)' that little T' formed a peculiar impression of the university. 'Last winter, T' understood from Pavel that everyone at the university is Chinese', Tanya noted in the red journal in summer 1909 in Kannuka.[7] 'She also knows that Chinese people eat rice. Suddenly, in the summer, she remarked, "There is a lot of rice at the university"'.

But while Paul and Ioffe slogged through a mountain of mathematics material, his often sharply phrased criticism also worked against him. Not everyone heard it with the same affectionate ears as Tanya or his friend Ioffe. And not everyone took criticism as phlegmatically as Lord Rayleigh, conceding how 'terribly easy' it was to make mistakes. In any case, the people Paul needed to secure long-term employment didn't appreciate his critical remarks.

Tanya also found herself caught up in bureaucratic red tape. Alongside her jobs, she had been studying physics at the St Petersburg State University for two years. And now a misunderstanding and bureaucratic formalities thwarted her graduation. 'To my deep disappointment, I learnt that at the last faculty meeting my name was not on the list of people who had passed all the [mathematics] examinations', she wrote in June 1909 from Kannuka to the 'Most honoured Dmitry Fyodorovich!'[8] Dmitry Fyodorovich Selivanov had also been her teacher at the Bestuzhev Courses. Tanya reminded him he had given her an exemption for Probability Theory, a subject she had already taken with him. 'Before registering for the Probability Theory examination, which was also examined by you, I again approached you to ensure I understood correctly that you would give me a grade without an examination', she wrote. But 'upon inquiry' she was shocked to see her name wasn't on the list of graduates, due to 'the absence of a grade' for this subject. Everything could still be put to rights if Selivanov sent a signed statement before 1 July. 'If you do not do so, issuance of the certificate of state examinations passed, and thereby also admission to the magister's examination, shall be delayed for a very long time. Please be so kind as to help me, most honoured Dmitry Fyodorovich, to obtain the certificate of examinations passed before July of this year'.

It was an unpleasant situation, especially as no one knew where Selivanov was. Paul, who was still in St Petersburg at the time, had already sent two letters on Tanya's behalf, one to St Petersburg and the other to the Frisian Island Juist. Tanya posted this third letter to Göttingen, hoping that Selivanov might be there. But the response Paul received shortly afterwards in St Petersburg wasn't what they hoped for. Selivanov refused to send the statement, even after a persistent Tanya wrote to him again on 20 June.[9] 'Postponing the examination until the autumn, as you suggested in your letter to Pavel Sigizmundovich, is not a good solution for me, firstly because I could easily have passed the examination last year, and secondly because of the things you said'. Postponement would put her in a 'very uncomfortable situation', she explained once again.

But Selivanov was unyielding. In Germany, he had studied under renowned mathematicians Weierstrass and Kronecker. He had even been in cafés with Sofia Kovalevskaya. Later, he married one of his students from the Bestuzhev Courses. Why was he so obstinate? In the end, Paul pinned the university insignia on her only on 22 December. 'T' naturally asked what it was for', Tanya noted in the red journal.[10] And when she heard that the insignia indicated that her mother had now learnt everything there was to learn at university, the bilingual girl wanted to know if you could 'also learn to speak good Russian' there. 'Yes, you can'. 'Then why hasn't Papa learnt to speak Russian at university?' Whether she was impressed by the gold medal her mother again received for outstanding performance, wasn't recorded. For Tanya, it was scant consolation: she had missed the opportunity to start the magister's examinations at the end of the summer and pursue an academic career like Schiff and Gernet.

Fortunately, her contributions to the discussions on mathematics education at the Twelfth Congress of Naturalists and Physicians in Moscow during the Christmas holidays of late 1909 were well received. 'Tanya influenza—*Encyclopaedia* finished—little desire to go out', Paul had scribbled shortly before in their journal.[11] 'Leave in the evening, Lopukhinskaya—frost—snow—cold, cold ... Station: Ioffe, Shaposhnik [?], Gernet, Biron, Bulgakov, Rosing'. And a day later in pencil: 'Arrival in Moscow—suburb—omnibus—frost!—Sun—...—Hotel—*Kremlin in severe frost!*—the big square—the red monastery, the lines of white facades—green structures—violet shadows—snow on the roofs—on the right blue domes—the blue sky—all the way to the left the old church with the pointed domes—in front the big tower with black and gold trim—across the water a sea of houses with 1000 cheerful chimney plumes'.

In the ensuing days, Tanya presented her views on geometry education at a lecture and during discussions related to the international debate that had resulted in reforms to mathematics education in Germany, France and other countries. In France, a parliamentary inquiry had already painted a grim picture of lycée students doing sums without understanding what they were doing in 1899. In Germany, Alois Riedler had started the engineering movement, which advocated teaching students primarily practical and applicable mathematics even a little earlier.

In Russia, mathematicians at the Pedagogic Museum had been studying the work of mathematics reformers such as German Felix Klein, Frenchman Charles-Ange Laisant, Irishman John Perry and Englishman Alfred Lodge since around 1907. They also followed guidelines published by Swiss mathematician Henri Fehr in the journal *L'Enseignement Mathématique*, which aligned with the ideas of an international commission for mathematics education established in 1908.[12,13] Tanya now fearlessly defended her own ideas in the discussions, opposing some of those of such prominent figures. She contrasted them both with the ideas of 'practitioners', who aligned with the engineering movement and with those of 'logicians', who wanted students to dive directly into Euclid's geometric proofs. And she emphasised that her 'geometry laboratory' was distinguished by its focus on self-exploration and learning to think independently.[14]

But her husband had the most success. Enthusiastically, Paul outlined the work of Dimitri Rozhdestvensky on stage, one of the loyal attendees of his kruzhoks. He continued undeterred even as the audience started shuffling in their seats: Paul hadn't yet mentioned Rozhdestvensky's name! Was this plagiarism?! Only at the last moment did Paul explain that he was speaking on behalf of Rozhdestvensky, who couldn't attend because he had to prepare for his magister's examination at home. And with a sense of drama, he then chalked a lengthy list of what he considered the stultifying and largely useless examination subjects on the blackboard.

That this literally moved Paul to tears moved most of the audience, Ioffe wrote later (Ioffe 1967, 44). Paul recalled the progressive Muscovite professors roaring with laughter, while their conservative St Petersburg colleagues felt put to shame (Klein 1970, 89). Even more than before, he became the hero of a group of younger researchers. His lecture on Einstein's theory of relativity was even described as 'the highlight' of a congress that had started boringly, Ioffe wrote to his wife, dentist Vera Kravtsova (Frenkel 1977, Ch. 3). Both 'the content and the impression that his sincerity and enthusiasm made' turned Paul into 'the man of the moment'. 'Today he was elected honorary secretary of the evening session!'

But did the incident favourably influence the 'dangerous experiment?' Paul had been elevated and made new acquaintances, such as famous Moscow experimenter Lebedev and renowned mathematician Steklov.[15] However, his relationship with Khvolson and Borgman deteriorated further. It didn't help that in subsequent months he very explicitly stirred up the discussion about what in the eyes of young St Petersburg physicists were the overly demanding magister's examinations in mathematics and about the topics they should cover. When Sunday kruzhoks were held at the Polytechnic Institute, it had to remain a secret from the two professors, and Paul's contract wasn't renewed (Frenkel 1977, Ch. 3; Ioffe 1967, 44–45).

When he passed both magister's examinations in mathematics on 5 March and 10 April 1910, Paul thus had nothing to show for his pains. And although he claimed to feel strengthened in his didactic skills because Ioffe, as an experimental physicist, had scored even slightly higher than he did, he no longer had the opportunity to put these skills into practice in a lecture hall (Klein 1970, 87). 'Nowadays, I often feel I shall never gain a foothold in Russia', Paul had written in 1909, around the time Tanya faced opposition from Selivanov and he himself was somewhat disappointed with the enthusiasm for his kruzhoks.[16] A year later and unemployed once more, to him the situation seemed even more hopeless. When they departed for Kannuka in May, Paul was despondent. And Tanya was heavily pregnant.

14
Rest, cleanliness and regularity
A daughter and a spa

'On the night of 9–10 July at quarter to four, Galya arrived', Paul scribbled two months later in the red journal.[1] Tanitchka had a little sister. 'In the evening, Tanya was still in good spirits. Upstairs (Tanya' with me) I couldn't sleep until quarter to three. At quarter to four, I was awakened. Came—lamp was on—already heard crying—Tanya was joyful and smiled ... In the morning, Tanya' was very happy—her whole little face beamed—she looked curiously at her newborn sister. Galya lay like a bundle of laundry on the changing table and made endlessly funny faces'.

Twelve days later, Tanya added another entry.[2] 'At three o'clock, Pavel left for St Petersburg to take auntie to Wiesbaden and to take the waters himself in [Bad] Kissingen'. Auntie, baba Sonya, had been ailing for some time. Tanya had travelled to St Petersburg to support her in the weeks before the birth. 'And Paul...' Tanya wrote how she had explained to Tanitchka that it was sad but necessary for Paul to go.

Tanitchka stayed behind in Kannuka with her mother, Grandmother Katya, the cook and the maid, where she observed her little sister with her customary logic. 'Yesterday, baba Katya said: Galetchka is already looking at everything', Tanya noted in the red journal. 'Surely she thinks: why are there so many things. T': 'She doesn't understand things'. 'Well at least: why are there so many'. T': 'She doesn't understand "are" because she can't talk yet'.[3]

A stream of postcards and letters from Paul soon started arriving for his 'dearest, dearest darlings who sleep so far from me in a little house in the country by the sea'. First, they came from St Petersburg and from stations on the way, then from Wiesbaden on the Rhine, where he left Aunt Sonya at Villa Windsor, 'a nice (Swiss-style) guesthouse in a large, clean south-facing room'—in the good hands of '(Russian-speaking) Dr Simon' and in the vicinity of 26 hot springs.[4]

Paul continued travelling. In spring 1910, he had felt isolated. He barely had any contact with his brothers. His friends Tietze and Hahn were preoccupied with mathematical research at the universities of Brünn and Czernowitz. And his friend Walter Ritz, who brought him to Leiden in 1903 and with whom he walked and discussed so much in Weesen during later years, had died in 1909 from the lung disease that had blighted his health for many years (Forman 1975).

Paul's primary bridge to the West was now Herglotz, with whom he had resumed regular correspondence in 1909, and who had become a professor of astronomy at Leipzig in summer 1909. The quiet university city improved his spirits, Herglotz had written.[5] But melancholic and prone to drink, he had slipped back into despair in spring 1910. 'Outwardly I am healthy and prosperous, but psychologically I am terribly fatigued, unable to pick up anything, to think about anything for a long time. I hope taking the waters at Kissingen this summer shall restore me'.[6] This had given Paul an idea. If he accompanied his friend to this spa in western Bavaria, he too could escape everything that depressed him so much. Moreover, Herglotz could help him strengthen his ties with Germany, where physics was flourishing.

It was something of a culture shock when Paul, with a beard, tunic and a head full of Russian experiences, saw Herglotz again at Bad Kissingen—he barely recognised him. His friend was 'completely clean-shaven!!!!!!!!!, without a beard or moustache, [he wears] a gold pince-nez and [has] a glossy face'.[7]

On postcards and in lengthy letters to Tanya, Paul detailed their days at 'Villa Hedonia'. Their routine began at six-thirty in the morning at the baths and ended at quarter to eleven, when they went 'exhausted' to bed. The days were marked by a focus on rest, cleanliness and regularity. Everything happened at set times: bathing, walking, more bathing, resting, some talking, some walking. And the meals were more or less the same every day: juices in the early morning, a light lunch, buttermilk with bread in the late afternoon and a relaxed dinner at night.[8]

But post from Kanuka was less regular. 'Very unfortunate that I haven't heard from you', Paul wrote to Tanya several times.[9] Her letters only reached each stop after Paul had already moved on, it transpired after a week. Only in Bad Kissingen did he finally read in detail in the forwarded letters from St Petersburg, Warsaw and Wiesbaden that Tanitchka was diligently knitting and painting, Galya no longer looked so crumpled, baba Katya had grown thin from caring and Tanya was trying to delve into the article on the theory of relativity Max Born had sent from Göttingen.[10] 'I'm terribly curious how your trip was', she wrote.[11] And: 'May God grant that we all come together again soon and indeed in Russia'.[12]

Yet Paul's interest in Russia had waned. In a long response, he shared his reflections at the baths in Bad Kissingen with Tanya, 'and you must accept that you shall hear many unpleasant things. ... Since my encounter with Herglotz, I have been consumed continuously by the following feeling: it is extremely unlikely I can develop normally in Russia (work and livelihood), while on the other hand, I have lost every material and mental contact with Germany and Austria. If we didn't have children, ... it wouldn't worry me as much as it does now'.[13]

It wasn't only the tranquillity of Villa Hedonia that allowed Paul to clear his mind so well, he continued, but also 'because I am back in Germany and that I can now verify whether all the images in my recollection, that I work with are correct'. For the most part, it was so: the landscape was indeed unremarkable, unless it reminded him of Russia. The 'average person on the street, in a tram or restaurant' resembled those in his memory. The neat streets and houses still had the same ambiguous effect on him: 'pleasant and sometimes unpleasant—the Sunday afternoon feeling!'[14]

Yet he also noticed all manner of things in Germany, which he hadn't appreciated before. In Wiesbaden and Bad Kissingen, for instance, there were truly beautiful buildings, he wrote, 'beautiful without any ifs or buts' and certainly not marred 'by the senseless historical frills that so often made us laugh in Göttingen'.[15] Above all, as the entire letter revealed, Paul was in awe of his friend Herglotz, who lived in seclusion in Leipzig and valued this.

Paul continued that he 'could have easily portrayed' his old friend such that Tanya would 'shrug her shoulders with a pitying smile'. Herglotz had become chic: with a 'white waistcoat, yellow shoes, grey hat, all extremely comfortable and expensive'. He had 'an elegant suitcase, bag, etc'. He smoked 'only very good cigars'. He had 'a languid and passive bearing' and 'a drawling but witty manner of speaking' that allowed him to evade any topic of conversation uninteresting to him (such as a job for Paul at the University of Leipzig). Herglotz had abandoned his old pacifist ideas and read the Greek philosophers, delved deeply into Goethe's work to 'convince himself that great individual achievements are only made in times of struggle, when the lives of many individuals are literally at stake; that times without struggle always and everywhere lead to sickly, senile decay'.[16]

Paul still harboured some reservations about this, but he was also grateful to 'destiny for having such friends', as Herglotz had opened his eyes. 'You suffer from a disease that didn't affect me as deeply', Herglotz had explained to Paul, 'you see in all the people and things around you primarily how they relate to you—and not how they relate to each other. This robs you of all serenity and strength'. Herglotz, too, used to feel connected to everyone and involved in all manner of things, exhausting him. His remedy was to suppress any 'desire to influence anyone at all'. He even viewed and treated his students with aloofness.[17]

Ultra-cultivated Herglotz essentially recommended Paul live a detached existence. Study, develop yourself and publish—this was what life was all about. And he struck a chord, as it was beyond question that Paul's university career would only get off the ground if he published more scientific articles. 'What "justifies" [Herglotz] most is how his scientific work has developed steadily and progressively', Paul wrote to Tanya. But how could he put this advice into practice? Thinking about his own future, all he could see was chaos, he continued despondently. This was because of Tanya's motherland. Even if the political situation remained stable, Paul would only stand a chance of securing a position if he could match the achievements of Mendeleev or Pavlov, he speculated gloomily. 'What shall become of us in Russia? Perhaps our children shall slip back into the proletariat'.[18]

Even if he focused only on the short term, say until Christmas, he doubted whether he would be able to work in peace and with discipline. The only way to succeed was under 'extreme circumstances', he explained. They would have to decline all private visits, abandon the kruzhoks and he would no longer attend meetings of the Physics Society. 'You are free to continue your probability kruzhok, and we can arrange things such that you can go out in the evenings whenever you wish, but I can live in complete isolation—in a sense, as if I had been officially declared ill'. But then, he lamented in a remark revealing that he perhaps saw her as less equal in practice than they had envisioned before their marriage: 'the opposite shall probably happen, if only because you have no other reception room at your disposal than my study'. After some further grumbling, he concluded by noting that the waters seemed beneficial and expressed his hope that Tanya 'might soon also be able to take a suitable cure'.[19]

But Tanya needed no cure to pass judgement. 'Never in my life have I thought you so foolish: your beloved Herglotz surrounds himself with the utmost care with everything imaginable that only the most peaceful times can offer people…, meanwhile delighting in the old-fashioned "individuality" … that war and revolution are supposed to bring, and you let yourself be led astray by such a philosophy?'[20] That revolution brings progress is inherent in the definition of the word, she continued. But she found it absurd to conclude, by invoking Goethe, that many must perish to create better individuals. 'There are more ingenious ways to perish than as cannon fodder—for the benefit of speculators who wage war'.[21] Paul was making the same mistake that Herglotz lay at his door, she chided, always wanting to 'adopt something from others, like now from Herglotz, while your own temperament—thank God—is completely different'. Be yourself, she urged. 'You shall see you can also lead a full and meaningful life without being clean-shaven'.[22] As for herself: "I don't see life as a gaping void before me. On the contrary, I see there is more to do than I have time and strength for'. And it was the same for Paul, of course. Only: 'you never tackle what is right in front of you; now you have convinced yourself once again that you must first study the Greeks to become like Herglotz'.[23]

Only one thing did she welcome: Paul's suggestion to work hard and with discipline until Christmas. 'Then you can complete your magister's degree. You shall see how getting up on time and taking regular walks shall help you. And you can walk with Tanitchka!!!!!!!!!!!!!!!!!!!'[24]

While Paul struggled along at the baths, Tanya set to work. In the discussions on mathematics education at the Twelfth Congress of Naturalists and Physicians, participants had repeatedly referred to their practical experiences. But these were subjective. Success in the classroom was as closely related to the quality of teachers as to the teaching method, Tanya pointed out to them. She proposed putting the matter to a large group of students. Which area of mathematics did they find interesting? What had stuck with them? Had this knowledge proved useful?[25]

The Pedagogic Museum was eager to conduct such a survey. The idea was to send questionnaires to 10,000 first-year students at various institutions, ranging from technical colleges to art academies and from universities to conservatories. Tanya was tasked with formulating the questions and Maksheyev, the director of the Museum, was pleased with her initial draft in May. 'I looked through your survey and found it interesting and instructive', he wrote on 4 May. The list didn't need to be shortened, as Tanya had anticipated, 'rather it needs to be expanded in some areas, and the questions should be clarified with examples'. Maksheyev found the survey 'eminently' suitable 'for the purposes of the Museum' and wanted it printed by July or early September at the latest. 'I leave the decision up to you, including how many copies should be printed. Let us know whether it should be indicated that the survey was prepared by you and in what form this should be presented. I believe your name should be mentioned'.[26]

But as so often, other mathematicians wanted to have their say too. It wasn't until she was in Kannuka that Tanya could put the final touches on the questions, which were to be answered 'as accurately and impartially as possible' and 'without consulting textbooks'. She felt 'very cheerful and relieved' when, a month after Galya's birth, she finally had the survey 'completely ready to send out'. 'Now I can think about other work again', she wrote to Paul. 'I hope I can also finish the history of probability some time'.[27]

She was trying to complete an article on the role of probability in physics. Physicists were careless with probability and randomness, she believed. With great bounds, they skipped over countless subtleties and higher-order effects in probability theory and, even in their conclusions about nature and the cosmos, they were cavalier with the outcomes of probability. All major physical theories, such as statistical mechanics and kinetic theory, assert that nature (or the part of which lends itself to investigation) tends to move towards the most probable state. But was this justified?[28] 'It is assumed that this tendency existed long before the appearance of organic life on Earth and shall continue to exist for a long time. But it also follows that in all the intervening time, the world is still far from reaching this most probable state. We seem to accept that the world primarily undergoes the most probable changes, and yet, that the state of the world is improbable. If randomness truly drives nature, then nature presents us with a wonderful analogy, with a long series of draws yielding only black balls from an urn containing more red than black balls', was her response in a nutshell (Ehrenfest-Afanasyeva 1911Ra, Ehrenfest-Afanassjewa 1958).

The cosmos as a magnificent accident versus all those theories in which states of equilibrium are the outcome. But for the time being, she was too preoccupied with people coming to see the baby, the girls and baba Katya, writing and reading letters or friends dropping by in Kannuka.

Her cheerful postcards describing this world were viewed by Paul with growing detachment. 'For three years I hardly laughed, and now I often laugh with Herglotz for half an hour at a time', he wrote back.[29] It wasn't kind, but it was clear:

Paul no longer shared Tanya's ideal of a life in Russia, full of long conversations and walks with friends, much thinking and science. He had abandoned the notion that his ambitions and dreams could be realised in Russia. Eager to make his mark in the academic world, he grew increasingly sceptical of their 'Russian' life. The divide between the autocratic elite and the radicalising revolutionaries was widening. Even stern yet reform-minded Prime Minister Stolypin appeared to be losing control. It seemed as if assassinations of ministers or notable figures occurred almost weekly in the vast empire, while reactionary and anti-Semitic sentiments were growing louder. How could such a climate nurture a thriving and progressive scientific culture?

It was therefore hardly surprising that towards the end of his 'cure', Paul found himself captivated by a lecture on Zionism. His day had started with a service in a 'little Russian church', after which he attended the lecture and then had dinner with a few of the speakers, he wrote to Tanya.[30] Palestine was located extremely favourably, he concluded a few days later in a subsequent discourse. 'Damascus to the north and Egypt to the south are rapidly developing. The Mecca-Damascus railway running through Palestine shall soon become very significant because construction of the Baghdad railway is beginning, which shall connect Europe with Central Asia (via Damascus), and in Egypt, a railway shall be built connecting Suez to South Africa [sic]'. And in Palestine, the Zionists wanted 'to create an area where Jews ... are the absolute majority—and in a manner that all Jews and non-Jews who come there feel they are on Jewish soil'.[31]

It wasn't mere wishful thinking. Most doctors, agronomists and engineers in Palestine were Jewish, and a growing number of Jewish farmers were reclaiming swampy areas and greening parts of the desert. Paul described how the speaker incidentally mentioned how such developments contradicted prejudices about Jews being 'physically degenerate and incapable of hard work ... —even a fleeting visit shows this is not true'. To Paul's delight, there were even concrete plans to establish a polytechnic college in Jerusalem! This contrasted with the situation in Europe, he continued. Of the roughly twelve or thirteen million Jews in the world, between eight and nine million lived in Poland and Galicia alone. An enormous mass for whom 'the watchword "assimilation" ... was empty and meaningless'. This would almost inevitably result in 'liberalism being followed by antisemitism'. And 'this problem could assume horrific forms in the next hundred years', he feared.[32]

It was only natural that Paul was struck by this analysis. Assimilation and anti-Semitism had marked his youth in Vienna. The Ehrenfests had been an almost perfectly assimilated family in Favoriten, but on the street, Paul was constantly singled out as a Jewish boy. Perhaps it was precisely this schizophrenic situation that he had wanted to end when he rejected and ridiculed the religion as a teenager after his mother's death. Yet he did cherish memories of his large Jewish family, and unlike his brother Arthur and friend Ioffe, who had become Lutheran after marrying, he hadn't converted to another religion. Still, although he had become openly non-religious instead, his environment often still regarded him as Jewish and an outsider.

It wasn't overly surprising that he felt rootless and constantly stressed that he hadn't found a place where he felt at home. 'I have been uprooted three times', said his contemporary, composer Gustav Mahler, who had converted to Catholicism. 'As a Bohemian among Austrians, as an Austrian among Germans and as a Jew everywhere in the world (Beller 1991, 207)'. And almost the same applied to Paul. 'Our spirit can be cosmopolitan, our heart can't', he scribbled at the front of his notebook during his time in Leiden.[33]

In fact, this quotation from Dutch literary scholar Carel Vosmaer was just as applicable to 'homeless' Paul as it was to all those who welcomed him into their circle yet still regarded him as an outsider. Paul almost always had to relate to an environment where his presence wasn't a matter of course, where he was given an identity or was equated with all those other millions of orthodox, Hasidic or assimilated Jews who were conveniently seen as one and the same. Was it to escape this that Paul described the other Jewish visitors to the spa so negatively to Tanya? 'There are a great many Jews here—Russian, Polish(!), German ... and they behave so unpleasantly that I get very depressed: if a Russian were to come here, he could easily say: Thank God our barbaric exclusionary rules keep these people away from us'.[34] He hoped to show Ioffe during a joint visit to Vienna in September that things weren't much better there than in Russia, he wrote, and that he hadn't exaggerated when 'vividly describing antisemitism [in Vienna] to him as well as the conditions that justify this antisemitism'. In any case, he was glad the whole issue 'was resolved for him personally by his children being non-Jews'.[35]

Another question was whether they should remain unaffiliated with any religion, a topic he had also discussed with Herglotz. Eloquently, Herglotz had argued that Paul and Tanya had put their children in an exceptional position. Ioffe too had raised this issue before. Their children would always have to justify their lack of religious affiliation and suffer from it. 'We must baptise the children (Orthodox— if you want; Protestant—if I may choose). I no longer need to elaborate on all the arguments Ioffe and Herglotz gave. Please think seriously about this point. ... I certainly believe we committed a sin [sic] by not having them baptised', Paul concluded.[36]

But Tanya again refused to question their life. 'For every decision we make, we can fear a future objection. This leaves us no choice but to choose to the best of our knowledge what seems best to us', was her rebuttal of Paul's request. To which she added a hint: '[pecuniary] problems are best addressed by working as much as possible, thereby making the girls materially independent'. Moreover, people would probably deal better with the whole religious issue in the future. She ended the letter more optimistically than Paul and at the same time somewhat tartly:

'People are becoming more sensible, even if Uncle Herglotz still felt the need to be cleanshaven!'[37]

It wasn't until late September that Paul, after stopovers in Munich and Vienna, returned to his 'dearest company' and could embrace his daughters and his 'dearest darling', his 'lovingly kissed' and his 'dear, faithful friend, you!' And Tanya could do the same to her 'dearest', her 'God be with you' and sometimes so 'poor, sad' Paul, who had gained '3¼ pounds'. They were together again in Russia.

Would it turn out well there after all? 'Galya can sit and then looks like a pitiful chimpanzee', Paul noted a few weeks later in the red journal.[38] Tanya was working on her article on probability. Together they had nearly finished the article for the *Encyclopaedia*, and when Einstein, who had risen to professor in Prague, called Paul's article on the rigid disk 'elegant', Paul started to enjoy his work again (Einstein 1911). With renewed enthusiasm, he prepared his lectures at the Physical Society, together with Russian physicist Stephan Timoshenko, a lecturer at the Electrotechnical Institute on Aptekarsky Island. 'We decided to meet in the morning at the Botanical Garden, opposite my house', Timoshenko later wrote. 'At this hour [ten o'clock] there were no other people, and we could discuss scientific problems in peace. If figures were needed for clarification, we could easily draw them with a stick in the fresh snow (Frenkel 1977, Ch. 3)'.

Yet gradually, hope for a successful outcome of their Russian adventure began to fade (see Figure 14.1). Disturbances became more frequent, such as those in Moscow, where students gathered to commemorate the recently deceased Tolstoy in the cold month of February of 1911. In late 1910, Aunt Sonya read aloud at the table the newspaper eulogies for the writer who had ended his life as a messiah preaching nonviolence.[39] His calls for a simple life, antiauthoritarian education and critical thinking had also profoundly inspired Paul and Tanya. But the new, ultra-conservative Minister of Education, Lev Kasso, had little regard for nonviolence and antiauthoritarian education. He pandered to conservative and ultra-conservative factions in the Duma, such as the anti-Semitic Black Hundreds movement, which sought to rid universities of 'undesirable elements'.

The police, sent by Kasso to the university campus in February 1911, cracked down on the students. The university rector, Alexander Manuilov, was promptly dismissed after voicing his protest against this. Approximately 25 professors and 75 scientific staff members subsequently resigned in solidarity, including physicists Alexander Timiryazev, Nikolay Umov and Pyotr Lebedev, whose famous laboratory was closed to the dismay of Ioffe, Paul and Tanya (Josephson 1991, 14; Kassow 1989, 356–357).

In St Petersburg, Paul, together with Ioffe, Khvolson and Friedmann, drafted a petition for the Russian Physical and Chemical Society, which had been silent hitherto. They underscored their position in Moscow on 8 March with a speech

Figure 14.1 Paul and Tanya Ehrenfest with Russian colleagues. Seated from left to right: Paul Ehrenfest, Vladimir Chulanovsky, Harald Perlitz, Tanya Ehrenfest. Standing from left to right: Leonid Isakov, Abram Ioffe, Dmitry Rozhdestvensky, Georgi Weichardt, Victor Bursian, Yadwiga Schmidt-Chernysheva. Standing behind from left to right: Yuri Krutkow, Alexander Dobiash, Karl Baumgart.—*Photograph Ehrenfest Family Archive.*

that Paul, by his own account, had 'wasted infinite [he wrote "∞"] amounts of time' preparing.[40] During a tumultuous session, the cautious Society 'eventually adopted everything' 'step by step' and after 'chaos'. 'We cannot accept that the oldest university in Russia is deprived of its exceptionally valuable school of physicists', the Society declared (Tsipenyuk 1973).

After the long meeting, Paul slept 'on a plank bed in the dining hall', much to his friends' appreciation (Frenkel 1977, Ch. 3).[41] But by this time, he saw events primarily as a hopeless repeat performance. 'Plague in Silesia. Student uprisings. We must leave', he noted in the red journal on 3 February.[42] And he definitively abandoned his plan to write a magister's thesis, the final step towards a magister's degree that would have paved the way to a career at a Russian university.

15

It could have been so wonderful here

Inaccessible German-speaking universities

As Paul's morale steadily waned, a learned company gathered in the conference room of the Hôtel Métropole in Brussels. In the first week of November 1911, chandeliers with milk-glass shades spread their even light over the attendees' heads. Teak panelling, heavy wallpaper and central heating lent the room an air of sumptuousness. It wasn't an environment physicists were accustomed to. For Ernest Rutherford, who had travelled from Manchester, this was even more reason to attend the meeting in Brussels. 'Some wealthy man in Brussels pays a thousand francs each for our expenses. This is the sort of congress I have no objection to attending', he said (Heilbron 2015, 2044). Arnold Sommerfeld from Munich, however, thought it 'stupendous pretentiousness. ... A lunch not under five courses. Silly!' he wrote to his wife (Eckert 2015, 2066).

Ernest Solvay, the benefactor, had no need to worry about the costs. A patent on the production of soda ash had turned his family business into a multi-million-franc enterprise and made him a wealthy man. After relinquishing daily responsibilities at the company in the 1880s, Solvay chose to devote part of his wealth to his 'fifth child', the natural sciences (he had two sons and two daughters). The closed sessions at the Métropole perfectly aligned with this vision. The participation of scholars from five European countries also aligned with Solvay's broader mission of fostering good international (or at least western European) relations.

Yet, the meeting came about serendipitously. The topic of discussion was quanta, which Planck had introduced in 1900 as a mathematical 'device' and which were later identified by Einstein as true parcels of energy, possibly suggesting that the universe had a granular nature. But by 1908, the Nobel Committee still found Planck's work and its potential implications too speculative to award a Nobel Prize. Walther Nernst, director of the Institute of Physical Chemistry at the University of Berlin, had largely organised the Brussels meeting to finally make sense of quanta (Lambert 2015, 2025).

This was partly out of self-interest. As a true experimentalist, Nernst studied specific heat and heat capacity of materials. He was impressed by Einstein, who had elegantly described Nernst's measurements using the concept of quanta. Nernst undoubtedly harboured hopes of winning a Nobel Prize, as his work seemed to confirm these latest insights elegantly. Yet, as quantum theory was still speculative, its falsification could also potentially 'taint' his work. When he realised that Robert Goldschmidt, his young colleague in Berlin, was scientific patron Solvay's right-hand man, things started to move quickly.

With Solvay's support, Nernst drew up a list of learned participants who, during a meeting in Brussels, could help definitively resolve the question of quanta (Lambert 2015). Goldschmidt and Solvay then slightly tweaked the list by removing some Germans, while adding a few French scholars. Marie Curie, who had just been awarded a second Nobel Prize and was the only woman in the group, was one of them. Besides her, the French delegation also included renowned mathematician Henri Poincaré, and physicists Jean Perrin, Marcel Brillouin, Louis de Broglie and Curie's colleague Paul Langevin. The only Britons to accept the invitation were Ernest Rutherford and James Jeans. Both were relatively young and were silent during the discussions. Rutherford was even reticent about his latest, brilliant experiments, which were the first to show that atoms had a dense and positively charged core, the nucleus, which was surrounded by mostly empty space through which negatively charged electrons travelled, orbiting the nucleus at relatively large distances (Heilbron 2015).

The German delegation was fully represented, with Planck and his superior Heinrich Rubens, director of the Physics Institute at the Royal Friedrich Wilhelm University of Berlin, along with Wilhelm Wien, Emil Warburg, Sommerfeld and, of course, Nernst. Representing Belgium were Solvay's aide Goldschmidt, Georges Hostelet and Edouard Herzen, grandson of renowned Russian philosopher Alexander Herzen. Versatile Viennese physicist Fritz Hasenöhrl was present, while young Martin Knudsen represented Denmark. Dutchman Heike Kamerlingh Onnes also journeyed to Brussels. His liquid helium made his laboratory in Leiden the coldest place on earth in 1908, and just before the Brussels meeting, he demonstrated that electrical superconductivity could occur at such extremely low temperatures. In terms of subject matter, he was somewhat out of place, but he accompanied Lorentz, who was indispensable.

From their very first encounter, Solvay had been convinced that Lorentz should chair the meetings (Lambert 2015). All the participants later agreed that he was ideally suited to the role. He was courteous, considered impartial, sharp-witted, composed, well-versed in the latest developments in physics and fluent in French, German and English. French physicist Brillouin found Lorentz's 'perfect' command of French 'truly astonishing (Brillouin 1926)'. The second-youngest participant, Albert Einstein from Zurich, called Lorentz 'a living work of art'. This was a genuine compliment as Einstein was, as always, critical of others: he referred to the gathering beneath the chandeliers as 'a delight for diabolical Jesuit priests' in a letter to a friend.[1]

For Einstein, the question wasn't whether quanta should be taken seriously, but rather what they signified. Unfortunately, he hadn't gained any new insight in this regard, he complained in the same letter. For him, the best outcome of the Brussels sojourn had been the opportunity to meet French and other non-German physicists. For many of these physicists, the best result was the article Poincaré published just over two months after the conference in which he demonstrated for once and for all that Planck's quanta weren't merely a mathematical device. Poincaré didn't go as far as Einstein, who suggested light and radiation were granular by nature, but he did observe that matter seemed to absorb energy from radiation in discrete parcels, implying granularity was an essential aspect of nature (Poincaré 1912).

How bitter all of this was for Paul! Far away in Russia, he had independently reached precisely the same conclusion by an alternative method. Of course, he had noticed physicists' increasing interest in quanta. He must have been aware of Einstein's work using quanta to describe the thermal properties of materials—their so-called 'specific heat'. This work illustrated that quantisation was a much more fundamental trait of nature than perhaps initially thought (Klein 1965; Duncan and Janssen 2019, Ch. 3). And perhaps this was even what had prompted him to revisit, clarify and refine his own earlier ideas of 1906 (Ehrenfest 1911b). Were quanta a necessary condition for describing the spectrum of a black body? In other words, was working with them one of perhaps several potential ways to describe the spectrum? Or were they also a sufficient condition, and was quantisation exactly what was required to describe nature? These were the questions he returned to.

As in 1906, Paul deployed his deep knowledge of Boltzmann's work giving a statistical basis to thermodynamics to once again meticulously compare Boltzmann's concepts and methods with those of Planck. He used a different route than in 1906. He now looked at the a priori weights that are given in such statistical calculations to, for example, gas atoms with a certain kinetic energy or black body radiation carrying an energy within a certain energy interval. Building on the work of others, he then showed step by step that, in the case of black body radiation, such weights must be proportional to the ratio of the radiation energy and frequency, if the well-known and monumental second law of thermodynamics were to hold. Such weights, he went on to show, then lead directly to Wien's famous displacement law, which states that the spectrum of a black body simply shifts from reddish to bluish with increasing temperature.

Paul next investigated two specific cases: the long wavelengths—infrared and longer—at the low-energy tail of the distribution, and the very short wavelengths—ultraviolet and shorter—at the high-energy end of the spectrum. He discussed how working with these weights helps explain the shape of the 'long tail' (very low frequencies). And he argued how to avoid extremely short wavelength radiation (extremely high frequencies) from carrying away all an object's energy—essentially by imposing a zero weight for the limit where the ratio of energy to frequency approaches zero. In fact, he implicitly introduced the now familiar term 'ultraviolet catastrophe' for the latter phenomenon, when he named this section of his paper 'Avoidance of the Rayleigh Jeans Catastrophe in the Ultraviolet'.

Before ending his article, he even argued—unfortunately, without providing the full mathematical proof—that Planck's distribution describing the entire spectrum can only be obtained if these weights are integer multiples—in other words, their energy must not only be proportional to frequency, but also quantised (Klein 1970, 245–251). And all of this ineluctably led to the same conclusion as Poincaré: matter, or electrons within it, somehow absorbed and emitted energy only in the form of specific chunks of energy (or colours), and these quanta were an indispensable (necessary and sufficient) trait of nature. It was an article with which he could have made a significant impact, perhaps even more than with his paper on the Ehrenfest paradox. If it had been noticed, later physicists judged, it could have helped accelerate the development of quantum theory (Klein 1970, 245–251; Navarro and Pérez 2004).

Except, it wasn't mentioned anywhere—neither in the proceedings of the Solvay Conference on quanta nor in Poincaré's subsequent article. Paul had imagined things rather differently. In the weeks when his eminent colleagues were preparing for the conference and later enjoying their five-course lunches in Brussels, he was in St Petersburg exploring opportunities for university positions in the West. He wrote to his old friend Hans Hahn, who had been appointed Professor of Mathematics in Czernowitz. He corresponded with Heinrich Tietze, who was an Associate Professor in Brünn and, of course, sought advice from Herglotz in Leipzig. The latter quickly devised a solution. Unfortunately, Leipzig University didn't recognise Viennese doctorates, he wrote. But if Paul pursued a habilitation in Leipzig, he could become a lecturer. Furthermore, Sommerfeld in Munich had already indicated that a revised compilation of Paul's earlier work could serve as a dissertation (Huijnen 2003, 62–67).

Should Paul try this? His brother Hugo, who had made a successful career as a gynaecologist in St. Louis in the United States, urgently advised him in a letter to secure a stable position. From afar, Hugo had diagnosed his younger brother. Paul suffered from overstrained nerves, neurasthenia, he concluded—in accordance with the spirit of the times. The uncertainty in Russia and the lack of stability further aggravated his frayed nerves. A steady job would be therapeutic. 'I can imagine that with a "constant" joy in work your entire life, your successes, and with it your mood and (through a virtuous cycle) finally your joy in work itself will ultimately bring you complete satisfaction'.[2] A stable position, he added, would allow Paul to finally support his family and share his knowledge and skills with the rest of the world. In fact, Paul had to find such a place outside Russia—perhaps Switzerland or the Netherlands. Paul had chosen Russia as his adopted homeland, but it didn't want him, mainly for reasons of religion, Hugo wrote. 'Your choice was unhappy'.

Yet Paul soon shelved the Leipzig plan once more. Neither the prospect of another doctorate nor Leipzig appealed to him. Instead, he planned a tour of German and Austro-Hungarian university cities—to the bewilderment of Herglotz, who had come to feel extremely Prussian. 'I am beginning to truly believe that you have crossed the boundary into pathological territory', Herglotz wrote indignantly.[3] He accused Paul of '*inertia foeda*' (revolting laziness). 'It is merely a matter of writing down [the dissertation]. You cannot excuse yourself with nervous disorders or anything else if you can still work on other matters, as is the case. It is simply Austrian-Slavic laxity'.

This time, Paul paid Herglotz no heed. At the beginning of the new year, he boarded the train in wintery St Petersburg and was instantly overcome with relief. 'The sun has been shining with unbroken strength since this morning', he wrote to Tanya as the train steamed through the white plains towards the southwest.[4] 'It greatly lifts the spirits. Isn't it both a sin and madness that we, through frivolity or neglect, condemn our children to fog and indoor air during their early years?' It was the start of a new stream of postcards and letters between German university cities and St Petersburg.

A few days later, Tanya read about Paul's first stop, Berlin. Shortly after arriving, Planck himself welcomed him into his home. The 'youthful looking' scientist with a 'very finely chiselled, delicate face' and 'graceful, intellectual glasses' had invited Paul to dinner, and it was an animated evening. Paul, who was still blissfully unaware of Poincaré's article, discussed his work on quanta and felt that Planck was impressed. 'He was very surprised and highly interested that I had proved Wiens' formula is also based on energy levels', Tanya read. She gathered from the letter that Planck had been 'astonished and delighted' by Paul's nuanced analysis and the 'insights I reached by meticulously working through his [Planck's] ideas'. Planck reportedly said, 'I had never so precisely understood which aspects of my theory caused this or that; it is completely clear now'.[5]

Even better, Paul wrote, was that an enthusiastic Planck had promised to try to have him habilitate in Berlin. Paul went to great lengths to swiftly organise a lecture that Rubens, head of the Physics Institute at the Royal Friedrich Wilhelm University of Berlin, could attend as he had the final say in such matters. He received help from old acquaintances like Gustav Hertz in Göttingen and Lise Meitner, who had moved to Berlin at the end of 1907 and since then, living off an allowance from her father, had been conducting experiments with radioactive isotopes—first in a basement (as women were not allowed to enter university buildings) and then above ground, together with chemist Otto Hahn.[6] Once again elated, Paul informed Tanya that Rubens had praised his lecture highly. 'Planck now needs to take the initiative, and he [Rubens] shall not object'.[7]

But it wasn't going to be that simple. Paul was already a bit old; a backpedalling Planck explained the next day. Paul was also a foreigner and had no religious affiliation. And actually, the position had already been promised to someone with a German doctorate.[8] Paul was left deeply disheartened. Adding insult to injury, on a brief visit to Herglotz in Leipzig, Paul stumbled upon a note by Henri Poincaré in the scientific journal *Comptes Rendus*. The French scholar announced his 'quantum work', without any reference to Paul's published article. 'What will become of me?' he lamented in his notebook (Klein 1970, 174).

Compared to his friends, who were already professors or well on their way, Paul's career did indeed appear to be stagnating. Even Paul Epstein, his host at his next stop in Munich, seemed to be surpassing him. Epstein and Paul—both Pavel Sigizmundovich in Russian—had met at the Twelfth Congress of Naturalists and Physicians in Moscow in 1910. Epstein, a former student of experimental physicist Lebedev, had nearly finished his magister's thesis at the time and was working as an assistant professor of experimental physics in Moscow. However, Paul so impressed him that he abandoned all his plans for experimental work to become a theoretical physicist as well. He went to Munich, where he trained as a doctoral student under Sommerfeld, honing his skills in 'calculating', as physicists liked to refer to working with mathematical formulas (Epstein 1965, 69–71). And this was where he received a downcast Paul.

Fortunately, Minna Epstein, a well-known Russian pianist, immediately played the piano for two hours after Paul's arrival. 'Tchaikovsky, Rachmaninoff, Brahms and Beethoven ... *very* beautiful', Tanya read.[9] Later that day, Epstein expressed comprehension of Paul's frustration, although he sympathised with Planck too. Offering someone an unpaid position for his habilitation shouldn't be taken lightly, especially when, as in Berlin, there was no prospect of a subsequent permanent position, Epstein remarked apologetically. But Paul couldn't yet see things in perspective. 'Perhaps I should just lie down quietly for a whole day and sleep for a long time', he wrote to Tanya. He added that his health was good: 'I eat a *gigantic* (egg) pancake with compote every afternoon and often in the evening as well. In addition, two cups of yogurt (every day!), eggs, cheese, bread, butter. Sometimes something sweet. My stomach is perfectly fine'.[10]

But his state of mind was not. The letter Tanya had sent ahead to Munich made him even more 'upset' a few hours later. Had Tanya sensed that things wouldn't turn out as Paul hoped? In her letter, she advised Paul not to abandon the Leipzig plan right away. But it was exactly that advice which had made him 'very depressed', he replied. 'Just look at the situation: a postgraduate from University A [Vienna] looks for a PhD post at University B [Munich] to - - - - - do his habilitation as a *Privatdozent* [untenured senior lecturer] at University C [Leipzig]. You really must be very young to be able to do this calmly. It is truly discouraging!'[11]

He extensively listed all the disadvantages of the plan again, concluding that living in Leipzig would be 'absolutely impossible'. 'Legendary coal dust, perpetually gloomy and dreary, an indescribably desolate city, truly indescribable and very expensive to boot! You cannot imagine how dreadful Leipzig is'. In short, it was 'evident' that habilitating in Munich was 'nonsense'. 'And that you have suddenly changed your mind now ... is really good only for making me feel despondent'.[12]

Paul's mood fluctuated between highs and lows during the trip. The low he underwent became even deeper when Sommerfeld—a professor of theoretical physics at Munich; an X-ray expert and an esteemed participant at the Solvay Conference—showed him a proof of Poincaré's article (Klein 1970, 174). But with lengthy discussions with Sommerfeld and 'inspiring' encounters with renowned Wien during a two-day excursion to Würzburg, Paul gradually began to recover. Munich looked fresh under a blanket of snow. 'If only you could see how *rosy* all the children look here!!!!!' he wrote to Tanya. And: 'Tell T' that dogs and horses here understand only German, not Russian'.[13] Buoyed by his improved spirits, he even began sending packages containing his quantum article and their joint *Encyclopaedia* article to Poincaré and Langevin in France, to Rayleigh and Jeans in Britain and to Lorentz in the Netherlands.[14] In Munich, his lectures on his further developed critique of the Le Chatelier-Braun principle were well received (Ehrenfest 1911a). In extensive discussions with Sommerfeld, Paul also 'explained the finer points of my work on radiation'. Overall, he regained confidence in his own abilities. 'It is clear Sommerfeld is quite reluctant to see me leave'.[15]

But Tanya was mistaken if she thought that Herglotz's plan of a doctoral degree in Munich and a habilitation in Leipzig was back on the table. Paul was captivated by an alternative path, sparked by news about Einstein, the rising star in the physics firmament. 'It is certain that Einstein is going to Zurich, and I shall definitely do the same', Tanya read in a postcard in early February.[16] 'Then Einstein and Debye shall be there. You shall then complete your doctorate there'. When the newspapers confirmed the news of Einstein's appointment the next day, for Paul 'any doubt vanished. I am absolutely determined to go to Zurich, even if I can't become a lecturer there. ... I shall travel there tomorrow afternoon and promptly and decisively pursue a habilitation at the Polytechnikum [Federal Institute of Technology Zurich, ETH Zurich]'.[17]

In the jolting railway carriage to Zurich, he reiterated how the city was soon likely to become a 'phenomenal centre of physics'. 'If the Zurich dream comes true, Zurich shall soon become for theoretical physics what Göttingen was (and no longer is!!) for mathematics'. This thought lifted Paul's spirits immensely, he wrote in his long letter. And the idea of possibly working in Zurich, surrounded by 'excellent, hardworking people', made him even happier—'which is to say, I am now in such a mood that I would greatly regret having to die'.[18]

But his mood remained volatile. On the morning after his first night in Zurich, in a 'shabby' little pension arranged for him by the mother of Paul's late friend Walter Ritz, he visited the physics library at the Federal Institute of Technology Zurich, commonly known as the Polytechnikum. It was disheartening. Paul opened the latest issue of the French *Journal de Physique*, which contained Poincaré's 'quantum work', confirming what he already knew: Poincaré drew the same conclusions from Planck's work after the Solvay Conference as Paul had already done before that conference. 'If I had been at the quantum conference, I could have told him everything. To hell with it', Tanya read in another long missive.[19]

In the afternoon, the sunny university district lifted his spirits. 'The Polytechnikum university quarter is superbly situated. ... As the streets rise steeply, there is light and sunshine everywhere'.[20] He outlined his plans once more. He intended to focus his efforts on two goals. First, he placed his hopes on the university, where Dutchman Peter Debye was his contact—'he is already a full professor and likely shall accept an offer from Utrecht nonetheless'.[21] Furthermore, he visited the Polytechnikum that same afternoon, where he was given a '4½-hour' tour by Frenchman Pierre Weiss, 'a slender, thin, Nordic-blond Frenchman who speaks German with an accent', although he quickly tempered any expectations: 'If he does not help me, it shall be because he simply *cannot*'.[22]

And he could not as the end of the letter revealed. Paul sounded dejected once more. 'I have just spoken to Weiss. The prospects at the Polytechnikum are virtually nil because Einstein is coming. There is no room left beside him'. Paul assured Tanya he wouldn't give up. 'As soon as Einstein arrives, you must submit a request', Weiss had said. 'I shall wholeheartedly support your request, and Einstein shall do the same. Then we shall ... find suitable work for you'. And once Paul was in Zurich, the path to a professorship might still be open. 'Zurich is very beautiful!' he declared in the final lines. Still: 'My dear, dear darling, I was *very* sad many times today. How much I needed you today!! Oh, dear, dear darling!'[23]

A few days later, Paul wrote to Tanya that his further efforts had also stalled. He had visited Weiss, who had invited Debye as well. He had also met with other professors 'to get to know them in advance'. On Sunday afternoon, he had walked with Debye among 'beautiful coniferous trees' not far from the university district, and they had discussed Debye's highly mathematical work at length. The Dutchman had promised Paul to raise the habilitation issue one final time at the university. But all this effort proved futile after the weekend. 'Debye told me that he spoke to Professor Kleiner [the head of the institute], and that Kleiner rejected it out of hand: it would be too crowded. Good heavens, it is truly unbelievable! What is that old bloke afraid of?'[24]

Kleiner was known for viewing theoretical physics as a rather speculative endeavour and was said to have appointed Einstein mainly because he could not avoid doing so (Uffink 2006). Or was it Debye who feared Paul mightn't fit well in Zurich? Two months later, Debye wrote to his mentor and friend Sommerfeld in Munich using unflattering and even blatantly anti-Semitic language about Paul and his background. 'If you are considering bringing Ehrenfest to you [and giving him a position in Munich], I must express some reservations. A Jew, as he clearly is, of the "high priest" kind, can exert an extremely harmful influence with his bewitching Talmudic logic. Many fresh, yet undeveloped ideas, which one would otherwise work out with enthusiasm, can easily be stifled by him. And in this regard, I find dealing with him dangerous'.[25]

Unawares, Paul attributed all the resistance he faced in Zurich to Kleiner. He felt down. 'It is so extraordinarily beautiful here, and I would so love to stay', he wrote to Tanya.[26] 'You could complete your doctorate here, and it would be ideal for the children'.

16
A big, dear boy
Meeting Einstein

What was Tanya to make of Paul's letters, even after he left Zurich behind? Cheerfully, he described his stopover in Weesen, where they spent their summer vacations during their early years of marriage: 'In the morning, the mountains high above the clouds. Do you still remember, my dearest? How beautiful Switzerland is. If the children grow up here, they could feel at home—even in Zurich'.[1] But on the way to Austria, he already started grumbling again: 'You should see the Swiss and Austrian border guards. Simply kind young and older farmers there [on the Swiss side]. Here, arrogant, over-zealous mugs of indeterminate age—unattractive both sexually and ethnically'.[2]

It was hardly surprising that Austria brought Paul little joy. He found Innsbruck provincial. Viennese physicists' work on radioactivity was a poor fit with his theoretical interests. The capital itself made him feel gloomy. 'Greet all the Viennese for me, but don't stay with them too long', Tanya had presciently written to him.[3] And indeed, when Paul drove through his old neighbourhood Favoriten with his brother Emil and his friend Heinrich Tietze three days after arriving, the streets and squares struck him as 'incredibly desolate, ugly places'.[4] The same evening, he clashed with Arthur, who had been his favourite brother. 'I'm already fed up with Vienna'.[5]

Bright spots included meetings at the Vienna physics laboratory, such as that with Hasenöhrl, one of the Solvay Conference participants. After Planck, Wien, Sommerfeld, Lorentz, Langevin and Jeans, Paul also directed this Austrian scholar to his work on quanta. But most of all, he looked forward to leaving for Prague, where he would meet Einstein, the most promising physicist of his generation. Einstein had participated in the Solvay Conference, had worked as an assistant professor in Zurich and as a professor in Prague and was soon to return to Zurich. He had everything Paul lacked in terms of career: a permanent job, a stable income and even prestige.

Paul thoroughly prepared for his visit to Prague. Tanya had already sent a package from St Petersburg on his behalf. 'This morning, I sent three books of Russian fairy tales with illustrations by Bilibin', she wrote from St Petersburg, where it was 'a beautiful day, one degree, sunshine and much snow'. Tanitchka had gone 'for a walk with Parasha [the nanny]'. And: 'I eagerly await news from Prague'.[6]

Paul, too, had sent some post ahead. In Zurich, he had written letters to Einstein and his boss, Anton Lampa, Tanya's witness at their wedding. He had explained his situation, and Einstein's responses showed that he had Paul's best interests at heart.[7] Naturally, Paul hoped Einstein could help him secure a position in Zurich. He also contemplated the possibility that Lampa might offer him the professorship in Prague, which would become vacant upon Einstein's departure; however, he swiftly dismissed the notion. 'Of course, I would have to accept that offer unconditionally. But really, that would be a big mistake'.[8]

With whom would Paul discuss ideas in Prague if Einstein left? The local environment was no more conducive to flourishing than in Innsbruck, he explained to Tanya on the journey between Innsbruck and Vienna. 'At a university like Innsbruck, one must be an exceptionally gifted physicist not to sink completely. The situation is terrible: no physics community, no seminars, no scientific society, very few journals and so on—just think, for instance, of Prague'.[9] Brünn, his last stop before Prague, seemed just as inconsequential to him a few days later. It was pleasant to stay with his friend Tietze, who worked at the university. 'His wife is very beautiful and kind ... sings very well'. But 'unlike previous weeks, I haven't learned anything new this entire past week'.[10] Paul wanted to move forward, to push the boundaries.

'Prague must now serve that purpose! In a few hours, there will be a twist in my world line: a meeting with Einstein. I have only heard enthusiastic praise about him as a person—especially also from an antisemite'.[11] The 'world line' he was referring to is an element of Einstein's theory of relativity and describes the path of a particle through spacetime. Did Tanya immediately know which anti-Semite this was among the many German, Swiss or Austrian professors for whom anti-Semitic remarks were as commonplace as putting on a shirt or belittling women's intellect?

Such lofty expectations, so much preparation. It almost seemed to invite disappointment. Instead, Paul's heart leapt when, on Saturday, 24 February at ten to three, he saw a man standing on the platform in Prague under a station clock, wearing a large overcoat over a wrinkled suit and a thick head of black hair above. There was no mistaking him. "I recognised him from the photos. He recognised me 'because that is how he had imagined me'", Paul wrote that evening in a long letter to Tanya (see Figure 16.1).[12] Einstein's study contained only 'a few books', he observed, somewhat intimidated: 'Feels "miraculous" that I am sitting in Einstein's room as if it is nothing. Yet I am very aware that his name will still be mentioned long after I have been forgotten'.[13]

Paul gave his first impressions of that afternoon. 'He: broad, simple, cigar in his mouth, magnificent brown eyes ... we immediately talked about physics, without stopping. He doesn't have a single unpleasant trait'.[14] On the way to Einstein's house, they visited a coffee house, after which Einstein showed Paul around the

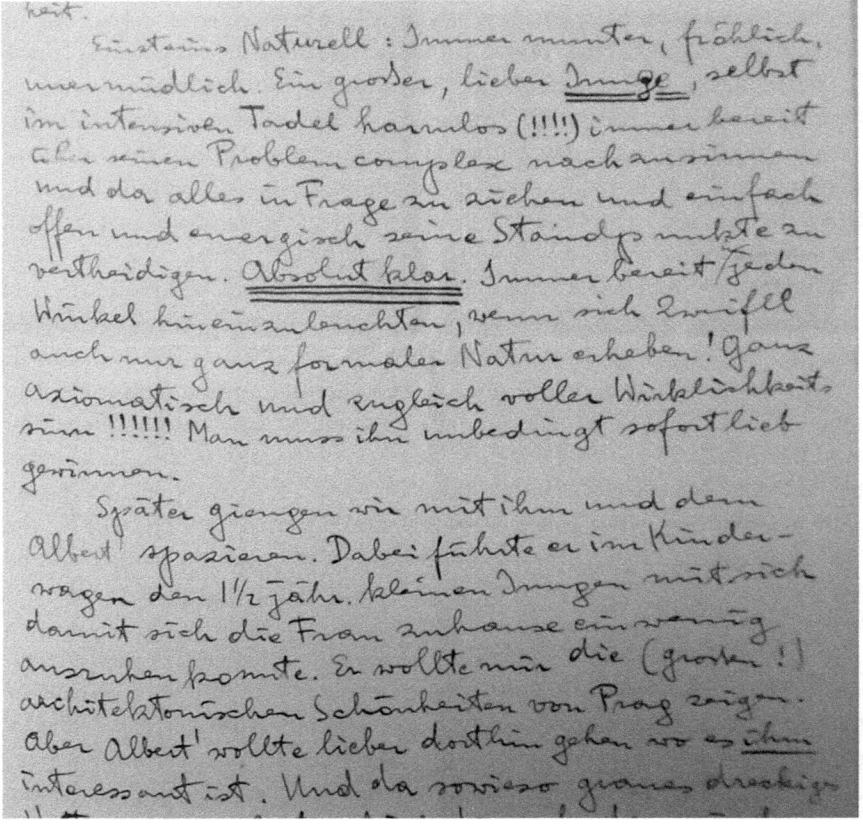

Figure 16.1 Excerpt from a long letter from Paul to Tanya in which he describes the first two days of his stay with Einstein in Prague.—*National Museum Boerhaave, Leiden.*

Physics Institute, and they discussed 'everything related to radiation theory and kinetic gas theory' all the while.[15] Even when Einstein returned from his string quartet after midnight, they picked up where they had left off.[16] Then the conversation immediately continued about the new theory of gravitation, until half past one. ... He said: 'It's quite common for someone to say to me, "I don't believe that", but it's rare for someone to actually substantiate it'.[17]

Einstein, the new scientific star, took insecure Paul seriously. But Paul also felt another, deeper kinship. Was it their background, both from a Jewish family with a love of knowledge? Assimilated, yet maintaining their own traditions and somewhat distanced from Austro-German class-based society? 'That I have opened myself up to him so completely is because I already know that, irrespective of physics, he loves me too', Paul was sure by Sunday, a day that 'was among the most beautiful and pleasant in my life'.[18] They went for a walk, he wrote to Tanya. Einstein pushed the pram with one-and-a-half-year-old Eduard in it. Seven-year-old Hans Albert scampered along, and together they watched trains being shunted behind the station. At home, they played music—Einstein the violin and Ehrenfest the piano. Brahms sonatas in the morning and Händel in the afternoon. 'So different from when I play with one of the Germans. Instead of pedantry, there was simplicity and naturalness. ... If there had been any doubt we would become good friends, it vanished at that moment'.[19]

Einstein discussed his work on gravitation and relativity with him, including the tensors he intended to use to properly shape that work. Paul discovered a small error in one of Einstein's proofs and then found and discussed an alternative proof—all 'in great harmony'. He described Einstein as 'always cheerful, happy, and tireless. A big, dear boy'.[20]

The days flew by. Even Paul's fantasy of taking over Einstein's position in Prague quickly materialised—or rather, could have come true in an instant. By Monday, just two days after his arrival, Lampa and Einstein presented him with a proposal. 'If I declare myself Jewish, I can immediately become Einstein's successor here in Prague', Paul wrote to Tanya. Conversion to a religion was crucial. 'Even Einstein had to give up his non-religious status. ... He says: I didn't get baptised. Returning to Abraham's bosom—it was nothing at all. Just a signed piece of paper: you must do the same'. But despite their kinship and friendship, this wasn't something Paul could do: 'You can imagine how bitterly I reproached him [Einstein]'.[21] Paul accused Einstein of setting a precedent by signing and stood by his non-religious stance. In turn, Einstein called Paul's principled stubbornness just as foolish a whim as his vegetarianism. 'He said: if you give up this utterly foolish non-religiosity now, I shall even allow you to subsist entirely on vegetables'.[22]

Paul stood firm. Not even the fact that Lampa and Einstein gathered 'all the old pistols [from Mach's speed of sound experiments]' the day before Paul's departure could change his mind. They 'held them to my chest, making me swear I would accept the professorship in Prague', Tanya read a few days later.[23] What Einstein said afterwards, referring yet again to Paul's Tolstoian, vegetarian and anti-religious lifestyle, didn't sit well with her: 'Out of religiosity, he doesn't eat meat and out of love for cattle, he doesn't adhere to any official religion', even though Einstein immediately softened his words. 'A terribly foolish man—just like my dearest friend Besso'.[24]

But was it really about principles for Paul? Or was it more about the other option which seemed to excite him more—the possibility of habilitating in Zurich? This option, which he discussed extensively with Einstein, would also please Einstein, Paul wrote to Tanya. Einstein thought there was even a good chance that Paul could become an assistant professor. 'There is no doubt that I must now pursue this path without hesitation'.[25] Rather than risk languishing in Prague, Paul gambled on such an uncertain position in Zurich, close to Einstein. For 'one thing is incredible. How much courage and joy in life my contact with Einstein gives me. I certainly do not want to die now. You shall be pleased by how Einstein plays music'.[26] And 'Oh', he wrote wistfully after Albert Einstein and Mileva Marić had seen him off at the railway station: 'if only Ioffe could also be sent by Röntgen to Zurich as Debye's successor'.[27]

Yet aside from Tanya, few people understood Paul's stubborn refusal to 'return to Abraham's bosom' and accept the offer in Prague. Only Herglotz was sympathetic. 'I wouldn't have done it myself'. Although he added that his circumstances were, of course, different to Paul's, for whom 'other prospects should perhaps weigh more heavily—such as the ability to provide advantages for your children with a higher income'.[28] His other friends tried their best to persuade him to take the position. Hans Hahn even sent a fiery letter to Czernowitz, asking Tanya if she couldn't bring Paul to reason (Klein 1970, 181). Einstein also turned to her.

But Tanya made no attempt to steer Paul. In a postcard, she responded to the incident with Einstein, Lampa and the pistols, as well as their remarks about religion and vegetarianism: 'If even such men possess only limited chivalry—what shall become of the world? ... That E., with all his logic and freedom to think, doesn't see beyond his own nose, is very disappointing to me'. She offered Paul no further advice, only noting that Einstein's and Lampa's attitudes 'reveal the problematic side of the Zurich plan in all its grimness—everyone there shall likely be the same as they are. Still, I wish you success'.[29]

While this stern postcard travelled in postbags across Europe, Paul was in good spirits. In Lemberg (present-day Lviv), he spent pleasant days with versatile and warm-hearted Polish physicist Marian Smoluchowski and his wife Zofia. 'There is no doubt that the Slavic disposition suits me much better than the "Viennese"'. They 'shall most certainly visit us next year in Zurich'.[30]

Afterwards, in Czernowitz—an 'impossible backwater'—Hahn 'taught him quite a few things, about which I shall tell you about in more detail later'.[31] Paul stayed longer than planned because his passport had expired a year earlier, preventing him from obtaining a visa for Russia. 'Oy oy oy, sweetheart, you are going to scold me now, oy oy oy!!!'[32] Fortunately, the Smoluchowskis pulled strings in Lemberg, arranging a new Russian visa in five days—an impressive outcome for an unaffiliated Austrian of Jewish descent without a valid passport!

Perhaps to win Tanya over, Paul showed that his concerns weren't solely focused on his own position in Zurich: 'In a small town [as he evidently saw Zurich], university people are closely reliant on one another. For you, such interactions shall only be worthwhile if you can keep uninteresting professors' wives firmly at bay. A woman like Smoluchowski's wife certainly wouldn't bore you ... I hope the same shall apply to Professor Weiss' wife'. He was less impressed with somewhat downcast Marić: 'Unfortunately, [the same] can only partially be said of Einstein's wife, but then again, we have Einstein himself!!!'³³

Conversely, Tanya was slightly less reticent about Paul's plans in a letter than in her earlier postcard. 'Last night, the Ioffes were here. I told them everything about you. Ioffe believes that Zurich is truly the best thing you could wish for now. He pointedly remarked: if [Paul] doesn't necessarily want to become a professor, he could just stay here. He could also earn up to a thousand roubles in St Petersburg with various "lectures". But he is a *Mensch* who needs to be a professor'. However, Tanya continued, 'I must add to this, my dearest, that I don't agree with Ioffe's ironic stance; I don't like the idea of earning money from work in which you only half believe'.³⁴

Tanya was well aware that Paul's friends in the West expected her simply to follow her husband. All too easily, they dismissed her work because she was a woman and because Russia was beyond their field of view. The centre of gravity for mathematical and scientific research was undeniably in Western Europe. Yet, Tanya didn't believe she could build a career at Western universities. 'There is hardly any point in me taking a doctoral degree. I won't get a job as a lecturer anyway', she scribbled on one of her postcards after Paul expressed hope for a doctoral position for her.³⁵ And somewhat stoically, she continued focusing on her mathematical work in St Petersburg during Paul's trip.

When she wrote in frustration about the didactics discussions at the Pedagogic Museum, Paul candidly replied that 'it naturally works in my favour that the lack of understanding at the Pedagogic Museum repels you'.³⁶ Her enthusiastic conversations with mathematician Veniamin (Benjamin) Kagan were probably less auspicious for Paul. Kagan studied at Kiev and St Petersburg, then was exiled to southern Ukraine for a time due to his involvement in the democratic student movement, after which he spent most of the year teaching at the University of Novorossiysk by the Black Sea. He also taught at Jewish schools and women's colleges in this city. But he returned to St Petersburg, where Tanya and he discussed 'entropy, probability theory, and mechanics without resorting to empty space. ... I am convinced I would have learnt mathematics much better if he had been in St Petersburg during my studies. He thinks it would be wise for me to pursue my magister's in applied mathematics'.³⁷

The significant difference between Tanya's and Paul's visions for their future became evident when Paul expressed his desire to use the final leg of his journey to visit Yasnaya Polyana, the Tolstoy family estate near the city of Tula, some 200 kilometres south of Moscow. Tanya struggled to understand: 'what is there to see at this time? And how are you supposed to reach it at the end of February? If you do go, you must dress very warmly. If I wanted to see Russia, I would go to Vesyegonsk. Or in the summer—along the Volga! You could also travel up the Vyatka to Kukarka'.[38] But Paul was committed to his plan.

First, however, he visited Kiev, Tanya's leafy birthplace. Undoubtedly 'beautiful in spring', but, unfortunately, it was 'very grey, dark and dirty' at this time of year.[39] He read her postcard about Lampa and Einstein and wrote back that she had 'formed a mistaken impression of Einstein'. He met some pleasant local professors but still found Kiev repugnant: 'streams of filthy water cascade through all the streets, unfathomably dirty. Apart from a few *very* beautiful, very old and very new buildings, all the other houses [are] sagging, dreary timber-framed houses'.[40]

He found snow-covered Kharkov, where he travelled a few days later, even more dreadful. 'Uglier than anything I have seen on this journey so far. Filthy, devoid of poetry, tasteless, unfriendly', he wrote shortly after arriving. 'Already, I can see it would utterly destroy me if I had to live in a provincial Russian university city'.[41] A day later, after visiting the university: 'Living here would be utterly, utterly impossible for me. First and foremost, because of the grotesque ugliness of the city itself, and secondly, the lack of a scientific atmosphere'.[42]

In his final, lengthy letter on this journey, Paul described his visit to the Tolstoy family estate.[43] After leaving his suitcase at Kozlova Zaseka station 'at 9:58 in the morning', he set off under a bright sun. Initially hesitant, he soon found his way following a sleigh track. 'Fields of snow to the left and right. Parallel to the road (beyond) walls of birch, lime and pine forests. Everything covered in gleaming, fresh white snow. Up and down hills'. The two 'pillars' he spotted after a while turned out to be 'two utterly filthy and dilapidated gatehouses' surrounded by 'straw and manure' and two 'dogs that induced respect'. And since the park around the estate was closed, he approached a 'very shabby peasant's hut with a sleigh in front', sinking to his knees in the snow. A peasant woman invited him in: 'very low, very small windows, a loom, a spinning wheel, pans for dyeing wool, a large samovar, three children above the stove and even two cats'. Paul spoke with a nephew of the peasant woman about the deplorable conditions in the village and about the two armed Circassians who chased the women of Yasnaya Polyana off the estate when they gathered dry wood. Meanwhile, the peasant woman set off with a scrap of paper: 'Dr Ehrenfest from Vienna courteously requests permission to view the house'.

A little later, Paul walked with her down the old tree-lined avenue and past the pond to the house. 'Greyish white with a green roof. First, the terrace with a wooden balustrade, in which animal shapes and such have been sawn out. Then around the back to a door covered in oilcloth. A slack bell cord ... A servant opened the door. Staircase, bookshelves. Old mirror. The servant asked me to wait as he had to serve the meal just then. I examined the book spines: books in every language, religious-scientific works and related subjects. The servant returned. Up the stairs to the first floor. Everything dreary and dilapidated, like in a neglected old dacha'.

The drawing room—with 'family portraits and (poor!) reproductions of the famous Tolstoy paintings'—gave a view of the park. 'The windows swarmed with flies!!!' In Tolstoy's study ('see our Tolstoy stereoscopic images'), he noted: 'Writing desk. Books. Chair. Everything very much alive'. Finally, he viewed 'his' bedroom. 'Very small. Cot. Nightstand. Bookshelf (containing the French translation of Tolstoy's last books). Travel chest, washstand beside the bed, belts, a very beautiful photograph of eldest daughter. I get the impression that the family has displayed and hung many items in this room that do not belong here'. Afterwards, the countess appeared to be 'expecting me' and invited Paul to the dining room, but 'I had no desire to let her spoil my mood'. Instead, he walked with the peasant woman roughly half a verst to Tolstoy's simple grave in a young lime forest, 'unfortunately surrounded by a factory-made wooden fence'.

While Paul drank tea with the peasant woman afterwards, the peasant collected Paul's suitcase by sleigh and then took him, suitcase and all, over 'glistening snowfields lit by the sun' to a station six versts away. 'Went up and down hills. The peasant cursed the horse. The horse kept sinking deep into the snow … Alexandra Lvovna's [Tolstoy's youngest daughter, who lived nearby but was not at home] dogs roamed through the snow like wolves. We rode and rode'.

The peasants here, Paul wrote to Tanya a few hours later from 'the godforsaken station', knew almost nothing of Tolstoy and were deeply superstitious. 'Is it true that the newspapers say it is freezing 140 degrees in the east?' they asked. 'Could this be a Russian legend related to Kamerlingh Onnes' liquefaction of helium?!' he wondered in his letter. 'Tomorrow morning, I shall be in Moscow'. He was almost home in St Petersburg, with a good story about Russia. But his future was elsewhere.

17
Coincidence and new opportunities
Farewell to St Petersburg

Tanya's trip along the Volga did come about. At the end of April, she travelled with Paul down the river from Rybinsk all the way to Kazan, the 'melting pot' where East meets West (Klein 1970, 185–186).[1] Oh, if only Paul could love her motherland as she did. 'Don't forget that a person who has seen Russia and has come to know the best Russians is in a different position to a German for whom a slightly idealised vision of Greece evidently represents the highest measure of life, emotion and beauty', she had written two years earlier when Paul was captivated by Herglotz's ideas about ancient Greece, Goethe, and 'Prussian' militarism.[2] But Paul's heart stayed more closely tied to the German-speaking parts of Europe than she had hoped. Instead of Herglotz, he placed Einstein on a pedestal, and Zurich beckoned. However earnestly Tanya tried to draw his attention to the good aspects of Russia, Paul seemed to see only fog, wind, cold, bureaucracy, chaos and his lack of prospects.

The days since his return drifted by torpidly. 'Darling, dearest. Forgive me for not taking stronger control of my fate. I should do so but cannot. ... I am simply amazed how it is becoming easier and easier for me to neglect everything I wish to neglect, and how impossible it is for me to do what I must do', Paul had written from Munich.[3] In St Petersburg, he found it utterly impossible to revise his articles into a dissertation for a habilitation, to resume work on his magister's degree or even to secure a part-time teaching post. He needed the rhythm, respectability and security of a job, his brother Hugo would have counselled; but that was easy to say far away in America.

Only the girls were wholeheartedly glad about his return. Not just cheerful and cautiously chattering Galya, who danced around when he came home, but more serious Tanitchka as well. 'Papa doesn't miss us. Why does he stay away for so long?' she had complained during Paul's trip.[4] She missed him and had wanted to show him that she had learnt to write. Through acquaintances, Tanya had also hired a tutor experienced in teaching writing, chemistry and music. He gave lessons three times a week and charged 50 copecks per half hour, she informed Paul in a letter. 'Do you agree? You are of the opinion that we should not teach the children things we do not understand ourselves, and also I thought that Sweetie would be better educated in the idea of systematic work.'[5]

Paul was perfectly content with this arrangement. He had unpleasant memories of Viennese schools and, like Tanya, shared many of Tolstoy's views on education. Once, after visiting a school in Kissingen, Tolstoy also expressed his horror at the German education system: 'It is terrible! Prayers for the king; blows; everything by rote; terrified, beaten children' (Simmons 1949, 205). This was in sharp contrast with the education Tolstoy—as well as Paul and Tanya—believed in, with educational material catering to children's talents that would naturally make them yearn for more knowledge.

Yet Paul was still restless, however much he enjoyed their lively household and conversations with his daughters or friends. He couldn't cast off this restiveness even when he finally sailed with Tanya down the slow-moving Volga. It could have been so beautiful. Cool fields stretched out beneath a pink morning sky. The spring sun warmed fragrant steppes in the afternoon. Lonely farmhouses accentuated the vastness of the land and the green, blue and golden onion domes of the fairy-tale monasteries in Kostroma, Novgorod and other cities shone. But at every stop, Paul hurried to the post office to see if perhaps post from Germany or Switzerland had been forwarded from St Petersburg. And although thousands of stars twinkled at night and the water smelled soft and sweet, he still noted in his diary: 'depressed'. His friends looked on with incomprehension as he sank into a mire of gloom and the opportunity in Prague slipped through his fingers.

'Your stubborn refusal to acknowledge any religious affiliation really bugs me; drop it for your children's sake', Einstein had fumed just before their boat trip.[6] An income, status and regularity at a university just within the European network of universities—what was holding Paul back? By then Einstein knew that Paul could forget about a position in Zurich: the Swiss had firmly rejected the request to create an assistant professorship for him. This only made Einstein more insistent: 'after becoming a professor here [in Prague], you could revert to this strange hobby horse of yours—the whole thing would take only a short time. Let your wife work you over'.[7]

It didn't happen. While the Volga lapped and the fields smelled of spring, even Tanya could do little to change Paul's mind. He was adamant about going to Zurich—or perhaps Munich first—but in any case, his sights were firmly set on the future 'physics paradise' of Zurich. Perhaps she wasn't the right person at all to persuade him. To her, what could Prague offer that St Petersburg didn't? They had many friends, lived in a sunny flat near the botanical gardens and she had her job at the mathematics group of the Pedagogic Museum.

Debate among Russian mathematicians, such as S.A. Bogomolov, S. Shokhor-Trotsky and Andrei Kiselyov, still pitched 'theorists', who wanted to adhere strictly to Euclid's teachings in geometry education, against 'practitioners', who emphasised applicability. Flamboyant shipbuilder and mathematician Aleksei Krylov, for example, who had gained world fame with theoretical publications in both fields, belonged to the latter camp. He believed that overly theoretical and out-of-touch students should finally learn to apply their common sense through application-oriented assignments, such as those in shipbuilding. This annoyed Tanya. 'It is indeed beyond doubt that someone without common sense and an eye for detail is helpless', she complained in a letter to Paul a year earlier, 'but someone who has an eye for detail yet does not appreciate why something must be so, who cannot explain to others why he is right, is also helpless—and unbearable, besides'.[8] At the same time, she enjoyed putting forward her own ideas during these debates.

It was also gratifying that over 10 per cent of the 10,000 first-year students from various disciplines filled out her extensive questionnaire on mathematics. This was a satisfying percentage for such surveys; the leadership of the Pedagogic Museum assured Tanya. Some respondents had grown somewhat restive with the occasionally challenging questions (Ehrenfest-Afanassjewa 1961, Introduction). 'The devil only knows', they had scrawled next to assignments like: 'a. What is the converse of the theorem "If a triangle has two equal sides, the base angles are equal". b. Is this theorem correct? c. Does this follow straight from the first theorem? d. Are the last two questions the same?' Or questions like: 'As is known, $\sqrt{2}$ can only be expressed as an infinite decimal fraction. Do you think it will one day be possible to express the square root of two with a denominator other than ten?'[9]

Nevertheless, over 800 students approached the questionnaire with genuine seriousness. Many respondents provided extensive answers to a series of detailed introductory questions, such as question seven: 'Were you interested in mathematics in secondary school? (*Yes, no*)', or question eight: 'How do you explain your attitude towards mathematics? (*Due to inherent traits or due to education*)', or question nine: 'When did you become interested in mathematics? *Before you started school* (before the start of school lessons), *from the lower grades, from the introduction to a particular area of mathematics* (which? ----), *with certain applications of mathematics* (which? ----)'. Incidentally, some questions stemmed directly from Tanya's own interests, such as question 33: 'Do you have the habit of looking for starting points in every argument or dispute, considering the logic of the proof and not simply being interested in the results of the reasoning?' or question 34: 'If so, do you attribute this habit to the influence of geometry?' The answers could teach Tanya something about her own idea of using geometry education as a vehicle for learning logical thinking.[10]

Undoubtedly, her thoughts turned occasionally to this while they sailed down the Volga. Just as they must have turned to the question of how all the answer sheets would be processed. This would require more people than just herself. Earlier in the year, she had floated the idea of financial support from the West on a postcard to Paul in Vienna: 'If only Solvay could help!!'[11] She had recently learnt from Paul that the Belgian patron wanted to establish a physics institute that would include a fund for 'needy' physicists.

And while Paul jotted down formulas and ideas in his notebook on the deck of their boat, Tanya probably also thought about Veniamin Kagan urging her to pursue her magister's degree. She had become close friends with this mathematician, who shared her interest in the foundations of geometry. She had previously written to Paul that she considered Kagan 'an exceptionally gifted and energetic person'.[12] His life, however, was burdened with adversities. In St Petersburg and during his exile in Ukraine, Kagan had faced vehement opposition due to his political activism and Jewish heritage. Only recently, his good friend Mikhail Iglitsky, founder and director of the Jewish gymnasium in Odessa and co-founder of mathematics publishing house Mathesis, had shot himself at his son's grave, who had been murdered by right-wing students. Tanya wrote to Paul that Kagan was someone 'who scientifically speaking, has certainly been unable to achieve everything he wants due to financial difficulties'. Almost in the same breath, she added: 'We are exceptionally fortunate, dearest! And, of course, sometimes we too, like everyone, must make compromises'.[13] But while Tanya counted their blessings, Paul mostly saw the other, darker side of Russia. When would the post set him free?

Not during their boat trip. But the day after returning to St Petersburg, a letter arrived from Munich. Sommerfeld had paid no heed to Debye's warning that Paul's 'bewitching Talmudic logic, would exert an extremely harmful influence'. Quite the contrary. In February, Sommerfeld had been so impressed by Paul's teaching abilities and his talent for uncovering the essence of a problem that he offered him an alternative to Herglotz's plan. Instead of completing his doctorate under Sommerfeld and then moving to Leipzig, Paul could habilitate directly under Sommerfeld and remain as a *Privatdozent* (lecturer) in Munich, the city where children looked so rosy and where Sommerfeld spent his days off hiking with his students in the mountains. Paul understood it was a game of musical chairs. As expected, Debye was appointed professor in Utrecht. Sommerfeld's talented assistant and privatdozent Max Laue took the vacant professorship in Zurich. And Paul could now take Laue's place as a Privatdozent in Munich. His income would depend on the students attending his lectures, but everyone knew that appointment as a privatdozent led to a permanent appointment (Klein 1981, 186).[14]

Yet, he wasn't altogether pleased. Was he afraid of failing? Sommerfeld, who had trained as a mathematician, was regarded as the most mathematically inclined theoretical physicist, while Paul regarded complex formulas with a certain amount of dread. Or was he daunted by the prospect of ending up in an environment he knew only too well from Vienna, with barely concealed anti-Semitism and without the protective presence of someone like Einstein? Laue's move to Zurich certainly served as a painful reminder of Einstein's failed attempts to bring Paul there.

What is more, it was doubtful Tanya would be happy in Munich. She hadn't taken well to the formality, conservatism and hostility to women wanting to develop their talents in Göttingen and Vienna. It was unlikely a city like Munich—probably much the same—would warmly welcome a nearly 36-year-old mother with ambitions.

They didn't have to agonise about it for long. For years, Paul had felt isolated in St Petersburg, despairing that his work was overlooked. And now, just days later, yet another offer arrived. He was astounded, as he later recounted, and dashed so swiftly to Tanya that he nearly knocked Aunt Sonya—their financial support in tough times—down the stairs. His hands clutched a letter from Lorentz, the Nobel Prize laureate from Leiden, whom he had visited with Walter Ritz ten years earlier, and the chairman of the Solvay Conference to whom Paul had sent a package of publications from Munich.[15] And in this letter, Lorentz had written something almost unbelievable: he was looking for a successor in Leiden and 'because I hold your work in the highest esteem for the depth, clarity and insight you demonstrate in it, I have also considered you'.[16]

Paul's earlier correspondence with Lorentz suddenly took on a different light. In April, Lorentz had written to Paul requesting information about the former Lebedev Laboratory, which had closed during the Kasso Affair. Lebedev and his colleagues were trying to continue their laboratory work outside the university with funds from sponsors, and Lebedev sent a letter to Lorentz in February inquiring whether Solvay's new foundation might function as a sponsor. Lorentz must have been surprised that an academic in Moscow was aware of this yet-to-be-established foundation, about which only a few people were aware. But his April letter to Paul revealed that he had soon grasped the situation.[17] 'I suspect you encouraged Professor Lebedev [to request a grant from the Solvay Foundation], after hearing something about the new foundation from Professor Sommerfeld [when Paul visited him in early spring]'.[18]

Sadly, Lebedev had since succumbed to a heart condition. Could Paul, in his stead, provide information about the current state of the laboratory in view of a potential future grant? Lorentz seized the opportunity to thank Paul for the 'excellent and thorough *Encyclopaedia* article' he had sent from Munich. 'Since the pleasure of briefly seeing you here with your friend Ritz years ago, our paths have not crossed again'. Could Paul tell him how he had fared since their meeting in 1903 and what his current academic position in St Petersburg was? 'I did not realise at the time that you were Russian'.[19]

Unsuspectingly, Paul had clarified the matter of his nationality and informed Lorentz in detail about the Lebedev Laboratory.[20] How could he have known that Lorentz also had an ulterior motive? During the academic year 1912–1913, Lorentz planned to exchange his professorship in Leiden for a position as director of Teyler's Physics Laboratory, housed in the former Haarlem residence of silk manufacturer and banker Pieter Teyler van der Hulst since 1784. The cabinet had a large collection of art, minerals, fossils and other natural objects, and there was also a laboratory which Lorentz planned to completely modernise. He wanted to combine management of this laboratory and cabinet with secretaryship of the Hollandsche Maatschappij der Wetenschappen (Holland Society of Sciences), freeing him from the many, to him irksome, bureaucratic duties in Leiden University. As an extraordinary professor, he would continue to lecture only on Monday mornings in Leiden, granting him sufficient time for all Solvay's plans, with whom he shared the ideal of promoting international relations between scientists, which 'foster valuable feelings of appreciation, solidarity and good fellowship ... and therefore promote peace' (Lorentz 1913, 6).

But first, he needed to leave the theoretical physics department in Leiden in capable hands. Lorentz sought a successor of international repute with a keen intuition for physics, someone able to keep pace with the rapid developments in the field. To his disappointment, Einstein, the candidate closest to his heart, had just accepted the prestigious post in Zurich (Berends and van Delft 2019, 323–330, Kox and Schatz 2021, Ch. 7). 'Indeed, I consider him as one of the very foremost, who sees deeper and farther than others, while I also highly value his personal qualities and his clear and simple presentation. I hold him in such high regard that I consider it my duty to attempt to win him for Leiden', Lorentz had previously written to university trustee Johannes van der Waals.[21] Incidentally, Einstein found it 'inexpressibly oppressive'[22] to measure himself against his predecessor in Leiden. He even wrote to Paul that 'when Lorentz offered me a job, it really gave me the creeps'.[23]

The suggestion by Sommerfeld and others to approach Debye was subsequently disregarded by Lorentz. He didn't wish to offend Utrecht University, which had appointed Debye just recently. But Lorentz probably also preferred someone from outside, with a fresh perspective and fewer connections in the small, close-knit community of Dutch physicists (Berends and van Delft 2019, 323–330, Kox and Schatz 2021, Ch. 7). It was exactly during this period that Paul's package of publications landed on Lorentz's doorstep, followed by Lebedev's letter. Paul was suddenly catapulted into Lorentz's line of sight.

Clearly, Lorentz must have been struck by Paul's article on quanta, the subject of the first Solvay Conference. Likely, he remembered how he had first introduced Paul to Planck's work in 1903, which later inspired the concept of quanta. Perhaps he even recognised elements of his own approach to Planck's calculations in Paul's article. Undoubtedly, Lorentz appreciated that Paul conducted his quantum analysis independently; Lorentz had also worked largely in the relative isolation of his study in his younger years. Furthermore, it is certain Lorentz was impressed by the *Encyclopaedia* article on statistical mechanics. And who knows, perhaps the fact that Paul had co-authored the article with Tanya moved him. Lorentz's daughter, Berta, was putting the finishing touches on a dissertation on Brownian motion (De Haas-Lorentz 1912).

When Lorentz discreetly sought colleagues' opinions, they supported him. In late April, for instance, Sommerfeld painted an enthusiastic portrait of Paul. He was a 'masterful' teacher whose notes on the blackboard were 'as clear as could be. ... He can make the most complex matters concrete and intuitively clear. Mathematical arguments are translated by him into easily understandable images'.[24] Additionally, Paul appeared much more the true physicist in person than in his articles, which mostly portrayed him as a dialectician and logician. 'My personal contact with him gave me the impression, more than his publications, that he is concerned with physical facts. Mathematics is not an end in itself for him, and this is as it should be'. He also proved to be a 'far more versatile' thinker during their meetings, Sommerfeld noted, capable of selecting the right problems: 'He only builds upon experimental results when they touch upon fundamental questions'.[25]

The correspondence shows that some Solvay attendees had indeed noticed Paul's earlier articles. 'Planck regards him as highly insightful, as he confided to me in Brussels [during the Solvay Conference], although he has only recently become personally acquainted with him, and then only briefly and more superficially than me', Sommerfeld wrote.[26] Lorentz must even have gained the impression that Paul was highly sought after as a physicist in German-speaking university cities. 'He is looking forward to the possibility of habilitating in Munich, Berlin or Zurich', Sommerfeld added, expressing his hope that Paul would come to Munich, 'especially since I learned from his visit that he ... has a strong intuition for physics'.[27]

Did Lorentz know that in the previous weeks Einstein had put forward Paul's name for a position in Prague and later as Debye's successor in Zurich? 'Ehrenfest is a man of a lucid and critical mind who has few equals in his ability to extract what is essential in a theory, and who is completely independent vis-a-vis contemporary endeavours. Both in his lectures and in conversation, Ehrenfest knows how to present even a difficult topic in a clear and lively manner; the members of the commission had the opportunity to see this for themselves a short while ago, during Ehrenfest's brief stay in Prague', Einstein wrote in a letter to the selection committee for the professorship in Prague.[28] Yet it is doubtful that this characterisation reached Lorentz. What is known is that Lorentz dismissed the characterisation of Hasenöhrl in Vienna, who couldn't understand why Paul so stubbornly maintained his lack of religious denomination. Hasenöhrl found it 'rather too idealistic, I believe almost too much', which could be interpreted as a polite way of saying 'overzealous'.[29]

But perhaps it was enough for Lorentz to hear from other leading German physicists that Paul was an inspired teacher, well-versed in the latest developments in physics and highly regarded. He wasn't deterred even by Sommerfeld's remark that Paul was a committed vegetarian nor his observation that Paul sometimes appeared constrained by bouts of melancholy and dissatisfaction.[30] In a letter to the university's trustees, Lorentz predicted that Paul would one day 'be counted among the foremost'.[31] And in his letter to Paul, he was direct: he was seeking a successor. There were some caveats—including the trustees' approval—but would Paul be willing to remain available for this position?[32]

Was Tanya as astounded as Paul? Did they walk home through the leafy streets of Aptekarsky Island in stunned silence? Or did they engage in animated conversation, gesticulating as the girls listened intently beside them? In the ensuing days, Paul drafted a long reply. He provided an extensive account of his life since his first encounter with Lorentz ten years earlier. Instead of merely listing diplomas and jobs, supplemented with a few personal details, he described a search for a place where he finally could thrive for an extended period.

'My experiences over the past ten years have been defined above all by an involuntary lack of a homeland. I have always been convinced that, apart from the exceptionally talented, it is an indispensable condition for the development of one's abilities that one does not perceive those one normally interacts with as strangers. In contemporary Vienna, I felt and feel more a stranger than anywhere else. I felt incomparably more "at home" among the group of comrades in Göttingen, just as I did later in German-speaking Switzerland. Indeed, however ridiculous it may sound, even during a few weeks spent among the fishing community on the island of Schiermonnikoog, I soon felt more at home than I ever could in Vienna'.[33]

He also wrote about Russia. 'Undoubtedly, Russia would have become my homeland in the deepest sense of the word had I been granted a teaching position here. Despite my poor command of the language, I feel completely at ease among all sections of the population (naturally excluding political officials), and remarkably, Russians from all walks of life (including the few with whom I have had conflicts) do not treat me as a foreigner'.[34]

Paul wrote that as a 'substitute for formal teaching positions', he and Tanya organised other activities: a weekly student seminar, mostly held at their home, and an informal two-weekly discussion club with young physicists and physical chemists, also primarily at their home. He didn't mention that the heads of the Physics Institute, Professors Borgman and Khvolson, weren't informed of these activities 'due to their hostile attitude towards the new physics of Einstein, Planck, the theory of relativity and Ehrenfest personally', as Ivan Obreimov later recounted (Frenkel 1977, Ch. 3). Yet, he did write that he and Tanya had recently decided—'the catastrophe at Moscow University being the decisive factor'—that a teaching position in Russia was impossible for Paul.[35]

'Without such responsible activity [an academic position], I feared I would stagnate. I already clearly felt the increasing disorganisation of my activities, and for me, "work" is inextricably linked with the oral communication of my ideas'.[36] Moreover, Paul and Tanya were responsible for their two children, and their inheritances weren't sufficient to continue supporting all of them. So, he had started to look abroad, encountering various obstacles, but a meeting with Einstein had encouraged him to take concrete steps.

In conclusion, he wrote—contrary to the truth—that if he were offered a position in Zurich, he would need to carefully consider Lorentz's offer in Leiden, particularly given 'the language barrier in the Netherlands'. Otherwise, he was certainly willing to remain available for the position in Leiden.[37]

It was an open and candid letter, and apart from his feelings about Vienna and 'political officials', he struck a positive tone. Paul's bouts of melancholy and periods of lethargy and confusion were neatly concealed. What must Lorentz have thought of it? The punctilious Dutchman revealed his inner thoughts so rarely that some wondered whether he had any. On this occasion, too, he gave no intimation of his thoughts or feelings. Had he already taken his decision when he realised Paul had befriended Einstein and was appreciated by Planck and Sommerfeld? Did he have any doubt that Paul, as an outsider, would bring a fresh perspective to the physics community in Leiden, dominated by Kamerlingh Onnes and the otherwise so staid institutional culture? In his polite reply, he asked only for a list of publications, which Paul promptly provided.[38]

Then the long wait began in Kannuka (see Figure 17.1). Of course, Paul had read that Lorentz was convinced that any university—including Leiden—would be fortunate to have him. Yet in his uncertain mind, he undoubtedly also heard another voice, reminding him that Lorentz in his first letter also wrote that 'our faculty might ultimately decide in favour of a young Dutchman'.[39] Would it be a rerun of what happened in Berlin, with Planck and Rubens? In the long, white nights, the breakers in the Baltic Sea murmured. Again, Paul and Tanya had long conversations with friends. Again, Paul walked with his daughters along the coastal path to Sillamägi.

A year and a half earlier, Tanya had recorded a conversation with ever-so-logical Tanitchka, who had been reflecting on the belief of some that Jesus had risen from the dead. 'Mama, have you ever died?' she began the conversation. 'No, if I had died, I wouldn't be alive', Tanya replied. 'But there are people who believe they'll rise from the dead. Do you believe that?' the girl urged. 'I'm not sure', her mother responded, 'but I'd like it to be true'.

'I don't believe it, and it's good that you don't feel any more pain when you're dead', Tanitchka stated decisively. Tanya's weak counterargument, that you wouldn't experience any pleasure either, failed to impress her. 'But I certainly won't need to because I'll no longer feel anything', Tanitchka asserted, 'I won't even know the word "feel!"'[40]

Her father, however, felt his emotions even more strongly. They coursed through him. 'Lorentz, Sommerfeld, Prague?'; 'Leiden or Munich?' he scribbled in his notebooks and the red journal. 'Will Holland say yes or no?' 'When will Lorentz finally write?'[41] And his doubts grew as the days passed. He took little heart from Einstein's letter congratulating 'Dear Mr Ehrenfest' in June: 'no one would be happier than I if you were called to Holland. You are one of the few theoreticians who has not been robbed of his common sense by the mathematical contagion!'[42] Nor did a hike with Ioffe calm him. Even when Tanya fell ill with a high fever, Paul was still preoccupied by the matter. 'Lorentz????' he noted.

Finally, after two anxious letters from Paul about the possibility of lost post or misunderstandings and two reassuring replies from Lorentz, a telegram arrived on 29 August. It was immediately forwarded to Kannuka by an acquaintance in St Petersburg: 'Ehrenfest appointed professor leiden heartfelt congratulations letter following lorentz'[43]

Figure 17.1 Tanya with her daughters Tanitchka and Galinka, as most people called Galya, in Kannuka, Estonia, 1912.
—*Photograph Ehrenfest Family Archive.*

And, of course, the entire Ehrenfest family, including Aunt Sonya and the nursemaid, moved to Leiden as quickly as possible. 'We hope to arrive in Leiden between 15 and 20 October', Paul wrote to Lorentz on 5 October (Western or New Style).[44] The house in St Petersburg had largely been emptied for their summer stay in Kannuka (see Figure 17.2). Their suitcases were packed 'presently'. Nursemaid Masha wrapped her meagre belongings in a sling quickly enough. They just had to bid farewell to their friends in St Petersburg.

Figure 17.2 Paul and Tanya in Kannuka, Estonia, in the summer of 1912.
—*Photograph Ehrenfest Family Archive.*

When they arrived from the coastal town, Ioffe awaited them at the station with congratulations. Paul had received what Ioffe thought he so desperately needed: a professorship—and at a prestigious university no less. He would have regular working hours, a steady salary and, with any luck, finally more peace of mind. Already, the new prospect afforded Paul enough relief to love Russia once more. 'These last years have been so wonderful that it can hardly get any better', he wrote to his friend Herglotz, who responded with warm-hearted mockery.[45] Meanwhile, their friends in St Petersburg sang a rhyming farewell song, lamenting their departure

PART 2

18
The man in the empty sphere
A flying start in Leiden

On 4 December 1912, Paul made his debut as a professor, less than two months after arriving in Leiden. 'Dr P. Ehrenfest, having arrived from St Petersburg as the successor to Prof. Lorentz, assumed his position today as professor of theoretical physics in the Faculty of Mathematics and Physics at the university, delivering a lecture in the Foyer of the City Hall on "The Crisis in the Hypothesis of Luminiferous Ether"', the *Leidsch Dagblad* reported.[1]

In the months before, Lorentz had made every effort to welcome Paul and Tanya warmly. During their stop in Berlin on the journey from St Petersburg, they had been warmly welcomed by Lorentz's daughter, Berta—who held a doctorate in physics—and her husband, physicist Wander de Haas, who were visiting the city at the time. Paul and Tanya also spoke to Max Planck—how different Paul's position was to six months earlier!—as well as to old acquaintances such as Gustav Hertz and James Franck. A pile of congratulatory cards had awaited them at the hotel. And with a Meccano set his brother Emil had sent for his nieces, Paul built a Dutch windmill for his daughters (Klein 1970, 193–194).

Their destination Leiden must surely have seemed a little insubstantial in comparison to St Petersburg and Berlin. In any rate, Hotel Pension Futura, at the beginning of Groenhovenstraat, was in Paul's words 'impossible'. Fortunately, Lorentz had enlisted the help of Margarethe Nieuwenhuis, a professor's wife born in what is now Lithuania as the German-Baltic Baroness Von Uexküll Guldenband and a doctor in biology, who helped Tanya look for a flat. Paul and Tanya swiftly exchanged their hotel rooms for an upper-floor flat in one of the red brick houses on Groenhovenstraat, which was soon equipped with electricity, a telephone and a toilet (Klein 1970, 195).

A day before they moved, Lorentz sent them a short note: 'I forgot yesterday to ask you whether Madam, if she is not too occupied setting up home, would care to attend the colloquium tomorrow evening. It would give me great pleasure.'[2] And with yet another thoughtful note two days later, as well as a bouquet for the new flat, he reassured them that it hadn't mattered at all that Paul and Tanya, in the chaos of moving house and still unfamiliar with Leiden's twisting layout, had arrived late at the colloquium. He was glad they had come, and the attendees had only waited a few minutes for them, 'for some time is always taken up with drinking tea'.[3]

Lorentz was constantly at their side. He introduced Paul and Tanya to colleagues and acquaintances, informed them about Dutch academic customs and advised Paul to buy a dark overcoat and hat for The Hague, where his appointment as professor needed to be finalised after government approval. After the event, little Galya cheerfully put the hat on her head.[4]

In turn, Lorentz was the first person Paul and Tanya invited to their home: 'My workroom is in order, to a first approximation', Paul wrote. 'Permit me to enumerate its virtues: it has five windows, but it can be kept quite warm, nevertheless, with a gas stove. There are blotting paper, writing paper, pen and ink; also, two tables, two chairs, a couch and an electric table lamp—even a blackboard. ... If you are willing to put up with this inventory and with the rather camplike meals (in an utterly bare room, decorated only by your flowers!), well, if you are willing to put up with this, we should be very, very, very pleased to have you. My room, that is really quite tolerable, but I am afraid that the meals will, unfortunately, be very bad' (Klein 1970, 196).[5] They hadn't yet found a cook.

In many ways, Leiden took some getting used to. For Paul, it was a renewed acquaintance, whereas every street was unfamiliar to Tanya. Instead of the mighty Neva and its tributaries, there were murky canals. Instead of magnificent palaces with neoclassical and rococo façades, there were narrow, tall houses with paneglass windows along the quays. Instead of the Winter Palace, there was the old Burcht van Leiden fort, a circular, thick brick-walled shell keep atop a motte in the middle of the diminutive city.

There were no schools where girls learnt to walk like ladies. There were no rebellious female students. And while Paul made his first acquaintance with Leiden in spring, when the tulip fields around the city bloomed and a fresh breeze blew over the Katwijk and Noordwijk beaches, Tanya first saw the city in autumn, when wind drove through the streets, drizzle made the cobblestones slippery and heavy raindrops streamed down the windowpanes like tears. It would forever colour how she saw the city.

But Paul also had to find his way. He still remembered Kamerlingh Onnes' famous laboratory from 1903. He recalled the maze of passages, the instrument workshop where young men in blue smocks learnt to blow glass and build instruments, and perhaps even the attic where students extracted helium from monazite sand. Undoubtedly he had heard that the British scientific journal *Nature* considered the efficient laboratory 'amongst the best provided (and, one may add, most productive) research laboratories'.[6] Yet being a direct colleague of the slightly authoritarian head of this laboratory, Onnes, was a strange sensation.

Paul didn't immediately take to Onnes, who had been a member and even the principal of a student fraternity during his studies at Groningen and had added his second name, Kamerlingh, to his surname out of a sense of refinement (Delft 2005, 91–101, 159). He reminded him of Khvolson in St Petersburg,

Paul noted.[7] This wasn't exactly a compliment, as he had always found Khvolson rather authoritarian and rigid in his views, not particularly sympathetic towards lower-ranking staff and, moreover, condescending towards women—from the outset Khvolson had never taken Tanya's work seriously (Frenkel 1977, Ch. 3).

Professor of Experimental Physics Johan Kuenen was more to Paul's liking. After completing his doctorate under Onnes, Kuenen had risen to become his right-hand man, while as the son of a politically well-connected professor of theology, he was also a helpful ally when Onnes had budgetary troubles. But Paul soon realised that Kuenen had little influence. It was Onnes who set the course, wielded the sceptre and set the tone. And it was Onnes who would receive the Nobel Prize for liquefying helium and discovering superconductivity a year later in 1913.

Did Leiden offer Paul—with Onnes, Kuenen, several younger physicists and, somewhat removed, Lorentz—an environment where he could flourish? Shortly after arriving, he contacted Onnes' mentor, Johannes van der Waals, in Amsterdam, who had been awarded a Nobel Prize in 1910 for his work on the equilibrium states of gases, which was closely related to Boltzmann's work. In Amsterdam, he also visited experimentalist Pieter Zeeman, who had shared the 1902 Nobel Prize with Lorentz. Taken together, Amsterdam and Leiden had more Nobel laureates than Berlin, with three laureates; Göttingen, with one; Munich, with three (including nearby Würzburg) or his beloved 'physics Mecca' Zurich, where chemist Alfred Werner wouldn't win the city its first Nobel Prize until 1913.[8]

At the same time, the atmosphere in Leiden and its university was staid. It was out of the question for a professor to take students on long walks along the beach, the Dutch equivalent of the hiking and ski trips Sommerfeld undertook with his students. Let alone for a professor to cycle in shirt sleeves to his girlfriend, with flowers from his garden on the handlebar, like Hilbert in Göttingen. There were no bustling coffee houses like in Vienna and no dimly lit taverns like in Göttingen, where one could go over matters again with colleagues. And neither did professors have the habit of inviting students over for more-or-less informal meals or tea in their gardens.

Indeed, most professors did no more than 'run into a class, give the lecture, and disappear', in the words of mathematician Dirk Struik (Powell and Frankenstein 1999, 323). Onnes emerged from his lab with 'visible displeasure' to read out his dry notes, other students said (Hollestelle 2011, 44). As a student, it 'was up to you' whether you turned up. Whereas students in other disciplines were regularly members of a fraternity—smoking, drinking, singing loudly or worse, much to Paul's dismay—most mathematics and physics students took the train back to their family homes after lectures and lacked any further connection to the city—which was also to Paul's dissatisfaction. And while academic life in Leiden wasn't as provincial as in Innsbruck, Prague or the 'insignificant backwater' Czernowitz, he believed that, at the very least, a reading room and a colloquium ought to be set up (Hollestelle 2011, Ch. 1).

But first he had to formally present himself as a professor. 'I am not yet in a state in which I see myself delivering this speech every night in my dreams, but this moment will come soon enough', he wrote to Lorentz in November.[9] Tanya read the text, provided input, stayed up late into the night when needed, straightened Paul's suit and calmed him. On 5 December, she sat with Baba Sonya in the audience in the auditorium. Relieved, she noticed that Paul's nerves disappeared as soon as he stepped onto the stage, after walking with the trustees from the Senate Room on Rapenburg Canal to the auditorium.

Rightly so, as his lecture was beautifully composed. 'Allow me', he began, 'to speak about a crisis that currently poses a serious threat to a fundamental hypothesis in physics—the ether hypothesis'.[10] (Ehrenfest 1913a) And while discussing ether and the 'threats' to it, he lavished praise on his predecessor Lorentz, highlighted the work of his friend Einstein and his late friend Ritz and captivated his audience with vivid examples on subjects that would ordinarily have lulled them to sleep.[11]

'Imagine a large hollow sphere in front of us', he began abruptly. 'Much bigger than the Earth, much bigger than the Earth's orbit. ... An experimenter sits precisely at the centre of the hollow sphere'.[12] What, he asked rhetorically, would happen if this experimenter, equipped with a very bright lamp, switched on this lamp for a moment and switched it off straight away? The vivid, almost Jules-Verne-like image immediately caught his audience's imagination. This was no dignified, reserved and slightly soporific professor. This was a stout man with sparkling eyes, lively gestures and a thick Viennese accent, pouring his heart and soul into his argument.

As long as the man was in the dark in the hollow sphere, he couldn't tell whether the sphere was stationary or advancing at a constant speed—just as a stowaway in the dark hold of a ship couldn't say whether the ship was moving. But what would happen after he emitted a flash of light? Here opinions differed. If light consisted of a stream of particles, as Newton had proposed in the seventeenth century, these particles would hit the sphere's wall and rebound like bullets. Suppose that in a stationary hollow sphere the particles took one hour to reach the wall and another hour to return. The man would see the wall of the sphere briefly light up two hours after the flash. And since everyone knows that a ball game follows the same dynamics on the quayside as it does on a steadily moving ship, it was easy for the audience to grasp that the man would see the same thing in a moving sphere. All of this was perfectly in line with the principle of relativity, which requires the laws of physics to produce the same results in both systems.

But then Newton's idea that light consists of a stream of particles was supplanted in the nineteenth century by the concept that light propagates as waves. Frenchman Augustin-Jean Fresnel theorised that these waves were carried by an ether that filled the cosmos, anchored, so to speak, to a few fixed stars. Unlike light waves that were carried by the ether, the other stars and planets drifting through this 'fixed' ether were then subject to an ether wind, much as a cyclist feels the wind blowing through their hair. And although alternative ethers were proposed, such as that suggested by Irish physicist George Stokes, observations and measurements ultimately showed that such a fixed ether was most plausible. At this point, it was fitting that Paul mentioned Lorentz's hero, French experimental physicist Hippolyte Fizeau. After which he, of course, cited Lorentz, whose convincing reasoning seemed to resolve the issue definitively in favour of fixed ether.

'What does this fixed ether mean for the man in the sphere?' Paul asked. He used the example of a stone dropped into a pond. In still water, a splash produces a circular pattern of waves, which spreads out at the same speed in all directions. If these waves were to hit a symmetrical circular edge, they would return at the same speed. Translated to the sphere: the man in the stationary sphere would see the walls light up evenly after two hours. Yet if the man in the sphere was speeding through fixed ether, and so encountered an ether wind, he would see something different. The pattern of the light waves would be as asymmetrical as the wave pattern created by a stone dropped into a flowing river. The walls of the sphere would be illuminated unevenly, as Paul made clear. In other words, in this case the principle of relativity no longer held.

Which picture was correct? The great thing was, Paul promptly continued, that we know exactly what the man in the speeding sphere would observe. His sphere was essentially an exaggerated, cartoonish version of the famous experiment by American physicists Albert Michelson and Edward Morley. Their experiment demonstrated that the man in the moving sphere would see the walls light up evenly. The principle of relativity had been upheld.

The big question now, though, was how to proceed with the fixed ether, which underpinned almost all optical measurements so perfectly, yet conflicted with this result. Lorentz had developed a solution in 1904, introducing his famous formulas that slightly 'contracted' fast-moving objects in the direction of their motion. He interpreted this outcome as ether wind pressing the molecules in an object, such as a sphere or measuring instrument, closer together. And he went on to explain that ether wind disrupted the even illumination of the sphere by either impeding or boosting the emitted light, but that it also distorted the walls of the sphere, or any other measuring instrument, such that it exactly compensated the first 'disruption'. Paul's admiration for Lorentz's achievement was evident in his remark that it would be 'immodest for me ... to try to assess the value of this method with any sort of qualification'.[13]

This was an excellent moment for Paul to switch to Einstein's 1905 interpretation of Lorentz's formulas: the measuring instruments weren't distorted but rather space itself was elastic, as was time, for that matter. In other words, it was space and time that stretched and contracted according to Lorentz's formulas, causing the walls of the sphere to light up perfectly evenly. And in this interpretation, it was no longer necessary to invoke the concept of ether at all.

To complete his lecture, Paul then discussed the work of his late friend Ritz, who had also declared ether redundant. Moreover, Ritz had objected to Einstein's idea that time and space are elastic. He had revisited Newton's ideas and conducted calculations on the particle nature of light, an ambitious endeavour cut short by his premature death in 1909. But Paul did know how to determine whether Einstein and Lorentz's or Ritz's view was correct: by measuring the speed of light from a star travelling away from us and from a star travelling towards us. According to Einstein's theory, which was in practice indistinguishable from Lorentz's, the speed of starlight would be the same in both cases. Ritz's theory suggested that these speeds would differ.

Perhaps the most remarkable aspect of Paul's lecture was that it immediately inspired a colleague in Leiden. Willem de Sitter, a congenial astronomer, had listened attentively in the audience and realised that such precise measurements could be made using binary stars. When he did just so in 1913, he confirmed Michelson and Morley's results and provided the strongest evidence to date in support of Einstein's theory (Sitter 1913a, 1913b, 1913c, 1913d).

But that was later. That evening, Paul, in a tailcoat and top hat, danced with Tanya on the bridge over the Witte Singel. Everything had gone well. Lorentz had taken him aside and spoken to him with 'full, full sympathy' (Klein 1970, 199).[14] Even though Paul had forgotten in all the excitement to thank trustee and Nobel laureate Van der Waals, Lorentz reassured him about this too. In a friendly note afterwards, Lorentz wrote that the omission could easily be seen as a sign of modesty. That Paul had not immediately remedied this during the reception afterwards was perfectly fine, too, as 'one does not say much to a "king" either'. A kind note to Van der Waals with the printed copy of the lecture would surely suffice.[15]

19

Water to the left, water to the right and water in between

Cosmopolitans in a provincial town

'This is what I want', Paul had written to Tanya less than a year before their move to Leiden.[1] 'To live in a way that I don't lose contact with advances in physics and on the other hand that I also do something myself... and everything such that you don't lose your life'. And now Tanya was in Leiden, and she mightn't have lost her life, but she didn't live as she had in St Petersburg.

Dutch sounded like gibberish to her; Dutch customs were unfamiliar to her. Sometimes she felt as if she unwittingly committed faux pas. For instance, at a tea party, when her cup was not refilled, after she had politely said 'thank you' in her heavily accented Dutch, mistakenly believing she had accepted a second cup, when in fact she had inadvertently declined the offer. 'Can you make sense of that?' Tanya still wondered decades later.[2]

It was almost just as difficult to define the position of women in the Netherlands. Berta de Haas-Lorentz, who had defended a clear dissertation on the Brownian motion of electrons that made her one of the first scientists to look into the phenomenon of thermal noise (De Haas-Lorentz, 1912, p. 82),[3] appeared more intelligent than her husband. Yet she limited her professional activities to incidental contributions to textbooks, contenting herself further with a life as a dutiful spouse and housewife. Rather like her mother, Aletta Lorentz, who stood smartly beside her husband at formal gatherings, warded off unwelcome visitors to the house, had kept the children calm and quiet and, in short, was the strong woman behind a successful man. This was expected of professors' wives, Tanya noted: well-bred and married women didn't have a paid job and stood at their husband's side. And more so than in Russia, it seemed that not only older women but also younger women acquiesced to this unwritten rule.

Tanya was therefore not only suddenly cut off from all her activities, including in the field of geometry education or probability theory, but she also had little prospect of work to replace them, at least for the time being. In the mornings, she brought Tanitchka to 'Miss' Hellema, whom she and Paul had charged with teaching the girl to read, write and do arithmetic. Afterwards, she often withdrew at home into the Russian atmosphere created unmistakably by the tableware, rugs and reproductions.

Sometimes she took Galinka for walks along the Witte Singel and Rapenburg canals. She enjoyed it when friendly chemist Hilmar Backer called on them in the evenings, or tall physicist Adriaan Fokker, who was completing his doctorate under Lorentz. Gradually she became familiar with Dutch rain, which could drizzle or tap gently but could also lash and sting. Fortunately, a series of foreign visitors made their way to Leiden. Yuri Krutkov, one of the participants of the colloquium in St Petersburg, stopped by. Epstein and Herglotz visited; and even Ioffe paid a visit at the end of April 1913. Tanya and Paul took them to the beaches at Katwijk and Noordwijk, to Delft and the Dutch masters at the Mauritshuis in The Hague. With Ioffe and Herglotz, they also visited Lorentz in Haarlem (Klein 1970, 201–203).[4] But Tanya missed the company of someone like Kitty Miloradovich, a quadratic-equation enthusiast, or Valentina Doinikova, who would come storming in to discuss mathematics.

What mostly kept Tanya going, alongside her daughters and Paul's enthusiasm, was the house she and Paul wanted to build at the end of the Witte Rozenstraat. Partly thanks to Tanya's railway shares, also a popular investment among the Dutch, they purchased a sizeable plot of land. The property was bounded to the south by the waters of the River Schie and separated on the west by a small ditch from the neighbouring plot owned by Professor Nieuwenhuis and his wife, Margarethe Nieuwenhuis—von Uexküll Guldenband. The house Tanya and Paul wanted to build was to be large: with space for themselves and the children, for Aunt Sonya, for Baba Katya when she came to visit and for other guests. In addition, they wanted to organise colloquia there.

On large sheets of paper, Tanya sketched a series of designs. 'Good', Paul wrote beneath one of the few sketches featuring Art Nouveau-style decoration. Tanya clearly preferred neoclassicism, which reminded her of St Petersburg. 'Her' house would have plastered façades and clear lines.[5] To keep out the damp and inclement Dutch weather, it needed the same sturdy walls that kept severe cold at bay in buildings in St Petersburg. Inside, she wanted stout heating pipes running along the walls of the rooms, just like in Russia. The windows would be fitted with double glazing, a novelty in the Netherlands at the time.

Thus, she gradually developed a sketch of a house with a largely closed frontage on the north-facing street side (see Figure 19.1). Unusual by Dutch standards was a shutter at one end of the frontage, next to the kitchen windows, through which suppliers could deliver goods directly into the kitchen. At the other end of the frontage, where the study was located and far from the front door, was the letterbox. She designed the easterly and westerly exterior walls with large pediments featuring semi-circular windows. At the centre of the westerly wall, she placed the front door. Through a hall and a small blue stone staircase, it led to a straight, wide corridor ending at a large study spanning the entire width of the house. On the south side, facing the garden, the side room, dining room and conservatory,

Figure 19.1 One of Tanya's many sketches for the house on Witte Rozenstraat.— *Ehrenfest Family Archive.*

which jutted out at the study, were fitted with large windows. She sketched a raised terrace with steps. Above all, she ensured that light would travel in long, straight lines through the rooms. For example, light flowed from the stairwell, with its narrow windows in the north wall, through a glass door into the dining room.[6]

That Dutch contractors could build a Russian-style house was something Paul and Tanya could see just around the corner. From their temporary residence in Groenhovenstraat, they only had to take a few steps to view the Leiden Observatory across the Witte Singel canal. Astronomer Frederik Kaiser had it modelled after the neoclassical Pulkovo Observatory just outside St Petersburg (Icke 2011). The plastered building, built precisely on an east-west axis by architect Henri Camp in 1860, had the same clean lines and symmetry that Tanya aimed to achieve. Her pleasure designing it was evident in the pile of sketches. The result was a house larger than that of almost all professors in Leiden. The 'sturdy domestic servant' and the potential 'second maid for half days', Tanya and Paul had already advertised for in the local newspaper, the *Leidsch Dagblad*, before arriving in Leiden, would indeed be necessary.[7]

At this time, Paul was trying to shake up the stuffy physics faculty. On his way to Leiden, he had promised Lorentz: 'I really want to do my utmost to justify, at least to some extent, the confidence placed in me by the men who appointed me to your chair'.[8] His students gave him this soon enough. In the lecture hall, Paul captured their imagination with vivid language, expressive gestures and an unorthodox approach to handling formulas. 'The most important formulas were presented on the blackboard almost as individual aesthetic creations, rather than merely as links in a coherent argument. He avoided lengthy calculations and often paid little attention to numerical factors. "4π" could mean almost anything' (Casimir 1983, 65).[9] It was about the crux, the essence of a problem, Paul believed. His mission was accomplished when students grasped this.

He also livened things up outside the lecture hall. It bothered him that many students who had to live at home caught the train home straight after lectures, often carrying a book borrowed from the university library. A reading room would change this, like the one at the Mathematical Institute in Göttingen, where Paul and Tanya had so often come across interesting work while browsing in the bookcases. Like in Göttingen, visitors would have to read the books and scientific journals there. Paul hoped of course that students and staff working on the long reading tables in the new reading room would share their knowledge with each other and begin to form a community (Hollestelle 2011, 43; Lunteren 2003). Together with mathematics professors Jan Kluyver and Pieter Zeeman—nephew of the eponymous Nobel laureate—he founded a society called the 'Reading Room for Mathematics and Physics'. Onnes reserved a lecture hall for this purpose. A member of his staff, Claude Crommelin, had bookcases installed, and the university library was willing to transfer academic literature to the reading room. 'The librarian, prof. Ehrenfest, along with his staff, created a card catalogue, and with everything ready, this reading room was officially opened this afternoon', the *Leidsch Dagblad* reported on 20 May. All mathematics and physics students who were a member of the faculty association were automatically enrolled in the new reading room, the newspaper continued.[10]

Despite such successes, Paul still felt uncertain. The academic traditions, the language and his new phlegmatic colleagues; he always seemed to be on the back foot. Moreover, his position gradually dawned on him. He lived in the shadow of a predecessor who seemed to possess an overview of the entire field of theoretical physics, had produced brilliant work, was a Nobel laureate as well as a linguistic prodigy. The thought had sent shivers down his friend Einstein's spine. Paul sometimes anxiously wondered what he had been thinking when he accepted the position.

Just before Christmas, he vented his frustrations in a letter to Lorentz (Klein 1970, 14).[11] 'Please have patience with me—with my blunders, my frittering and squandering my energy. Give me two, three years to decide whether I ought to retain this professorship, or whether I must ask you to give it to someone else'. And, no, Lorentz shouldn't address this cri de coeur, 'neither verbally nor in writing'. Paul would rather Lorentz deferred his reaction for two or three years. 'Only one thing was necessary for me now: to be able to say to you frankly that I look upon these years as being completely a probation period (for myself!)'. He was glad he had expressed himself: 'You cannot imagine how relieved I feel now I have said everything'.

The biggest problem was Paul had no clear line of research in mind. His mentor Boltzmann had had atoms; Onnes had his cryogenics; Lorentz electrons; Debye his work on molecules and Einstein had sunk his teeth into gravitation and relativity. But Paul drifted, as he had done through cities and countries, between topics in physics: enthusiastic, passionate and critical, but without a clear direction. It troubled him that he still couldn't present Lorentz a well-defined research topic. Attached to the candid letter to Lorentz were a couple of sheets on which he presented a small matter related to the 'Quantenhypothese' from the perspective of the adiabatic and reversible processes that he had so often studied, also with Tanya. But he refrained from writing that this could be a starting point for further research into quanta and concepts in thermodynamics, in line with his quantum articles from 1906 and 1911. He had, his letter to Lorentz explained, mainly written this because 'I felt I could not send you the list for the reading room without sending something new in physics (no matter how small)' (Klein 1970, 14).

Yet the ideas Paul outlined weren't ridiculous at all. Besides, nobody, and certainly not Lorentz, expected him to produce a brilliant insight a few months after arriving. It was chiefly Paul who was unsure whether he was the right person in the right place. 'Lectures, lectures, lectures—you know how I prepare for them', he wrote to Ioffe around the same time early in 1913. 'I know nothing clearly enough to be able to teach it—everything must be relearned, recalculated, etc. Discussions with students. The horribly bad French translation of the *Encyclopaedia* article must be put in shape. A number of reviews for the *Physikalische Zeitschrift* are due. Lectures, lectures, lectures. In short I work on nothing, read hardly anything ...' And with a sense for the dramatic: 'I work on nothing, I am stagnating, while all others—Einstein, Debye, Laue, you, and all the young fellows go forward, forward' (Klein 1970, 15).

Paul missed Ioffe, with whom he had endlessly discussed subjects such as Planck's quanta, and who always helped him get back on track when he became stuck. Apart from doctoral student Fokker, Leiden lacked young theorists or even experimental physicists to spar with. 'Tanya does as best she can'. But the real problem—the 'worm'—was that he felt like a failing toiler in comparison with impeccable Lorentz and powerful Kamerlingh Onnes. He became increasingly convinced 'that Lorentz should have taken Debye as his successor and not me. And I feel that Lorentz already perceives it. And I know that even you dare not say absolutely anything against it', he wrote to his Russian friend (Klein 1970, 15).

The sense that he must prove himself had a paralysing effect '[F]or everything that I've managed to get out at all has come from the impulse to play, from interest in some paradox, and not from striving to accomplish something significant'. And although Paul tried to put his misery into perspective by facetiously talking of his 'hysterical despair', there was real pain in his sombre closing sentences. 'I know very well: Leyden is a little university in Holland—to which, by chance, Lorentz came as a high school teacher, and then just stayed. His successor had to be a really first-rate young fellow; they couldn't get one to come here, and so they had to take a second rater, and I know that I am really in the forefront of the second raters now'.

Although his mood brightened again, his uncertainty was revived when a couple of weeks later he realised that he had not been invited to the Wolfskehl Lectures in Göttingen in mid-April. 'That I wasn't invited as a speaker, I consider fair (if also bitter if I have to admit it to myself), but they really could have invited me as a listener', he complained in a letter to Ioffe (Klein 1970, 202). The Wolfskehl Lectures were an informally organised counterpart to the Solvay Conference, also open to uninvited attendees, so Paul went anyway (Schirrmacher 2015, 2118). But it stung him that his fellow train passengers, Lorentz and Debye, had official invitations.

Fortunately, he was cheered by the spring sun in Göttingen, where he met mathematicians Hermann Weyl and Richard Courant and reacquainted himself with physicist Max Born, played Brahms on the piano with Planck and had long and comforting conversations with Ioffe. Despite everything, Göttingen is 'a dear little spot on earth to both of us', he wrote to Tanya.[12] On the lectures by Hilbert and Klein, he was even so enthusiastic that he wrote: 'You absolutely must come here—immediately!'[13] Although he knew this was of course impossible, he repeated it during Planck's lecture the next day: 'What an incredible pity you are not here—not yet here (?)'.[14]

Hence, Paul went homewards in a better mood than when he had left. And his spirits were lifted further when Lorentz wrote to him in early July that wealthy tea planter Ru Bosscha, son of Leiden physicist Johannes Bosscha, was to donate 2500 guilders towards the reading room.[15] This allowed Paul to acquire almost all the titles on his wish list. The large bookcase contained the complete works of Cauchy, Stokes and Thomson, alongside journals such as *Annalen der Physik*, *Physikalische Zeitschrift* and *Philosophical Magazine* (Burgers 1962, 48).

The highlight was perhaps that Paul and Tanya had by then already stayed more than two weeks in Zurich, where they spent almost every day, bar one, with Einstein.[16] Meanwhile, Tanya was making the final corrections to a short paper about the proper statistical formulas to describe concentration variations in radioactive solutions (T. Ehrenfest, 1913).[17] But above all, they immersed themselves in lively debates. 'We have now been in Zurich for three days and are indulging in the latest theory of relativity, quanta and other matters', Tanya noted at the start of their trip.[18] Einstein introduced them to his friends Michele Besso and Marcel Grossmann, and after these weeks, Paul was no longer addressed as 'Dear Mr. Ehrenfest' in Einstein's letters but rather 'Dear Ehrenfest' or simply 'Dear E'. When Paul and Tanya went to the seaside in summer 1913—not to Kannuka this time but to the sandy white beaches of the resort and fishing village Noordwijk aan Zee—they had plenty to be satisfied with.

Yet Tanya wasn't entirely content either. Just as Paul had felt in Russia that, so far from central and western Europe, he was missing the latest developments in physics, she felt disconnected from the discussions on mathematics and her work in mathematics education there. So, when she received an invitation in late summer to speak at the Second All-Russian Congress of Mathematics Teachers in Moscow at the beginning of 1914, she accepted without hesitation. Paul was taken aback. 'The opportunity arose for my wife to make a significant contribution to the reform movement in the field of mathematics education', he had written to Lorentz about their Russian years.[19] But that Tanya wanted to return to her motherland for this was to him absurd—a 'beer idea'.[20] Why such a long journey? Why invest so much energy in Russia, where chaos was followed by more chaos?

Imperturbable, Tanya boarded the train in the first weekend of January 1914. 'You must be deep in Russia by now', Paul wrote on Monday on the first of an avalanche of postcards to St Petersburg and then Moscow. 'I envy you for the snow. Things are capital here. Everything is all right (inwardly too)'.[21] He described how the girls played ('healthily') with the 'enormous collection of dolls and animals they [had] collected'. 'T' found it most intriguing to walk with me through a dim Leiden so early in the morning' to Miss Hellema. 'Galya insisted I cleaned my nails. She is now sitting at your place at the table'.[22]

Work on the new house on Witte Rozenstraat had gone smoothly in the preceding days. 'All three of the pediments have been painted provisionally, but the shade of yellow contrasts too little; carefully examine the colours in St Petersburg', Paul scrawled on a card on 5 January.[23] The first ceilings were plastered, and in the garden, work had started on the steps leading to the terrace: 'I have reiterated that they must work exactly according to your instructions'.[24] A few days later, De Sitter came to view the progress: 'He examined it intently from all sides ... and thought it very good. Then I showed him the inside. He noticed the flatness of the steps and the rounding of the wooden contours, which he found very appealing'.[25]

Only the weather was 'rubbish'. It was 'cold' for a while and then 'gloomy-miserable weather'. There was 'rain-wind-cold. So bad that we did not send T' to the teacher'.[26] 'Ghastly weather; water to the left, water to the right and even more water in between'.[27]

From Tanya's letters—too few and not well dated, in Paul's opinion—and from Russian newspapers, he tried to glean as much as possible about her trip and the congress. 'From your letter, I gather you haven't heard anything from Ioffe. Naturally, this saddens me greatly, but it also vexes me intensely. Try to make a note every day so that later you can recount everything to me well. It is very good that you can blow off steam', he advised her on 11 January by postcard.[28] And on a second card later that day: 'Formulate your final conclusion very carefully in your lecture. The province [teachers from provincial towns] wants guidance on how to proceed!'[29]

Some time during these days, Tanya was presenting the results of the mathematics survey and showing 'statistical tables', aided by a 'projection lantern'.[30] Paul was afraid that Tanya would be too obstinate and single-minded: 'Today or yesterday must have been your presentation', he wrote on 17 January. 'I hope you don't argue too much—don't forget that the people have in any case far more practical experience than you, and this: that they approach you kindly and that fighting is of no use if it only leads to annoyance'.[31] He constantly gave Tanya unsolicited advice on other topics as well. 'Don't catch a cold. Eat well'. 'Don't engage in unnecessary conflict. Don't react to angry words'. 'Take care of your health and watch out especially for trams and automobiles'.[32] Also important: 'Please, don't accept any obligation from anyone to write something'.[33] In Moscow, Tanya would undoubtedly, be asked to write something about Russian mathematics education, Paul thought. 'I suspect you might be imprudent enough to agree to this. But it is essential that we first work on something together in the new house—for orientation on the use of the rooms and the layout'.[34]

His own work was progressing poorly. 'I realise I have let this holiday slip away as well', he wrote on the weekend of 10 January. His brother Emil had suddenly arrived: the successful businessman and his partner Max Eisler were on their way to an art auction in London. The girls were both gifted a 'scarlet cloak' with 'glass buttons' and had tried to communicate 'in all possible ways' with their Viennese uncle. But Paul was exhausted after a trip to The Hague with the two men in a car that they had arranged. 'You shall understand that walking and driving around was tiring and depressing for me'.[35] And: 'I am ashamed, for your sake, that I have nothing interesting to tell you. Explanation: I am not working on anything at all, not thinking about anything either and there is no point in moaning to you while you should be happy right now (yes—sweetie, that is what I wish very much for you)'.[36]

The next Wednesday, after giving his 'crystal lectures' in the early morning to a rather empty hall, he was still complaining. 'All in all, only I am extremely discontented, and I am doing nothing. But as you know, this passes quickly with me, and by the time you read this card, I shall probably feel much better already . . .'[37] The next day, he indeed described in somewhat better spirits a visit to Onnes, who had just collected his Nobel Prize and spoke 'very cheerfully about his experiences in Stockholm'. Onnes had 'received whole piles of letters. Incredible how many' and was incidentally 'quite kind and cheerful *these days*'.[38]

Interaction between Paul and Onnes was evidently not yet altogether smooth. Perhaps Paul was also still getting accustomed to the phlegmatic Dutch character with its reserved etiquette. Friendly chemist Backer and astronomer De Sitter also maintained some distance, despite visiting often enough. When the canals froze in mid-January, Paul watched them swish over the ice amidst the throngs of merrymakers. 'Backer and Coseyn skated to the Hook of Holland. De Sitter [skated] with his eldest son to The Hague', Paul wrote to Tanya on 18 January.[39] 'That I turned 34 today is no reason to celebrate either', he added dejectedly.

Not even the fact that the reading room would now also be open in the evenings cheered him up. This was thanks to its daytime success, yet Paul's expectations remained gloomy. Who would come, he grumbled. 'Probably a few old physicists. But it is the young people who hardly use the reading room at all—they all work poorly anyway. The eternal focus on examinations, sheer laziness, no one lives in Leiden and so on and so on ...'[40]

Likely, he missed Tanya most of all. Even in Göttingen, he would have preferred her by his side. She was his anchor when his moods fluctuated, as they did at this moment. The fact that she was in Russia made him uneasy. 'You sense that I fear the conference fever will turn you completely against life here', he had written just a few days after her departure.[41]

'Don't think too harshly of your Western European exile', he added late that evening. 'Consider three things: 1) there is naturally an endless amount of urgent work to do in Russia—but why? Because "no one lets anyone else" do it. 2) Even you, and especially you, can do almost nothing there. 3) If only, if only, you were willing to approach the matter from the right angle, you could do much good here too, while developing yourself and exerting enduring principled influence. Especially by educating your children and through reading and writing.'[42]

Between 1907 and 1912, Paul had tried, on his own steam, to develop his expertise and exert 'principled influence' in Tanya's beloved Russia. Did he think it was Tanya's turn to try the same in the Netherlands?

20
War and friends
Leiden 1914

On 1 October 1914, Jan Burgers Sr. took his son, Jan Burgers Jr., to Leiden by train. Burgers Sr. was a postal employee in frail health, with a meagre income and a passion for the natural sciences. Undoubtedly, he glowed with pride: his son Jan was going to study physics in the city where famed physicists Lorentz and Kamerlingh worked. The two Nobel Prize laureates were a regular topic of discussion in the Burgers household. Their home in Arnhem was a small temple to the natural sciences, with cabinets filled with crystals, minerals, magnets, electromagnets, induction coils, galvanometers, small electric motors and microscopes. The house filled with interested townspeople when Burgers Sr. gave lectures on electricity, microscopes, astronomy or geology between these displays. Jan Jr. often assisted him in operating the devices and maintaining the 'collection' (Burgers 1962, 20–24).

So it was unsurprising that Jan Burgers at once recognised Lorentz a few days later. A boy from Kamerlingh Onnes' workshop had brought him to the lecture hall, through 'corridors filled with all sorts of complex machinery and large bundles of pipes, across a small courtyard'. When Lorentz walked past the lecture hall in the Physics Laboratory shortly afterwards, just before his first Monday-morning lecture, Burgers immediately noticed the two lumps on his forehead. They were the consequence of deep contemplation, it was said in Arnhem, where Jan—like Lorentz—had been to an HBS secondary school.[1] But who was that man—'dark looking' and of 'smaller bodily posture'—next to Lorentz? It wasn't long before Burgers discovered that this new 'Russian' professor was Ehrenfest. Paul soon spotted Burgers, too, as there were only a handful of physics students. Burgers stood out for his keen intellect and the 'awe for the wonders of nature' that had been instilled in him from a very young age. Somewhat overwhelmed, Burgers was invited at the very start of the semester to Paul's Wednesday evening colloquia at Witte Rozenstraat (Burgers 1954, 47–48).

Burgers was astonished by the cavernous house with plastered walls and an expansive garden. Years later, he still recalled the study perfectly: 'a large room with three windows in one of the walls, looking out on a part of the garden, and a large couch at the other wall, where I have been sitting so very often. On the shorter wall there was a large blackboard, a strange form of decoration even to me' (Burgers 1962, 50). In the lecture hall, too, Paul deviated from the traditional mould. Most professors in Leiden were aloof and expected students to manage on their own. How students went about studying or taking notes or whether they also wanted to spend a semester studying Egyptology, so to speak, was their affair (Burgers 1962, 48). 'The contact with Lorentz at that time was non-existent. You went to the lecture, but he was, so to say, the god', one former student concluded (Uhlenbeck 1963, Session III).

But Paul taught as he had seen Boltzmann and Exner do in Vienna as well as Hilbert and Sommerfeld in Göttingen and Munich. Rather than providing detailed calculations, he sketched the physical principles, research methods and mathematical techniques in broad brushstrokes. He made—truly excellent—students dive into the deep end as quickly as possible by having them attend colloquia, prepare presentations, engage in discussions and encouraging them to publish something, preferably sooner rather than later. In this way, Paul selected a small group of students from the broader pool, much as he had seen Boltzmann do in Vienna and Hilbert with his *'Wunderkinder'* in Göttingen. Burgers was one of these young men—they were all men—whom Paul sought to nurture into physicists after his own heart.

Burgers was fortunate. Many of his contemporaries were 'deprived of scientific and cultural distractions in a garrison on the border'. But as Burgers was rejected for military service 'due to an overly narrow chest girth', he could absorb the latest developments in general relativity and quantum theory in Paul's study, along with all the other knowledge Paul generously shared with his students (Burgers 1962, 46). Here, and later in the dining room 'which so often had full sunshine coming from the garden', he was welcomed by Paul and Tanya 'as a close friend' and even 'almost as a close relative' (Burgers 1962, 51).

Paul and Tanya had moved into their spacious house at the start of July (see Figure 20.1). 'Last night Pavel and I spent the night in our own home for the first time', Tanya wrote in their journal on 7 July. 'The girls were with Nadya Plekhanova [the nanny] in Noordwijk. May God grant all of us good fortune. And may everyone who visits us enjoy themselves'.[2]

On the walls between the windows in the study, they hung portraits of Boltzmann, Maxwell, Tolstoy, Dostoevsky and their old friend Walter Ritz. Next to them hung views of St Petersburg (Burgers 1962, 50). On the blank wall to the left was a large blackboard, and a milk-glass ceiling light spread a soft glow at night. In the dining room, slender Thonet chairs surrounded the dining table, and a servant lit the stove in the kitchen.[3] The entire house still smelt of plaster and freshly planed wood. This was a blissful time. Paul dug up sea holly in the Noordwijk dunes for the garden.[4] The girls performed tricks on rings under the trees. Tanitchka practised cycling round the house. Sunflowers blossomed and the neighbours had 'dahlias in magnificent colours', Paul wrote in the journal in early September.[5]

Figure 20.1 The house on Witte Rozenstraat shortly after construction, photographed from the garden of the Nieuwenhuis family's mock castle on Jan van Goyenkade. The garden in the foreground is part of the Nieuwenhuis estate; there are already swings in Paul and Tanya's garden.—*Photograph Ehrenfest Family Archive.*

The contrast with the rest of the world grew all the more unreal. Following the assassination of Austro-Hungarian heir presumptive Franz Ferdinand and his wife Sophie in Sarajevo in the simmering Balkans, war spread in one week in September over almost all of Europe, like a 'cyclone of death'.[6] In the west, German troops marched straight through Belgium into France, while in the east, Tanya's motherland became a war zone. 'One would like to know how things will be in two years time, but perhaps it is good that we do not know', Paul wrote on that same September 2 in the family journal.[7]

'First panic over money-goods [sic], then more and more soldiers here. No one is allowed to sit idle', he had noted in early August, just a week after the outbreak of the war.[8] Since then, the echoing voices of mobilised recruits rang out against the façades in Leiden, and newspaper headlines screamed of death and destruction. Shortages and hoarding increased the price of bread, provisions and gas.[9] But for Paul and Tanya, the war felt most real when thinking of their friends and family in Vienna and St Petersburg.

Almost inevitably, Paul didn't support Austria-Hungary and its powerful ally, Germany. 'Vienna and German-Austria can never, never be anything other than "hostile strangers" to me', he later observed.[10] Thinking of Vienna, he felt the loneliness of a Jewish boy among Catholics, a middle-class son among working-class children and a teenage son of a nouveau riche family among the old elite. And the malice of the Viennese towards their Jewish fellow citizens still wounded him to the core.

Russia chose to side with Serbia against Austria-Hungary and for the Slavs, and it counted Britain and France as allies. But would the Russian army, with its many illiterate and innumerate recruits, be strong enough? Just before the outbreak of war, Nikolay Maklakov, the Russian minister of Internal Affairs, was concerned that a war was 'not very popular with the people of Russia' as 'the masses set more store by revolutionary ideas than by a victory over Germany. But you cannot escape fate' (Lieven 2015, 378). This fate manifested itself as a well-oiled war machine that German newspapers had been stirring up for months against 'rapidly modernising and aggressive' Russia (Lieven 2015, 354–355).

While grim reports of the war rolled off the printing presses, Paul sometimes withdrew to the university library. He read what newspapers had written about the Napoleonic battles: propaganda, distortions and lies, he concluded (Hollestelle 2011, 241). With renewed focus, he then read how the German, British, Belgian and occasionally Russian newspapers reported on the battles and acts of terror: a relentless stream of 'endless agitation'.[11]

More practical Tanya offered their flat on Groenhovenstraat, which they had kept, to Belgian refugees. Belgians fled across the Dutch border in large numbers, some also reaching Leiden. Taylor Willem de la Court of the Roman Catholic Housing Committee kept Tanya informed in the ensuing months about the welfare of the families temporarily sheltering there. They could do little else.[12]

In the preceding academic year, Paul had tried to establish himself in Leiden. It hadn't been easy. In Russia, he had also been surrounded by gifted physicists, but he had been able to distinguish himself with his network in western and central Europe. This and his extensive knowledge had made him a guide in theoretical physics, a field that developed more slowly into a strong and independent discipline in Russia than elsewhere. Things were different in Leiden. Lorentz had a bigger network than Paul, and theoretical physics was already firmly established thanks to Lorentz. As much as his colleagues in Leiden had valued Paul's oration, it was still an overview of existing work. What they expected from him was his own, original work.

For the time being, Paul gained recognition and respect primarily from students in the lecture hall, and he cautiously focused on quanta, thermodynamics and Boltzmann's work. The war and its effects on international relations between scientists didn't make it any easier for him to maintain this focus. Lorentz tried to maintain a modicum of pre-war relations by sending diplomatic but slightly naïve letters to German, French and Belgian colleagues. He corrected errors in national newspapers, countered propaganda and sought to rectify incorrect conclusions (Otterspeer and Schuller tot Peursum-Meijer 1997, 68–73).[13] From time to time,

Paul assisted him, but their efforts had little effect. It was hard to imagine that Germans, French, Belgians and Britons had recently sat together brotherly during the Solvay Conference in Brussels. Or that Lorentz, as chairman of the Institute for Physics founded by Ernest Solvay, had awarded scholarships to forty researchers from seven European countries in recent years (Berends 2015).

At the same time, Leiden was disconcertingly calm, and the university was 'unmoved. ... as if the neighbour's house was not in flames'. (Otterspeer and Schuller tot Peursum-Meijer 1997, 61). Paul too stuck to his routines. In the Bosscha Library, he insisted that books weren't lent out and that it was quiet in the reading room. Of participants of his Wednesday-evening colloquia, he demanded 'great seriousness and diligence'. Students who 'oscillated'—showing up occasionally but not consistently—were promptly removed from the attendance list (Casimir 1983, 76). Unsurprisingly, foreign guests were scarce. Gunnar Nordstrom from Finland was one of the few to spend some time in Leiden. Most other attendees were local physicists, such as Lorentz' post-doctoral student Adriaan Fokker and Paul's colleague, experimental physicist Kuenen. And there was the small group of excellent students, the youngest of whom was Burgers. It also included somewhat older Dirk Coster, highly intelligent Hans Kramers and sharp mathematics student Dirk Struik.

Tanitchka and Galya were fascinated by all these guests engaged in discussing complex issues. They helped set up the chairs in front of the blackboard in the study beforehand and occasionally peeked around the corner during colloquia.[14] It was a respite from their long days at home, as the girls didn't attend school. Paul and Tanya had persisted with homeschooling, which was still given by Miss Hellema. From the lengthy list of subjects Paul believed the girls must be introduced to, she hadn't checked off many yet. Nevertheless, Tanya felt that Hellema's curriculum, with additional input mainly from herself, was better than what Dutch primary schools had to offer.

'They see plenty of children in our garden, and little good can be expected from the schools here', she wrote in the journal that summer. 'From a very young age children here are drilled with the names of all the Dutch canals and the types of soil in the various Dutch provinces and such things, and they have no idea about the rest of the world. The reading material they receive consists of saccharine stories specifically created for children, completely devoid of any poetry. At least Tanitchka has time to understand and appreciate the Russian language and literature at home, and she shall have time for her own observations and thoughts'. In short: 'I thank God every day that we haven't sent Tanitchka to school'.[15] And this view was shared by Paul, who despised 'cheap hand-me-down clothes for the mind', as he called school curricula (Casimir 1983, 21). Five-year-younger Galya most enjoyed the art teacher's weekly visits (see Figure 20.2). She 'draws really lovely things: a girl pulling a little cart, with a parasol in her hand; trees, the sun, stars, houses, boats', Tanya noted. 'Nowadays, she tells us every morning about the beautiful dream she had and promises to draw it'.[16]

Figure 20.2 Tanya with Galinka in front of the house on Witte Rozenstraat in 1915.—*Photograph Ehrenfest Family Archive.*

As for Tanya, she tried to resume her work on mathematics education. Lorentz introduced her to Rommert Casimir[17], the founder and headmaster of the modern Nederlandsch Lyceum[18] in The Hague and also an enthusiastic educational reformer. Casimir advised her to translate one of her essays on mathematics education into Dutch to contribute to discussions on educational reform in the Netherlands. With a little help now and then, she produced an essay, which was published in Dutch as 'The Role of Axioms and Proofs in Mathematics' in *Weekly Journal for Preparatory and Higher Education* in 1915.[19] As she had done in Russia, Tanya opposed the idea that geometry could best be taught by following Euclid's work. His principles weren't so absolute, she wrote. At one time, his five axioms appeared to be indisputable truths, but it was understood by the nineteenth century that these propositions were, in fact, chosen. With a slightly different set of axioms, a different geometry could be constructed through successive steps of logical reasoning—one which was equally valid. Such a geometry could describe surfaces like that of a saddle or the Earth's surface, where the three angles of a triangle add up to less (a saddle) or more (the Earth) than 180 degrees.

Einstein used the descriptions of such curved spaces as a tool in his work on his general theory of relativity. Tanya employed them as a rhetorical device. If the axioms didn't have a unique status then teachers might as well 'take any proposition that pupils found self-evident as a starting point, as an axiom'. (Ehrenfest-Afanassjewa 1961, 97–109). This would make it possible to align teaching material with what pupils intuitively understood—which was Tanya's hobbyhorse. Only afterwards, when pupils had practised logical reasoning, should teachers have them think about choosing a unique, minimal set of axioms, which could serve as a foundation for geometry. This would 'exert a beneficial, emancipating influence' on their 'way of thinking'. (Ehrenfest-Afanassjewa 1961, 97–109).

Tanya's other enduring hobbyhorse was that mathematics—and especially geometry—is such an important school subject because it stimulates 'pure thinking'. Yet it was difficult to convince staid secondary school teachers of her idea to teach geometry in a different way as they proudly pointed to five Dutch Nobel prizes between 1901 and 1913. It didn't help that Tanya wasn't employed as a teacher, nor that she was a woman and yet freely spoke her mind. In the end, only a small group of mathematics teachers, encouraged by Casimir, attended her colloquium on mathematics education at Witte Rozenstraat in autumn 1914. Over tea and biscuits, they listened to Tanya explain her ideas in German, with a smattering of Dutch words, drawing from the discussions she had held in St Petersburg and Moscow.

At the sixth sitting of the Congress of Mathematics Teachers in Moscow in early 1912, for instance, Tanya had discussed the 'geometry laboratory' that she had introduced at her girls' gymnasium in St Petersburg. She had warned attendees at length that 'the term laboratory method [can] easily lead to misunderstandings'. In laboratories in the 'so-called experimental sciences', researchers test hypotheses which they then use as the basis for further research once sufficient evidence had been obtained. But the 'laboratory method' in geometry lessons was only about exploring space. In other words, the idea wasn't for pupils to test the basic principles of geometry—the axioms. There was absolutely no reason to do so as these axioms were 'so simple that anyone going to secondary school is convinced of their correctness'. Above all: 'why teach pupils to conduct experiments when they could also learn to think?'[20]

Thinking is what it was always about for her. In her laboratory, simple observations, practical measurements and playful tasks were intended to 'focus pupils' attention ... and open their eyes ... to something already embedded in their consciousness', she wrote in her 1915 essay. In a playful and more or less intuitive manner, pupils were meant to explore geometrical insights, rules and patterns. After this, geometric postulates would fall into place more easily and pupils would gain a keener appreciation of Euclid's systematic thinking. Ultimately, his thirteen works on arithmetic and geometry were of course a wonderful example: 'regarding the accessibility of both the material and the goal they set for themselves, the propositions in the geometry course are derived from a few axioms which are accepted without proof'.[21]

Tanya was certainly not opposed to Euclid, as this essay and debate evenings clearly show. Yet even the participants in her colloquium on mathematics education were hesitant, wondering whether her ideas weren't too radical. The heated discussions at Witte Rozenstraat sometimes lasted so long that participants almost missed the train home. 'Nothing will come of mathematics education reform', Paul once remarked after a session. 'Just look how they can't agree on anything'. But Tanya wasn't so easily discouraged: that mathematicians were at least engaging in discussion was a victory in itself to her (Ehrenfest-Afanassjewa 1961, 11).

Once again, despite his insecurity, Paul's work was received most warmly. He was praised for his crystal-clear explanations of complex concepts and theories in lectures and colloquia. 'First the proposition then the proof!' was his motto. Then he expounded the proofs using only the most essential formulas. He illustrated them with vivid examples that left a lasting impression on his audience, such as the empty sphere or the dog with fleas. The only drawback was that some students were terrified whenever they had to give a presentation themselves (Casimir 1983, 77). Had they read the publication critically enough? Had they unearthed the hidden assumptions? And the key point of the argument? Paul called this key point the 'springende Punkt' or the 'Witz'. Or: 'That's where the frog jumps into the water'. (Casimir 1983, 68; Powell and Frankenstein 1999, 422). Students knew that Paul was liable to subject them to a nerve-wracking Socratic grilling if they failed to clearly indicate where exactly this frog landed in the water.

Moreover, his presentation style was almost unsurpassable. 'The blackboard begins at the upper left corner', he complained as students fumbled at the blackboard where he accurately chalked all the crucial formulas from the top left to the bottom right. When a student stood in front of the blackboard during a lengthy explanation, he joked: 'you may have many good characteristics, but transparency is not one of them'. (Casimir 1983, 67). It was a great relief that he usually made amends for his occasionally razor-sharp criticism. And when students managed to clarify a confusing issue, he also gave them generous praise. 'Herr ... has finally pulled the whole rat out of the soup', he would say sometimes, perhaps with an image of the soup kitchens in Vienna or St Petersburg in mind, where rats might indeed be found in the soup (Casimir 1983, 68).

During colloquia, they discussed the work of Viennese Philipp Frank, Einstein's successor in Prague; Max Abraham, who had his own ideas about electrons and, for example, Walther Nernst, the man behind the first Solvay Conference. Students capable of handling such an intensive approach felt valued. They learnt, Struik later observed, what physics entailed: 'He made science a living thing. All the science I had learned before was static. Ehrenfest showed me how science is a living and growing field'. (Powell and Frankenstein 1999, n. 16).[22] During their daytime visits to Paul's home, conversations often expanded to include Russian literature, art or economics and sometimes 'wonderfully lively' Paul would even play the piano for them in the study (Burgers 1962, 52). In doing so, he forged a close-knit group that partly replaced the 'circle of comrades' in which he had felt so cherished and happy in St Petersburg.

21

Oasis in an ugly world

Work, war and Witte Rozenstraat

If the mathematics congress in Moscow had inspired Tanya to visit Russia more often, these plans now lay in ruins. Shortly after the outbreak of the war, some 'experts' still believed that the Russians, advancing towards Vienna, would swiftly end the war by causing the Austro-Hungarian Empire to collapse.[1] But from 1915 onwards, retreating Russian forces were repeatedly trounced by the Germans, equipped with new supplies of ammunition and often brand-new armoured vehicles. Russian soldiers, by contrast, were 'deprived of ammunition and weapons', as newspapers reported more and more often. Meanwhile on the Western Front, German and Allied soldiers were ensconced in trenches stretching from the Flemish coast to the Vosges mountains. Stranded in the Netherlands, Tanya was filled with anxiety for her family and friends.

In Leiden, she had no choice but to adapt to Dutch customs. She had grown up among women who had claimed their right to education. She was accustomed to her aunt Sonya travelling autonomously across Europe. She had been taught by women pursuing academic careers. At the Bestuzhev Courses, she had sat in lecture halls alongside female students from diverse backgrounds. And in Leiden, she had designed her own house and largely financed it herself. Yet under Dutch law, Tanya, as a married woman, had the same legal status as children and those with a cognitive disability: she lacked legal capacity and couldn't independently open a bank account, withdraw money, purchase property or a car or make other significant decisions.

On a smaller scale, Tanya had to find her place among the professors' wives. This was perhaps less challenging for her than for her neighbour, Margarethe Nieuwenhuis-von Uexküll Güldenband, who once literally became unwell during a lengthy coffee conversation among the professors' wives about cleaning carpets (Steen 2013b, 10). Yet Tanya didn't truly make a meaningful connection either.

After their visit to Zurich, in March 1913, she published the article in *Physikalische Zeitschrift* that linked her earlier ideas on probability theory to the random motion of radioactive particles in a fluid (T. Ehrenfest 1913). This touched on work by aristocratic Smoluchowski, who had studied such 'drunken movements' in Lemberg [Lviv] and Krakow, and who had delayed publishing for so long that Einstein stole his thunder with his groundbreaking work on this kind of movements in 1905. 'The day after tomorrow, my wife shall speak about her latest publication during the colloquium', Paul announced to Smoluchowski in 1915.[2] But all this wasn't as matter-of-course as it sounded. Just as during her studies in Göttingen, Tanya was caught in a no-man's-land between the sexes during this lecture. Only occasionally did a female student attend a colloquium, such as the sisters Jo and Nel van Leeuwen who were just beginning to work on their dissertations with Lorentz and Kamerlingh Onnes, respectively. But there were few other women Tanya could identify with, and her contribution could easily be dismissed as merely an inevitable accessory that came with collaborating with Paul.

Did this make her insecure? 'That you at least find my efforts worthwhile gives me immense pleasure', she had written to Smoluchowski who had read the proofs of the article.[3] He had sent them back to Zurich, where Tanya and Paul were visiting Einstein, along with—to her relief!—a warm, complimentary letter. Tanya admitted freely that she had hesitated unnecessarily long collecting and making corrections for final publication. 'The thing is I am always so unsure about whether to have something printed … Incidentally, you may understand me better than others because you, too, published fine works much later than you could have'.[4] Did she also dare to write so openly to Smoluchowski because he was progressive and had called the notion that women were unsuited to the natural sciences hopelessly outdated in a speech in 1912 (Smoluchowski 1912)?

Such views were rarely aired in the Netherlands. The emancipation movement hadn't entirely passed over the country, and the professors' wives weren't completely inactive. Aletta Lorentz, for instance, had been engaged in charity work all her life and had had some involvement in feminist activities (Kox and Schatz 2021, Ch. 3). Yet her approach was often different to Tanya's. When the ladies of Leiden addressed these issues, they rarely questioned underlying social structures. For example, at the end of the nineteenth century, Aletta Lorentz collaborated with textile manufacturer Krantz and his wife to start a nursery school where female labourers, forced to work out of necessity, could bring their children. Nearly ten years later, she also initiated a 'community playground' where poor children could play in fresh air. In both cases, she and others stressed the importance of child development instead of just supervising the children. Yet these efforts did little to change workers' impoverished lives, with meagre wages and poor housing.

The same was true of Tesselschade, an association which supported impoverished women of respectable standing unable to seek employment because of their 'class'. Tesselschade enabled them to supplement their income by making handicrafts at home. But by selling these handicrafts anonymously, Tesselschade simultaneously upheld the notion that earning money was indecent for women, especially upper-class women.

Even in their most political activity, the struggle for women's suffrage, Aletta Lorentz and the women around her were moderate. In 1907, they broke away from the suffrage movement led by Amsterdam-based Aletta Jacobs, a renowned physician and advocate of women's suffrage in the Netherlands. Instead, they founded the Dutch League for Women's Suffrage (Nederlandsche Bond voor Vrouwenkiesrecht), which allowed men to join the board and didn't explicitly mention the goal of women's suffrage in its statutes. Some members of this rather genteel league preferred to extend suffrage gradually to groups deemed sufficiently 'educated' to vote (Steen 2011).

This approach differed from Tanya's, who wanted to educate every child to become a critical and independent thinker and who demanded more immediate equality between the sexes. For Tanya, the best aspect of the initiatives undertaken by the moderate Leiden feminists, aside from their struggle for women's suffrage, was their Leiden Cooperative Kitchen. Tanya had a particular aversion to cooking, so it was convenient that for a small sum she could order hot meals, with or without meat, from this collective kitchen, which were delivered by horse and cart or by carrier bicycle. This saved her time, which she could dedicate to more interesting pursuits—which was in fact the very purpose of the kitchen (Steen 2013a).

In this same period, Paul cautiously began to develop what might become his own line of research, building on his previous work. In his 1905, 1906 and 1911 papers, he had compared Boltzmann and Planck's statistical methods step by step to determine whether 'energy quanta' were indeed a necessary (and sufficient) condition for describing the light spectrum of a black body. Now, he took the discrete, 'granular' character of the process as a starting point, and added an ingredient from thermodynamics: adiabatic processes.

In his 1911 paper, Paul had already explored such adiabatic processes in which a system changes so slowly that it is barely noticeable. He had toyed with Planck's model of a black body as an almost completely sealed cavity, with walls at a constant temperature and covered with oscillators, which absorbed and then re-emitted electromagnetic radiation in all directions. And he had worked out what would happen if this cavity shrank incredibly slowly. He showed that the balanced mixture of all possible colours (or frequencies) of light would turn slightly bluer (slightly higher frequencies) as the temperature in the shrinking cavity increased almost imperceptibly. However, the energy of the radiation remained proportional to its frequency, and the constant of proportionality remained exactly Planck's constant (or an integral and non-zero multiple of it).

This led him to think that these extremely slow processes might shed light on deeper, underlying questions, even if the exact nature of the interaction between light and matter was still unknown. For example, if quantisation were indeed a feature of nature, would systems other than Planck's 'oscillators' also exhibit some kind of 'granular', discrete behaviour? To find an answer, Paul began to study a variety of systems which, just like the imaginary, electrically charged 'oscillators' in Planck's work, were 'periodic' and repeated the same motion at the same rate over and over again.

In his notebooks, he worked out examples such as molecules bouncing back and forth between two walls, a pendulum and rotating or vibrating molecules. Helped by a crucial remark by Tanya,[5] he subsequently showed that the ratio of kinetic energy (of the moving molecules) to the frequency (of their motion) stayed the same for such systems under adiabatic transformations (such as one wall moving very slowly towards another) (Klein 1970, Ch. 11).

The most striking example proved to be dipole molecules (i.e. molecules with an uneven charge distribution resulting in one slightly negative and one slightly positive charged pole) which vibrated like Planck's oscillators in the presence of an electric field. By extremely slowly increasing the strength of this external electric field, such vibrations were converted into rotations, Paul showed, with a rotational speed ('angular momentum') that was also 'quantised' and determined by the amount of energy (the 'quanta') the dipoles had previously absorbed. In other words, the adiabatic processes of (classical) thermodynamics seemed to offer a recipe for moving from one discrete, 'granular' process to another, potentially representing a bridge between classical descriptions of nature and 'the future mechanics of energy quanta' (Ehrenfest, 1914a).

It was Einstein who subsequently coined the term 'adiabatic principle' (Einstein 1914; Mehra and Rechenberg 1982–2000, 439–440). 'I cannot get your idea of adiabatic transformations off my mind. This may be our most valuable resource in our state of general hopelessness', he wrote to Paul.[6] The principle could serve as a guide to chart the mysterious quantisation of nature, he thought. But he was one of the few to react enthusiastically. Paul's other colleagues barely noticed his articles in the Proceedings of the Royal Netherlands Academy of Arts and Sciences (Ehrenfest 1914a, 1914b). In Munich, Sommerfeld even seemed simply to disregard them (Klein 1970, 284; Navarro and Pérez 2004; Pais 1991, 189–190).

This lack of recognition may have been partly because Paul limited himself to hypothetical processes that could be expressed in the simplest mathematical formulas. Moreover, he formulated his conclusions rather cautiously. But probably more importantly, Paul's work was completely overshadowed by that of a man who had audaciously given a concrete example of quantisation in nature. Niels Bohr of Copenhagen was five years Paul's junior and looked like a friendly bear. He was inspired by Rutherford, whose sophisticated measurements in Manchester showed that atoms have a small, positively charged nucleus. Negatively charged electrons swarmed around them, Rutherford suspected. The only difficulty was that Maxwell's theory insisted that these moving electrons continuously emitted small amounts of electromagnetic radiation. This would cause them to lose energy and spiral towards the atomic nucleus, causing atoms to 'collapse' according to this model. Building on the idea that matter absorbs or emits energy only in precisely measured parcels, Bohr had proposed a solution in 1913.

He had begun by describing the hydrogen atom, which with only one electron is the simplest atom. The atomic nucleus was like a miniature Saturn, around which was an imaginary disc composed of progressively broader circles. Each of these circles was a potential orbit in which the solitary hydrogen electron could revolve around the nucleus, provided it had the necessary energy. These circles actually corresponded to 'energy levels', ranging from a low electron energy close to the nucleus to a high electron energy in orbits farther outwards.

The crux was that, in a bold mental leap, Bohr subsequently made this atom 'discrete'. He separated the energy levels like the rungs of a ladder. This explained why atoms don't collapse: while collapsing, electrons would plummet towards the nucleus by continuously giving off small amounts of energy. But in Bohr's model, electrons could only absorb or emit specific chunks of energy, allowing them to jump from one rung to another. What then reinforced his model was that the 'energy chunks' enabling electrons to make such jumps corresponded exactly to the colours of light emitted or absorbed by hydrogen atoms during measurements (Bohr 1913a); for a much more detailed and nuanced discussion, see: Kragh (2012); Aaserud and Heilbron (2013); Duncan and Janssen (2019, 145–201).[7]

It was thus unsurprising that physicists in various university towns were examining this model closely. But in Leiden, Paul called Bohr's idea brazen, sweeping and poorly substantiated. No one could explain why the electron orbits were only in certain positions while other positions were excluded, he grumbled. And how could electrons 'know' in advance which orbits they could jump to and where they would fall between two stools? If it had to be like this, then 'I must throw all of physics in the bin (and myself along with it)', he wrote 'perturbed' to Ioffe in August 1913 (Moskovchenko and Frenkel 1990).[8] Perhaps it also rankled a little that he had proceeded with such caution and tried to justify each step of his thought process so precisely, while Bohr, with his blunt assumptions, was the one in the limelight.

Would Paul have changed his mind if he could have met Bohr in person? For Paul, who thrived on discussions and conversations, the war was likely a greater hindrance than for Bohr. Post was slow, travel was virtually impossible and relationships were strained. Shortly after the war broke out, for instance, German philosopher and biologist, Ernst Haeckel returned all his honorary English university titles. Paul found this 'totally nonsensical and wrong'. 'As if they were military decorations!'[9] And he sadly looked on how the free exchange of ideas ground almost to a halt.

Paul and Tanya were even more shocked by the 'Manifesto of the Ninety-Three', published by eminent cultural figures in German newspapers shortly after the war broke out (Klatt and Lorenz 2010, 113–134). Men they held in high esteem, such as Klein, Röntgen and Wien, also signed the text, which denied that 'Germany is guilty of having caused this war'. According to the manifesto, the situation was different. 'Not till a numerical superiority which has been lying in wait on the frontiers assailed us did the whole nation rise to a man'. And: 'Were it not for German militarism, German civilisation would long since have been extirpated'.[10]

Signatories included even Nernst, who had persuaded Belgian industrialist Solvay to organise and finance the famous Solvay Conferences in Brussels in 1911 and 1913. And Planck, the man with 'the finely chiselled and delicate face', with whom Paul had recently played the piano.[11] 'In the current chaos of the world, it takes a very particular temperament and maturity to know with absolute certainty on every occasion what our conscience demands of us', he lamented in a letter to Lorentz.[12]

The situation became even more troubling in the ensuing months when several eminent cultural figures from the Kaiser Wilhelm Institute for Chemistry in Berlin developed poison gas. Led by Fritz Haber, director of the institute and head of a research department at the Ministry of War since the outbreak of war, they were turned into practically deployable lethal weapons. Haber personally helped to deploy them for the first time near Ypres in Belgium on 22 April 1915, and Paul's colleagues in Berlin, James Franck and Gustav Hertz, were among those who oversaw the proper burial of the long rows of cylinders filled with liquid chlorine. When the wind blew towards the French trenches and the cylinders 'automatically' opened, they saw with their own eyes a whitish yellow cloud of chlorine gas slowly roll into trenches with thousands of French and Algerian soldiers. Barely two weeks later, Haber travelled to the Eastern Front, where poison gas would also be used against Russian troops.

Yet Paul and Tanya didn't lose their faith in the power and beauty of science. During their colloquia, they continued to welcome like-minded individuals. They felt supported by Lorentz, who tirelessly persisted in his efforts to maintain or restore international contacts (Kox and Schatz 2021, Ch. 9; Berends and Van Delft 2019, 364–376). Perhaps their greatest comfort was their friendship with Einstein.

In March 1914, Einstein had left Zurich for Berlin, and while his household belongings were transported from Zurich to Berlin, he stayed with 'Dear E' and his family in Leiden. 'I accepted this odd sinecure because giving lectures gets on my nerves in such an odd way & there I do not have to lecture on anything', he had written about his 'Berlinerisation' in a letter to Paul at the end of 1913.[13] Exemption from teaching was one of the privileges with which the grand men of the Kaiser Wilhelm Institute—Nernst, Planck and Haber—had enticed Einstein to Berlin. With a prestigious and relatively 'free' position as a member of the Academy, they had, not long before war broke out, overcome his aversion to Prussian class society with its inclination towards militarism and nationalism.

What Einstein hadn't told Paul was that his marriage to Marić was falling apart. While he increasingly withdrew into the remoteness of his mind, his once so clever and unconventional 'little witch' languished amid laundry racks, children and errands. And as the two drifted apart, Einstein had felt at home with his cousin and second cousin, Elsa, during a visit to Berlin in 1913 to discuss becoming Van 't Hoff's successor at the Academy. He looked forward to seeing her more often when he moved to Berlin for this new position. 'I rejoice at the thought that I will soon be coming to you. We will probably already move in the fall. Seeing you regularly will be the nicest thing that awaits me there!', Elsa read in summer 1913.[14] And a couple of days later: 'I do not have to teach any courses there—I am totally free to do whatever I wish. And one of the main things I wish is to see you often, to gad about with you, and to chat with you'.[15] It led to further deterioration in his already fragile marriage, even though the move to Berlin was postponed to spring 1914. Marić had changed, he wrote mercilessly to Elsa in December 1913, into a 'humourless creature who does not get anything out of life and who, by her mere presence, extinguishes other people's joy of living'.[16]

This issue was too delicate for letters to Paul. Einstein corresponded with him about the world and especially about his work. About six years earlier, with his Ehrenfest paradox, Paul had highlighted a limitation of Einstein's theory of special relativity. Einstein elegantly used Lorentz's equations for space and time in this theory to translate descriptions of mechanical and electromagnetic phenomena from one system to another, but this only worked as long as such systems moved at constant speeds relative to each other. When a system accelerated, as the rotating disc in the Ehrenfest paradox did, things broke down. Besides, neither could gravity be transformed from one system to another.

Einstein's brilliant insight that acceleration and gravity—that is, accelerated falling in a gravitational field—are equivalent would eventually make everything fall into place. It would result in his general theory of relativity, which provided the correct time, space and equations always and everywhere, and was a theory of gravity; it described how masses, such as the sun and other stars, shape and distort spacetime, creating 'dents' which cause time and space to stretch or contract locally.

But such clear concepts hadn't only to be wrested from nature, which often yielded murky observations, but they also had to be expressed in formulas that unruly mathematics didn't readily provide. In these years, Einstein was labouring over the geometry able to describe a more-or-less curved spacetime in mathematically precise terms. And although mathematician Riemann had already laid the foundation for this, Gregorio Ricci-Curbastro and Tullio Levi-Civita's resulting calculation methods were immensely complex.

In Zurich, Einstein's friend Grossmann had helped him with all the formulas and mathematics (Pais 1982, Ch. 13). But after moving to Berlin in late March 1914, Einstein was on his own—although he couldn't have imagined just how alone he would feel there, while he was still in Leiden in March of that year. His arguments with Marić escalated that summer, and despite Haber's mediation, she returned to Zurich with their sons a day after the war broke out. The war thundering across Europe left its mark on Berlin. 'At such times one sees to what deplorable breed of brutes we belong', Einstein wrote to Paul after its outbreak.[17] 'I am musing serenely along in my peaceful meditations and feel only a mixture of pity and disgust'.

Subsequently, Einstein, 'all by myself in my large apartment', withdrew to the abstract high plains of space, time and relativity. Lorentz and Paul were among the few for whom he occasionally descended.[18] From time to time, he wrote them letters about his progress, often with new work for their perusal, sometimes with specific questions—and more often towards the end of 1915.[19] Einstein valued the considered judgement of Lorentz, upon whose work he was building. He held great respect for Lorentz' extensive knowledge of theoretical physics and his sharp, adaptable mind (Kox and Schatz 2021, Ch. 8).[20] But he relied just as much on Paul's honest critique, as Paul scrutinised Einstein's letters, formulas and work down to the smallest details (Pais 1982, 271–272).

By contrast, Einstein's colleagues in Berlin were more critical of his approach. Like Einstein, most of them adhered to Viennese philosopher Ernst Mach's 'positivist' ideas. They followed his principle that natural scientists should restrict their work to descriptions of what can be counted, measured, characterised— in other words, sensory observations and empirical data. Only when formal mathematical descriptions subsequently described such observations as accurately as possible were meaningful statements about interpretation and concepts relevant, they believed (Pais 1982, 282–288; Sandner 2019). But Einstein's deep insight and vibrant creativity escaped this straitjacket. Despite his admiration for Mach, Einstein constructed his theories time and again on well-chosen assumptions. Only afterwards did he consider whether his formulas accurately represented counted, measured and characterised phenomena and whether they could make predictions. That Lorentz and Paul, although more cautious in their assumptions and methods, appreciated this approach, deepened their bond with Einstein.

Letters travelled back and forth. Paul and Einstein needed each other. Einstein, brimming with ideas, had the overview, saw the big picture and the broader context, but was sometimes careless with details. For this, he could rely on Paul, who wasn't particularly fond of mathematical subtleties but did want to understand concepts and ideas down to the finest details and examined every part of a theory to see whether it withstood scrutiny. And Paul, often frustrated by his inability to rise above such critical analyses with his own original work, could feel tremendously useful in this way.[21]

For Paul and Tanya, it was comforting to know that a 'man of culture' of Einstein's calibre was also in Berlin. Conversely, Einstein, with Elsa but without his children in sombre Berlin, must have seen Leiden as a sunny oasis. Rightly so, for Paul and Tanya's bright and sunny house had increasingly become a cheerful haven. Students like Burgers, as well as colleagues such as Lorentz' doctoral student Jan Droste and Finnish gravity specialist Gunnar Nordström, who lived farther along Witte Rozenstraat, enjoyed attending colloquia or just dropping by. Absent-minded astronomer De Sitter occasionally visited and professors from other disciplines in Leiden, such as paediatrician and biochemist Evert Gorter, would stop by as well. Philip Kohnstamm, a physics professor and educationalist from Amsterdam, had become a close friend. And there were also Russian friends in The Hague who loved spending time in Paul, Tanya and their children's world, which seemed like a joyful oasis amidst the flat Dutch polders.

Paul had started introducing ten-year-old Tanitchka to 'grown-up books'. In the family journal, he listed writers such as Nikolai Gogol, Charles Dickens and Jules Verne.[22] Tanya described her eldest daughter above all as a keenly observant girl with a sharp mind. That her father claimed that there were more women than men on earth 'is not true', Tanitchka had remarked for example. 'It turns out that she spent several days counting the men and women walking down our street. Usually, there were far more men', Tanya noted. 'But perhaps the women are simply at home', her daughter observed and of course she was right.[23]

Her mother also took delight in Tanitchka's love for numbers and mathematics. The girl came up on her own with the trick to quickly calculate the sum of all numbers from 1 to 4, 1 to 6 or 1 to 10, and she presented her 'discovery' to her mother as a riddle: 'What do 4 and 10, 6 and 21, and 10 and 55 have in common?' Tanya almost at once understood that her daughter had realised you didn't need to add the numbers $1 + 2 + 3 + 4$ individually to reach 10, or $1 + 2 + 3 + 4 + 5 + 6$ to reach 21, or 1 to 10 to reach 55—instead, a clever shortcut could deliver the answer in an instant.[24]

Galinka, by contrast, had little interest in numbers and arithmetic. She preferred drawing, thinking up stories and was lively, skilful and could sing beautifully. Paul noted that both Galinka and Tanitchka loved classical music, especially Beethoven's *In questa tomba oscura*, in which a voice from a dark grave curses a woman, Beethoven's *Mailied*, Schubert's *Heidenröslein* and Haydn's Symphony No. 94 'Surprise'.[25] Tanya sounded more matter of fact when she wrote that Galinka had eagerly sung *Volga-Matushka* when a Dutch lady asked if she knew any Russian songs.[26]

All in all, there was music, scientific discussions, frequent visitors, drawings everywhere—and in spring 1915, a little brother joined the family. 'Until yesterday, *Sweetie* still worked a few hours each day in the garden', Paul noted in the family journal on 11 May. 'The entire house was bathed in sunlight'. The apple trees and strawberry plants were in bloom; it was a beautiful spring. The previous evening, Paul had chatted with Tanya until half past midnight. 'Then Tanya grew anxious', and Dr Hellema came to prepare everything for the birth.[27]

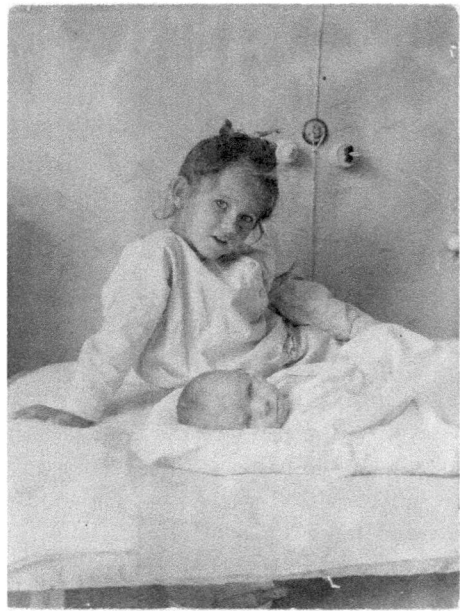

Figure 21.1 Galinka with her newly born brother Pawlik.—*Photograph Ehrenfest Family Archive.*

The baby, 'expected to be named Anouchka or Lenotchka', so a girl, was born the next afternoon at half past three, just after Paul returned home from the funeral of mathematics professor Pieter Zeeman, nephew of the eponymous Nobel laureate. Together with their daughters, Paul and Tanya chose the name Pawlik, little Paul. Out of nervous excitement, Pawlik's sisters both had a fall that afternoon.[28]

In the ensuing weeks, Paul took his daughters to Ermelo, a small town in the heart of the Netherlands, for an extended stay at their friends Philip and An Kohnstamm's wooden summer house, Schapendrift, nestled among woods, heathlands and drifting sands.[29] In Leiden, Tanya wrote in the family journal how Pawlik was learning to smile, lift his little head and, later, after Paul and the girls had returned, even began to babble (see Figure 21.1).[30] The boy was born into a continent showing its ugly, dark side, yet in a house bathed in light.

22

The red professor and the Russian princess

Happiness in wartime

Almost immediately after Einstein had expressed the formulas of his relativity theory in their definitive form at the end of 1915, he declined Paul and Lorentz's invitation to come to Leiden for a couple of days. 'Your tempting invitations make it difficult for me to stay here', Einstein wrote to Lorentz in early January 1916.[1] But he felt that he had to stay, so he exchanged several long letters about his theory with Lorentz instead. With good reason. 'If anyone had to relive exactly the same struggles here in the considerations on general relativity, I'd ardently wish that it be you', Einstein wrote to Lorentz in January 1915.[2]

But with equally good reason, Einstein sent his first complete exposition of the theory to Paul at the beginning of 1916.[3] Einstein 'needed Ehrenfest since, though Ehrenfest was a great physicist himself and did some very good creative physics, Ehrenfest could tell the other fellow, Einstein, what his theory really meant', Paul's student Dirk Struik later recounted (Powell and Frankenstein, 1999). And although this statement is undoubtedly coloured by Struik's admiration for his mentor, it was certainly true that Paul's enthusiastic reading and thinking, also in the ensuing months, helped Einstein succeed in his Herculean task.[4]

It was good fortune for them both that an exhausted Einstein was issued a Dutch visa in the autumn of 1916 after struggling through a mountain of paperwork. He travelled to Leiden, where he played music with Paul in the study on Witte Rozenstraat and debated at the blackboard. He slept above in the attic under the beams. Unlike other guests, he was allowed to smoke cigars. And although he didn't necessarily look forward to the meals from the cooperative kitchen, he enjoyed sitting in the dining room with the whole family, where light poured in from the high stairwell like a waterfall and where German, Russian and Dutch mingled (see Figure 22.1).

With Paul, he visited Lorentz in Haarlem, and Lorentz and his son-in-law De Haas reciprocated by dining at Witte Rozenstraat. De Sitter and Nordström also visited often. 'With Lorentz, Nordström, De Sitter, Droste, gravitation (rotating coordinate systems, at night on the dark road, the starry sky above)', as Paul summed it up with keywords in the journal. 'Discussion with Tanya about dimensions'. 'With Kapteyn, Heymans and then with Tanya about the fundamental concepts of gravitational theory'.[5]

Figure 22.1 Einstein with about one-year-old Pawlik in the dining room of the house on Witte Rozenstraat.—*Photograph Ehrenfest Family Archive.*

Paul and Einstein strolled with Nordström in the dunes near Katwijk, dined at Herberg De Zwaan and discussed 'German politics (Treitschke, Bismarck)'. Einstein attended the colloquium and visited Willem Keesom, Heike Kamerlingh Onnes' right-hand man, and Belgian physicist Jules Verschaffelt. He accompanied Tanitchka and Nordström to the doctoral graduation ceremony of Paul's student Dirk Coster—a great honour for Coster—after which, they rang the bell in the tower on the roof of the Academy Building. On the final day of Einstein's visit, Paul and Einstein walked together from Noordwijk to Rijnsburg, where Einstein's hero Spinoza had worked on his *Ethics* (see Figure 22.2). And even though Einstein wasn't yet the renowned scholar he would later become, he and Paul were quite something to behold, walking home along the serene Leiden canals with their billowing coats, fluttering ties and animated gestures.[6]

Their discussions, and especially debates with De Sitter, would have profound consequences for international developments in physics. They had already inspired De Sitter to work out the astronomical consequences of Einstein's theory, something he did independently and from scratch. By then publishing the results in three articles on 'Einstein's theory of gravitation and its astronomical consequences' in the English journal *Monthly Notices of the Royal Astronomical Society*, De Sitter introduced Einstein's work to scholars in English-speaking countries, particularly the United Kingdom. This in turn stimulated astronomer Arthur

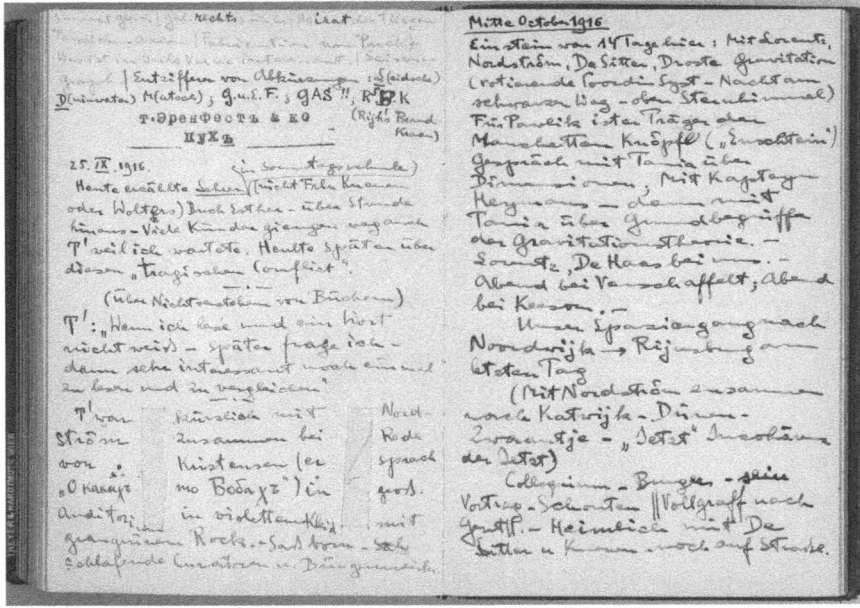

Figure 22.2 Page in the red diary about Einstein's visit in 1916.—*Ehrenfest Family Archive.*

Eddington to carry out one of the most important tests of the theory after the war—but that was later.

More important now was that the trip did Einstein 'an indescribable amount of good, physically and mentally; I am much more refreshed and cheerful. Solitude can be tolerated only up to a certain limit, you know', as he wrote to Paul.[7] In another letter to both Paul and Tanya, he added that 'the reinvigorating days spent with you have melted into a beautiful dream which I relive tirelessly in my imagination'.[8] Delighted, Paul then threw himself into the music of J.S. Bach, which Einstein had introduced him to but which impressed Tanya less. Einstein had teased her about this in the same letter to 'Dear (P + T)', 'in you [Paul] I see developing a new admirer of these magnificent things [Bach's chorales]. T also will join, despite the alarming straightness of her intellectual world line (exception to the law of motion?)'.[9] It was a reference to his theory, in which a worldline describes how a particle moves through four-dimensional spacetime. Did he, with this 'alarming straightness', mean to characterise Tanya's steadfastness in her love for Russian composers and Russian church music, for Russia itself, for geometry teaching and for the natural sciences?

Tanya wasn't like the women in Einstein's milieu. She wasn't like Mileva, who languished amid laundry racks and dishes. Nor was she like Elsa, bustling about and fussing as she awaited Einstein. In a sombre mood, Einstein had described them in a letter to his friend Besso as impossible: 'these women' who 'wait for someone to come along who will use them as he sees fit'.[10] Tanya wasn't like this. She charted her own course. And although Einstein preferred Elsa's goose cracklings with onions to the meals Tanya ordered from the cooperative kitchen, and although he didn't love Russia or Russian music, his tongue-in-cheek remark about her straightness also seemed to betray a certain admiration.

Tanya's affable imperturbability was equally clear when she walked the streets of Leiden, stately and upright as she had learnt at school. 'A Russian princess', people whispered, and they regarded with some awe the expansive garden where Tanya planted bulbs, made strawberry beds, and where her daughters frolicked, often in geometrically patterned dresses.

Which of them could have guessed that, amid her daily activities, Tanya had regularly sat down at her desk in the study to write an article for the prestigious *Physical Review*. It was a response to American theoretical physicist Richard Tolman's work. What would happen, Tolman had wondered, if you shrank the universe to the size of a water droplet? This question was in keeping with the fresh wind blowing through physics, carrying Bohr's quantum theory and Einstein's relativity theory with it. Suddenly everything seemed possible, and so too Tolman's unconventional idea. Could you build a miniature cosmos using precisely the same components—with the same physical laws and constants—as our universe?

If the cosmos could indeed simply be 'scaled', Tolman postulated, such scaling would almost naturally yield fundamental physical laws such as the ideal gas law in thermodynamics (Tolman 1914). But Tanya made short work of the physical significance Tolman attributed to this scaling, and she was not alone in doing so. Tolman had merely applied 'dimensional analysis', she observed, which had nothing to do with actual physical principles (Ehrenfest-Afanassjewa 1916).[11] And only few people would have expected such rigorous reasoning—logical, formal and analytical—from a 'princess' or indeed any woman.

Meanwhile, in Leiden, Paul began to earn a rather different nickname: the 'red professor'. This was not least thanks to his students, Burgers, Dirk Coster and Dirk Struik. They were ordinary young men from backgrounds where pursuing higher education was still uncommon. Struik was the son of a Rotterdam schoolteacher, Coster the son of a blacksmith and Burgers' father worked as a postal clerk. All three had, like Lorentz before them, taken advantage of the opportunities provided by the *HBS* (higher secondary school) to pursue further studies. And all three were also active in the Social Democratic Party, which later became the Communist Party of Holland (CPH).

Struik, who came to Leiden as a mathematics student in 1914, was introduced to socialism by a physics teacher in his hometown Rotterdam. He had even considered becoming a 'professional socialist' but found himself too drawn to 'bourgeois science'. Instead, as a 'socialist professional', he regularly contributed to *De Tribune*, the Social Democratic Party newspaper. In addition, Struik also founded a rather liberal chapter of the socialist youth organisation *De Zaaier* (The Sower) in Leiden (Alberts 1994).

That Coster and Burgers joined Struik was not surprising. Burgers had seen his mother's struggle to supplement his ailing father's meagre income with needlework (Burgers 1962, 28–29). As the son of a blacksmith, Coster knew firsthand the travails of making ends meet (Brinkman 2013). It was no wonder that they advocated a fairer distribution of income and better education for all.

At the same time, they arrived in Leiden with reverence for the unwritten rules and customs of academics and the upper middle class. Somewhat overawed, they must have entered the large house on Witte Rozenstraat—a house so spacious that it could have fitted their parents' homes with room to spare; where the traditional Dutch meat-and-potatoes of their upbringing was replaced by vegetarian dishes and where Dutch gave way to a cosmopolitan mishmash of languages. And how liberating it must have been that Paul took all these rules and conventions with a grain of salt. Even their socialist ideals were open for discussion there; Paul had never forgotten how his father had impressed upon him that Viennese workers were 'tired men' who 'spent the entire day doing hard and dirty work' and therefore 'deserved great respect'.[12]

But what mattered most to Paul, of course, was that these young men were intelligent. Developing their talents was his priority, as he had promised the students in his audience after delivering his inaugural lecture in Leiden. 'I must apply all my knowledge and skill to help you find, with as little harm as possible, the path suited to the essence of your talent', he had said, adding that it was 'absolutely necessary that I also come to know you as individuals' (Klein 1989, 29).

And so it went. Paul guided Coster, Struik and Burgers through their studies almost like a father. He spared them complicated practice calculations, reasoning that 'the most interesting problems are those you discover yourself' (Hollestelle 2011, 53). He only set examinations around Christmas, rewarding success with a bar of Kwatta chocolate.[13] He wrote an impassioned letter when Struik nearly lost his scholarship in summer 1914: 'Professor Lorentz! This is really *completely* impossible!' He highlighted Coster's precarious financial situation: 'He is a former teacher who is living on his savings, which, so far as I know, will just last him through the year' (Klein 1970, 212). He opened new worlds for them by, for instance, taking them to the Museum of Ethnology, where 'a set of five wonderful statues of the Buddha [were] standing under a magnolia tree in full flower' in the sunny garden (Burgers 1962, 53) and receiving them in his study with Einstein. But most of all he made them feel their worth by encouraging them to lead tutorials, deliver colloquia and contribute their ideas.

When Paul, alongside Lorentz, became one of the very first people in the world to grasp Einstein's theory of relativity early in 1916, his students' respect for their mentor deepened. This esteem only grew when it became clear later that spring that the new rising star in the physics firmament, Niels Bohr, was intrigued by Paul's work. In his efforts to solidify the foundations of his atomic model, Bohr had 'extensively used Ehrenfest's idea on adiabatic transformations' and had described the adiabatic principle as 'very important and fundamental', as Sommerfeld conveyed to Paul in a letter from Munich.[14] And Paul was so gratified by this belated recognition of his work that he took back all his earlier criticisms of Bohr's 'cannibalistic' atomic model (Pérez and Valls 2015, 274–275). With renewed energy, he prepared a lecture on his principle for the Royal Netherlands Academy of Arts and Sciences, ensuring that the English-language version in the *Proceedings* was crystal clear. For good measure, he submitted a detailed German account to *Annalen der Physik* in June: Planck, Epstein and the other 'quantum men' wouldn't be able to ignore his work this time (Ehrenfest 1916b, 1916c; for a much more detailed and nuanced analysis, see Pérez 2009; Pérez and Valls 2015; Valls and Pérez 2016).

'In a great number of physical problems, the foundations of classical mechanics (and electrodynamics) are used together with the quantum hypothesis, which contradicts them', he wrote in the article, which was eventually published in October (Ehrenfest 1916b). 'Naturally, it remains desirable to arrive at a general perspective that can consistently delineate the boundary between the classical and quantum realms'. Could his principle indeed help demarcate this boundary and better chart the quantum world?

For his students, it certainly felt as if they were witnessing trailblazing developments firsthand. Their admiration for Paul reached a peak after Einstein's stay that autumn when they saw the warmth and camaraderie between the two men. 'Sometimes he played the piano for us', Burgers later wrote of the period after Einstein's visit. 'It was apparently Einstein who opened Ehrenfest's ears to Bach's preludes and fugues. Ehrenfest played several for me ... I also remember Beethoven's bagatelles, which he often played. ... A few times, he asked students to perform Bach's fugues with each voice played by a different instrument, so you could better hear the individual melodies' (Burgers 1962, 52). Struik and the other students began to see 'that little fellow with his little black beard', who enthusiastically 'danced in front of his class and began to yell', in a new light (Powell and Frankenstein 1999).

Paul seemed to be at the height of his abilities. He exuded an 'electrifying energy', swiftly grasping concepts others found obscure and conveying them with razor-sharp clarity. 'No longer it was my father's maxims which took the first place in my thinking: Ehrenfest's influence became the stronger one', wrote Burgers about this period around late 1916, when he had just begun his doctoral research (Burgers 1962, 52). 'Professor Ehrenfest's example and friendship had a decisive influence on my life in Leiden', said Coster, who started a position as an assistant at the Technical College in Delft in autumn 1916 (Kramers 1951, 2). And much later, an elderly Struik still vividly recalled 'a good thing we did before the doctoral examination [in 1916,] which can I recommend to all students. ... With my friends, the Kramers, we rented a boat and went canoeing, and then to visit our professor, Ehrenfest, who treated us to strawberries' (Powell and Frankenstein 1999).

Hans Kramers was the fourth member of Paul's talented group of students. He wasn't involved with socialism or communism. The son of a general practitioner, he had joined a collegiate fraternity upon arriving in Leiden instead. With his somewhat hesitant and indecisive nature, he hadn't immediately given in to physics in the months thereafter nor did he warm straight away to Paul either. Alongside studying physics, Kramers enjoyed playing the cello as well as reading English literature and poetry. And despite admiring the Leiden physics professors, Kramers occasionally skipped Lorentz' Monday morning lectures and 'oscillated' at the colloquia—sometimes attending, sometimes not—to which Paul had invited him relatively early on.

For Paul, this was almost unforgivable. Many years earlier, his friend Herglotz had once advised him to suppress any 'desire to influence anyone whatsoever' and to keep distance from his students, believing that intense personal relationships drained energy better spent on science (see Ch. 14). Yet Paul poured his heart and soul into his students. He presented his best self to them, using his network for their benefit, keeping them informed of the latest developments in physics and engaging all his enthusiasm and sharp intellect to encourage them to think. But this came at a price, as Kramers soon learnt. In return, Paul demanded unconditional dedication to the discipline and loyalty to physics and himself.

It sometimes seemed as if the affection and respect Paul's students gave him were necessary to bolster his ego. While he certainly felt validated in Einstein's company or indirectly through Bohr's work, the proximity of these great men, not to mention Lorentz, also underscored what was lacking in his own work, as he had written to Ioffe in 1912/3. 'Of course, you already know exactly what the rub is. Unlike you, Ritz or Einstein ... I have no central, powerful set of ideas, nothing I can call "my problem"—just crumbs—and so I remain stuck in the mud'.[15] And this was probably why he seemed to take Kramers' reserve—which could easily be mistaken for arrogance—as an almost personal affront.

Initially, Kramers was unaware of these sentiments. Despite his aloofness, he was impressed by Paul. 'Such intelligent enthusiasm was unusual in these low countries, with their phlegmatic inhabitants. ... Anyone who saw and heard Ehrenfest could not escape the sense of being swept up by a whirlwind that brought promise and renewal to every corner of their soul', he wrote years later (Dresden 1987, 92). Yet Paul regarded Kramers with scepticism and distrust.

During Kramers' first year in Leiden, it was especially his membership of a student fraternity that augured ill. Paul despised the drunken rowdiness, misplaced arrogance and braggadocio that characterised fraternities in his view. Undoubtedly, memories of his own student days in Vienna played a role. Anti-Semitism had been unmistakable in Viennese student fraternities, and even in the student associations for physics students and for science students in general, which had 'nice entertaining meetings, no fighting, no immense almost compulsory drinking of beer'. Paul's younger Viennese colleague, Erwin Schrödinger later described how, around 1906, during one such gathering 'in such a *Kneipe* (meeting, with beer, in the club, singing pretty songs etc.)' 'poor Hasenöhrl', Boltzmann's successor, turned 'anxiously from his right neighbour to the left and back again, asking timorously, but we aren't truly antisemitic, are we?' His neighbours 'found some way of saying neither yes nor no', according to Schrödinger, who added that general student fraternities were far more anti-Semitic and malicious 'hotbeds of National Socialism'.[16]

Even after Kramers exchanged his fraternity for the Christiaan Huygens Study Association, Paul's mistrust lingered. Kramers was one of the brightest members of this group of around twenty, which met fortnightly in one of the students' rooms to discuss mathematics and physics. The evenings were enlivened with tea, biscuits and lemonade, and friends remarked that Kramers had become a veritable 'nerd' (Dresden 1987, 90–94). Yet Paul continued to resent him. He considered it a waste of time that Kramers, who had a gift for mathematics, would solve mathematical problems with no practical application simply for pleasure. He was equally unimpressed by Kramers' study of the work of American physicist Josiah Gibbs, who approached thermodynamics with far more abstract mathematics than Paul's mentor, Boltzmann, had done. Above all, Paul appeared unable to stomach how Kramers steadfastly kept his own interests and friendships, making him elusive. Unlike Burgers, Paul couldn't take Kramers by the hand and lead him through physics. Even years later, he wrote to Tanya that Kramers looked 'as if the whole world belongs to him, but he finds it too shabby to claim' (Hollestelle 2011, 79).[17]

Only after the canoe trip with strawberries in spring 1916 did Kramers suddenly realise Paul's opinion of him. Contrary to his usual habit, Paul had set Kramers a carefully worded and highly demanding final examination. And after a dumbfounded Kramers struggled through the barrage of questions, Paul was blunt and direct: a lack of passion made Kramers unsuitable for science, he declared during the concluding discussion. As a result, a disheartened Kramers became a teacher at a gymnasium in Arnhem—where, incidentally, he was again described by some as arrogant. When confronted about his frequent tardiness for example, Kramers reportedly retorted that his pupils learnt more in half an hour with him than in an hour with other teachers. Yet those close to him knew how much Kramers also struggled with self-doubt (Dresden 1987, 94–96).

It was Kramers' salvation that the headmaster of the Arnhem gymnasium did recognise his talent and that Kramers' father didn't take Paul's judgement too seriously. Together, the two men encouraged him to attend the International Student Conference in Copenhagen in August 1916. And the postcard Kramers subsequently sent to Bohr arrived at exactly the right moment in the Danish capital.

Bohr, overwhelmed by the administrative demands of his new post as professor in Copenhagen, was in desperate need of an assistant. So, almost immediately, Kramers immersed himself in Bohr's work on the atomic model for helium, a heavier atom than hydrogen. He also investigated why hydrogen atoms emitted or absorbed certain colours of light more intensely than others, as if electrons were more likely to make some jumps than others. And with his elegant mathematics and acute intellect, he soon became Bohr's right-hand man (Dresden 1987, 99–101).

It was a successful year for the Leiden circle in other ways too. Paul's three other students also secured positions: Burgers as a doctoral candidate in Leiden, Coster as an assistant in Delft and Struik as a teacher in Alkmaar. Paul was still hopeful for the success of his adiabatic principle and shared fond memories of Einstein's visit with Tanya.

Little Tanitchka was delighted by a different kind of visitor at the end of the year. Shortly after Einstein's departure from Leiden, Aunt Sonya managed to escape war-torn and turbulent Russia and joined them in Leiden. Tanya happened to be visiting the Gorter family with the children when she arrived, and Paul called her at once with the Bakelite telephone by the stairs. 'I'm telephoning. They're coming. From the garden, through the glass door in the dining room, T' can see Aunt Sonya sitting. She won't move for fear of "shattering the dream"'.[18]

23

Ups and downs

Bankruptcy, peace and another son

No person has a 'straight worldline'. Not even Tanya, and certainly not Paul, whose life and emotions were in constant flux. The elation he experienced in late 1916 gave way to profound despondency in early 1917. His colleagues' response to his publications on the adiabatic principle disappointed him. Particularly Sommerfeld, who had eagerly anticipated the work, responded with reserve: he found Paul's presentation of the principle rather too sketchy. Could his rough examples be expressed in formulas that would make possible concrete predictions about quantised—discrete—systems (Navarro and Pérez, 2004)? It was a pity that Paul's late friend Ritz could no longer apply his imagination to this issue, had been Sommerfeld's answer to this rhetorical question (Klein 1970, 291). And Paul subsequently foisted the quest for such formulas on Burgers, whose mathematical skills exceeded his and who would need a topic for his dissertation near the end of 1917.

In the meantime, the ongoing war didn't improve Paul's mood. Contact with colleagues abroad was slow or downright difficult, and travelling was still complicated. Lacking Dutch nationality, Paul couldn't leave the Netherlands at all. Peevish and suffering from jaundice, he spent much of spring 1917 in bed and complaining in letters.[1] Einstein, exhausted from his intense work and his divorce, living in a dejected Berlin, reminded Paul of his good fortune. 'Just think how insignificant it will be twenty years from now how one knocked about on this Earth, provided no wrong was done. It is all the same whether you or someone else has written this or that paper. You are certainly never mindless, except for when you happen to be contemplating whether you are mindless or not. So, away with the hypochondriac! Enjoy yourself with your family in the fair land of the living!'[2]

But above all, Tanya succeeded in this. In the family journal, she described how precocious Tanitchka crossed the garden on stilts while pondering geometric problems, magic squares and a formula for the number of combinations of two or three tones that could be made from the twelve tones (black and white piano keys on the piano) of an octave. At Christmas, the girl delved into astronomer Anton Pannekoek's astronomy books and, according to Tanya, made 'very valid comments on the points she did not understand'. Galinka, five years her junior, had learnt to crochet, drew a great deal and started violin lessons. 'How does it work, the way we think?' she wondered. 'Is there a separate tube for that?' Finally, gentle Pawlik had started singing songs, Tanya noted (see Figure 23.1).³

Such familial enjoyment 'in the fair land of the living' wasn't enough for Paul. The distraction physics couldn't give him for the moment, he sought instead in economics. In February, Paul, Tanya and Aunt Sonya read in newspapers how demonstrations on International Women's Day in St Petersburg had escalated into riots over food shortages. They learnt that a new, provisional government had ousted the Tsar, and that chaos proliferated in the country. It became a recurring topic of conversation around the dinner table on Witte Rozenstraat, along with the distribution of wealth, goods and labour. Paul and Tanya had read Marx and Engels, as well as works by English-language economists, and Paul took pleasure in using his quick mind to ponder how physical concepts might give the field of economics a firmer foundation.⁴

Figure 23.1 Tanya with Galinka, Pawlik and Tanitchka in 1917 at the Witte Rozenstraat.—*Photograph Ehrenfest Family Archive.*

Now his student club had flown the nest, or were about to do so, he clung more firmly to Burgers. It even led him to take the rather unusual step of visiting Burgers' parents in Arnhem in summer 1917 (see Figure 23.2). And in their small house, filled with crystal models, galvanometers and other apparatus, he offered plenty of advice.

The devices and models must have reminded Paul of the items his eldest brother Arthur had once so impressed him with. Now the roles were reversed, and Burgers Sr., donning a traditional cap, listened respectfully while Mrs Burgers hovered in the background. Before the visit, Paul had already instructed the couple in a postcard to ensure 'that Jan absolutely does no calculations or theorising during this [free summer] period. He may do anything else as intensely as he likes (although perhaps not too much reading).'[5] On a separate postcard, he enjoined Jan not to write Zs and Ts with such unnecessary flourishes. 'Forgive me for this interference—but I hope to read much more of your handwriting in my lifetime.'[6] On cards after the visit, he also addressed Jan's younger brother Willy, who was beginning his studies in Leiden that autumn. 'Ask Willy to think about which points in the HBS textbooks he finds ineffectual, difficult or superfluous—as preparation for the didactic colloquium'.[7]

Figure 23.2 Jan Burgers' father, Jan Burgers and Paul at the Burgers family home in Arnhem.—*Photograph Ehrenfest Family Archive.*

Paul even arranged accommodation for the brothers. Farther down Witte Rozenstraat, the Van Leeuwen sisters' studio had become vacant. That summer, the younger of the two sisters, Nel, had married Finnish gravitation expert Gunnar Nordström, working with Paul in Leiden. Around the same time, the sisters' father also died, necessitating the eldest of the two, Jo, to live with her mother again in The Hague (Blaauboer and Van der Heijden 2025). Wasn't this studio suitable for the brothers? Then they could have a study at Paul's house across the street, 'especially nice during the cold winter. ... You understand it would give me great pleasure to have you close to us'. In fact, Paul had already reserved the Van Leeuwen's studio, and if the boys had already paid rent for another place, 'I can reimburse it anyway!'[8]

Unsurprisingly Paul's meddling began to weigh on Burgers now and then, and it was probably good for him that Lorentz offered him a job in Haarlem. Almost immediately after obtaining his master's degree, Burgers began working as curator at Lorentz's Physical Laboratory in this town.[9] He also began writing a long essay that he submitted to the annual competition of the Teyler Foundation's 'Second Society', whose theme in 1918 was 'Rutherford-Bohr's atomic model'. One year later, his analysis of the latest insights and controversies surrounding the model wouldn't only be awarded the prize (it was the only submission) but would also be the lion's share of his dissertation—with only a relatively brief discussion of Paul's ideas on adiabatic processes and their relation to quantum theory on the last pages (Burgers 1918).

Nevertheless, Burgers could still often be found in Leiden, studying in the reading room, attending the colloquia and also regularly enjoying the lively household on Witte Rozenstraat—in no small part thanks to Tanya. 'Mrs Ehrenfest had (and has) a very pleasant way of treating us, and of talking; she was almost as inquisitive as Ehrenfest was himself, and she often engaged us for work in the garden of which she was very fond', Burgers wrote much later (Burgers 1962, 51). He didn't mention Tanya's scientific work. Viewing her from his traditional Dutch perspective, he focused more on what Tanya *wasn't* than what she *was* or did. She wasn't submissive, modest, emotional or nurturing. Yet with her upright posture, sharp mind and amiable composure, she was Paul's anchor in life—even in these turbulent years.

For shouldn't Tanya be the one who ought to have been shaken? She had been deprived of her status as an independent woman. She heard nothing from her family and friends in Russia, and news from the country grew ever grimmer. 'The brilliant expectations' of the February Revolution had 'darkened', the *Algemeen Handelsblad* wrote in September, and 'the collective energy of the people' had been 'stifled'.[10] Anyone 'without a heart of stone' must surely suffer alongside the Russian patriots 'of whichever persuasion'. And when in October the Bolsheviks took advantage of the chaos to seize power, any hope of a peaceful resolution of the revolution soon faded.

Did Tanya welcome the Bolsheviks' pursuit of equal rights for men and women? Their drive for education for all? Was she one of many who supported the Bolsheviks' aim of negotiating a peace agreement with Germany as soon as possible? Yet she could hardly have forgotten some of her fellow students' fanaticism during the strikes at the Bestuzhev Courses and what she had once described as 'comrades' courts' (see Ch. 4).

Paul regarded the revolution with even greater scepticism. It was admirable that Russian socialists recognised science and technology as pillars of modern society. Before the war, chemistry and physics discoveries too often were developed and patented abroad, forcing Russia to import products it could have manufactured itself (Kojevnikov 2004, 5–7). Yet he disagreed with the Bolsheviks' fundamental premise. 'I simply cannot assume that all men are equal', he once said (Alberts 1994, 282).[11] 'I could think', observed his former colleague James Franck from Göttingen, who would be awarded the Nobel Prize of Physics in 1925 and emigrate to the United States in 1933, 'that some of the conservatives here would call him pinkish, or what. That he certainly was, because he had a very strong social feeling' (Franck and Sponer, 1962, session V).

At the same time, Paul and Tanya lived rather carefree lives of relative affluence. Of the scientists in Leiden, probably only Onnes had a larger house—elegant villa 'Huize ter Wetering', just outside the city and surrounded by chestnut trees (Van Delft 2005, 415–418). True they hadn't yet bought much furniture, wallpapered the plastered walls or laid parquet on the wooden floor. But they employed nannies, Miss Hellema for homeschooling, and Baba Sonya lived with them. It was a lifestyle largely enabled by the railway shares Tanya's Uncle Pyotr had so wisely invested in—shares many Dutch investors also regarded as excellent exposure to industrialising Russia. They also counterbalanced the mortgage Paul and Tanya had taken on the house and would one day hopefully fund their children's higher education.

At least, until the Bolsheviks unleashed financial catastrophe. In their eyes, shareholders like Tanya were simply 'members of the capitalist class', whom they no longer intended to pay. Overnight, Tanya's shares become worthless, and their financial situation uncertain.[12] For Paul this was an even bigger shock. Years earlier, he had come to terms with the idealistic decision to refrain from eating meat. He had dutifully eaten meals from the cooperative kitchen without complaint. But the prospect of returning to the austerity of his youth, with meals of rice and garlic, filled him with horror, and under no circumstances did he want to leave his children worse off than he had been at the start of his life.

That Tanya again fell pregnant around the time of the Russian Revolution was thus likely unplanned. As her belly grew, the Russian Tsarist empire collapsed. In March 1918, Lenin signed a peace treaty with the Germans in Brest-Litovsk, on behalf of the Bolsheviks. By ceding Finland, Poland, the Baltic countries, the Caucasian republics and Ukraine, the new Soviet Union lost nearly a third of its farmland, a third of its population, half of its industry and almost all its coal mines.[13] A power struggle between Lenin's 'red' Bolsheviks and the 'whites'—a hotchpotch of moderate socialists, liberals and Tsarist conservatives—then plunged the country into civil war.

In the Netherlands, ration cards and anxiety about the war in Europe characterised daily life. Yet despite everything, Tanya refused to neglect her ideals. Together with Tanitchka and Galinka, she translated texts on vegetarianism by Tolstoy, who had condemned eating animals as immoral and described the cruelty of a 'horribly revolting' visit to a slaughterhouse (Tolstoy 1892; LeBlanc 2001). 'People can be healthy without eating animals that have been killed. When they eat meat, they take part in killing animals just because they find meat tasty. That is wrong', wrote Galya, just seven years old, on sheets of paper. Then skipping along Leiden's cobblestone streets, she brought them to local butchers.[14]

Her father, meanwhile, had moved 'miles away' from physics. He was 'sitting over problems of politico-economic theory (with Papa Lorentz's shoulder-shrugging blessing). When am I going to return to physics again with qualms of conscience and remorse? At the moment it is unspeakably odious to me', he wrote to Einstein in May 1918.[15] He had even turned down Hilbert and Planck's 'remarkably kind' invitation to the Wolfskehl Lectures in Göttingen, although his refusal was mainly due to his aversion to the war. Paul couldn't travel to the country whose troops had 'raped' Belgium, as he wrote to Hilbert.[16]

This led to a rare disagreement with Einstein, who found it incomprehensible that Paul hadn't just politely declined the invitation. Did Einstein fear that Paul's comment might come back to haunt him after the war? In a letter to Hilbert, Einstein tried to draw the sting of Paul's remark. 'His understandable bitterness over politics has swollen to such a degree that now he must give vent to his heart on such an entirely unsuitable occasion. I am convinced that you see the decent fellow behind this clumsy, almost childish conduct and do not hold this blunder against him so much. Such a character is surely a thousand times more welcome than the many miserable fawners who populate our faculties in the main.'[17]

Then Einstein reprimanded Paul directly. 'Imagine if you had been appointed to Göttingen a few years ago instead of to Leyden and were now held to account for the great reprehensible machinery!'[18] It was one of the few times Paul disagreed with Einstein. 'I did not "blow off steam" at Hilbert', he replied. 'I wrote to him in a way that merely expressed how I felt, in loyal attachment to the atmosphere pervading my Göttingen student days. ... besides, I did what I had to do.'[19] Hilbert didn't hold it against him. He was one of the few Prussian scientists discontented with German militarism and hadn't taken 'E's honesty badly, though'.[20] All three men naturally agreed that the war was a 'delusion of epidemic proportions', as Einstein had written to Paul the year before.[21] A delusion that 'after producing immeasurable suffering, will vanish again, so the generation after the next can gasp at it as something completely monstrous and incomprehensible'.

And so 1918 limped on. To Einstein, Paul tried to put a brave face on things. 'Personally, we are doing very well, and the joy the children give us is beyond description', he wrote to him in spring 1918.[22] Yet he betrayed his disappointment elsewhere in the letter: 'my not writing [to you, Einstein] stems from: my chronic discontent with work.—But no more of it!'

Tanya must have occasionally felt lonely amid Paul's ups and downs, with her growing belly and a head full of worries. Perhaps it was even a relief that he left for Kernhem that summer. At the family estate of his student Carel de Ridder, with a 'driveway 250 to 300 metres long', Paul taught De Ridder and his friend Gijs van Aardenne 'what higher mathematics is and how its applications look in mechanics' at the end of July.[23] In return, he received board and lodging as well as a stipend, which Tanya and he greatly needed.

Paul also enjoyed the rest of the entourage at Kernhem. The young men took him to meet Dutch writer Arthur van Schendel, where he also met painter Jan Toorop.[24] With a pianist staying at the estate, he had long conversations about Beethoven's compositions.[25] And even though the 'two *malchiki* [boys]' weren't excellent students, they were interesting in their own way. 'Van Aardenne has a love for all things old and beautiful and enjoys rummaging around old attics and cellars', Paul wrote to Tanya.[26] And: 'I am sorry you are stuck in Leiden, so overworked and tired, while I am having such a wonderful time. But it would be wrong not to make the most of what is offered to me here'.[27]

In Leiden, Tanya's days were more monotonous. 'It makes you lethargic when you can barely move around', she wrote back to Paul.[28] She kept him updated on the children's activities and reminded him that he had forgotten to send a card or gift for Galinka's birthday. 'I thought she was four days younger', he apologetically replied.[29] Paul was more preoccupied with the outside world than with his family. The boys wanted to come to Leiden at the end of their workweek at Kernhem, he mentioned to Tanya. Could she place an advertisement in the *Leidsch Dagblad* for 'two or three quiet, airy rooms for two gentlemen?'[30] And could the boys stay at Witte Rozenstraat in the meantime? 'Let them stay with us; they shall bring their own bedding. Please arrange the rest'.[31]

By this time, Paul had cycled to Ermelo, where he stayed with the Kohnstamms. 'Thank you for your package (shirt, undershirt, socks, stockings)', he scribbled from their home, Schapendrift, 'but now I *urgently* need: bread coupons, fat coupons (send by registered mail) and perhaps two or three extra bread coupons as a gift, if it doesn't leave you too short'.[32] Food had become even scarcer as British mines and German submarines disrupted the supply of imported agricultural goods by sea. When Tanitchka also arrived at Ermelo station a few days later, Tanya had sent her with cheese, eggs and sugar.[33]

This left Tanya, Aunt Sonya, Galinka and Pawlik in Leiden, where the summer wind carried a 'sickening reek of rotting meat and fish' through the streets, penetrating even 'the living rooms of houses'.[34] The stench had angered the residents of Leiden since spring, resulting in a flood of letters to the *Leidsch Dagblad*. Yet Van Calcar's animal feed factory near Zoeterwoude continued its activities unabated. Barges laden with rotting fish, abattoir waste and condemned sausages docked there every day for processing into animal feed in heated drying chambers.

Gynaecologist Carl von Winning occasionally mentioned the cloying smell when he parked his gleaming black automobile at the end of Witte Rozenstraat to visit Tanya. Often the gaiter-wearing German aristocrat would find her in the garden. If he brought up the stench, she would shrug mockingly. Dutch people! Did they think they could solve the problem by writing letters and lodging objections? 'What else should they do?' he once asked. 'Have a match handy?' she replied with an ironic smile. And indeed, on Friday, 30 August, in the dead of night, the wooden factory complex burnt to the ground. But Tanya had an alibi. While Von Winning believed she was 'capable of such an act', she had simply been in bed that morning.[35]

Twelve days earlier, after Paul and Tanitchka had returned to Leiden, she had given birth to a son. Wassily would be his name, as the other three children had decided. A week after the birth, Von Winning visited the family again on a Sunday morning. To his surprise, they were all gathered in the study by the blackboard, immersed in mathematical puzzles, a family emanating joy. 'Dear Ehrenfest', Einstein wrote shortly afterwards, on 4 September.[36] 'I congratulate you heartily on little Wassily. Your will for life fills me with admiration. I also am full of hope for the future again, even if I don't give it concrete expression'.

Two months later, on 11 November, the warring parties in Europe laid down their arms. During the same week, a half-hearted revolution declared by Pieter Jelles Troelstra of the Social Democratic Workers' Party came to nothing in the Netherlands. And just before all these events, on 7 November 1918, Burgers defended his dissertation on the Rutherford-Bohr model of the atom and also a bit on Paul's adiabatic principle. Needless to say, Paul had inundated him with advice—about the cover of the dissertation, the acknowledgements and the ceremony. He had applauded Burgers' decision to defend his thesis publicly, with an audience and a speech for laypeople but, knowing Burgers' fervent sympathy for the Bolsheviks, added: 'For heaven's sake, don't show up in a Bolshevik uniform!'[37] And even Burgers' fiancée, physics student Nettie Roosenschoon, received unsolicited advice from Paul: 'Perhaps it is mistaken—but you know my wife and I are convinced that the key to a lasting marriage is a shared passion for a third matter. Usually, this would be children, but in the case of a man like Jan, who finds such happiness in intellectual work, I think it is excellent that his wife is a true physicist at heart! Even more reason for me to hope you can continue your studies with determination'.[38]

But Burgers had long grown weary of such interference. Partly, this was because Paul had previously expressed pronounced scepticism about Roosenschoon. But it was also because Paul exhausted those around him with such instances of meddling. 'His analytical mind stirred up everything, so that at times it looked as if nothing would be left as it was. On the long run this pushed his students somewhat away from him .... There were things which we did not like to have analysed ... in several cases it was an instinctive protective reaction from our side', Burgers would later write diplomatically (Burgers 1962, 53). Paul was 'a vampire', Coster concluded more bluntly (Klein 1970, 210). At the same time, it was almost inevitable that Burgers, after four intense years, would distance himself from the man who had taught him so much. Now he had Roosenschoon, who made him see relationships, emotions—and Paul—in a different light.

'I remember a meeting in the spring of 1918, which Ehrenfest had arranged to make physics teachers acquainted with recent discoveries', Burgers later wrote. 'He asked some of us to give talks; I believe I spoke about the work of Franck and Hertz. This meeting naturally gave him great pleasure, but what was strange to us was that he said it had given him more pleasure than the birth of his youngest child in that same year. This surprised us greatly. I have spoken about it with Lorentz and asked him whether he could talk with Ehrenfest and help him to find a way back to feelings which looked more normal to us. But even for Lorentz it was too difficult to penetrate into the deeper recesses of Ehrenfest's mind' (Burgers 1962, 53).

Paul didn't follow a straight path in life. Yet was his remark so odd? With Burgers' move to Delft, where he became a professor in aerodynamics and hydrodynamics on Lorentz's recommendation, all the students from his close circle had dispersed. Such a loss would make anyone melancholic, and certainly Paul, who lamented in a letter to Burgers at the end of 1918: 'I feel that I am now *rapidly* losing all contact with the younger people and am growing old ... I am always being reproached for having hurt younger people by my uninvited interference' (Klein 1970, 210). And perhaps there was something Burgers overlooked—something which wasn't openly discussed but which Paul and Tanya had fearful suspicions about. Little Wassily had Down's syndrome, or as was said at the time, the boy was an 'idiot'.

24
Physics curator
Bohr as Rembrandt and Einstein as Holbein

It was an entire year after the revolution before Paul and Tanya received any sign of life from Petrograd. A letter from Krutkov arrived in the post in November 1918. Paul immediately informed Burgers: 'They suffer from hunger and cold, often to the point of deep despair, but all my friends are still living, working and even publishing (!). Krutkov, for example, in line with your work on adiabatic invariants. "My" colloquium is still going vigorously, even though people seriously do not know whether they shall survive this winter. Oh, you know, the enthusiasm of the Russian youth is utterly unique. I cannot tell you how much warmth these messages convey towards me—*Heimat* [home/homeland in German]'.[1] He thickly underlined the word four times. Did he still feel out of place in the Netherlands with its commuting students and its phlegmatic inhabitants?

Diligently he set to work on behalf of Krutkov. During Krutkov's lengthy stay in Leiden before the war, they had already discussed the first steps towards the 'adiabatic principle'. With an empty stomach and wearing too thin a coat, Krutkov had further developed these ideas in Petrograd. His elaboration paled in comparison to Burgers' treatment of the principle, but out of loyalty, Paul still tried to get his work published (Krutkov 1919).[2]

Almost effortlessly, he reverted to the role he so enthusiastically fulfilled in St Petersburg: mediator, coach and linchpin. Once again, he could serve as a bridge between theoretical physics in St Petersburg—Petrograd since the beginning of the war—and in Western and Central Europe. The war had slowed the expansion of his network, but he had strengthened his connections with the Dutch Nobel laureates and there was his good relationship with Einstein. Moreover, recently Paul had met another great physicist: Bohr of Copenhagen.

After a series of postponed and stalled attempts, Bohr finally wrote down the most recent insights in quantum theory and the spectra of various atoms in the first of two thin books in May 1918 (Bohr 1976). Sure enough, he mentioned Paul's adiabatic principle in them. 'As you will see, the considerations are to a large extent based on your important principle of "adiabatic invariance"', he wrote to Paul in mid-May 1918 (Klein 2010, 308). It took some time for Paul to grasp Bohr's work fully. When Bohr spoke, his lengthy arguments seemed to lead his colleagues through an unfamiliar and increasingly erratic landscape, until, at the very end, all the earlier steps fell into place. And 'he knew well what he wished not to say when he strove in long sentences to express himself in his scientific papers', as young Viennese physicist Wolfgang Pauli later characterised it (Pauli 1945). Understanding Bohr took time.

This time, one of the questions he tackled was where the 'quantum model' of the atom transitioned into a 'classical' model. In other words, when were quantum calculations necessary to describe their behaviour, and when could classical formulas be relied upon? The questions were related to Paul's research into the limits of and possible bridges between the quantum and classical worlds. And Bohr, more philosophically inclined than Sommerfeld, had been inspired by Paul's adiabatic principle to formulate the 'mechanical transformation principle' which answered such questions.

As it turned out, the interpretation of this principle would evolve both in Bohr's and others' eyes over the years as quantum physics became quantum mechanics. It is hard to say whether this was because it was so vaguely formulated or because it was so profound, but in 1918 it was a guiding principle for mapping the quantum world. At which scales did electrons in an atom no longer need to jump between energy levels, as if jumping from one rung of a ladder to another, and instead glide back and forth as if on a slide for example? The 'mechanical transformation principle' (later renamed the 'correspondence principle') demonstrated that electron transitions between successive energy levels become so small at large distances from the nucleus that they approximated the classical model—as if the steps converged, making the space available to electrons seem smooth and 'continuous'.

Bohr had approached the matter from a different angle to Paul. He had 'therefore not used the same terminology', he wrote to Paul.[3] But when Paul finally understood 'where the frog jumped into the water' at the end of May, he was nonetheless pleased to have inspired Bohr. Added to which, Bohr had lavished praise on his skills as a mentor. More specifically, as the mentor of the young Dutchman about whom Bohr had already written favourably to esteemed colleagues such as Rutherford: 'Kramers, whose stay here has brought me much pleasure in these sombre times. I consider him a very promising physicist, and he has been enormously helpful to me in my work' (Bohr 1976, 671).[4]

Kramers! Paul had never congratulated him on his position in Copenhagen. In 1917, he had even withheld from him the announcement by the Teylers Foundation in Haarlem regarding the essay prize on 'the atomic model according to Rutherford and Bohr'—a topic ideally suited to Kramers.[5] Yet his former student in Copenhagen evidently spoke highly of him. 'I very much hope to meet you once the war is over', Bohr wrote to Paul. 'Kramers has told me so often about the interesting scientific life in Leiden, and I should like so much to come to Holland once when it is possible to travel again'. Paul replied enthusiastically: 'I impatiently await the day when you enter our house as our guest' (Klein 2010, 310).

Not long after receiving Krutkov's letter and after the dreadful war, Paul began arranging this visit. In a letter, he invited Bohr to the seventeenth Netherlands Congress for Natural Science and Medicine in Leiden at the end of April 1919. Bohr could meet renowned physicists such as Lorentz, Kamerlingh Onnes and Zeeman, who all lived nearby, Paul wrote on 13 January. Moreover, he could stay at Witte Rozenstraat 'for as long as you can stand it here (bad service, bad food—well, Kramers can prepare you for everything)' (Klein 2010, 310).

It would have been even better if Einstein had come too, but having just finalised his divorce in Zurich, he needed rest. Fortunately, Bohr accepted the invitation. He also suggested that Kramers accompany him. A lengthy review article by Kramers on the intensities of the spectral lines of the hydrogen spectrum (related to the atomic model), that was due to appear in the proceedings of the Royal Danish Academy of Sciences, could serve as his dissertation, enabling him to obtain his doctorate from Leiden University in late April or early May (Kramers 1919).[6]

So it happened that on 24 April, Bohr was seated in Leiden's venerable city hall, where the mayor welcomed the assembled scholars. It was fortunate, observed the mayor, that science hadn't come to a standstill in the 'sorrowful years of the war', and it was regrettable 'that science and its inventions in some cases perhaps served war more than humanity'. But neither Bohr nor Paul had reason to consider themselves the subject of this remark. And, given his aversion to student fraternities, Paul probably saw no cause either to drag his guest to the Minerva student society after 'tea and lemonade' at City Hall, where many others continued the evening 'in conviviality'.[7]

Did Paul stick to his draft notes when introducing Bohr to the audience at the Inorganic Chemistry Laboratory on Hugo de Grootstraat the next morning? According to his scribbles, he first apologised for speaking German. In English, he would simply sound 'completely incomprehensible'. Then he thanked Bohr for making the journey, which was so 'tiring and time-consuming' in these times and expressed his belief that a 'personal meeting' would lead to a much 'deeper understanding' of Bohr's 'groundbreaking work on the structure of atoms'.[8] The time when Paul had found this work troubling or even cannibalistic was definitively over.

He and Bohr even got along extremely well. During walks and at home at Witte Rozenstraat, they had long conversations about art, the world, philosophy and physics. Bohr enjoyed the children, who frolicked around the house, and Paul ensured that Bohr could regularly withdraw from company. To show him another side of the Netherlands, Paul also took Bohr for two days to the country house in Maarn belonging to the family of Gijs van Aardenne, one of the two '*malchiki*' (boys) whom Paul had taught the previous year.[9] And at the end of the visit, on 8 May, Kramers successfully addressed all the critical questions posed by the PhD committee in Leiden, which included, in addition to his promotor Paul, Bohr, Lorentz and Onnes (Dresden 1987, 113) (see Figure 24.1).

Figure 24.1 Bohr visiting Heike Kamerlingh Onnes' (far right) cryogenics lab, with Hendrik Lorentz (left of Bohr) and Paul (far left).—Photograph National Museum Boerhaave, Leiden.

Back in Copenhagen, Bohr tried even harder to express himself as precisely as possible in his letter of thanks to Paul than in his scientific articles. He felt 'terribly inadequate', especially writing in English, a foreign language to him: 'You do not know how miserable and stupid I feel when writing this letter. I am sitting and thinking of all what you have told me about so very many different things, and whatever I think of I feel that I have learned so much from you which will be of great importance for me; but, at the same time, I miss so much to express my feeling of happiness over your friendship and of thankfulness for the confidence and sympathy you have shown me. I find myself so utterly incapable of finding words for it' (Klein 2010, 313). The joy of their newfound friendship was mutual. Paul wrote, if Bohr wished to visit again, whether for a few hours or a few days, all he had to do was show up at the door, and—'even at midnight!!!!'—he would be 'welcomed jubilantly' by every member of the household (Klein 2010, 313).

Three weeks later, Bohr sent a package of gifts: for Tanitchka, an etching of the Danish Renaissance castle Kronborg; for Galinka, a porcelain bear and for Pawlik, a wind-up toy. Bohr had been extremely busy, he wrote. 'However, I have taught and told so much of the wonderful time in Holland, that it seems to me that it was only yesterday that, together with your wife and van Ardenne, [sic.] we were looking at the woodcuttings of Hokusai', he wrote, along with much else.[10]

In Leiden, Paul was still writing a detailed reply to Bohr's earlier letter, tackling the subjects they had discussed one by one. And, as in his letters to Einstein, occasionally grousing passages made their way into this missive. 'I have unfortunately done practically no work at all lately. That is to say—I do go to Delft every day as one of the [external] examiners [at the Institute of Technology] five hours a day of examining!!! This is the first time in my life that I am earning money for work that has absolutely no value in principle for me—earning money by doing hangman's work. Besides that I also do it badly in just the technical sense, because I have a particularly poor knowledge of the subject matter on which I am examining' (Klein 2010, 314).[11] To add to this, all kinds of paperwork and arrangements for his students and colleagues kept him from doing his own research and thinking.

'But the older I get', he continued, 'the more clearly do I see that the essence of my personality is really expressed by all these "accidental", "temporary" "diversions" from the straight and narrow path of work. That is a depressing acknowledgment because success in scientific work could give me joy that would be somewhat lasting, but although running around on organisation business or some other "human" matters can indeed fascinate me temporarily and even fill me with intense joy temporarily it is soon all vanished and then comes "the morning after"' (Klein 2010, 314). It sounded almost like his childhood when he had so often felt torn between the bustle of the shop and the comings and goings of his brothers on the one hand, and the quiet landing with Minna and his maps and other projects on the other.

His letter was candid. 'Please don't be annoyed over my perpetual schematising and my wanting to put everything into words. I don't know myself what drives me to it—I am unspeakably restless, dissatisfied, and can't by any means achieve a mood, even if only for a short time, in which I might feel—now for once I have the right to set aside a few peaceful hours for something. Then I have only two remedies—sleep and hasty rushing to and fro' (Klein 2010, 316).

Yet there was growing respect in Leiden for the diminutive professor with a rolling r, a Russian wife, adorable children and a mind the greatest intellects from abroad came to draw upon. Only a few months after Bohr left the country, Einstein knocked on Paul's door. 'Indeed, I have a great desire to visit Ehrenfest after an interval of almost three years', he had written to Lorentz.[12] Once again, Paul began making arrangements. They would accommodate Einstein's diet for his 'tyrannical stomach [severe stomach problems]' in every way, he wrote to Berlin. If Einstein granted him 'all the necessary authority', Paul would turn away anyone who threatened to disturb his peace. 'Besides that, you can on the spur of the moment isolate yourself completely for a couple of days on the estate of a very, very dear friend of mine in pine woodland, with most excellent accommodations, and peace!'[13]

This was the Van Aardenne family estate in Maarn, where he had also taken Bohr.

Anticipation grew when Lorentz picked up big news shortly before Einstein's visit.[14] During an expedition to observe a solar eclipse over the African island of Príncipe, British astronomer Arthur Eddington—inspired by De Sitter's papers on the cosmological consequences of Einstein's theory—had captured on photographic plates light from a distant star skimming past the sun. Normally, the star's faint light would have been drowned out by the sun's intense brightness. But as the moon had blocked the sunlight, the light appeared as a halo on Eddington's photographic plates. And this halo beautifully showed that the starlight hadn't travelled in a straight line. It had passed through a curvature in space caused by the sun, which had distorted its course, much like water follows the curve in a riverbed. This was proof that spacetime bends and curves under the influence of gravity, precisely as Einstein's theory of relativity had predicted. 'Eddington found star displacement at the edge of the sun', was the somewhat garbled telegram Lorentz sent him on 22 September.[15]

When Paul's 'Dear, dear Einstein!' greeted Lorentz, Paul, Tanya and the children on 18 October in Leiden, they thus all knew that he had been right with

Figure 24.2 Paul and Einstein with Pawlik, address unknown (not the Witte Rozenstraat).—Photograph Ehrenfest Family Archive.

Figure 24.3 Ibid.—Photograph Ehrenfest Family Archive.

his theory.[16] It added even more brilliance to the days in which they immediately picked up the threads of their discussions, conversations and musicmaking. The garden framed the house in autumnal colours, snatches of music rose up the stairwell, Tanya burst into laughter as she always did, Baba Sonya bustled around with tea and trays of food and the children romped about with Einstein (see Figures 24.2 and 24.3). 'Never have I taken part in such a vivacious home setting before; it simply comes from two independent persons not being bound together by compromises!' wrote Einstein afterwards.[17]

He had learned 'to consider you all a part of me and me a part of you all', he continued and sent 'to you, your wife, and to the touchingly solicitous aunt, each a separate, hearty kiss—as meets gender, tradition, and temperament, of course—in thanks'. And 'no less must Tatya, by nature so wondrously dreamy, although somewhat infected with the germ of the classical school and literature (the smiling enigma), suffer a kiss naturally, also the shoo!-shoo! lizard Galinka with the animated pencil, the little blue tomcat Paul, and finally the patient little crawlikins [Wassily]'. 'From now on', he wrote, addressing Paul directly, 'let's stay properly in close contact with each other. I know that it does both of us good and that each is made to feel by the other less of a stranger in this world'.[18]

This, in the flat Netherlands and with all his feelings of uncertainty, must have lifted Paul's spirits. So much so, in fact, that for once he limited himself to a relatively restrained reply, in which he enclosed a drawing by nine-year-old Galinka. The girl had of course heard at home that Einstein's name was in all the newspapers after the Royal Society publicised Eddington's results more widely in early November. For that matter, Lorentz and her father had so often spoken about the work of the pacifist professor who had spent the war in Berlin upending the concepts of space and time, while others violently redrew the map of Europe.

Galinka had drawn an African astronomer in a tiger skin gazing at the light of the stars behind the sun because Einstein had explained 'very far away ... in his house' that this light bent. In her drawing, Einstein was tossing calculations out of the window, while a squirrel skipped by, birds abandoned their nests and people excitedly ran towards his house, where he muttered: 'Oh well—I know it—I did calculate it'.[19] In his reply, Einstein wrote that her drawing was 'inimitably prettier than all that cackling by the startled flock of newspaper geese'.[20]

Had Galinka inherited her talent for drawing from her mother, who had designed their house with such pleasure and once wrote that she would rather have been an architect?[21] Or did she get it from her father too? Years later, Paul would compare Einstein to the painter Holbein, who portrayed his subjects with clarity on a well-composed canvas. Bohr, he felt, was more like Rembrandt, who illuminated a single spot brilliantly against a dark background.[22] These comparisons didn't spring out of thin air. Others also noticed how viewing art could put Paul 'in a gentle mood'. Swedish theoretical physicist Oskar Klein, for instance, remarked: 'I remember with what quiet intensity he—who else would discuss without intermission—on a sunladen day of early spring, some months after the death of Lorentz, would show me Buddhas and Chinese landscapes in the ethnographic museum of Leiden'.[23]

But Paul was no painter, even if he did have a keen eye for art. And the same was true of his work in physics. He was astute, driven, witty, charismatic, stimulating, observant and an associative thinker but not a 'creator' like Einstein and Bohr. Paul had a knack for identifying significant work in physics and could accurately place it within specific theoretical movements. He could pinpoint its strengths and weaknesses flawlessly. He had the gift of inspiring a wide circle of enthusiasts for new work, and with his humour, acuity and insight into human nature, he could bring even the most independently working individuals together. And although he quietly—and sometimes not so quietly—regretted never producing anything truly spectacular, with these qualities, all the overnight stays and blackboard sessions, he nevertheless made the house on Witte Rozenstraat flourish as an international 'physics gallery', over which he presided as its charismatic 'curator'.

Even Paul's old dream of Zurich as a phenomenal paradise for physics returned, albeit in a different guise. Paul hadn't succeeded in securing a position alongside Einstein in Zurich, so what if he could now bring Einstein to Leiden? After all, Einstein had openly expressed his irritation at Prussian arrogance. Moreover, the collapse of the German mark made it increasingly difficult for him to provide Marić's alimony in Zurich, on top of running his household in Berlin with Elsa. And not unimportantly, he always enjoyed his time in Leiden immensely.

After sharing these considerations with his colleagues in Leiden, Paul immediately sent Einstein an enthusiastic letter in early September, pre-empting all the practical obstacles that could arise between dream and reality. Come to us, is what it came down to. The Dutch guilder was a strong currency. 'You'll get as much remuneration as is necessary for you, your children, and whatever else—you have to calculate that yourself. (Our remuneration's maximum of 7500 guilders is your minimum)'.[24] No one in Leiden would impose strict demands on him: 'Absolutely no lecturing duties will be imposed on you—the only real obligation will probably be to choose Leyden (or its surroundings) as your principal place of residence'. As for the language, 'you will learn whatever Dutch you need for practical living in a fortnight'.[25]

To add strength to this proposal, Paul listed the Leiden physicists who would love to exchange views with Einstein. 'Lorentz, de Sitter, me, Kuenen, Droste, de Haas, his wife [Berta de Haas-Lorentz], Fokker, Burgers, Julius, Zeeman, and [you] will regularly see intelligent and impressive young fellows of pleasant personalities. Moreover, guests like Nordström [who had worked on an alternative theory of gravitation and had joined De Sitter, Lorentz, and Paul in examining Einstein's work in Leiden in 1916], Bohr, etc'. In short, Paul exhorted Einstein to 'consider that here you would join a group of people who are very fond of you personally and not just of the brain power you exude!'[26]

But once again, his dream didn't materialise. Einstein had indeed been annoyed by the 'megalomania and the voracious appetite for power that the economic boom had formerly brought' and which his German colleagues hadn't relinquished after the war, as he had written to Lorentz.[27] Nevertheless, his colleagues in Berlin still showed him 'much warmheartedness' and not only those simply wanting to 'lap up the droplets I sweat from my brainy brow'.[28] Above all, he wanted to be loyal to Planck. The esteemed German scientist, recently awarded the Nobel Prize, paid him a generous salary despite a tight budget and cutbacks. Furthermore, Planck had expressed some regret for signing the deplorable manifesto denying German responsibility for the war.

Yet, Paul's dream wasn't entirely lost thanks to enthusiastic Onnes, diplomatic Lorentz and jurist and social scientist Cornelis van Vollenhoven, who presented the Leiden University administration with a plan that took greater account of Einstein's position in Berlin (Berends and Van Delft 2019, 402–403). They proposed offering Einstein a visiting professorship, an idea that fitted well with a potential role of the Netherlands as an international peace negotiator. Already in 1898, Tsar Nicholas II had given an impetus to this role when, to general astonishment, he invited the ambassadors of twenty-six countries to an international conference on peace and disarmament, to be held in The Hague. The meeting between the representatives of all those countries at the Dutch seat of government led, a year later, to the establishment of a Permanent Court of Arbitration, where states could submit disputes for resolution. And since the Netherlands had been neutral during the war, Van Vollenhoven and like-minded people believed the country could now expand this role. This would also go some way to repairing Dutch prestige, which had diminished as the country declined as a colonial power.

International science, which strengthened ties between peoples and nations, aligned with this role, Van Vollenhoven and Lorentz had previously argued already (Otterspeer and Schuller tot Peursum-Meijer 1997, 52–58). And Einstein the pacifist, who worked in Germany but wasn't just German, perfectly embodied this spirit of international science. Moreover, 'connecting such a light to our Leiden Alma Mater would greatly increase her glory', as Onnes remarked (Berends and Van Delft 2019, 403). With support from the university administrators in Leiden, Lorentz, could indeed offer Einstein a visiting professorship in Leiden, four days before Christmas. Diplomatically, Lorentz underscored that he wanted this offer in no way to conflict with Einstein's position in Berlin. The only requirement was that he come 'to Leyden once or twice a year for a few weeks' for an annual salary of 2000 guilders.[29]

Lorentz had of course always wanted to have Einstein in Leiden. He hoped Einstein's fame would reflect positively on the city. 'So long as one can say: "Einstein is at Leyden—in Leiden there's Einstein"', as Paul had put it in an earlier letter to Einstein.[30] In earlier letters, he had also kept Einstein informed about the discussions between Lorentz, Onnes and university administrators. 'Oh, it is splendid how cheerfully everyone is working together on this affair, but foremost Papa Lorentz and Uncle Onnes and Brother Kuenen', he disclosed in early December. 'And every step of the way we'll emphasise that there can be no question of any rivalry with your Berlin position. ... Forgive me for constantly harping on this matter instead of waiting until everything is absolutely set in stone. But you understand that I am just enthusiastic about it, like a child—and that it's hard to keep my mouth shut.'[31]

And Tanya? '[But] perhaps not even I have been revived quite as much by your stay as my wife!!!' Paul had written to Einstein at the end of November.[32] Tanya was tired. Sleepless nights and caring for Wassik, as they usually called little Wassily, weighed heavily on the 43-year-old. The boy was developing more slowly than his siblings. He had babbled 'Mama' in the summer, clapped his hands in the autumn and managed to sit up by himself and pull himself up on the edge of the playpen just after Einstein's visit—'which is very difficult because the playpen is smooth and high, reaching up to his mouth.'[33] Tanya still had to get up several times every night for him.

Did Paul and Tanya discern in Wassik's face the features of 'Mongolian idiocy', as Down's syndrome was called? At the time, doctors speculated that 'mongoloid' features indicated that a child had stalled in early foetal development. Some thought the condition was caused by the mother's age or exhaustion, especially since it often affected the youngest child in larger families. Others suggested the condition was associated with other mental defects in a family (Halbertsma 1922). But these were hardly topics for open discussion.

It must have weighed on Tanya, but she wasn't one to sit back helplessly. With Paul, she debated the relevance of the number of spatial dimensions—three, four or more—in physics. She contemplated economics and the organisation of the state. And she oversaw the children's home education. Yet in this same house—without a study of her own, without an income, without colleagues, without all those informal connections and conversations that help a person progress and with all her worries—she must have often felt despondent. Perhaps it was a telling sign that, in a letter to the *Leidsch Dagblad* in early 1919, she not only defended striking bakers but also referred to herself as a 'housewife'. A relationship between a housewife and a professor was never what Paul and she had envisioned.[34]

For the first time, their roles seemed reversed, with Paul more optimistic than Tanya. Thanks to Bohr, he had rediscovered his pleasure in physics, and Einstein's visit had energised him. He gave presentations at the Netherlands Academy of Arts and Sciences on topics like how 'does it become manifest in the fundamental laws of physics that space has three dimensions?' (Ehrenfest 1918b, Ehrenfest 1920). And his joy was complete when Einstein accepted Lorentz's offer to become a special professor in Leiden in early 1920. Einstein too, was grateful for the offer that would give him 'the pleasure of keeping fresh the comforting bonds of friendship linking me with my colleagues ... in Holland'. Although in his letter of acceptance to Lorentz, he also wrote that 'I blush, though, at the reasons behind your fine offer. I do know that kind destiny allowed me to find a couple of nice ideas after many years of feverish labour. Nevertheless, unlike you, nature has not bestowed me with the ability to deliver lectures and dispense original ideas virtually effortlessly as meets your refined and versatile mind'.[35] Earlier he had written something similar to Paul. But Einstein's lectures and teaching were probably a concern Paul thought they could deal with later.

In the same cheerful mood, Paul described his life in a long letter in early July 1920 to his old friend and colleague Karl Baumgart, who received it more than a month later in Petrograd after a circuitous passage through the Bolshevik bureaucracy. 'I cannot tell you how happy [your letter] made me, and not just me—it was a "Hello" to the entire Physics Institute and the Polytechnic', his Russian colleague replied at once. 'The letter was read aloud at the Optics Institute and the X-ray Institute. It truly seemed to me that I could hear your voice. ... Now I hasten to answer all 45 of your questions quickly'.[36]

In his lengthy reply, Baumgart soon lost track of these 45 points. '1) Your wife's mother lives in Moscow, is healthy … and has been informed of your letter by Fräulein Föhringer. 2) Ioffe, professor at the polytechnic, director of the X-ray Institute and member of the Academy of Sciences is writing to you himself. 3) Fräulein Föhringer is a librarian … and an assistant at the university (the higher women's courses and the university have now merged) … gives lectures at the Pedagogic Institute and the university on natural sciences history.— Krutkov is making great progress. … He sent you earlier work. Have you received it?—Rozhdestvensky, professor at the university and director of the Optics Institute, is working very hard. … Bursian is working diligently. … Khvolson is writing to you himself. Has published work here on the atomic nucleus. … Friedmann and Tamarkin are writing themselves'.[37]

And on it went, from one name to the next. Baumgart had lost contact with some; others had emigrated. A few were dead, drowned or missing, but many on his list were still putting energy into their research in Petrograd, 'even though it is difficult to work at all'. They were desperate especially for English and German books and journals, which hadn't arrived in Petrograd since the 1917 revolution. 'Since early 1917, we have only received foreign news by chance.... We learnt about the eclipse and relativity from the newspapers, without details'.[38] Could Paul send articles? 'We are particularly interested in: everything achieved in optics, both experimental and theoretical, the work of Paschen and his school .... Lorentz's work. Everything about quanta. Your work and that of Burgers. New work by Bohr. ... Everything Sommerfeld has accomplished. Sommerfeld's book *Atomic Structure and Spectral Lines*, if possible'.[39]

The letter continued, as if Baumgart kept remembering more things, until he finally concluded: 'I am so glad that you are happy in Leiden, can live such an intense scientific life and [that you] are having success and are getting recognition from truly competent people. ... How I wish I could be with you'.[40]

25

Physics at the highest level

Just not for women

Only much later did Paul and Tanya learn that stern Khvolson had ensured that lectures and examinations in Petrograd continued as much as possible, revolution or not. Not until then had they understood that their friend Ioffe had assumed the Bolsheviks' popularity would be temporary because they promised peace (Frenkel 2001, 18). It was also only then that they heard Baumgart had delivered well-attended public lectures for the Free Association for the Development and Dissemination of Positive Science during the revolutionary year.[1] And that mathematics professor Krylov and optics expert Rozhdestvensky participated in this association, which aimed to combat poverty and illiteracy by sharing knowledge. As was Pavlov, whose dogs they had heard barking on Aptekarsky Island (Josephson 1991, 42).

How many of these ideals had dissipated in vodka and gunpowder smoke by then? In the Netherlands, the *Nieuwe Rotterdamsche Courant* quoted in early 1920 a reporter from *The Times* who had travelled undercover in Russia. He told of atrocities, deprivation, unbearable living conditions, accusations followed by show trials and death sentences and a reign of terror as the government's only tool were the order of the day everywhere in Russia.[2] Even Dutch people who viewed the revolution idealistically couldn't ignore such reports.

The greatest revolution in the placid Netherlands had been the introduction of women's suffrage in 1919. This was enabled by progressive liberals such as Philip Kohnstamm but most of all by the unwavering efforts of steadfast physician and feminist Aletta Jacobs (Steen 2007/2008). 'Men shall have to descend from their throne as "lord of creation" and henceforth be satisfied with a place beside women, who, as political equals, as friends and as collaborators in all fields, shall stand by their side', Jacobs declared during the victory rally at the Concertgebouw in Amsterdam in September. 'Thrones are out of fashion in our time. We have repeatedly seen their holders willingly abdicate in recent years' (Jacobs 1919).[3]

But Jacobs' conviction that the 'political liberation of women [would] exert a beneficial influence' on men and 'their character development' proved to be as lofty as the Bolsheviks' socialist ideals.[4] With the perspective of a Russian and an outsider, Tanya saw that women's voices were almost invariably reduced to whispers. It was often unwritten rules which restricted them. Berta De Haas-Lorentz, Johanna Reudler and Eva Bruins, who earned their doctorates under Lorentz between 1912 and 1918, had all conformed to such rules and devoted themselves as 'housewives' to their husbands and children. Nel van Leeuwen left the field already before completing her doctoral research under Kamerlingh Onnes' supervision, because she married the gravitation specialist Nordström. She soon became a housewife and mother for a short while in the Netherlands and then in Finland where she had two more children (Blaauboer and Van der Heijden 2025).

Or in other words, outside the country, conditions weren't necessarily much better. Most successful was Marie Skłodowska-Curie, who won two Nobel Prizes and was the only woman invited to the Solvay Conferences in Brussels in 1911 and 1913. In 1911, she had also visited Leiden, where she had studied the effect of extreme cold on radioactivity in Onnes' cryogenics lab (finding no effect) (Van Delft and Kes 2016). Three years later, at the start of World War I, she equipped twenty light trucks with X-ray machines to take X-rays of soldiers wounded by shrapnel and bullets at the front. Skłodowska-Curie manned one of the 'petites curies', as they were known, as did her seventeen-year-old daughter Irène. Mother and daughter Curie also trained at least 150 other women in Paris to examine soldiers at 150 stationary X-ray stations around the country. After the war, all this led to Skłodowska-Curie becoming the first woman to have her own research institute, the Radium Institute in Paris, where over 30 per cent of the researchers were women (Dumancic and Enger 2024).[5] Still, even Skłodowska-Curie faced prejudice: When her love affair with married Paul Langevin became public in 1911, major French newspapers launched a smear campaign against her, and stones were thrown through the windows of her apartment. And for all her success, she would never be admitted to France's highest scientific forum, the Académie des Science.[6]

In Berlin, Lise Meitner was the exception that proved the rule. Initially, she had worked unpaid in the basement of the Kaiser Wilhelm Institute for Chemistry, with her sole source of income a monthly allowance from her father. As Planck's assistant, she later marked student work so she could continue her increasingly important, yet still unpaid, research on radioactive isotopes alongside Otto Hahn. Her situation improved in 1917 when she was asked to establish and head the institute's new Radiophysics Subdepartment, with Hahn heading its Radiochemistry Subdepartment. And at the age of 39, she was appointed the first woman physics professor of Germany in 1926 (Lewin Sime 1996, 79, 110). Notwithstanding, she and the handful of other women at the science faculties in Berlin stood no chance of securing a place in the highest bodies of science.

Most disadvantaged was formidable mathematician Emmy Noether. In 1915, she demonstrated the connection between certain symmetries in physical theories and conservation laws with what is known as Noether's theorem. It also upheld the crucial 'conservation of energy' in Einstein's theory of relativity, providing the theory with a solid foundation and enhancing its credibility (Noether 1918; Neuenschwander 2017). This was just one aspect of her prolific work. Noether, who took students 'like chicks under the wings of a mother hen', shared her ideas so generously that according to many, it is likely that half of them were published by others.⁷ Yet for a long time in Göttingen, she was only allowed to teach under Hilbert's name. Only in 1919 did the Georg August University of Göttingen grant Noether the right to lecture under her own name, although even then only as an unsalaried Privatdozent (Noether 2005).

The small community of mathematicians and physicists in the Netherlands hadn't produced a single woman of this calibre. Combativeness wasn't in Dutch female students' culture. Certainly not in Leiden, where most professors' wives regarded Aletta Jacobs with some suspicion, perceiving her as overly revolutionary, and where social conventions constricted everyone like a tight corset. It meant the meetings of the Association of Women Students Leiden, founded in 1900, were calm affairs. After one of the association's annual, festive meetings, not long before Tanya's arrival in Leiden, for example, the professors' wives noted with satisfaction that everything had 'gone off without a hitch', thanks in no small part to the 'petite string ensemble music' and the 'first-year students entrusted with serving tea and biscuits' (Steen 2013b, 11).⁸

Tanya was fortunate to be able to discuss thermodynamics with Berta de Haas-Lorentz and Philip Kohnstamm, a subject which had occupied her since the article in the *Encyclopaedia*.⁹ She had less contact with Jo van Leeuwen, who received her doctorate under Lorentz in 1919 for her research into the origin of magnetism. Van Leeuwen was the last of Lorentz's four women doctoral students, and also the only one to stay in the field. In 1920, she became an assistant at the Technical University of Delft, where she led the practicals for physics and electrical engineering students (Blaauboer and Van der Heijden 2025). From Delft, she also attended the colloquia in the Witte Rozenstraat from time to time. But her work was too far removed from Tanya's interests, and her relationship with Paul was rather distant. Whenever she attended his Wednesday evening colloquium, he would always greet her somewhat formally as 'Miss Van Leeuwen'. A woman wishing to pursue a career in public service in the Netherlands had to be unmarried—and therefore 'Miss'—until her death or retirement.

Despite the presence of several women students, the world of mathematics and physics thus remained a male bastion, although even men were less inclined to choose the rarefied field of physics shortly after the war. Nevertheless, two students chose this field in 1919. One was George Uhlenbeck, who, after a year at the TU Delft, had decided that he would rather delve into theory. The other was seventeen-year-old Sam Goudsmit. Both were mostly diffident railway commuters, travelling between Leiden and their hometown, The Hague. Uhlenbeck's father, a retired and penniless major, had settled in the city with his family after returning from the Dutch East Indies. Goudsmit's parents both had businesses in The Hague—his mother sold hats; his father sold bathroom products: their son Sam was the first in the family to go to university. And like Uhlenbeck, in Leiden he didn't, by his own admission, belong to the group of 'law students [who] were always rich, ... used to buy their lunch [in a cafeteria]', and who, like their relatives, wanted 'to go into business' (Goudsmit 1963, session I).

They hardly saw Paul in this period. He only tutored graduate students[10] and even then 'only two courses. The one year it was Maxwell's theory, which always ended up with a little bit of special relativity; the other year he gave statistical mechanics, which ended up with a little quantum theory and, atomic structure. And that was essentially all that he taught. It was four hours a week'. He left first-year and second-year physics lectures to his colleague Kuenen (Uhlenbeck 1962, session II). And to inspire students to engage with the subject independently, Paul had established a sophisticated system by this time (see Figure 25.1).

A crucial pillar in the system was the Christiaan Huygens Study Association, where about twenty astronomy, chemistry and physics graduate students met every other week in one of the student rooms. At the portable blackboard, one of them would open the evening with 'a little talk'; then there was usually a second little talk, the main talk, a little one, and then what you call improvisations and so on', Uhlenbeck later recalled. In the early 1920s, Huygens members also helped

Figure 25.1 Paul Ehrenfest at the blackboard, c. 1925.—*Photograph Archive P. Ehrenfest Jr., Noord-Hollands Archief, Haarlem.*

set up the Leiden Jar for younger students. In this club, older Huygens members would delve deeper into 'aspects of the lectures or questions about the lectures' with younger undergraduate students during Friday lunchtimes. There was 'a hierarchy, you see, and the best of the Leiden Jar became members of Huygens; that was the scheme' (Uhlenbeck 1963, Session V). The students propelled each other forward, while Paul occasionally dropped in.

The 'highest' achievement was partaking in the Wednesday evening colloquia with 'tea beforehand', attended by researchers and professors, sometimes from abroad. The meetings in the early 1920s took place 'every Wednesday, religiously', according to Uhlenbeck, and it was Paul, in particular, who started discussions. 'He always asked questions, and he did it really marvellously, because he often simply interrupted the speaker, and especially one of the younger ones. He said "Now you stand aside. Now I will talk". And then he summarised it, you see. I know it has happened with me and certainly with many people, that at that time you understand it all!' (Uhlenbeck 1962, Session III).

While Uhlenbeck, Goudsmit and, from 1920, talented Gerhard Dieke took their first steps in this system, Paul received his first personal invitation to a Solvay Conference. The meeting in the first week of April 1921 was the first after the war. German scholars weren't yet welcome. Only Einstein, as a 'non-German German', had been invited, but he was on his way to the United States with Chaim Weizmann to raise funds for the Hebrew University of Jerusalem (CPAE 12, Introduction). And while thousands of people cheered for him on the quay when he arrived in New York aboard the steamship *Rotterdam* on 2 April, Lorentz, Kamerlingh Onnes, Langevin, Marie Curie, Rutherford and their non-German colleagues gathered almost as they had of old beneath the milk-glass chandeliers of Hotel Britannique in Brussels.

The subject of the conference was 'Atoms and Electrons', which Planck, Bohr, Einstein, Sommerfeld, Kramers and Paul had been pondering in recent years. How were atoms structured, and what did the 'granular'—discrete—behaviour of electrons within those atoms mean? It seemed a more pressing issue than the relativity theory for which Einstein was being celebrated. In fact, Einstein's theory hadn't gained much traction in Leiden. Lorentz, the first person in the world to have lectured on general relativity in 1916, was mostly occupied in Haarlem with the Solvay Conferences and the Physics Cabinet. His former doctoral student, Fokker, who had also published on relativity, focused on other, more applied work in Delft. De Sitter had made Einstein's work accessible to English-speaking scientists during the war years and had shown in passing that his equations described the evolution of the cosmos. He had even demonstrated that these equations permitted the existence of an imaginary cosmos without stars and planets—a 'toy cosmos'—which could perhaps offer deeper insight into the fate of our own universe. But he had become so absorbed in his duties as the new director of the

Leiden Observatory that he seemed to have 'forgotten' his earlier groundbreaking ideas about relativity and the universe (Guichelaar 2016). Finally, Nordström, who became an expert on gravitation during his stay in Leiden, had returned to Finland, taking in tow his alternative idea for describing gravity with extra dimensions (Halpern 2004).

Paul, who in his fine article in *Annalen der Physik* had argued that both the solar system and atoms could be stable only in three spatial dimensions according to Bohr's model, maintained his interest in the exploration of spatial dimensions, as did Tanya (Halpern 2004, 2007). But the relativity theory itself was, Uhlenbeck later remarked, 'on the horizon of our attention. The thing which we had to work on, if there was anything, was statistical mechanics and quantum theory' (Uhlenbeck 1962, Session III).

Paul also enthusiastically immersed himself in quantum theory in spring 1921, partly because he unexpectedly replaced Bohr at the Solvay Conference in April. Setting up and organising a new institute for theoretical physics in Copenhagen had exhausted the Dane. He had written about this to Lorentz and Paul at the end of 1920, while Kramers also reported that Bohr was close to being overworked. 'Cast away everything that is not essential!!!! ... If, for example, you regret having accepted the Solvay invitation just calmly retract your acceptance. No one will understand your action better than Lorentz himself', Paul wrote, trying to reassure him (Klein 2010, 321). However, Bohr, who weeks later in his reply called Paul's friendship 'one of the greatest strokes of luck' in his life, couldn't bring himself to cancel (Klein 2010, 322). And as so often, Lorentz found another solution, suggesting to Paul: why not offer Bohr your help with the preparations?

It proved to be an excellent move. Bohr had brilliant ideas but found it incredibly difficult to express them clearly on paper. Paul, perhaps the wittiest and most insightful speaker in theoretical physics at the time, again had lacked a solid idea for some time. Now this collaboration would give Bohr some rest and himself a mission. All their other colleagues would also be pleased because Bohr's long-winded explanations, full of meticulous details, would finally be comprehensible once Paul laid out the framework with surgical precision. And although Paul wrote somewhat affectedly to Einstein that Lorentz had 'commissioned me to write on behalf of Bohr the Bohr-atom-presentation for Brussels. Result: suicidal thoughts', the collaboration and the prospect of seeing Bohr in Brussels gave him great joy in January 1921.[11]

Tension mounted when first Einstein cancelled in February, followed by an overworked Bohr at the end of March. Just half of Bohr's contribution had been sent to participants in preparation. Would Paul present the other half in Brussels, Bohr asked. Paul could do little other than encourage Bohr once more—this time to send as much information as quickly as possible. 'Send 100 letters if necessary, one after another, sending a letter off whenever you have written down one or two remarks. There is no harm done if I get some of the remarks after that report has been given!!! Just write freely and with pleasure!! Also send any quite spontaneous remarks that you feel may be useful to the Congress. I will always begin by saying "assuming I have correctly understood Bohr and Kramers", so you will be fully covered' (Klein 2010, 324). Yet just a single letter arrived on 28 March. And with this long letter filled with additions to what Bohr had previously discussed with Paul, the latter had to make do in Brussels.

It wasn't the first time Paul presented someone else's work. Eleven years earlier in Moscow, he had brought Rozhdestvensky's work into the limelight and moved his audience to tears (see Ch. 11). This time he had the good fortune of being able to speak in German. Paul was in any case masterly at initiating and leading discussions. It helped, moreover, that the discussions were about the correspondence principle, which was loosely based on his own earlier work.

Once again, Paul made an impression. The 1921 conference was defined by the correspondence theory. 'Bohr ... was unable to come, and Ehrenfest was in continual contact with him', French theorist Langevin recalled twelve years later. 'I remember those days when Ehrenfest arrived after receiving a letter from Bohr and brought us the echo of the thoughts being developed at that very moment in Copenhagen. He brought a joy and an ardour to that task that deeply moved us all' (Klein 2010, 325).

Paul's zeal extended to Bohr after the conference, as he continued to inundate him with good advice. Bohr certainly shouldn't worry about the text on his contribution for the Solvay Conference proceedings. It had to 'contain some of Bohr', but 'all of Bohr' really wasn't necessary, he wrote. For the future of physics, it was 'much more important' that Bohr would be 'not only completely healthy but also happy and free of cares'. Bohr surely understood that '100 years from now, people will not care whether one of your articles was wrong, but they would indeed if at the age of almost 36 you were threatened with a breakdown. (And then you wouldn't be blamed, but those who harass you!!!!!!!!!!!)'. And neither should he be worried about disappointing Lorentz for at most he would react 'the way a father does when his very dear little boy tears his pants while he is climbing: he makes an "angry face" while stifling a laugh!!!!! And you are still very far even from that with Lorentz!' (Klein 2010, 326-327).

Paul strung together many more recommendations and explanation marks. Couldn't Bohr declare bankruptcy? Couldn't he appoint a business minder who would send away anyone who burdened him with unimportant issues? And so on and so forth, until Bohr decided to submit only an improved version of the first half of his report. Lorentz then asked Paul to produce a supplementary report on the discussion on the correspondence principle. And many colleagues were pleased that he was the author, as it was concise and limited to the main points (Ehrenfest 1921a).

The short trip brought Paul recognition and some fresh air. Paul's passport had expired during the war—as had Tanya's—making most travel impossible, while the naturalisation process in the Netherlands was excruciatingly slow.[12] 'If only you could at least make it to Brussels', Einstein had written to his stranded friend who led a 'cage-life' in the Netherlands in a letter in early 1921.[13] Einstein knew from experience how slow Dutch bureaucracy could be. His appointment as a visiting professor in 1920 had been delayed by months on account of Dutch civil servants confusing him with a Communist namesake (Van Dongen 2012). Only in autumn 1920, during another stay at Witte Rozenstraat, had he been able to deliver his inaugural speech (Einstein 1920).[14]

Perhaps it was because of this stress about red tape that Paul had shown some irritation in a letter to Einstein in early spring 1921 when Einstein had just paid a lightning visit to Amsterdam. In the Dutch capital, he had spoken in support of making international contacts with peaceable human-rights organisations on behalf of the pacifist German League for Human Rights—something Paul didn't quite agree with. 'Your trip to Amsterdam depressed me very, very much', he wrote. Did Einstein really have to make this caper, just after Lorentz and Onnes had smoothed over the whole situation with his Communist namesake? 'If you do have to raise dust in all n−1 countries of the world, do at least stay good as a newborn baby in our nth little land.'[15] But 'my short tour to Amsterdam was extremely harmless, you scaredy-cat', was Einstein's riposte.[16]

Not long after the Solvay Conference, with Einstein still in the United States, Paul travelled to Berlin, again with special permission for scientific purposes. He met Planck and Nernst as well as Ioffe, who had travelled to the German capital from Petrograd. From Ilse and Margot Einstein, he then collected one of the two violins Einstein had ordered for him in December 1919 from his own violinmaker, 'a craftsman in violinmaking' whose instruments he was 'very enchanted with'.[18] The Hupfeld grand piano which Einstein had also bought for Paul—and had depleted 'virtually my entire cash flow'—had been in the study at Witte Rozenstraat since January 1920. And the other violin, for Galinka, Einstein himself had brought to Leiden. With the second violin, for Tanitchka, Paul slowly made his way back to the Netherlands, accompanied by Van Aardenne. His former summer student had become a family friend. He lent Paul a hand when the heating system broke down, assisted him when he had to fill in forms for the bank or a notary and was one of the few people who could get Tanitchka, a surly teenager, to laugh. Now they added several weeks to their journey: before actually returning to Leiden, they called in in Göttingen, where Paul met even more of his former colleagues. Their friendship was strong enough to withstand something like this. What is more, Lorentz appeared to find Einstein's trip as unproblematic as his fund-raising trip to 'Dollardia', as Paul called it,[17] which had obliged him to cancel not only the Solvay Conference but also his trip to Leiden in spring that year.

Tanya, meanwhile, was permanently in Leiden. She was there early in the Dutch spring when there are often four seasons in one day; she was there in summer, and she was there when autumn brought new rains. She was also there when, at the invitation of Bohr, Paul visited Copenhagen at the end of 1921. While acquaintances and friends dropped in, Galinka painted dreamy scenes in oil paints, Pawlik built castles and houses out of matchboxes and Tanya endlessly followed Wassik, who stumbled through the rooms in his wooden walker and called for his mother at night.

'I'm in Copenhagen with Tanitchka and Van Aardenne since 3 weeks (T' and I are living with the Bohrs)—how wonderful it is. Everything', Paul wrote from the Danish capital to 'Dear Einstein' in the United States. 'He's a mighty physicist—I feel like I'm a joke gone wrong whenever I open my mouth in his presence. ... Oh, if only I could witness both of you together (and if possible with Ioffe there!)'.[19]

He wrote how 'little T' has gone away with van Ardenne to the mountain snow' in Norway. And 'I haven't felt so happy as in Bohr's house in a long time—he has such a wonderfully kind wife ... three very dear little boys and a quite special mother. ... If my poor wife weren't so miserably stuck in the dumps, I'd now be in bliss and want nothing better than to be Bohr's assistant'.[20]

26
Einstein, Wassik and Russians
Concerns far and wide

It was kind of the Einsteins. In mid-September, they hosted Tanya and Tanitchka for a week and took them to a concert by Rudolf Serkin, a Russian-Jewish pianist.[1] Sirkin had become part of the circle around violinist, conductor and composer Adolf Busch, a friend of the Einsteins whom Paul had also come to know.[2] On a postcard, Paul had urged Tanya to arrange a meeting with Busch directly upon arrival in Berlin. He gave detailed instructions for the trip to his house, pressed her not to come late under any circumstances and to bring a box of chocolates for Busch's daughter.[3]

Did he hope this would cheer up Tanya? Just as he hoped, Margot and Ilse Einstein—who sang, did needlework and were cheerful—would be an example to Tanitchka? The girl was extremely clever but also closed and sullen. 'Why isn't there another girl like Tanitchka! If only she wasn't so cross all the time, but perhaps that has something to do with it—if she weren't so cross, she couldn't paint so well. Now she is always cross when she paints because she wants to do everything on her own, in her own way', her younger sister Galinka once observed.[4]

Tanya was downcast in Berlin, despite the delight of travelling again. She, Paul and the children had finally become Dutch citizens that spring.[5] But this first trip without Paul wasn't for pleasure: Tanya and Tanitchka brought little Wassik to the eastern German city of Jena on 4 September. For more than a year, he had been on the waiting list for a place at an educational institution in the Netherlands, but chances of him securing a placement were slim. For this reason, she had travelled with the four-year-old to Germany: he would be cared for at the Trüper Institute in Jena from now on. He was one of the youngest of over one-hundred children, who learned a little, did chores and craftwork, ate, slept and hopefully were well looked after in the spacious buildings.

'I often think of Wapsili', Paul wrote a few days later to Tanya in Berlin.[6] A few weeks earlier, he had prepared Einstein for her visit and described the reason for it. 'At the energetic advice of doctors and especially also of my brother [Hugo], my wife finally decided—provided it is possible—to place Wassik in an institution. Everyone is unfortunately unanimous that he is a member of a very well-known type of idiot ("mongoloid") that between the ages of five and fifteen still possesses some capacity for development if appropriately treated. In favourable cases the final state then corresponds to the intelligence of a normal five-year old child'.[7] (See Figure 26.1)

Would Wassik really never learn what his father and mother found so important: clear and logical reasoning? Paul had resolved to follow the advice of his brother Hugo, the gynaecologist in Saint Louis, who he had seen in Vienna that summer, when the five Ehrenfest brothers had met for the first time in almost twenty-five years.[8] On his way back to Leiden, Paul stopped by at the clinic in Jena. 'Darling, dearest dearest!' he scrawled on a postcard to Tanya on 23 August 1922.[9] 'I've just come back from the educational institution—was there for three hours—very favourable impression of the people and good impression of the facilities which are important for Wapsili'.

The Trüper Institution was renowned. Johannes Trüper had worked his way up from being a teacher to a medical doctor with a doctoral degree and then allegedly became the first person in the world to bring doctors, psychiatrists and educationalists together to examine and supervise children with behavioural and developmental issues. At his institution, children—they were called 'psychopaths'—were taught and supervised at their own level. Particularly in the years preceding the great war, doctors and educationalists from far and wide came to admire the beautiful buildings, large garden and modern methods through which these

Figure 26.1 Tanya with Wassik, probably in the summer before departing to Jena.—*Photograph Ehrenfest Family Archive.*

children—from across Europe and mainly from distinguished families—were nurtured (Gerhard et al. 2008).

Paul noted on a slip of paper that Wassik had been assigned to a group of five or six children of similar age, each with its own supervisor. If Tanya quickly registered Wassik on her passport, immediately booked the overnight train to Jena via Halle and packed clothes and things for the first few days—the institute would arrange the rest—she could bring him there as soon as 5 September. Paul explained his motive for hurrying along Wassik's move in the same letter to Einstein, a few weeks previously.[10] 'Whether it is yet too late to save my wife's health, I do not know'.[11]

Ilse and Margot, Einstein's daughters, then recommended Tanya and Tanitchka visit Hiddensee. The small Baltic island, fringed with sandy beaches and swaying marram grass off Germany's north coast, was a favoured retreat for artists and scientists from Berlin. Paul's colleague and future Nobel laureate Gustav Hertz had a holiday home there; writer Thomas Mann would later enjoy visiting the island, as did Sigmund Freud and famous actors and painters. 'Very clear sky, calm sea, as I like it. ... The evenings also glorious at sunset and at full moon', Tanya wrote a few days later to Paul.[12] And he responded as always with a litany of advice, as if Tanya might irrevocably fall apart at any moment. 'Stay very calmly there until 1 November and really rest well now for once. And don't be frugal with food! Once you are well-rested, you shall benefit from taking walks again'.[13]

The postcards sent by Tanya, who visited artists and studios on the island and enjoyed the beach with Tanitchka, suggested she was anything but fragile. Were Paul's concerns mainly an expression of his own sense of unease, disquiet and dissatisfaction? Whenever Tanya travelled there seemed to be an epidemic—cholera, the Spanish influenza, typhus—raging somewhere, a war about to break out or an economic crisis flaring up. And Tanya jeopardised her health, he felt, by taking too many walks and doing too much, while she also bought overpriced railway tickets. Her plan to take Tanitchka from the island back to Berlin unnerved him. Tanitchka, who finished secondary school early by passing state examinations at sixteen, wanted to take the train there to be in Leiden in time to take part in the Leiden Association for Female Students orientation days.[14]

But 'I absolutely fail to comprehend why you want to take T' to Berlin', Paul objected. A little further in the same letter: 'Why do you want to travel with her to Berlin at all costs? That is simply pure sentimentality!' And again: 'Don't travel with T' to Berlin ... Besides, there is a great risk of influenza'. Paul even devised an 'escape route' in case an Anglo-Turkish war broke out. 'Don't travel via Berlin but come directly: Stralsund, Rostock, Hamburg, Bremen, Groningen ... (and, if necessary, leave behind some of the suitcases ...)'.[15]

Paul certainly found life in Leiden without Tanya difficult. 'Yesterday I was at the professors' club. It's appalling how methodically conversation is limited to chitchat there'. Unsurprisingly, it rained in Leiden 'in all conceivable ways and variations'.[16] And when his colleague Kuenen unexpectedly died of a stroke, Paul felt vulnerable.[17] 'If you still doubt whether it was right for me to take out a very large life-insurance policy then all your doubt would vanish if you knew how many troubles the Kuenen family have now', Tanya read a few days later.[18] Suddenly, Paul was burdened with added responsibilities. Just the day after Kuenen's death, he went to Delft to oversee examinations in his stead. How would this work when lectures resumed, and he had to care for the children too?

'I feel very clearly that you've had enough of playing the family man', Tanya wrote back.[19] 'Shall I come back at once?' But he would manage, Paul replied apologetically, allowing Tanya to stay on Hiddensee for another two weeks. Without Tanitchka, who had indeed returned to Leiden via Berlin.

Tanya had a second mission too. In Moscow, her mother had finally obtained all the visas and stamps needed to travel to the West. She was supposed to let Ilse and Margot know when she was on her way, and they would then notify Tanya on Hiddensee if necessary. But when Tanya arrived in Berlin two weeks later, her 'heroic' mother had already been there for several days. Self-reliant Katya Afanasyeva had stayed with acquaintances and spent her time in Berlin sorting out her travel documents for the Netherlands. She had 'aged', Tanya observed, but hadn't been 'worn out'. Tanya didn't even need to travel with her mother when she boarded the train to Amsterdam, armed with a passport brimming with stamps and a suitcase holding all her belongings. Her eldest granddaughter Tanitchka would pick her up.[20]

Tanya again visited Wassik in Jena, and Paul thought it ridiculously sentimental that Aunt Sonya wanted to accompany her. Did he fear they would bring the boy back to Leiden? They didn't. Tanya travelled on to Tübingen, where she called in on German experimental physicist Friedrich Paschen at Paul's request. 'Tell him with how much respect I've told you about him', Paul instructed her.[21] The week after she stayed with her Russian acquaintance, Alexander Yanitsky, who had left Kiev and now worked in Frankfurt at the 'Institute for the Physical Foundations of Medical Science'.[22] En route she tried to soothe incessantly grumbling Paul. 'Darling, don't take everything straight to heart. Not even the bills. Remember now, every professor receives bills. I very certainly want to try earning something myself. When Mama is here, that should—time-wise—be manageable'.[23]

Tanya had cared endlessly for Wassik. Countless nights she had risen for him. Hundreds of times, she had followed him around the garden on his wooden walker. She hadn't wanted to put him in an institution. But now that Wassik seemed to have ended up in a good place, and now that her mother was safely in Leiden, Tanya seemed to think about work again. Even the heating had finally been repaired, so the study was no longer constantly filled with coal dust. Perhaps like Kohnstamm, she could publish essays and articles for a broader audience, as she had suggested before.[24] The encounters, the sea and the sun had done her good, she wrote, after not 'having seen summer for ten years'.[25]

Did Paul realise how restrained Tanya often felt? He seemed preoccupied with his own career, their financial setback and repairing the damage to scientific relationships caused by World War I. Finally, international cooperation was possible again! Finally, he could spread his wings and do what he was so good at: bringing people together, initiating discussions and building a network. His friendships with Bohr and Einstein offered a wonderful starting point. It was also notable that Coster, who had obtained his doctorate with Paul in September of 1922, had taken the initiative to speak to Bohr about X-ray spectrums and what they revealed about the structure of heavy atoms. Not much later, Coster started research delving deeper into this question in Copenhagen (Hevesy 1962). And even if Paul's role in this was only indirect, it still represented yet another strand in the web he was weaving and with which he connected promising and well-established physicists to each other and to Leiden.

Connections with Russia also grew stronger as a measure of calm returned there and borders opened a little. The stream of Russian visitors to Leiden, halted by war and revolution, resumed. In early summer 1922, Paul and his colleagues strolled with Ioffe along the canals of Leiden and the tulip fields, and Ioffe bought piles of books and cases filled with measuring equipment. All of which greatly pleased Paul for as he had written to Einstein, Ioffe was 'one of the showpieces in my menagerie!!'[26]

In addition, George Breit, a physicist from Russia who had taken the American nationality, spent much of the year in Leiden on a grant from the Rockefeller Foundation.[27]

The phenomenon of foreign visitors and contacts was of course hardly new to Leiden. Lorentz had connections with physicists throughout Europe, especially since winning the Nobel Prize in 1902 and after his chairmanship of the first Solvay Conference in 1911. Kamerlingh Onnes' almost industrial cryogenic laboratory had hosted foreign visitors for years. Alexander Stoletov, a Russian expert in the field of electricity and magnetism Onnes met during his studies in Heidelberg, had been one of the first in 1895. When Stoletov brought along a colleague and then also sent a student, Onnes even learned a little Russian (Delft 2009, 322). Three years later, when Viennese Fritz Hasenöhrl investigated the electrical properties of oxygen in Onnes' laboratory, it was already famous across Europe. In 1896, the scientific magazine *Nature* ranked it as 'amongst the best provided (and, one may add, most productive) research laboratories' in the world.[28]

Ever since, visitors to the laboratory had regularly stayed at Onnes' beautiful villa, Huize ter Wetering, just outside the city. Their number grew when 'cold' became fashionable. Breweries, ice factories, the meat and dairy industry, hospitals, goods trains, steam ships: all needed cooling. This resulted in the creation of the International Institute of Refrigeration and the First International Congress of Refrigeration in Paris in 1908, which attracted an impressive 3000 experts.[29] Onnes' prominent role at this congress and the next one, in Vienna in 1910, profited him handsomely. Soon Pierre Weiss and Albert Perrier came from Zurich to study magnetism at low temperatures at his laboratory; Curie travelled to Leiden with a bag full of radioactive isotopes; father and son Becquerel researched uranium salts at low temperatures and Philipp Lenard from Heidelberg immersed himself in the study of rare earth metal compounds (Delft 2009, 323–324). Moreover, Onnes and industrialist Ernest Solvay immediately established a strong rapport during the Solvay Conference in 1911, resulting in a friendship as well as extra grants to keep Onnes' laboratory up to date and attractive (Delft 2009, 323).

Thus, Paul's—and Tanya's—'new tradition' of receiving theoretical physicists from far and wide at the lecture hall and at their home was in keeping not only with Lorentz's ideal of the Netherlands as an international mediator but also with the long tradition of hospitality at Onnes' laboratory and home. When Einstein's inaugural oration as a visiting professor on 27 October 1920 concurred—not by chance—with one of Onnes' congresses on refrigeration, attended by Frenchmen Paul Langevin and Pierre Weiss, newspapers in the Netherlands were enraptured. 'Leiden an International Centre of Science', was the headline of the *Algemeen Handelsblad*.[30]

As Van Vollenhoven and Lorentz had anticipated, Einstein in particular was a preeminent representative of this international science, for the broader public too. The absent-minded, violin-playing professor, whom Paul had first met ten years before in Prague, had become a celebrity. And while he had retreated to his study until halfway through World War I, he now regularly gave his opinion about subjects close to his heart, such as pacifism, democracy, human rights and Zionism. His trip to the United States to raise funds for the Hebrew University was an example of this, as was his earlier lightning visit to Amsterdam for the pacifist German League for Human Rights. And his endeavour, shortly before his visit to Paul and Tanya in early May, to bring about a cautious rapprochement between German and French scientists was extensively covered in Dutch newspapers.[31]

But not everyone in Europe shared this desire for good international relations. Einstein's Prussian colleagues were less impressed by his public appearances. His pleasure at this public interest caused resentment, and for some colleagues this displeasure extended to Einstein's work. Stark, whose work and experiments had once piqued Paul's interest in Göttingen, was one such person. Shortly after he had conducted Nobel-Prize-winning experiments in 1913, the physicists in Leiden had congratulated him on his results and invited him to deliver a lecture (Kox 2008, 386). But just over a month after Einstein's visit to Leiden, Stark of all people attacked the theory of relativity in a pamphlet entitled 'the current crisis in physics' (Stark 1922).

The pamphlet stemmed from a longstanding debate—or perhaps more accurately, a 'witch-hunt'—in which Stark was supported by fellow Nobel laureate Philipp Lenard. Both physicists referred to the positivist tradition, which insisted physicists use mathematical formulas to describe what can be counted, measured and weighed. They had expressed their concern about the other direction physics seemed headed: the abstract and counterintuitive theory of relativity and quantum theory. And during a large assembly and elsewhere, they had also attacked Einstein personally for his pacifism and democratic stance, which they interpreted as a lack of loyalty towards Germany. It contained an anti-Semitic malice, which Paul remembered from Vienna. He immediately wrote to Einstein when the latter responded vehemently in the *Berliner Tagesblatt* to a lecture by Lenard. Einstein shouldn't bring himself down to this level and should remain silent, Paul thought, who said he also spoke for Tanya (Klein 1970, 321–322).

It was a parlous time in Berlin, Einstein had already realised. The idealistic government in the post-war Weimar Republic hadn't succeeded in overcoming the destruction, disgrace and poverty the war had visited upon their country. Berlin had become a city where 'gaping disparities [are] painfully noticeable', a city stalked by 'terrible want and hunger', where 'the infant mortality is horrifying' and the state had 'sunk into extreme impotence'.[32] This was fertile ground for extremism. Ultraconservative men who pined for the old Germany—dominated by aristocrats, with precisely defined ranks, classes and rules and its militarist leanings—conspired with physicists like Stark and Lenard, who wanted to return to good, old physics.

'Yet the greater public and—politics—have long since taken possession of my theory and my person and have tried to make both somehow suit their purposes', was Einstein's summary of his situation in this sombre Berlin.[33] The true severity only became clear when right-wing extremists assassinated German foreign minister, Walther Rathenau, on 24 June 1922. Son of a Jewish businessman who had founded the industrial conglomerate AEG, Rathenau was an advocate of assimilation, a philosopher and a physicist. Not much later, it emerged that Einstein was also on the extremists' blacklist.

Yet even at this stage, Einstein refused to consider offers for full professorships elsewhere, such as in Italy, Spain and the United States. He did cancel lectures and other public appearances. And for a long time, he hesitated about the request by Curie and others to join the rather idealistic International Committee on Intellectual Cooperation of the League of Nations, which aimed to promote and strengthen international exchange between scientists. 'Not only on the occasion of Rathenau's tragic death, but also on other occasions I perceived that very strong anti-Semitism prevails among those I to some extent have to represent at the League of Nations', he wrote to Curie, 'and generally there is a mentality of a kind that makes me unsuited to be the representing and intermediary person.'[34] That he joined after all would have been particularly appreciated by Lorentz in the Netherlands. Paul and Tanya were far gladder that he had escaped Berlin for a long sojourn in Japan.

For all of this, nature continued to be its mysterious self. During Einstein's stay in Leiden in May, Paul and Einstein pored over German physicists Otto Stern and Walter Gerlach's baffling results. Stern, a theoretical physicist and former student of Einstein, had conceived the experiment. Brilliant experimentalist Gerlach had ensured its success, partly thanks to subsidies that they had wrangled from Einstein and invested in their mechanical workshop in Frankfurt. They built a measurement apparatus with an electric furnace that vaporised silver atoms, which were then directed one by one through an asymmetric magnetic field. The aim was to test Bohr's model on silver atoms. According to this model, 46 of the electrons orbiting the nucleus of the silver atoms should cancel out each other's effects and together form a constellation without properties. The outermost, 47th electron ought to define the character of the atom. As it orbited the nucleus, it should generate a weak current, which in turn should produce a magnetic field perpendicular to its plane of rotation. And because it could travel either clockwise or anticlockwise, the magnetic field could only point in two opposite directions with respect to the plane of rotation, as if this direction were quantised.

Gerlach and Stern's experiment demonstrated this quantisation. Or at least, that is what most physicists thought. In the well-chosen magnetic field, the vaporised 'magnetic' silver atoms lined up like iron filings in a magnetic field. Indeed, this magnetic field directed the atoms like a traffic controller in two neat rows. In one row, the 'silver magnets' pointed towards the 'north', and in the other line, they pointed in exactly the opposite direction.

But Einstein, along with Paul and incidentally Stern as well, once again 'looked farther than all the others'. The furnace, they reasoned, ejected silver atoms haphazardly, pointing in all directions just like iron filings outside a magnetic field are jumbled up. But how then did the atoms instantly line up in the magnetic field? It takes energy to align tiny magnets with such a field, and the only plausible way of extracting this energy from the surroundings, or this magnetic field, is by absorbing radiation. Paul had precisely calculated this in his notebook: such a process would take these silver 'magnets', which were extremely weak and little inclined to follow the field, about one-hundred years. A century! While Stern and Gerlach had seen them turn instantaneously (Unna and Sauer 2013, A16).

Only years later would it become clear that Bohr's 'old' quantum theory first needed a new footing before this could be understood. Or as quantum physicist Max Born put it in 1932: 'Stern and Gerlach's experiment is perhaps the most impressive evidence we have of the fundamental difference between classical and quantum mechanics'.[35] Their measurements showed something of the elusive nature of the quantum world, which began to reveal itself to physicists only in the 1930s. This quantum world is granular, erratic and can be interpreted in such a way that each measurement always selects a single outcome from a range of possibilities.

It was therefore hardly surprising that Paul continued to struggle with his and Einstein's conclusions. In Leiden, partly together with Breit, his guest at the time, he spent weeks trying in vain to think up alternative explanations for the phenomenon. 'If we cannot clarify it enough for the two of us to reach agreement, then it's better if we let it be', Einstein eventually wrote to him at the end of May.[36] Paul took the hint and wrote a short draft article, although he couldn't resist saying in a few final sentences that everything might perhaps be quite different to how they conjectured here (Einstein and Ehrenfest 1922). And Einstein didn't remove these sentences, even though he didn't understand their meaning (Unna and Sauer 2013, A17). 'It would be good if you could elaborate on this to appease my literary conscience.'[37]

Their friendship was a bright spot for Einstein in a dismal Berlin. Their shared love for physics was just one of the elements. Einstein could express himself freely about other affairs with Paul, and he could make music with him. How often in Leiden did he pick up one of the two violins he had ordered for Tanitchka and Galinka in Berlin, while Paul took his place at the grand piano Einstein had also arranged for a bargain thanks to the exchange rate. Incidentally, thanks to these same monetary conditions, Paul administered a bank account in the Netherlands where Einstein kept income from patents and royalties in strong guilders, and about which they corresponded in cheerful code language.

'As I already informed you, we now managed to raise the concentration of the Au ions up to 6.96×10^3, thanks to the coprocessing of a local preparation of 1.00×10^3 and one from Methu[en] 1.08×10^3', Paul wrote to Einstein in early February for instance.[38] 'I received the report by Villars just now that he manufactured for us a preparation of approx. 0.38×10^3, all in our units'. Thus, there were 6960 guilders on Einstein's Dutch account because Leiden had transferred him 1000 guilders and Methuen had paid 1080 guilders in royalties, while Gauthier-Villars had 380 guilders of royalties waiting. This 'high concentration of Au ions', Einstein wrote in his reply, 'is very welcome, considering that the state of our examination here makes such a high concentration appear very desirable'.[39]

Sometimes insecure, Paul couldn't help feeling that their friendship was less important to Einstein than to him. It came as a shock to him when Einstein wanted to cancel his trip to Leiden that spring. 'If you do come (twice a year) I can bona fide and happily recount all that came out of it in my progress report.—And nobody then mindlessly asks, "What did he do?"—If you don't come, however, then it would put me in a pickle', he wrote early in 1922. He continued: 'understand me correctly: Neither Lorentz nor Onnes nor anyone else here would permit themselves to say such a thing to you—indeed, I think they are so very Dutch that they even understand the art of not just not saying it, but even banishing it from their thoughts'.[40] Paul expressed his insecurity even more directly when Einstein left for Japan later in the year. 'I'm not going to be seeing you again for a very long time now', he wrote at the end of July. This was shortly before Tanya went

220   THE DELIGHT OF THINKING

to Einstein's daughters, and Paul seemed to feel as abandoned as when his eldest brother and hero suddenly moved to Hamburg. 'You don't need anybody—but I need you very much!'[41]

But Einstein gainsaid this. 'Dear Ehrenfest, I need your friendship as much, perhaps even more, than you do mine', he replied, 'for my personal relationships are much feebler and sparser than yours and I have difficulty finding human contacts that make me feel good'. Then Einstein told of his happy 'Indian life:' the four weeks he had spent with his sons in his holiday home in Spandau, outside Berlin. 'I told them a lot about you'. Especially his younger son, Tete, who had a 'softer' character than his elder brother Hans Albert and was 'comical, fine, flexibly minded, and sensitive; he would happily fit in well among you.'[42]

Repeatedly, Einstein emphasised his appreciation of Paul and Tanya's cheerful family life. And time and again, Paul relapsed into grumbling. Alongside three children and his wife's aunt, he now also had a caring but destitute mother-in-law at home. 'She's a very, very kind person and brings an element of well-being to the home. Also, she is so happy to be with us', he wrote to Tanya.[43] But now there wasn't any money for a new bicycle for T'.[44] And Paul felt abandoned by Tanya with her trips to Russian friends such as the Yanitskys and with all her requests for support of Russian scientists. Her love for Russia risked ruining them, he made clear in a missive that he would have her 'read some time at the right occasion'.[45]

'Today I again lost an entire day because I had to wait for the oven repairman and the heating to be put into service. I await the coming week in trepidation. You know, we must finally sort out things so that I don't constantly "have to" do something which keeps me from my official commitments and my scientific work. This includes the necessity of finally erecting barriers against the unlimited "having to help" with money or services. I want to pay off my debts, I want to be able to buy what we want for *us* [word underlined three times], I don't want to have constant money problems (especially with the coming salary reduction of 8–10%). *Perhaps* then I can finally work, and this shall enable me to make a specific difference for fellow human beings, while now this constant "having to" is starting to make me dislike all of them with an ever-greater intensity. If I must constantly torment myself and no longer get down to working calmly and playfully because I must constantly defend myself (something I have been feeling increasingly over the past two years) then ultimately neither can I still do what I certainly would otherwise do contentedly: face a Krutkov attentively and with love. But when more and more bills and sundry Russian "have to" letters (another letter from Hessen has just arrived) keep piling up on my bureau then I can do nothing but curse all of them and ask myself again and again whether I am so much more worthless than all of these letter writers that my development and my specific talents must be sacrificed for them. It is very understandable that [illegible], aunt, Dr Babkin, a young Russian Jew Weinstein (letter from Weyl in Zurich), Kasterin, Boguslavsky, Marushya, the University of Simferopol, the Hessen committee,

Kulischer, Bursian, Besicovitch, Friedmann, an asthmatic Russian professor and so on and so forth turn to us, but what an awful burden all of this imposes on me, who is totally unsuitable for it! I really want, want, want finally liberation from this witches' sabbath. Because truly, after a short period of hope because of Wassik's placement and me daring to hope that you might get back on your feet, I am now once more utterly dispirited. I can see increasingly clearly: we are fated to be buried unresistingly and irrevocably beneath this Russian detritus. And more than once, I have already thought this isn't my concern. This isn't what I live for.'[46]

27

Second rate and second class

Capitalism and socialism, men and women

Were the Russians in such dire straits? Not all of them. Paul's friend Ioffe became head of the new State Physicotechnical Radiology Institute in Petrograd, which did research into X-rays and other physical phenomena with potentially useful applications.[1] In the new Soviet Union, the institute expanded rapidly, not least because Ioffe by his own account 'ambitiously strived to make physics the basis of all future socialist technology' (Ioffe 1967, 45–46). As the focal point of physics research was in Central and Western Europe, he went there in search of the latest journals, the newest measurement instruments and useful contacts in Germany, the United Kingdom and Leiden as soon as the borders reopened (Eckert 1993, 116–117). But cautious Ioffe would have been careful not to ask to much of Paul.

More than Russian requests for aid or broken heating systems, it was of course his own changeable state of mind that threw Paul off balance. He was no longer young and promising. His bristly hair still stood on end, and when his eyes twinkled and he gestured with his hands, he looked ten years younger than he was. But it weighed down on him that he had passed forty—and not because of the belly he had grown. Even more often than before, he was assailed by a despondent feeling that he was at most an excellent second-rate physicist who would never truly become first-rate.

At such moments, Leiden wasn't the best place to be. Even his friendships with Einstein and Bohr couldn't cheer up Paul at such moments. He liked to say that he felt safe with the two men 'from all the ugliness of the world, like in the snow high in the mountains'.[2] But the high mountains could also be cold and lonely, especially for people—the majority—who did better in the rolling hills beneath. On some level, Paul was aware of this. When Tanya took Pawlik and Galya to visit Wassik that summer, he described in a postcard Epstein's American colleague De Kahler's visit to Leiden as a 'day off': 'because he was someone who I could teach something, while for the rest I must constantly teach myself with great effort what Bohr, Kramers, Pauli, Sommerfeld, Debye think up'.[3]

Paul came into his own on the slopes just beneath the mountain tops, where one could map out the best routes to these peaks. He was masterful at summarising these routes to the summit: teaching, transferring knowledge, grasping the crux of a magnificent theory—all the things crucial to sustain science and enable great achievements, just as the slopes support the mountain tops. The only problem was that the slopes weren't good enough for Paul, and he despaired increasingly at his inability to mount them. 'I've been racking my brains about the conversion of Boltzmann's microcanonical proof to quantum systems, but I haven't come an inch further. And if I do find it, I shall consider it clumsy, trivial, and it shall feel like a disillusion'.[4]

Forceful words. Enough physicists would have given their eyetooth to correspond and share their thoughts with Einstein, like Paul. Just recently, he had called in Onnes at Einstein's request to assess in his cryogenic laboratory whether superconductivity was brought about according to the quantum principles sketched by Einstein. This wasn't the case—'yet another glimmer of hope to understand is dashed', Einstein wrote with disappointment[5]—something Paul could do little about.

But the collaboration did enhance his reputation. Increasingly, foreign guests found their way to Leiden, and their visits boosted Paul's fame. The *Leidsch Dagblad* regularly announced the arrival of foreign professors, highlighting the impressive fact that 'Prof. Einstein, renowned throughout the world for his theory of relativity',[6] regularly stayed 'at the home of his friend and colleague Prof. Ehrenfest'.[7] Once Einstein even, by coincidence, 'accompanied Prof. Ehrenfest to pay a call to the prince'.[8] And when, engrossed in conversation, Einstein and Paul strolled through Leiden and suddenly crossed a street, the trams would respectfully stop.[9]

Paul's gift for delivering compelling lectures also became known in the outside world. The *Leidsch Dagblad* reported extensively on the lecture series, 'peppered with numerous witty comparisons', that Paul gave to the patrician Royal Society for Physics, Diligentia, in The Hague. 'When does something really exist?' a journalist noted when Paul was discussing the 'question of ether'. 'A mouse is not constantly the same mouse; the animal is constantly wearing out and at the same time is growing new molecules ... and when it is eaten by a cat, a zoologist shall no longer see a mouse, yet for a chemist ... the mouse's matter still exists. ... Merely another assortment of the molecules exists which no longer resembles a mouse'. Even more difficult, the journalist gathered, was to define a wave as something material, even though it consisted of particles going back and forth.[10]

Like Paul, this journalist then moved from the mouse and the wave to ether and 'the further our knowledge of phenomena expands, the more the firm ground of sure observation gives way, and now and then great minds come who can think further, developing new theories which can again encompass the entirety', the report concluded. But despite the public having the sense it had come closer to these great minds, Paul couldn't resist the temptation to trivialise his stories immediately. 'It is still a terribly pernicious activity, both for the speaker and the listeners', he complained in a postcard to Tanya after another of these lectures in Rotterdam.[11]

He didn't feel lost only between his talent for giving speeches and his inability to fully enjoy this, but sometimes he also lost himself a little between the pleasure of rubbing shoulders with men such as Einstein and Bohr and his inability to measure up to them. At such moments, he needed Tanya's unconditional support to keep going, along with students' admiration and devotion. It was therefore fortunate that he had gathered a new coterie of students in the early 1920s, even if this group was less monomaniacally focused on theoretical physics than his earlier set.

Uhlenbeck became, thanks to Paul's personal intercession, the private teacher in Rome of the eldest son of the Dutch ambassador, J.H. van Royen. Soon he started dressing almost as sharply as Italians, and surrounded by all the Roman and Renaissance buildings, he fell in love with history. If Paul hadn't kept him in line, Uhlenbeck might very well have forsaken physics for the Romans, Renaissance art and rococo. Paul adroitly averted such an eventuality by sending Uhlenbeck—shortly before his doctoral examinations in 1923[12]—with a letter to his contemporary, Enrico Fermi.[13] Gifted and shy, the highly academic atmosphere in Göttingen, where Fermi had studied for a while, utterly bewildered him. Uhlenbeck's relaxed response to this, and his stories of the far more casual atmosphere in Leiden, sowed the seeds of a long friendship, bound by a love for physics (Uhlenbeck 1962, session III).

Goudsmit was less under Paul's sway. When Goudsmit joined his father on a business trip to Tübingen in 1921, Paul sent him to visit Paschen, as Tanya had done. Paschen, who was famous for his measurements of the faint infrared lines in the hydrogen spectrum, didn't take him in 'as a novice but as a physicist'. And in 1923, Goudsmit returned to spend a summer learning the intricacies of measuring this eerily infrared light emitted by atoms, which gave Sommerfeld and Bohr insight into the fine details of their atomic models. But experimentally so adept Goudsmit didn't fully appreciate theoretical physics in Paul's estimation. When he even took the liberty to call a few Wednesday-evening colloquia 'not of much interest', an affronted Paul sent him away for some time (Goudsmit 1963). By comparison, Gerhard Dieke was a model student, who flawlessly sailed through his studies. But Paul saw perhaps the most in Jan Tinbergen, who passed his *kandidaats*, the equivalent of bachelor's examinations,[14] within a single year in 1922, becoming Paul's assistant shortly afterwards.

In a sense, Tinbergen was a variation on Burgers. Like Burgers, he came from a family that stimulated thinking and studying, although his parents were a little more educated, and the Tinbergens didn't focus solely on physics. Jan's mother, who had taught at a primary school before marrying, loved mathematics, while his father—a Dutch teacher at an H.B.S. in The Hague—had a love for art and languages. They brought up their five children—Jan was the eldest—with excursions in nature, drawing and strong social engagement. Jan was keenly aware of wrongdoing and injustice, and influenced by the love of his life, Tine de Wit, he became a socialist while still at secondary school.[15]

In this sense, Tinbergen was more comparable to Struik. Like Struik, he organised debates in Leiden for students and authored socialist articles, including a long piece about the effect of the economic depression on employment in the Netherlands in the early 1920s. But Tinbergen was more measured. The debating club was explicitly social democratic, and he published his articles in the moderate socialist daily newspaper *Het Volk*. Yet he was also more rigorous than Struik, choosing his study explicitly because of his ideals and regarding his favourite subjects, mathematics and physics, as stepping stones. His real goal was to help build a more just society 'as an economist'. After mastering mathematical formulas and learning how to construct solid models in physics, he aimed to become a better economist (Jolink 2003). And Tanya and Paul supported him in this endeavour.

Tanya had previously documented her ideas on an 'economic order' that gave greater opportunities for people to prosper. People allowed themselves to be 'subjugated' out of fear of 'starving to death', she noted around 1920, which was the reason they worked 'far more than eight hours a day often for their entire lives'. This enormous amount of work often didn't result in good or beautiful products, she argued, but instead also in 'unnecessary and even dangerous shoddy work', such as tobacco and weapons. Moreover, most people only earned 'just enough' to 'stay alive and to procreate'.[16]

Clearly, she wasn't fond of capitalism. But neither did she care for socialism. Its preoccupation with 'liberating people from fear of starving to death' meant it neglected the second great problem—giving people 'as much freedom as possible in the production of goods'. This brought her to the question: 'What would happen if every person was provided from their birth to their death with the strictly necessary things for a healthy life?' While this would content a small fraction of the population, a far larger proportion would 'have greater demands, whether material or spiritual, causing them to want to work more', Tanya suspected. They would also be 'fully entitled [to do so] as long as they did not cause any harm to the wellbeing of others with what they produced'.[17]

Subsequently, she produced an idealistic plan. The state would take charge of part of the agricultural land, energy reserves and factories, and provide the population with basic products such as energy and bread. In exchange, every citizen would be obliged to contribute towards the production of these basic products. But people would be given the freedom to decide when they wanted to put in these working hours, as well as enough time alongside this work to do other things or earn more money elsewhere. This would liberate them because 'someone would only have to spend a minimal part of his life doing obligatory work and his livelihood would be so secure that he could only be enslaved if, led by his own foolishness, he starts working excessively hard in order to acquire luxury products'.

Furthermore, 'the disappearance of many companies [would result] in less disfigurement of the land on which we live and a decrease in the number of hours that we must work. Both developments would elevate the population's spiritual level'.[18]

Did such ideas appeal to Tinbergen? In his later life, when he became one of the founders of econometrics and the first Nobel laureate in Economics, income inequality, poverty and the relationship between economics and environmental problems were always close to his heart. On a day-to-day basis, Tinbergen had of course most to do with Paul, who had also developed an interest in economics and in 1918 had drawn parallels between economics and thermodynamics, with its states of equilibrium, optimalisations and transformations.[19] This supported Tinbergen's notion that economics needed a mathematical foundation and offered hope that one could logically work with models in economics too (Alberts 1994, 299).

In any case, beyond their small circle, such ideas were considered rather outlandish at the time. At an evening debate of the Christiaan Huygens Student Association, Tinbergen was 'the first sacrifice ... led to the slaughter, that is, to the blackboard' after tea and biscuits. But when he started talking about 'mathematical economics', he was greeted with mockery and so ridiculed that no one at all listened to him towards the end (Jolink 2003, 21). He didn't much care. Even less so when, as Paul's assistant from 1922, he could take part in discussions which sometimes included four Nobel laureates—Einstein, Onnes, Lorentz and Zeeman. Which other member of Huygens could say that!

What Tinbergen might have seen less clearly, and what Paul with his moaning about his second-rate talent perhaps forgot, was that Tanya really was in a second-class position. In academia, even the most accomplished women weren't given the benefit of the doubt, and they met resistance at every step. Female suffrage and news about developments in Russia brought only piecemeal change. In Göttingen, brilliant Emmy Noether might have become *professor extraordinarius* in mathematics in 1922, but she received an exceptionally meagre salary, especially devised for her. And as much as men such as Klein, Hilbert and Einstein admired her, calling her 'Der Noether'—der is the masculine article in German—and ascribing her a lack of feminine grace, these same colleagues underscored with language that women weren't worthy of mathematics.[20]

Gifted Lise Meitner had a similar experience in Berlin. She created a furore in her laboratory and collaborated closely with her colleague Hahn and with Planck. But Planck never took back what he had once written: 'Amazons are abnormal, even in the intellectual sphere. In some practical situations, such as in healthcare for women, the conditions may be different, but in general it cannot be emphasised forcefully enough that nature itself destined women for their role as mother and housewife, and that the laws of nature cannot be ignored under any circumstances without causing serious harm, which in this case shall occur in the next generation' (Fuchs et al. 2001, 178).

In the Netherlands, Tanya sadly looked on as women's opportunities to work at all outside the house only diminished. Christian parties were formalising in law the unwritten rule that married women shouldn't work. They wanted female civil servants to be 'honourably discharged' when they married—unless they did so over the age of 45 (Linders 2018). But even without this law, Tanya must have felt that many people in the Netherlands expected her chiefly to look after her home.

She never aspired to scale the heights of academia. Rather she wanted to give purpose to her life by contributing to socially relevant debates. And the way she did so during her geometry colloquium impressed participants. 'I believe this is when we understood that an original thought in the field of pedagogics could be as important as a scientific discovery', wrote Belgian Marcel Minnaert much later, who once was sent away from Witte Rozenstraat for his support of the pro-German Flemish movement (Molenaar 2003, 217).[21] He praised Tanya's 'love for education, unfettered by an inclination to pedantry, and vitalised by the fresh spirit of university science'.[22] At the same time, Tanya was still denied what many took for granted—a salary, a study and daily contact with colleagues who treated one as an equal and kept one up to date with the latest publications and test results.

At every turn, Paul encouraged her to keep working. 'It's impossible for you to go on walks with the children, the distances are far too great', he wrote in summer 1923, when she was staying with Pawlik and Galya at Bad Blankenburg and visited Wassik in Jena almost daily. 'Perhaps you can find a student or teacher!!! who is prepared to take them for a walk in the vicinity every day … that shall do the children much more good than when you drag them along with you'.[23] At least this would allow Tanya to study. But even this well-intentioned encouragement illustrated that work wasn't a self-evident, accepted occupation for Tanya. She was dependent on others' good will, and she could also feel that not everyone took her work seriously. 'I almost forgot to say that Tanitchka and I dined at the Onneses on the eighth—*tête à tête*', she wrote to Paul from Leiden a couple of months later, continuing somewhat mockingly: 'He even read my article on the theory of probability to have something to talk about with me!'[24]

Tirelessly she continued to seek intellectual nourishment to keep her life in Leiden interesting. 'On Friday I went with De Ridder to Huizinga and Kohlenbrander's [sic] colloquium on the development of parliamentarism in the Netherlands. Kohlenbrander gave a lecture, and the students then discussed it', she wrote to Paul around the same time.[25] Somewhat critically, she noted that Paul's colloquia were livelier, even though well-known historian Johan Huizinga told her afterwards that the colloquia at Witte Rozenstraat had inspired him. This would have done Paul good. But despite their mutual encouragement, their life together wasn't as equal as they had once hoped.

According to Dutch law, Tanya was the 'legally incapable spouse', and Paul the spouse who had to provide her with a 'housekeeping allowance'. It was hardly accidental that her discourse on economics, written in the period when Wassik interrupted her sleep, included the phrase 'nearly every married woman is on duty 24 hours a day'. And while the bills for the Trüper Institute kept coming, her railway shares had become worthless and Tanitchka's studies and Pawlik and Galya's homeschooling grew in expense, Paul's grousing about the bills for coal, clothes for her mother, shoes or teaching materials for the children, flower bulbs for the garden and so on started gnawing at her. She wasn't used to worrying about money, let alone having to ask for it.

In a letter to Paul at the end of the year, she emphasised that Nel van Leeuwen, who was left with two children after the death of her husband Nordström in Finland, was a 'spirited woman'. Nel, Tanya had gathered, managed to support herself and her children with translation assignments and language lessons. 'I would like that too', Tanya wrote.[26] 'Especially bringing in business'. But for now, the colloquia she organised and the articles she published were a labour of love.

The lectures she occasionally gave were also unpaid, such as the one to the First Leiden School Association on a languid summer's evening in 1923. On the school benches she had placed in the playground of this innovative primary school, the responsible parents listened to Tanya's story about her girls' gymnasium in St Petersburg. Approvingly, they learned that her gymnasium in Russia had 'spacious recreation rooms, two on each floor ... where the girls could run around for a quarter of an hour after each 45-minute lesson'.[27] They discovered that this was very sensible because it was difficult for a normal person to hold their attention for more than 40 minutes. Yet this sensible account didn't give Tanya any more economic independence than her work on her mathematics pamphlet or her plan, like Kohnstamm, to earn extra money as a pamphleteer.

In this light, it wasn't overly surprising that Tanya contemplated going to Russia to make herself useful. In summer 1923, she even received a visa for Russia. It would be waiting for her in Berlin. To Paul's consternation. 'It would be good for you to progress a bit further with the second L.T. [second law of thermodynamics] before travelling to Russia. Or this work, of which I had great hopes for you, shall come to nothing. (With a doctorate [evidently, he hoped that her work would result in one] even in Russia you shall have quite different opportunities than without one)', he wrote in one of his letters to Tanya in Bad Blankenburg.[28]

But the sentence he scribbled in one of his next postcards to Bad Blankenburg cautiously echoed his text ten years previously, when Tanya had travelled from Leiden to Moscow. 'I now see that I had actually hoped you wouldn't be granted permission'. This time, his greatest concern wasn't the hopeless situation in Russia. 'I fear most that you shall fall ill (colds and infections)—I don't feel entitled to stop you now, but when you are back here [in Leiden], we must thoroughly discuss everything once again. And understand me correctly: only my concerns that you return healthy and well and absolutely no affairs of secondary importance are unsettling me'.[29]

That Paul fretted eternally about illness was of course something Tanya was aware of. She didn't allow all his newspaper clippings about the malaria epidemic and other doom and gloom in Russia to distract her. 'What am I to do with you?' she responded. 'All the newspaper reports you send me simply increase my desire to do something somewhere to make things better and nothing else!'[30] Equally, she ignored Paul's exhortations to rest at Bad Blankenburg or somewhere else before travelling on to Russia. 'I don't know if I really can rest here—this is in fact impossible when the children are nearby, and I shall certainly rest best if I go to Russia absolutely alone'.[31] She hardly even responded to Paul's final attempt to stop her: that he was best able to work in Leiden when Tanya, too, was there. 'If you were here then I would be able to continue with the quantum interpretation of the microcanonical derivation of the II L.T. [the second law of thermodynamics]. But this has now unfortunately completely ground to a halt'.[32]

For Paul, it was thus quite a stroke of luck when Tanya's Russian plans fell apart. Indeed, this had to do with him. In Brussels in 1921 and during a trip around the United States in 1923, Lorentz had put in a good word for Paul with the brand-new Nobel laureate Robert Millikan of Caltech—the California Institute of Technology—in Pasadena in the United States. Millikan, who was the first American representative at a Solvay Conference in 1921 and chairman of the Executive Council of Caltech, wanted to bolster the institution's reputation by inviting talented European scientists as visiting professors. Paul fitted the bill Lorentz thought, and after intensive consultation with Millikan, with colleagues who would have to replace Paul in Leiden and with his brother Hugo in Saint Louis, who contacted the universities, a stay in Pasadena and a lecture tour across the United States was arranged at the end of 1923. 'Prof. P. Ehrenfest has been invited by the university at Pasadena (California) to deliver lectures', the *Leidsch Dagblad* wrote on 17 December.[33] 'And he shall also deliver lectures in several other cities in America. The professor expects to return to the Netherlands towards Easter'.

The newspaper's report was late. Ten days earlier, Paul had already embarked in Rotterdam and sailed to New York via Southampton. And although he would return from the western hemisphere, Paul and Tanya were starting to face in opposite directions, her looking east while he looked west. It didn't go as abruptly as the Stern-Gerlach experiment, but to some extent they did turn their backs on each other.

28

Two worlds

American girls and 'male logic'

On 15 December 1923, Paul disembarked in New York, following in the footsteps of Lorentz, Einstein and his mentor Boltzmann. 'I cannot make landfall in America without being a little jealous of Colombus', Boltzmann had written. Paul's more down-to-earth brother Hugo chiefly hoped that Paul 'could enjoy to the full' the skyline with the 'mighty buildings [standing out] against a blue sky', even though the weather was inclement.[1] In the last of three notes awaiting Paul in New York, Hugo also recalled his own arrival in his adopted country and assumed his brother must have found his sea legs. 'Vividly, I now recall my more-than-23-year-old impression of this mighty city—and how I staggered over Broadway and kept treading with one foot and then the other far too forcefully on the paving stones'.[2]

Hugo looked after his youngest brother well. A brother of Hugo's wife Sophie,[3] Arthur Schwab, collected Paul at the arrival hall in Hoboken and took him to his home. 'They insist you stay with them', Hugo had written and praised 'the interesting circle of people (artists, writers)' they socialised with.[4] Paul took a liking to them. 'With the boat over the Hudson to their small home, where I'm staying with him and his wife.[5] Impression from yesterday and this morning very, very overwhelming. Schwab and his wife and several friends with them are the nicest, best, *most lively* [underlined four times] people I have met in my entire life. His profession is: advisor on the improvement of working conditions (physical, psychological, economic) in the gigantic factories. ... his wife ... writer', he wrote euphorically to Tanya about his impressions.[6]

The new impressions and the company of Schwab's friends made Paul feel light. No tests, no examinations and everyone would understand if he didn't publish anything the coming months: he was busy with other matters. Another benefit was the extra income the coming leisurely months would generate. After visiting Hugo in St. Louis in Missouri, he would travel on to Pasadena in California and earn almost 600 dollars a month until the end of March for a loosely organised series of lectures about the latest developments in physics.[7] This would be followed by a tour of university cities, generating even more money. Hugo, who had thought through everything with him and corresponded on his behalf, would deposit the money for now.

Paul's arrival also made Hugo look with fresh eyes at the United States and his own place within it. He was completely integrated into American society: as head of the department of gynaecology and obstetrics at the Jewish hospital in St. Louis, an associate professor at the University of Washington and the editor of the important *American Journal of Obstetrics and Gynaecology* (Aswal 1990, 342–344).[8] The book he had just published about the injuries babies can sustain during childbirth had honed his reputation (H. Ehrenfest 1922). With Sophie, he had three children—Paul, Fritz and Ellen—and he would be 'very unhappy' if 'I *had* to go back to Vienna to live there', he wrote in one of the letters, mostly in English, he sent Paul during his journey.[9]

None of his other brothers had ever visited him, and Hugo indicated beforehand that he wished to introduce Paul to typical American public institutions as soon as possible. It would be good for Paul to see the public school system, the public library service and the welfare institutions. Hugo was also a little apprehensive of the impression his adopted country would make on his critical brother and vice versa. 'You're now in America. Be mild and restrained in your judgement. I know that you shall think better of us by May. We live in another world'.[10] But such concerns were unnecessary. Paul enjoyed his trip and described the United States in letters to Leiden—which he often sent via Hugo, who read them—in rosy terms.

Enthusiastically he described the colours, scents and flavours of the new places he visited, and he felt content when new forms of collaboration or relationships spontaneously arose. It had been the same during his earlier travels. Even though he had sometimes grumbled about 'grotesquely ugly' Kharkov or 'dismal' Tübingen, he loved new stimuli. In Leiden, too, he enjoyed foreign visitors' stories and was delighted when a nugget of information from one visitor corresponded with the knowledge of another.

This time, his opinion was perhaps already somewhat coloured by Einstein's reasonably positive assessment of the country after his trip with Weitzmann. In an essay, Einstein complained a little about Americans' materialism but commended their 'joyous, positive attitude to life'. Their invariable smile on photographs was to his mind a symbol of friendliness, self-confidence, optimism and a lack of envy. 'The superiority of this country in matters of technology and organisation' particularly impressed Einstein. 'Objects of everyday use are more solid than in Europe, houses much more practically designed [and] the marvellous development of technical devices and methods of work' saved labour, he wrote (Einstein 1954, 4). This allowed Americans to use their natural resources efficiently, despite their relatively small population.

Paul soon shared this admiration for the speed and ingenuity with which Americans converted knowledge into technology. The large scale of their enterprises filled him with awe. This was even more so after visiting his brother's family in St. Louis, when he travelled to California. 'Suddenly [one sees] endless orange tree plantations, endlessly flat and with identical orange trees, the size of men, lined up in a regular triangular grid, laden with golden oranges', he wrote in a long letter to Tanya about his railway journey to Pasadena in California.[11] And he couldn't help but contrast this modernity with the old—and often old-fashioned—continent.

'Sunday morning, early in spring in a small village in the Thuringian Forest or Wurttemberg, *Herr-Doktor*, the pharmacist, the schoolmaster, the postmaster, 50–60 years ago, in a light, sunny room, in front of the window the apple tree in blossom or else the cherry tree. In a moment they shall play one of Haydn's quartets together, first they drink a glass of cider, the schoolmaster's own produce, and smoke a pipe. Their cherry, apple and pear trees bear very different fruit, not always particularly good, rather "as God sees fit", but it all goes together: apples, Hayden, the Bible, a schnapps, the schoolmaster's snuff, perhaps Newton or Kant at the postmaster's house, a large dung heap, faith in God and a little typhus.'[12]

Paul sketched a nostalgic picture of this old Europe, with typhus as a nasty, old-fashioned element. He juxtaposed this in his next paragraph with orange plantations and other large-scale schemes in a dazzling, optimistic but somewhat overwhelming picture of the future:

'Take all of this apart [and in 50 years you shall get]: an orange plantation; a million-volt laboratory; a cathedral with wireless sermons, spread by telegraphy; a concert hall ... Develop the left-frontal quarter of mathematical brains, a lyrical poet's heart, a prima donna's vocal cords and a boxer's fist. Cultivate them "in vitro", ... and you shall get gigantic results which the "unscientific, old-fashioned" method couldn't even dream of—the Californian oranges are extra large, very good and absolutely uniform. Nothing can withstand this.'[13]

The combination of verve and power of observation revealed by such passages so touched Hugo that he felt his brother's letters should be published. '*Yes, yes*, to make some money!!', he continued teasingly, still astounded at the meagreness of Paul's Dutch salary.[14]

Yet for the time being, only Hugo and Sophie, and Tanya and the children a few weeks later, read the detailed descriptions in Paul's missives. They read about how he arrived at the station in Pasadena—that was 'as large (or small)' as the station at Ede, a very small Dutch town'. And that Epstein, who had once so looked up to Paul in St Petersburg and, like Hugo, had emigrated from Europe to America, had collected Paul in his 'beautiful big car'. On their drive to small but rapidly growing Caltech, everything had been bathed in a 'southern' light: from the bright-green canopy of the pepper trees and the brightly painted wooden houses, half concealed by rose bushes in the wide streets of the small town, to the distant mountains silhouetted against the deep-blue sky.[15]

In the ensuing days, Paul enjoyed the 'starry night sky' and the 'magnificent red morning sky' and the large 'oak of marvellous beauty' between the arcades and white buildings of the Institute. Only the Californians took him some getting used to: 'They don't know how to make a sensitive foreigner feel at home—a great difference with Arthur and Teddy Schwab in New York ... not to mention Hugo and Sophie in St Louis', he wrote to Tanya.[16] To which he added that during his lectures at the institute, he wanted to discuss quantum theory from a statistical perspective and to present her work on statistical methods that afternoon as a first introduction to the subject.

Three months later, Hugo wrote with satisfaction that the idea of staying in Pasadena until early April and then doing a lecture tour was working out magnificently. In Pasadena, Paul had become better acquainted with Tolman, the man whose idea of scaling the universe Tanya had made such short work of—a 'fine fellow', with whom he even published a short article about quanta (Ehrenfest and Tolman 1924). He had renewed his friendship with Epstein. And he had introduced the students to quantum work conducted in Europe and with Boltzmann's statistical mechanics. Yet Hugo thought the second part of Paul's trip, with the lectures, would be the most fruitful. 'You have that rare ability of smelling unusual personalities & to quickly approach them, so that even these hurried visits in these various universities offer you many chances to establish valuable friendships'.[17]

Paul did so, while also enjoying the landscapes, the meals, the cities and all the novelties. Not everything appealed to him. Modern American houses weren't in the same aesthetic league as venerable European buildings. He greatly preferred the naturalness of female European students to the 'made-up' faces of their American counterparts, with their 'painted lips and eyebrows'. And he wrote condescendingly about the 'football clothes' worn by male American students around the campus.[18]

Neither was he free of prejudice, despite being the victim of it himself, writing insouciantly of 'sleeping-car Negroes' in the trains.[19] And as like any traveller, he brought himself along. His abounding curiosity, which had put off Meitner and oppressed Burgers, was now unleashed on American students. Would they have thought of his fellow-Viennese Freud—if they had heard of him—when, during evening conversations, Paul—rather like a guru—went on and on about physics as well as about their background, their parents, their childhood and their aspirations?[20]

In St. Louis, meanwhile, Hugo received a stream of cheques in the post. Paul's visit to the Mount Wilson Observatory, where astronomer Edwin Hubble had just completed his measurements on receding galaxies, had earned him 100 dollars; in Minneapolis he had charged 125 dollars; a lecture in Toronto, Canada, had brought in 200 dollars; in Maddison he had been paid 300 dollars; Cornell University at Ithaca 250; Colombia University had even paid 350 dollars and Harvard had given him 300 dollars. Hugo invested all of it in short-term Treasury securities of 500 dollars at 4 per cent interest and revealed in his letters his low opinion of his brother's financial acumen.[21] Did Paul want to invest the money in a fund to cover the costs of Wassik's care? Did he want to pay off his mortgage? Over the months, impatience crept into these questions. When Paul complained about financial worries in a jeremiad shortly before his return to Europe, Hugo drew the line. 'Don't spoil your final days here with absurd and unhealthy worrying. Your current money worries are hallucinations!' was his tetchy reply on 1 May to Paul's 'peevish' letter.[22]

They were vastly different characters. There was an age gap of ten years, and an ocean had separated them for 23 years. On one hand, Paul and Hugo reverted straight to their roles as youngest and older brother. Paul was only too happy to take everything Hugo did for him for granted, and Hugo just as naturally managed Paul's earnings, booked railway tickets, took out travel insurance, lent him money when needed and more. On the other hand, they regarded each other with puzzlement. 'Together with this letter, a fat package goes to Arthur Schwab containing your pair of thick underwear, you left here—Paul [junior, Hugo's son] thought that such things are worn only by North Pole explorers', Hugo wrote a few days after his stern letter.[23]

Paul's attitude towards women also surprised Hugo. 'Personally, I was amused at your satisfaction of having finally found in Bryn Mawr [a famous women's college] 5 women really interested in studies and investigations—speaking disparagingly of all the women in all the other colleges & universities.—& the thought came to my mind what a horrible, terrible place to live in, this world or these United States would be, if all our girls would be just like those five. Is it really the duty or function or everyone going to a school to become an investigator or a scientist?? Is it not wonderful in itself that so very many girls in this country are willing to learn a little something while at the same time passing away, joyful and happy, those years necessarily lying between the end of public-school education & the beginning of their reproductive function?'[24]

Hugo held other ideas about fulfilment in women's lives to his younger brother. 'Many a times you mentioned in your letters an attractive, really charming wife of this or the other man—they all have gone to college—not to turn into scientists but to acquire a little more knowledge & much charm and physical health'.[25] Even though they had never met, Hugo must surely have regarded Tanya as a little too independent and serious, with her straight-cut hair, face without makeup and head full of mathematics.

But despite the pleasure the trip had given him, Paul was glad to return to Tanya and the children. How often hadn't he written to Tanya about his 'longing to see, hear and touch all of you'.[26] He lacked time at the end of his tour to call again on Hugo in St. Louis. But after Arthur and Teddy had brought him to the ship and after long weeks crossing the ocean, when he entered his home on Witte Rozenstraat, a card awaited him. 'Welcome home! Mailed today: Check on [sic] Rotterdam Bank Vereeniging'. 3600 dollars had been sent for Paul. 'Love Hugo'.[27]

Instead of Californian sun, cheerful Americans and their country of big cars, practical houses and gigantic orange plantations, there was Leiden, with its impoverished backstreets, still canals, cloud-filled skies, rain showers as well as a highly regarded university. Paul's travel journal was filled with new names; he had told his interlocutors a great deal about the latest developments in physics in Europe, and his role as teacher and knowledge broker energised him. Now he was almost equally delighted to fill in the details of all the impressions he had conveyed in his letters to his colleagues, students and Tanya and the children.

Hugo probably wouldn't have been surprised to hear that Tanya had thrown herself into mathematics during Paul's trip. Or rather the geometry education that she had occupied herself with in St Petersburg. In Leiden, Tanya hadn't had the opportunity to broaden her mind like Paul. But during discussion evenings with secondary school teachers, she had refined her thoughts on geometry education.

The ideas of the small club which regularly met at Witte Rozenstraat to discuss geometry education hardly reflected those of the typical secondary school teacher. Many of them were quite happy with the strict classical way of teaching geometry. That the abstract teaching method went above the heads of many pupils wasn't a drawback in their perspective. On the contrary, it enabled self-selection of an elite, which would serve as a model to the rest of society following further education at university (Bastide van Gemert 2006). This was underscored, in their view, by the stream of Nobel Prizes awarded to Dutch scientists at the beginning of the twentieth century.

Regular visitors to Tanya's colloquia, such as mathematician and physicist Willem Reindersma, educationalists Casimir and Kohnstamm as well as mathematician Gerrit Mannoury, saw things differently. They wanted to make education accessible to a larger group of pupils, certainly also to children who received less academic support at home. This had become even more important since the advent of universal suffrage in the Netherlands in 1919. At the same time, they each had their own emphasis. Reindersma who had taught mathematics and physics at Casimir's Nederlandsch Lyceum in The Hague was pragmatic. He wanted to use cutting, pasting, estimation and reflection to introduce new secondary school pupils to forms and concepts such as symmetry (Reindersma 1924). Communist Gerrit Mannoury, a mathematics professor in Amsterdam, mainly wanted to investigate what society needed from mathematics. Kohnstamm saw the importance of mathematics and physics in a broader perspective. He believed education in a well-functioning democracy must prioritise the development of balanced, emotionally intelligent pupils with a strong sense of responsibility.

Unrelenting secondary school teacher Eduard Dijksterhuis from the more conservative south of the Netherlands diametrically opposed this. He even thought it was lax that university applicants no longer had to take exams in Greek and Latin. The idea of tampering with geometry education, 'one of the most precious fruits of mathematics education', was anathema to him. Geometry was the field where students could get to know 'the purity and honesty of mathematical thinking and reasoning' and with the 'intellectual discipline, order and clarity mathematics strives for'. And according to Dijksterhuis, 'axiomatic reasoning' in geometry, with which new theorems were derived step by step from a handful of well-chosen assumptions ('axioms'), produced 'an edifice of such great beauty, and despite slightly shaky foundations here and there, of such harmonious solidity, that it must do every person good spiritually to tarry there for a while' (Dijksterhuis 1924/1925b). In other words, and as these quotations from a debate between him and Tanya show: it had been a rude awakening for him when Tanya published an essay voicing strong reservations about the way teachers led pupils through this magnificent edifice (Ehrenfest-Afanassjewa 1924).

To a certain extent, her essay in Dutch—with the title '"What can and must Geometry Education give to a Non-mathematician?"—agreed with Dijksterhuis. Tanya also considered "axiomatic reasoning" the conclusion of geometry education. But she thought it a bad idea to spur pupils blindly through fifty postulates and 250 exercises to instil them with what Dijksterhuis called "spiritual order and discipline". This taught most pupils nothing more "than parroting these postulates", she retorted to Dijksterhuis' objections (Ehrenfest-Afanassjewa 1924/1925). Instead, her method would allow a broad group of pupils to enter the edifice of geometry and—partly—to discern it (Ehrenfest-Afanassjewa 1924).

These ideas were driven by strong conviction. Tanya believed in the 'educational effect of geometry'. She was convinced that geometry was particularly good at developing and honing children's logical intellect. This was precisely why she regarded the subject as also having relevance for pupils who didn't want to become mathematicians—the non-mathematicians in the title of her pamphlet.[28]

Unsurprisingly her proposal for a broad teaching method included elements from her 'geometry laboratory', which she first assessed at the girls' gymnasium in St Petersburg. In the Netherlands, too, a laboratory like this could serve as the first phase of geometry education, Tanya argued. After all, it aligned with the first stage of thinking: the intuitive or perceptual period, in which exploration—in this case of space—takes centre stage. In the next, ordering phase, children learned to recognise more or less unconsciously seen patterns and to logically reason about them. Only in the third and final 'axiomatic phase' would they, together with the teacher, sum up all their knowledge about space in Euclid's five postulates and numerous proofs for other postulates describing space.

With this proposal, Tanya had attacked the sacred cows of Dutch geometry education. She suggested that in this second, ordering phase, pupils could take any postulate—statement about space—which seemed self-evident to them as a starting point for a logical reasoning, resulting in another postulate. As if they could use separate blocks to build bits of mathematics. Only after this, with the help of Euclid's postulates and corresponding reasoning, would they work out for themselves how geometry, this 'edifice of such great beauty', could be efficiently constructed.

Dijksterhuis was appalled. How could 'mathematics education still bear fruit when it becomes practice to accept without evidence postulates that an entire class finds obvious', he complained. Dijksterhuis thought little of the importance Tanya attached to active learning. 'A child must, after all, also learn to force themselves to concentrate their thoughts, and it is not always necessary for them to understand the "why" of the things they are made to do', he believed (Dijksterhuis 1924/1925b).

In a first reply, he had already tried to undermine Tanya's position with a great deal of rhetoric. He reproached her for vagueness and a lack of originality. If one replaced Tanya's term 'logical thinking' with 'power of judgement' and her term 'intuition' with 'perceptibility [*anschaulichkeit*]', Dijksterhuis wrote testily, one could hear Schopenhauer's 'grim voice'. This philosopher had already argued much earlier that everything started with visualising concepts and clarifying relations between them (*Anschaulichkeit*), after which the challenge was to convey everything that had become known 'correctly and precisely to the abstract consciousness'; and that everyone with good sense can then deduce the relevant postulates, that is, can 'calculate' in Mrs. Ehrenfest's terminology (Dijksterhuis 1924/1925a)'.

Nevertheless, Dijksterhuis had taken Tanya's pamphlet seriously enough to publish this reply in what would be the first number of *Bijvoegsel van het Nieuw Tijdschrift voor Wiskunde* or 'Supplement to the New Journal for Mathematics' in English. It would later become an independent journal on the didactics of mathematics and exists until this day under the name *Euclides*. At the time Dijksterhuis' response—'Must Geometry Education be changed? Comments in response to a Brochure by Mrs. EHRENFEST-AFANASSJEWA'—was the opening article of the first number, and thus in a certain sense also a manifesto for the journal (Berkel 1996, 121–142; Berkel 2000). Clearly, Dijksterhuis had the editors of the publication on his side.

Did he have a point? Tanya's definition of 'intuition' was a little vague. 'Without intuition no thinking is possible', she had written, and she quoted Gauss, the famous mathematician from Göttingen: 'I have already found the theorem; I just have not yet proved it'. With this pronouncement, she illustrated that 'axiomatic thinking' is the conclusion of a thought process, and that insights appear earlier, when people take in the world or when they logically order their observations. Perhaps she could have put this more explicitly. And perhaps she could have added weight to her argument by mentioning her experience with mathematics practicals in St Petersburg, by referring to Klein and Hilbert's ideas of visualised and conceptualised learning in Göttingen, the discussions and work of mathematics teachers in her colloquia or perhaps even Tolstoy's ideas on education that connect to the talents and interests of children.

By not doing so, she had made it easier for Dijksterhuis to dismiss her as 'naïve' and someone who didn't know the first thing about Dutch education. Tanya didn't know, Dijksterhuis claimed for example, that 'most HBS teachers' simply 'follow the secular tradition of their science and that their opinion at least had a fairly firm historical basis'. This tradition afforded scope, he wrote with much verbosity and solemnity, to various aspects of spatial theory during axiomatic reasoning—that is, for the spatial insight and thinking about space Tanya advocated. He didn't understand why she didn't want to see that this 'spatial theory' was almost ingrained in Euclidian geometry. Even worse, her alternative teaching method didn't meet 'the first requirement one might demand of a pedagogical method, namely that it has grown out of educational practice' (Dijksterhuis 1924/1925a).

That a new Dutch law, established in 1924, forbade Tanya, as a married woman to take part in educational practice as a teacher, didn't even seem to occur to Dijksterhuis. He focused on firmly advocating a geometric 'sphere' where the 'growing youth' learns that 'vague claims, careless expressions and misunderstood words are no longer acceptable, where any deviation from honest thinking is immediately evident and where they observe all knowledge being constructed from the start'. And he would oppose anything that, in his view, 'could harm the purity of the atmosphere breathed there, the solidity of the curriculum being built there' (Dijksterhuis 1924/1925a).

In her reply to his first reply, also published in the new supplement on the didactics of mathematics, Tanya stuck to her earlier standpoint. There was little point making pupils learn logical reasoning by rote if they had no idea what the relevance of it was, she repeated. In other words, what was the purpose of the art of formal logical reasoning if one hadn't learned to ask the right questions? Teasingly, in a footnote, she suggested calling all reasoning doomed due to an incorrect phrasing of the question 'male logic'—a counterpoint to the often-used term 'female logic' for illogical reasoning (Ehrenfest-Afanassjewa 1924/1925).

In a second and final reply in the publication, Dijksterhuis subsequently censured the 'lurid colours' Tanya used to portray existing teaching methods and 'that singular [sic] sense of superiority with which intellectual women in our time so gladly look down on the male sex', with Tanya 'unable to find any less esteeming qualification than "male logic"'. 'I still fail to comprehend why Mrs. Ehrenfest cannot concede that one can fully understand something geometric without the presence of a clear representation', he wrote offhandedly (Dijksterhuis 1924/1925b). The editors openly took his side by choosing a letter to the editor from a mathematics teacher, who wrote: 'I believe with one of the reviewers of the *Journal* that most colleagues ... in their conviction shall take the side of Mr. D' (Coster 1926/1927).

Much later, Tanya was delighted when a boy asked her why there were no straight lines in nature. What a magnificent observation! What a start to a thought process!²⁹ But for the time being, her ideas for adapting education to children's thought processes were rejected. When the superintendent of schools, Eibert Jensema, set up a commission to investigate whether secondary school mathematics education did indeed require reform, relatively moderate Herman E.J. Beth was appointed chairman and Dijksterhuis his deputy. In this position, he successfully quashed any serious reform of the teaching programme, with the support of *Bijvoegsel*, edited by mathematics teacher Piet Wijdenes.

'Why haven't I asked Gunning, Reindersma, Kohnstamm, Mannoury or Casimir as contributors? Because they are our antipodes and have already had enough to say in The Hague', Wijdenes wrote to Dijksterhuis. Smugly, he answered his own question. 'Our *Bijvoegsel* shall be the mighty guide in mathematics education and not idle chatter which does everything along the gently curving lines of feeble equality, synonymous with shoddiness, half-heartedness, laxity, toadishness' (Berkel 1996, 141; Bastide Van Gemert 2006).

Tanya wasn't mentioned in this correspondence. Neither was she asked to take part in the commission, even though she had started the discussion. Nobody questioned this: she simply didn't have any direct influence in The Hague, she wasn't a teacher with practical experience, she didn't have a position at a university, and neither would she get a job as a married woman for that matter.

29
Warm people and thermodynamics
Paul's network and Tanya's science

Guests staying at Witte Rozenstraat later remembered the cut-out scenes Galinka had glued to all the doors, the flag of the Dutch Social-democratic Youth Organisation with which she sometimes walked around the house with friends and the plays she sometimes performed with her brother and sister—with braids in their hair, streaks on their faces, wrapped in white sheets or coloured sheets. The children had a fantasy world in which they gave themselves and their parents other names: Tanitchka was called Saturna, Galinka was Aita, Pawlik was Aino, Tanya was called Twonella and Paul had the name Knan-Pans.[1]

Most likely, this fantasy world was partly connected to their relative isolation. Tanitchka had doggedly endured with her gymnasium schooling, and after doing her state exams early, aged sixteen, she went on to read mathematics and physics. But Galinka and Pawlik were still homeschooled. Paul was unimpressed with the Vrije School in The Hague, where Tanya looked with Galinka in 1923. Still small at the time, the school, whose founders included refined anthroposophist Hélène Droogleever Fortuyn, taught about twenty pupils in a house on Columbusstraat. But Paul had a low opinion of Rudolf Steiner's anthroposophy, the doctrine on which the school was based, and thought the school elitist and expensive too. It was fortunate for creative Galinka that she could express herself in drawing, for which she was given free rein, like cheerful Pawlik, who enjoyed craftwork. He liked tinkering with old radios, as his father had done with old clocks, or making little houses out of wooden blocks, more like his mother. The house was a cheerful 'laboratory' where the children could explore the world, even if adults mainly peopled this world.[2]

Paul and Tanya had made it a custom for visitors sleeping high in the attic, beneath the beams, to sign the wall (see Figure 29.1). James Franck from Göttingen (Nobel Prize 1925) had been the first during a stay in April 1923.[3] Einstein was next in June 1923 and then Patrick Blackett (Nobel Prize 1948), an English stockbroker's son who had just started working at Rutherford's laboratory.[4] Brilliant Hertha Sponer, who obtained her doctoral degree under Debye in 1920 and had become Franck's research assistant in Göttingen, added her name in August (Crull 2025). Einstein again graced the plaster with his name in September and December, between which stood his fellow-Berliner Lise Meitner's name (who was never awarded a Nobel Prize but should have been awarded one).

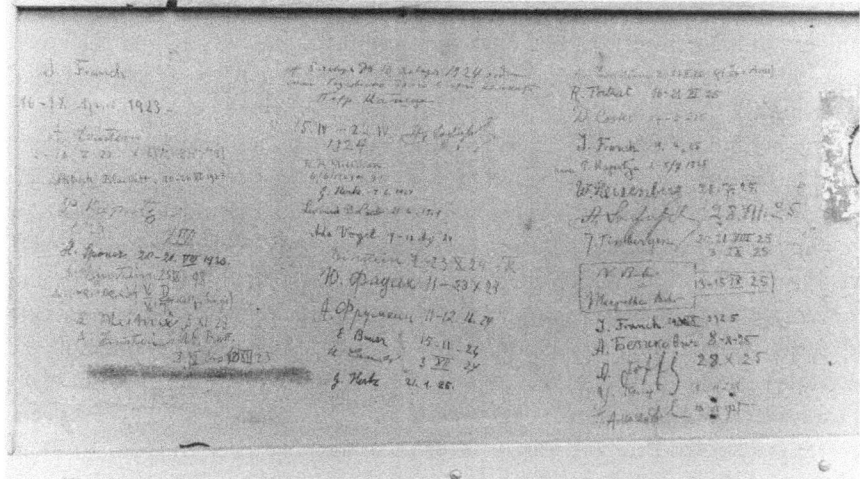

Figure 29.1 Part of the signature wall in the former guest room in the attic of the house on Witte Rozenstraat.—*Photograph National Museum Boerhaave, Leiden.*

In 1924, this was followed by the signatures of Pyotr Kapitsa (Nobel Prize 1978),[5] Paul's friend Gustav Hertz (Nobel Prize 1925)[6] and estimable Robert Millikan, who had come to Europe to receive the Nobel Prize he was awarded in 1923.[7] Ioffe and young Russian chemist Alexander Frumkin's signatures showed how travel grants and fellowships for Soviet colleagues had started to increase. The wall also included Fermi, who came from Italy on a Rockefeller grant in September 1924 and felt so at home in Leiden that he even published a report on 'spectral lines' in Dutch (Fermi 1924). And of course, Einstein once again visited Witte Rozenstraat in 1924, incidentally at the same time as the founder of Montessori education in Russia, Julia Fausek.[8] An inconspicuous name on the list was that of Ada Vogel, Galinka's friend from the Dutch Social-democratic Youth Organisation.

With all this hustle and bustle, it gave peace of mind that Paul had managed to earn enough money on his American trip to pay off most of their debts. Tanya had repeatedly asked him for money that spring and wrote several times informing him that she had borrowed 400 or 500 guilders from Kohnstamm, from future Nobel Prize laureate Willem Einthoven and from Onnes' deputy Eli Wiersma, who lived across the road from them. But Paul's American lecture series meant they could repay the borrowed money as well as the 10,000-guilder mortgage on the house at the end of 1924. A small remortgage meant they were out of the woods for the time being.[9]

That Tanya had started giving Russian lessons from home made less difference financially. 'De Ridder [one of the boys he tutored in summer 1918, along with Van Aardenne] has arranged a Russian pupil for me', she wrote to Paul in Pasadena in

March 1924. 'He described her to me as a penniless student, so I charged only three guilders an hour. But she turned out to be an immaculately attired and perfumed lady. We shall see. With one sheep over the ditch, others shall surely follow! I shall do my best'.[10]

But she undoubtedly felt relieved about something Paul wrote about Wassik around this time. 'Very important. On the last day of my visit to St Louis, a conversation with Hugo and two of his paediatrician friends ... made me realise that I completely misunderstood a comment Hugo made in Leiden. Wapsik's defect has absolutely nothing to do with a 'hereditary defect' on either of our parts, and conversely, Wapsik's defect won't have any consequences for our other children's offspring. Neither is the hypothesis that this defect can be traced to the mother's or father's fatigue or exhaustion supported in any way'.[11] This misunderstanding must have caused them a great deal of stress.

Things between Paul and Tanya weren't like they had been in Göttingen and St Petersburg when study, writing, work, music making, discussions and laughing with friends blended seamlessly. Instead of a Tolstoy shirt like then, Paul wore a suit. And whereas he had complained of the lack of structure and regular work in his life in St Petersburg, he was now weighed down by the responsibilities associated with a professorship. Peevishly, he waded his way through lectures, examinations and obligatory visits to colleagues with much small talk. His position alongside Nobel laureates Onnes and Lorentz seemed to oppress him more than before. He was almost more restless than at the end of his time in St Petersburg. In fact, scarcely four months after returning from the United States, he was already making his way to Russia—as he insisted calling the Soviet Union—while Tanya had applied for a visa to the country earlier.

Paul needed such trips. It allowed him to escape Leiden, where he was disheartened by never having become 'one of the foremost', and it energised him to meet friends who knew him when he was young and promising, who looked up to him because of his friendship with Einstein and Bohr and who themselves were good friends. It must have done his self-confidence good that he was asked to be the deputy-chairman of the Fourth Congress for Physicists in Leningrad, as St Petersburg had been renamed (Frenkel 1977, Ch. 7).

'What chiefly concerns me is the endeavour to pay attention only to the practical course', he remarked there about the course he believed Soviet science had embarked upon. Referring to his recent trip to America, he recounted often being asked 'how do physicists in Germany and the Netherlands (where I work) respond so quickly and, in many cases, so brilliantly to many practical problems'. Which problems exactly isn't mentioned in the record of the meeting, and it seemed to somewhat contradict the admiration he had for American pragmatism. Regardless, he thought that the only correct answer to this question now was 'this is because of the great attention paid to pure theoretical knowledge and especially to mathematics in these countries [Germany and the Netherlands]. People with a thorough command of theory can deal with problems when they occur. So permit me to hope that pure theoretical and practical avenues shall develop in harmony in the Soviet Union'. And like during the congress in 1912, shortly before his departure from St Petersburg, he again received loud applause. This was, according to congress chairman Petr Lazarev,[12] thanks both to Paul's ideas and his personality (Frenkel 1977, Ch. 7).

Paul was in his element. 'All of our problems arise from our habit of reflection', was his summary of a long discussion in the subsequent days. He defended Bohr's quantum theory with a witticism. 'As long as Bohr's theory can be sick [i.e. subject to criticism], it is still alive'. And his dazzling enthusiasm was rewarded when he was appointed a correspondent member of the Russian Academy of Sciences. At this event, Ioffe even called him a founder of Russian theoretical physics (Frenkel 1977, Ch. 7).

With complete devotion, Paul then made a new group of young Russian physicists enthusiastic about the discipline—one of whom had even come barefoot because he lacked money for shoes but still wanted to be there (see Figure 29.2).[13] And: 'I must tell you that Ehrenfest has won the hearts of all of the youth here and perhaps of older people too', young physicist Victor Frenkel wrote in a letter to his father. 'This is a man who combines a child's simplicity and directness with an

Figure 29.2 Paul with young Russian physicists A.N. Arsenyeva, B.M. Gochberg, V. Tomashevski, N.S. Usataya, B. Ja. Pines, N.A. Brilliantov, A.I. Shalnikov, V.S. Gorski in Leningrad.—*Photograph Ehrenfest Family Archive.*

extraordinary love for the people and with the inexhaustible ingenuity and intellect of a great and astute researcher. He brings inanimate objects, molecules, atoms and electrons into conversation with one another using, in terms of endings and cases, rather flawed but at the same time very sophisticated Russian, they love and hate each other and come completely to life, they change into microscopic inhabitants of an animated universe. For Ehrenfest, or rather for the physicist Ehrenfest, physics is not so much an exact science as a drama or comedy in the lives of atoms and electrons' (Frenkel 1977, Ch. 7).

Paul left a vivacious impression in Russia and, conversely, was himself deeply moved by his Russian friends. Not for the first time of course: when two young Russian physicists, Vladimir Chulanovsky and Aleksandr Arkhangelsky, visited Leiden in 1920 and discussed Ioffe's and Rozhdestvensky's new research institutes, he had responded just as enthusiastically in a letter to Rozhdestvensky. 'Dear Dimitri Sergeevich, Chulanovsky and Arkhangelsky were with us yesterday. I want to write to you at once. Above all I am proud of you, my dear friends, that you have been working with such exceptional dignity and harmony in these difficult times. Especially the latter greatly pleases me. With your intensive enthusiasm, you shall achieve results that shan't be outdone by work in Western Europe. Whenever Tsiolkovsky tells me of your successes, great and small, of things that have succeeded and things that haven't succeeded, my heart beats faster. ... The short blockade of Russia has resulted in your own strength flourishing. This is the historical significance of your accomplishments' (Frenkel 1977, Ch. 5).

Paul could no longer view Russia with any detachment at all after his trip in 1924, he wrote later that year to Carel de Ridder, who had sent him a critical analysis of the political and economic situation in the Soviet Union. 'For me and my family's problems, as for all problems arising from my relations with Russian physicists and mathematicians, an utterly subjective approach is the only productive one'. So, describing the Soviet Union in cool economic terms mattered as little to Paul as someone 'giving a strictly correct anatomical-physiological description of Tanya and Galinka'.[14]

That Tanya managed to publish despite all the guests, Paul's busyness and her Russian lessons, was thanks to Baba Katya, who enjoyed keeping an eye on her grandchildren. This allowed Tanya to study a subject that had engaged her for some time, alongside her work on geometry education: thermodynamics. A year earlier, Paul had regularly encouraged her to work on the topic: 'Nah, what about your thermodynamics?!'[15] And Tanya had progressed far enough in January 1924 to give Lorentz a draft article to read. 'I believe I have researched everything I could ever want to know in thermodynamics', she wrote to Paul. That she didn't first send the piece to him, she continued, was because he would undoubtedly be busy on the other side of the ocean. She 'greatly needed to talk to someone about it as soon as possible, and furthermore I would in any case like to hear Lorentz's opinion, and in this way, he shall have to give his opinion whether he likes it or not—and a well-considered opinion at that!'[16] During their first discussions on the matter, Tanya noticed that Lorentz approached thermodynamics very differently to her, with her strict logical reasoning. 'You can clearly feel that he doesn't feel the need to understand the things that so torment me. He isn't a philosopher at all, Puplik'. Yet she had learned something anyway, she wrote, 'namely how you must explain something so that physicists understand that it isn't 'mathematics'.[17]

Tanya's formal way of reasoning was indeed reminiscent of mathematical thinking. Her work on thermodynamics was in a sense perfectly in keeping with her ideas on geometry education. Thermodynamics arose when eighteenth- and nineteenth-century engineers played with glowing coals, hissing steam and clattering rods and valves of steam engines and observed with some astonishment the processes unfolding in them. What was this heat which flowed like a liquid from one part of the engine to another, they wondered. Did heat move around the world from object to object and from liquid to gas, perhaps even without being created or destroyed?

At least, this was the belief of Frenchman Sadi Carnot, named after a Persian poet, who served as an engineer in the French army. He noticed that when heat flows spontaneously it always does so in only one direction: from hot to cold. A hot cup of tea never gets hotter in cold hands, an ice cube melts in a glass of lemonade and a warm wind blowing over cool water cools down and never heats up. And just as flowing water can be used by a mill wheel in a water mill to saw wood or grind wheat, heat flowing from a heat reservoir to a cold reservoir can be used to turn the wheels of a steam engine. Carnot went on to demonstrate that a steam engine could never be 100 per cent efficient. He also thought that heat was never lost in the process, just as water doesn't disappear in a mill wheel.

British Brewer and amateur physicist James Joule then showed that this latter comparison didn't entirely hold true. He had actually wanted to investigate whether he could make his brewery operate more efficiently with electric motors, which had been invented recently. But upon discovering that electric wires generated heat when conducting electricity, further experimentation and contemplation led him to divert his attention from his brewery towards the idea that 'heat' and 'work' are manifestations of the same phenomenon.

In a very elegant follow-up experiment, Joule used a falling weight—controlled by a pulley—to set a series of paddle wheels in motion. This didn't only make the water in a barrel splash but increased its temperature a little, he discovered. In the language of modern physics: work (such as the paddle wheels) and heat (such as in the water) could be converted into each other. Prussian Rudolf Clausius and genial Englishman Lord Kelvin then conceptualised such observations into 'axioms', which they called 'laws'.

Altogether, this process was a wonderful illustration of the three steps each thought process consists of, according to Tanya. Engineers started by observing and discovering patterns and characteristics: exploration. People like Carnot and Joule organised such patterns, filled in gaps and made logical inferences: ordering. Then physicists such as Clausius and Kelvin formulated fundamental premises upon which the edifice of thermodynamics rests: conceptualisation. While Tanya had lauded intuition and playful research in her geometry essay, she now revealed another side of herself. Her article on thermodynamics focused on the final phase of the thought process: Clausius and Kelvin's 'axiomatic reasoning' and some loose ends in this.

Compared to quantum mechanics, thermodynamics was a rather obscure niche of physics. Yet it wasn't completely forsaken. Greek academic Carathéodory tried to give thermodynamics more rigorously conceptualised 'axiomatic' foundations in 1909, and mathematically minded Max Born in Göttingen attracted some renewed interest in Carathéodory's work in the early 1920s (Carathéodory 1909). It did Tanya good that Carathéodory was subsequently interested in her work, as incidentally was Kohnstamm, who immersed himself in thermodynamics as physics professor in Amsterdam.[18] It gave her the courage at the end of July to submit her work to the authoritative *Zeitschrift für Physik*, which in December published her 'On the Axiomatisation of the Second Law of Thermodynamics', in which she tried to supplement Carathéodory's 'exceptionally valuable attempt to axiomatise thermodynamics' (Ehrenfest-Afanassjewa 1925a).[19]

Her article focused on the second law of thermodynamics. The first law was undisputed. It states that energy can be transformed from one form into another—from electrical energy into heat into work and so on—and that the total amount of energy in an isolated system remains the same (incidentally, temperature is the measure of the internal energy of such a system, as Kelvin had shown). But with the second law, a confusing new concept had been introduced into thermodynamics. It supplemented the list of three variables—pressure, temperature and volume—with which physicists characterised gases and liquids.

Clausius initially needed this new concept of 'entropy' mainly for the definition of the second law, which states that 'in a circular process heat can never be transferred from cold to hot'. 'Entropy' allowed Clausius to distinguish between spontaneous processes in which heat flows and the kind of cyclic processes which occur for instance in a steam engine, in which hot steam moves the valves and rods in a repeated rhythm. When an engine performs work because heat flows from a heat reservoir (such as a boiler) to a cold reservoir (such as cooling water), a small remnant of heat will always flow into the cold reservoir. Or conversely, the heat reservoir always loses some heat which can't be recovered, even if the valves and rods only produced heat and returned all this heat to the hot steam. This allowed Clausius to show that entropy increases in spontaneous processes, while it doesn't change in circular processes.

Perhaps representing the concept of entropy as guilders sheds light on the matter. Paul and Tanya's bank balance would increase greatly if a rich man such as

Solvay were to deposit 10,000 guilders on their account. By contrast, the decline of Solvay's bank balance would be relatively small. In this case, entropy is the measure of the total of relative changes from transferring money. Or in nature: transferring heat from a heat reservoir (with a 'high' heat balance) to a cold reservoir (with a 'low' heat balance). In a well-designed, cyclic steam engine, the amount of heat flow converted into work is exactly such that the relative decrease in heat in the steam equals the relative increase in heat in, for example, the cooling water system in which residual heat is dumped, Clausius demonstrated in 1854. Only a system like this, without a change in entropy, can be continually repeated. In all processes that couldn't be repeated or reversed, entropy increased. And as the universe abounds with such processes, Clausius stated that 'the entropy of the universe is increasing to a maximum'.[20]

Kelvin formulated this second law slightly differently but with the same result when he established that a steam engine (or any other process that always goes through the same cycle) can never work when there is no difference in temperature. Did Clausius' 'tendency to maximum entropy' accord with Kelvin's idea that the entire cosmos shall become a uniformly tepid and lifeless bath because every difference in temperature shall be erased? Did this tendency to maximum entropy concur with what Boltzmann later seemed to demonstrate: that atoms and molecules, colliding and jostling, invariably move towards a state of equilibrium in which they all move at roughly the same speed and carry roughly the same amount of energy? Or was the word 'tendency' significant here? Because Paul and Tanya's urn model had long since shown that such a state of equilibrium wasn't necessarily definitive.

Such questions made Tanya think. What are states of equilibrium and disequilibrium, and what is the difference between reversible and irreversible processes? Were the definitions of these precise enough to draw a clear distinction between these cases, and was this distinction even justifiable at all? But in her article in 1925, she demonstrated chiefly that the axioms proposed by Kelvin and Clausius and supplemented by Carathéodory weren't entirely sufficient to underpin the 'edifice of thermodynamics'. She illustrated the lacunae in these laws with an engine with two boilers that tapered cleverly and were linked together through pistons. Even when the pressure was the same throughout the system, the pistons could keep each other moving with their clever shape. This enabled extraction of work (using heat) from a reservoir with uniform temperature everywhere—contradicting Kelvin as well as Carathéodory's elaborations (Uffink 2000; Uffink and Valente 2020). Incidentally, in her article Tanya thanked 'P. Ehrenfest', who had pointed out such clever trickery to her.

Tanya seemed a little finicky when she then emphasised that Clausius' and Kelvin's definitions of the second law weren't equivalent. Clausius' definition only applied under one extra condition: the absolute temperature was constantly positive (or constantly negative) in the systems he was talking about. Kelvin had defined temperature as a measure for the movement of atoms and molecules, and his 'Kelvin scale' went down to absolute zero (−273 on the Celsius scale) when atoms and molecules stop moving. This gave the impression that temperatures automatically had to be positive. Yet it wasn't mere pedantry that Tanya wanted to take negative temperatures into account. 'However, if one is compelled—such as by quantum theory—to deviate from this explanation, it leads to a distinctive statistical interpretation', she wrote (Ehrenfest-Afanassjewa 1925a). And 31 years later, in 1956, American physicist and Nobel laureate Norman Ramsey did indeed describe quantum mechanical systems which resulted in 'negative' temperatures (Ramsey 1956). He was unaware that he had fallen back on Tanya's old prediction.[21]

With the upheaval of quantum mechanics and atomic theory, interest in this field had waned by 1925. When Tanya wrote to him a year after her publication, even Carathéodory responded late and apologetically. 'With great joy I read in the programme of the British Association that your husband shall also be present. Hopefully, you shall come too. You shall probably be annoyed with me for having been out of touch for so long', he had scribbled on a postcard in summer 1926.[22] He said that he had started to write a book about thermodynamics but had abandoned the job halfway through because '[I] have again been doing pure mathematics in the last few months. When this little book is finished, I should like to ask you, if you are willing, to read it'.[23]

In Amsterdam, on the other hand, was preparing a book on thermodynamics (Bussato 2015, Ch. 3). The book was a revised edition of his two-volume Textbook of Thermodynamics,[24] in which he precisely and exhaustively gave account of the lectures his mentor, Nobel laureate Johannes van der Waals, on thermodynamics (Van der Waals and Kohnstamm 1912). 'Dear Mrs Ehrenfest', he wrote on a postcard on 6 September in which he thanked her for a manuscript she had sent as a supplement to her article. Was this manuscript an attempt to turn her article into something like a dissertation? In any case, Kohnstamm had sent it back, shortly before Tanya was supposed to travel, and after having 'read it with much interest and almost constant assent'.[25]

'I am sorry that we shall no longer see each other. ... I have, in fact, found reason in our conversation to send the enclosed note to Leipzig', he wrote. In the enclosed typed letter, he had asked the printer Ambrosius Barth in Leipzig to suspend preparation of the proofs for the time being. 'In the interim, discussions with *Frau Prof.* Ehrenfest-Afanassjewa have ... shown me that substantial points in the manuscript can be made clearer ... To avoid unnecessary corrections I would like to request you to immediately stop further preparations of the manuscript for printing and to advise me how far you have progressed with the proofs. Then I shall decide whether it is better to stick to the original plan and include the axiomatics in an appendix or to rewrite the manuscript and directly incorporate the axiomatics in the systematic organisation'. The intended changes went further than reorganising the manuscript. 'The changes appear to me in principle so extensive that the thought occurs to me that it would be considerably better for the title to be "Textbook of Thermostatistics"', Kohnstamm even wrote.[26] This book was indeed published under this title in 1927, with an acknowledgement to Tanya in the preface.

Was this praise sweet or bitter? Anyone wanting to count in academia needs publications and citations. The silent prompter who shares her ideas in the domestic circle or travelling with her partner to congresses will be forgotten, even if her name is included in a word of thanks. And even if this name is mentioned elsewhere, this gave women no guarantees: future Nobel laureate Alfred Landé's *Handbuch der Physik* in 1926 referred in the section about Carathéodory's axiomatisation to Tanya's work and called her additions 'important' (Landé 1926).[27] But the debate with Dijksterhuis showed that, certainly also in the Netherlands, few people were welcoming towards a woman, however highly educated, who made her presence felt.

'A woman who has attended university shall later, in society – even if she like thousands of other women, takes charge of a household – be able to create an academic atmosphere. ... She shall channel the upbringing in other directions and by doing so serve science, even if she does not directly contribute towards its advancement', sympathetic Zeeman had explained two years before in a newspaper interview in answer to a question about the point of having female students in a physics laboratory.[28] The chances of Tanya having some lasting influence through her work, as she had so very much wanted, seemed lost in the Netherlands. When her article was published in 1925, Tanya was almost fifty and a mother of four children in a world that defined women differently to how she did.

30
Physics curator 2.0
Passionately interested in people's fate

Paul answered better to what the world expected of him, a man of talent. It was difficult not to be impressed by him, many colleagues said. His network now extended from the American west coast to deep in Russia. Halfway in between, in Leiden, when he appeared on a colleague's doorstep, his twinkling eyes seemed to light up the world (Reinder 2018, 46). He would positively shimmer with delight whenever he put people in contact, successfully arranged funding or spoke about the latest publications, whose offprints had arrived in his letterbox in Leiden.

His trips to the Soviet Union and America had energised him and enlarged his network. 'My trip shall bear fruit. Perhaps not for myself but for the boys in Holland, Germany and Russia', he wrote to Tanya during his stopover in London en route to America.[1] He had just spoken to Dr Wickliff Rose, a 'fine old doctor', who assessed which European researchers deserved development grants and scholarships for the American Rockefeller Foundation. Paul immediately presented him a list of names of British, Russian, German and Dutch physicists and gave some unsolicited advice. 'I started off diplomatically, but after two minutes, forgetting everything, I railed against the high-pressure American scholarship system, and it was precisely with my incautious complaining that I won his confidence. I said: one must also plant forests and not just build factories. Give 50% of your money to people who dream'.[2]

Paul was one of these dreamers too. In America, he had been impressed by the enormous laboratories, such as those of General Motors, and by some research facilities, such as the magnificent telescope on Mount Wilson. Hugo praised the speed with which Paul discerned the 'unique way' in which the United States dealt with 'the problem of universities (education), materialism, [societal] utility' and issues such as 'who pays the universities; who determines the programmes?'[3] Several times, he reiterated in letters to his little brother the value of the broad education at American colleges and the many advantages of American 'materialism' and private individuals sponsoring research. Professors really didn't have to live like scholarly paupers, Hugo thought, who had gained a rather dramatic impression of Dutch professors' salaries from Paul. But Paul, having of course little influence on the matter, was chiefly impressed by American pragmatism and how Americans almost directly turned research into technology, even if this research wasn't yet at the same level as in Europe.

The Delight of Thinking. Margriet van der Heijden, Oxford University Press.© Margriet van der Heijden (2021). DOI: 10.1093/9780198927112.003.0030

In the Netherlands, he shared these impressions with leading lights in business such as Gilles Holst, the first director of the Physics Laboratory (NatLab) that Anton and Gerard Philips had added to their light-bulb factory in 1914. As a student in Leiden, Holst had become familiar with Onnes' cryogenic laboratory, which was a large-scale research laboratory for the time. Moreover, he had also heard about the latest developments in quantum and atomic theory and the associated light spectra during his postdoctoral research in Zurich under Pierre Weiss. And who better than Paul, well acquainted with Onnes, Weiss and many others, could explain the developments in these fields to NatLab staff?

An added attraction of Paul's lectures and classes at NatLab, which started in 1920, was the extra income. In addition, he made new friends there and chatted convivially for many hours with young '*Frau* Holst'.[4] Of course, Gilles Holst's interest in his vision of the relationship between science and business flattered him. Paul also shared these ideas with his former assistant Adriaan Fokker, who had left theoretical physics for a chair in Modern Physics and Technology at Delft, and whose cousin was aircraft manufacturer Anthony Fokker. And he discussed them with his politically engaged colleague and good friend Philip Kohnstamm, whose father-in-law, August Kessler, had made a success of the Dutch oil company in the Dutch East Indies which would go on to become Shell (Poley 2000, 93). Paul had added another 'space' to his 'gallery', where universities and companies could exchange knowledge.

But his true love would always be theoretical physics. In Leiden, he threw himself into it with Uhlenbeck, who had succeeded Tinbergen as Paul's assistant in 1925. Paul and Uhlenbeck worked together 'practically every afternoon during the week'. They discussed 'the problem on which he was working or recent papers in the literature which he wanted to understand in detail'. It was intense. 'I can personally testify that in the beginning ... at the end of the afternoon one was dead tired', Uhlenbeck recalled years later. 'Especially because one had to follow [everything] in detail. The greatest sin was to say that one had understood the point if it was not the case. And it was always found out!' (Uhlenbeck 1956).

Paul was 'really a maniac', others also said a bit jokingly. He 'hated' it when people tried to escape his lengthy expositions by saying 'Yes, yes, I see already', said future Nobel laureate James Franck. And it wasn't that easy to shake him off: 'So if you see, then tell me ...' Franck enjoyed recounting how Paul once started such a lengthy explanation in the middle of the street at a busy intersection, leaning against a lantern with one hand. 'Apparently, if he would have had a piece of chalk, he would have written down something on the lantern'. After this incident, Franck resolved to ask him questions only about issues requiring forensic examination (Franck and Sponer 1962).

Ardent Uhlenbeck still had the patience and unconditional loyalty as a student to endure such expositions and to profit from them. 'The wonder was that after a while the tiredness disappeared, and after a year one worked almost as equals.

In fact, as a student you often had the sneaking suspicion that you really knew the things much better. At that point one stood on one's own legs and one had become a physicist!' (Uhlenbeck 1956). Burgers, who was the head of a new institute in Delft for aerodynamics and hydrodynamics by then, had the same experience (Burgers 1962). Even Kramers, who had gained fame as Bohr's deputy, continued to praise his strict mentor. As did Marcel Minnaert, who Paul had dismissed from his colloquia in 1917 for his pro-German and Flemish-nationalist sympathies but who had become a prominent astronomer in Utrecht. In letters, he wrote that 'Prof. Ehrenfest' had always remained 'the main person in Leiden' for him—'the model professor and almost the model person' (Molenaar 2003, 109).[5]

It is true that the entire set Kramers, Burgers and Minnaert were part of had landed on their feet.[6] With the aid of Rockefeller grants, Struik was making a career for himself in mathematics in Rome and Göttingen. Coster made it into the newspapers with his discovery, together with Hungarian Charles de Hevesy, of the new element hafnium in 1923. And undoubtedly, they were an example to the new group of students with Tinbergen, Goudsmit, Uhlenbeck and Dieke. 'Dear Tinbergen and dear fellows on the other side, so far, far away. I think at least five times as often about each of you individually as all of you think about me *together*!' Paul wrote to the group from Pasadena in early 1924, according to the letterheading from his 'office in the Laboratory with a splendid variable airstream—*cold* or *hot* as one pleases and everything in between—unfortunately no divan'.[7] Was this really so? It was clear that, once again, Paul demanded unconditional effort, loyalty and admiration in exchange for all the attention.

Goudsmit was probably the member of the group least inclined to reverence. Suffering from an immense fear of failure, it was good for him that Paul, sometimes to the frustration of colleagues, dared to bend the rules. Pacing up and down the corridor discussing the subject matter also counted as an examination for Paul. Moreover, it had been a good idea for Paul to send Goudsmit three days a week to Zeeman in Amsterdam from September 1924. Goudsmit delighted in spectroscopic measurements and enjoyed the more casual atmosphere in contrast to stilted Leiden. 'The jokes you heard and could tell in Amsterdam, you couldn't tell them in Leiden. That was impossible. It wasn't respectable enough. It was really convivial in Amsterdam' (Goudsmit 1971).[8] Yet Paul's advice also instilled doubts in insecure Goudsmit: did he see a theoretician in him? But his self-confidence received a boost when Paul paired him with friendly and intelligent Uhlenbeck in summer 1925.

Uhlenbeck had left behind the ambassador's son in Rome, along with his dream of becoming a historian, and succeeded Tinbergen as Paul's assistant after summer. But his knowledge of physics could do with improvement, Paul thought, despite having attended lectures in Rome. 'There in Italy ... they only know classical physics' (Goudsmit 1971).[9] So he proposed that Goudsmit bring Uhlenbeck up to standard in the field of 'atomic structure and all this spectroscopical business' (Goudsmit 1971).[10] And while Uhlenbeck and Goudsmit went through Bohr's latest ideas on the structure of atoms and studied Sommerfeld's former doctoral students Alfred Landé and Wolfgang Pauli's work, they proved to be a perfect match. Uhlenbeck dealt with problems analytically and loved formulas; Goudsmit had keen instincts and approached his work like a detective (Pais 1989, 37).

Perhaps this caused Paul to underestimate him a little. In 1921, when Goudsmit 'guessed' the equation for how two chromatic tones can be close in the atomic light spectrum, Paul's response was tepid. He had Goudsmit write a brief note to *Naturwissenschaften* (Goudsmit 1921) and a longer text to *Archives Néerlandaises des Sciences exactes et naturelles* (Goudsmit 1922), a journal which, as Goudsmit later said sardonically, 'was published in French in the Netherlands to be really certain nobody would read it' (Goudsmit 1971, 387).[11] The result was that almost three years later American scientist Millikan published the same formula again, with new data as well, and received a great deal of praise. Coster delivered a lecture on the subject in Leiden in spring 1924.

'I put up my hand and said: "I talked about exactly the same thing three years ago"', Goudsmit recounted years later (Goudsmit 1963; Goudsmit 1971, 387).[12] 'And suddenly Coster remembered, "Oh, this poor man", he said, "he talked about the same thing three years ago"'. But from the United States, Paul matter-of-factly brushed aside the incident, described in a letter from an upset Goudsmit. It was something completely normal, which could have happened to 'any well-educated beginner', he wrote back, while seated in an armchair in the Men's Faculty Club at Columbia University in New York.[13] 'You're a blockhead. ... Don't you understand what a fine thing this is—what you discovered corresponds with what the best qualified, mature specialists discover almost at the same time?' And Goudsmit, never very self-confident, left it at this.

But now, in summer 1925, he could share all his knowledge with Uhlenbeck. He knew the atomic spectra inside out, and Uhlenbeck posed exactly the right questions. Soon they published two short papers together in *Naturwissenschaften* and Dutch-language journal *Physica* (Uhlenbeck and Goudsmit 1925; Goudsmit and Uhlenbeck 1926). What was even better was that their knowledge dovetailed with Pauli's intriguing publication on atomic spectra and on 'quantum numbers'.

It had been twelve years since Bohr had described the simplest of all atoms, the hydrogen atom, by assuming its electron could only move in 'discrete' orbits. In other words, orbits you can count with natural numbers (1, 2, 3, ...), while disregarding all numbers between them (between 1 and 2, 2 and 3, ...). As if the electrons in an atom can't skate fluidly through space like over a smooth icy surface but had to jump from ice floe to ice floe. In an extended exchange of ideas with Sommerfeld and partly also Epstein in Munich and others, Bohr subsequently refined this model and expanded it to heavier atoms, with far larger numbers of electrons travelling around the atomic nucleus. Spectroscopic data had been crucial, establishing the specific range of colours—or energy parcels—which atoms emitted or absorbed when electrons jumped back and forth between possible orbits.

In 1922, Bohr grouped these orbits into energy levels like separate 'shells' around the nucleus, separated by empty space. The first shell, closest to the nucleus, contained two electrons, each in its own elliptical orbit. A more distant second shell had eight electrons, followed by a third shell with ten electrons and so on until the outermost shell, which was often not completely filled. In principle, when electrons absorbed a specific light particle, they could jump to empty shells farther from the nucleus, and at a given time, each electron could be characterised with three quantum numbers: one specifying the 'shell' and two defining the elliptical trajectory in the shell.

Wolfgang Pauli then proposed the existence of a fourth quantum number, less concerned with the orbits of the electrons and more with the electrons themselves (Pauli 1925). Such a quantum number could perhaps explain the old results of Pieter Zeeman, who had demonstrated that a single colour in the atomic spectrum—a 'spectral line'—could split into two tones, or spectral lines, in the presence of a magnetic field. Lorentz' classical explanation of this phenomenon was no longer adequate in the new quantum era. So when Goudsmit and Uhlenbeck made a concrete proposal for this fourth quantum number in 1924, their idea was not entirely unexpected. Indeed, Ralph de Laer Kroning had already made a similar proposal but had withdrawn it after strong criticism from Pauli.

Goudsmit and Uhlenbeck proposed that, alongside their trajectory through three-dimensional space, electrons had another way of exhibiting motion: spinning around an imaginary axis. In fact, they argued that this spinning caused the electrically charged electrons to align in a magnetic field: if they spun leftwards, for example, they pointed northwards; if they spun rightwards, they pointed southwards. This meant this characteristic, which they named 'spin', was 'discrete', resulting in an enumerable number of possibilities. And with Goudsmit's extensive knowledge of atomic spectra, they soon demonstrated that the spectrum of hydrogen, the spectra of heavy atoms as well as the Zeeman effect could be thoroughly explained with the addition of this fourth quantum number.[14]

As was so often Paul's wont, he involved Lorentz. He asked his 'judgement and advice on a very clever idea Uhlenbeck has had about spectra' (Kox 2018, doc. 271).[15] And Lorentz, in a conversation with Uhlenbeck, produced a counter-argument Paul had apparently overlooked. The problem, according to Lorentz, was that a spinning electron would generate so much extra energy that the atom couldn't exist at all. But this time, Paul hadn't waited for Lorentz' advice, having already sent off the article to *Naturwissenschaften*. Was he compensating for his coolness towards Goudsmit's earlier work? Or had he thought that

Lorentz' objections wouldn't be insurmountable? He maintained that Goudsmit and Uhlenbeck would come out unscathed if it were a failure: 'You haven't yet built up a reputation, so you have nothing to lose' (Goudsmit 1971).[16]

Things went well. The article was well received by younger physicists when published in November (Uhlenbeck and Goudsmit 1925). Up-and-coming talent Werner Heisenberg from Göttingen called it courageous (Pais 1989). And by happy coincidence, a group of international physicists gathered in Leiden for the 50th anniversary of Lorentz' doctorate on 11 December 1925. They discussed the idea at once, before and after speeches by Prime Minister Colijn, Rector De Sitter and international guests such as Curie and, of course, Einstein.

It was a festive occasion, and it was the second time that year that Einstein took part in a procession of professors through Leiden. Earlier that year he had also been invited to the ceremony at which Dutch Princess Wilhelmina received an honorary degree, wearing what he mockingly called a 'pompous outfit'. 'It is all fine with me as long as I don't have to give a speech', he had written on a postcard to Tanya at the time.[17] But now, he was only too glad to deliver a speech to Lorentz because Lorentz meant more to him 'than all the others I have met in the path of my life'.

For Paul, the highlight of the festivities was afterwards. Despite his successes, he felt increasingly oppressed in the low-lying polder landscape of Leiden. He even considered asking Kramers to succeed him and transferring to Utrecht. But now, seated on the sofa in his study—in Leiden for Lorentz' celebrations, but still— were two giants in physics: Einstein and Bohr. He loved their more intuitive and conceptual approach to physics, which he found far more comprehensible than the mathematical approach of Sommerfeld in Munich and, these days, Max Born in Göttingen. With these two 'giants' slumping lightly there, he undoubtedly forgot for a moment that he had never reached the level of Lorentz, whose reputation abroad grew even further that day. Paul captured with his camera how Einstein smoked a cigar somewhat pensively, while Bohr looked at the smoke rings, perhaps looking for counterarguments (see Figure 30.1).

This was a crucial moment. Einstein had been the first person to suggest the granular nature of light in 1905. Then Bohr presented his atomic model with discrete electron orbits eight years later. But while Einstein wanted to take the granularity of nature as the starting point for further research, Bohr still harboured doubts. His 'correspondence principle' connected the granular inner world of atoms to the more fluid outside world, but Bohr wanted somehow to preserve this continuous and fluid world inside atoms too. In 1924, he, John Slater and Kramers, made a final attempt to lay a continuous description of nature beneath the quantum description of atoms (Bohr, Slater, Kramers 1924). Much to Kramers' dismay, he was even willing to sacrifice one of the most important fundamental assumptions in physics for it: the conservation of energy. Paul sided with Einstein, who thought this an exercise in futility. 'The "Bohr-Kramers-Slater company" is

Figure 30.1 Einstein and Bohr on the sofa in de study at Paul and Tanya's house on Witte Rozenstraat.—*Photograph Ehrenfest Family Archive.*

persistently accused of "wanting to buy something" with a false bill of exchange', he scribbled in the margin of a letter from Bohr in October 1925.[18] But by then, Bohr's old quantum theory had already largely been supplanted by a radically reframed description which no longer questioned the discrete character of nature at the smallest scales: quantum mechanics (see e.g. Duncan and Janssen 2023). And now, Paul looked on with inward satisfaction as the two discussed the matter.

That Bohr and Einstein both stayed with Paul and Tanya must have compensated for all the praise for Lorentz, whose achievements so often overshadowed Paul's life and work. Lorentz was a Nobel laureate. Lorentz was the much-vaunted chairman of the famous Solvay Conferences. Lorentz was a much sought-after speaker. Lorentz spoke fluent Dutch, German, French and English, along with a smattering of Italian and Spanish. Lorentz and Einstein had a relationship based on mutual admiration. Lorentz, like Curie and Einstein, was a member of the International Committee on Intellectual Cooperation, which aimed to restore good relations between researchers after the damage wreaked by World War I, even becoming its chairman. Lorentz' article on Einstein's theory of relativity was published in *The New York Times*. Lorentz gave something back to the people of the Netherlands from 1918 by endlessly calculating the water flows in the Zuiderzee as well as the proposed Afsluitdijk causeway.[19] Even the Dutch movement for women's equality was well disposed towards him! Paul seemed to lag him in every area.

Except perhaps as a 'charismatic curator' of science. A network like his, with so many young people in it, was something Lorentz lacked. Neither could he offer his guests a comparable decor: with a large 'Russian' house with talented children in it; where the doors were decorated with Galinka's cut-out tableaux; where Tanitchka was making a furore in mathematics and physics; where a violin on the grand piano in the study was ready for Einstein; where an elderly aunt and mother-in-law made conversation in French and Russian at the dining room table and where cheerful Tanya tended the garden as meticulously as her publications in thermodynamics. Together with his witty and sharp intellect and love for the subject, this somehow also made Paul an important player (see Figure 30.2).

With his genius, one-liners and absent-mindedness, Einstein affirmed the general public's idea that physics was the domain of great men, who occasionally emerged to present a groundbreaking formula. Lorentz stood as a brilliant mind above all the parties. But Paul stood for another aspect of science. 'It is not by discoveries only, and the registration of them by learned societies, that science is advanced', celebrated Maxwell said on this subject. 'The true seat of science is ... in the living mind, and the advancement of science consists in the direction of men's minds into a scientific channel; whether this is done by the announcement of a discovery, the assertion of a paradox, the invention of a scientific phrase, or the exposition of a system of doctrine' (Maxwell 1965, 401).

Figure 30.2 Einstein, Goudsmit, Uhlenbeck and Bohr at the blackboard in de study at the house on Witte Rozenstraat.—*Photograph Ehrenfest Family Archive.*

It was thus fitting that Maxwell had a prominent place in the row of portraits on the study wall in Witte Rozenstraat. He looked down at Tanya, who tirelessly participated in debates on thermodynamics and the didactics of mathematics, and on Paul who practised this living intellect in lectures, colloquia, picnics and trips with students, in working long afternoons with assistants, looking for paradoxes and metaphors that elucidated physics and starting fiery discussions with learned colleagues. Paul wasn't an erudite recluse or an isolated bookworm but was 'passionately preoccupied with the development and destiny of men, especially his students', as Einstein later wrote (Einstein 1934; Einstein 1976, 238).

PART 3

31
Falling behind
Mathematics in Moscow and mathematics as a blight

'Dear', 'Dear Mummy', 'Dear Pavlinka.' From 1926, a flurry of Soviet-franked letters arrived at Witte Rozenstraat. Tanya had taken the train to Moscow in early January to work there for a few months. 'A.F. Ioffe has already discussed my wish with you to work at the organisation for education and teacher training in Russia', she had written a month earlier in a draft letter to Albert Petrovich Pinkevich.[1] The Russian educator had played an active role in the 1905 revolution and had been a moderate Menshevik in 1917—causing him to reject the October Revolution—yet he had been prepared to work with the Bolsheviks. By this time, he was the dean of the Second Moscow State University.

'I appreciate that allowing me to hold a certain post raises all kinds of problems because I cannot move to Russia with my family - in any case for now, and I have no choice but to offer you my services for only part of the year', Tanya continued. She understood that it was difficult for him to invite someone he didn't know at all. And: 'the third possibility is undoubtedly that all existing posts are already occupied by people with an excellent understanding of their profession, rendering me a little superfluous. Personally, I would find this latter circumstance especially troubling because I have no desire for my appointment to curtail someone else's professional domain. Nonetheless, I am convinced that Russia needs workers at this time and that it would be wrong of me to refrain from devoting my capacities for Russia's benefit'.[2]

These complemented, she felt, all the developments in education in Russia, where educational thinking was 'brimming over' and 'where old conventions clash with the many totally new and sometimes still developing ideas'. For the best results in education, it was essential to clarify these new ideas and to make them 'practicable for the great mass of teachers'.[3] And Pinkevich and his colleagues agreed that Tanya was eminently suited to this task. She was issued a visa in early December 1925.[4] As a temporary professor, she delivered a lecture series on mathematics didactics at the Pedagogic Faculty of the Second Moscow State University from 1 February 1926.[5]

Paul acted as if this were commonplace. 'My wife has gone to Moscow via Jena-Göttingen-Berlin to work there for 3–4 months. Yours Sincerely, your P. Ehrenfest', he wrote at the end of a letter to Bohr in early January.[6] But their Dutch acquaintances were astounded: how could a professor's wife, with a magnificent house, so many interesting visitors and three growing children run off just like that? People in the Netherlands were less used to delegating childcare to governesses, nannies or family members than the circles in St Petersburg where Tanya grew up. Tanya had also been brought up by her aunt, with the aid of tutors and domestic staff. In a sense, it must have seemed rather fitting that Tanya's own mother—who had worked for years with children in the orphanage in Petergof—now kept an eye on Pawlik.

The ten-year-old boy, cheerful, clever and dextrous, was the only child still needing much supervision. Tanitchka was studying for her *doctoraal* degree in mathematics, having sailed through her *kandidaat* examinations in physics and mathematics, aged nineteen, at the end of 1924.[7] Sunny, fifteen-year-old Galinka had started training to be a pre-school teacher in The Hague later in 1926, at Cornelia Philippi's institute, inspired by educational innovators such as Maria Montessori and Jean Piaget. She was staying in The Hague with the family of conductor Peter van Anrooy, a good friend of conductor Adolf Busch in Berlin, and so didn't necessarily need her mother to be in Leiden.[8] And Tanya could travel back and forth from Russia to visit Wassik.

Yet many Dutch people found it difficult to grasp why Tanya exchanged her life in Leiden for an uncertain adventure in a country racked by hardship, violence and scarcity in almost every sphere.[9] And this while the 'Ehrenfest house' had grown into such a scintillating oasis, where scientists from the world over met. In 1925 alone, Einstein, Hertz, Franck, Werner Heisenberg, Ioffe, Tinbergen, Coster and Margrethe and Niels Bohr had left their signatures on the wall in the attic— and of these names, only Coster and Margrethe Bohr weren't destined to become Nobel laureates. Such guests minded little that the meals were only just adequate, the bathroom was shared, and alcohol and cigars were taboo. Just as most of them put up with the fact that the cordial reception, the informal atmosphere and Paul's cheerful banter went hand in hand with long, long discussions.

Franck understood why Einstein felt so happy in the relaxed household, he later said, where most guests walked straight from the garden into the dining room and where a pocket watch on a nail in the wall sufficed for a clock. Einstein could 'be himself and do what he wanted to'. Neither did Einstein care whether there were 'normal nice chairs there or what not' (Franck and Sponer 1962, session V). Just as he didn't care that there wasn't any wallpaper on the walls; not all the windows had curtains; by some mysterious logic, the letterbox was in the wall in the study, the bell was at the garden gate on the other side and the front door was farther along in the west wall.[10]

However, some guests also knew that Tanya wasn't very happy. The humid Dutch climate irritated her respiratory tract, and she found Leiden provincial. 'Until we came to Holland, we were accustomed, wherever we were, to find a circle of people with whom one could at any moment discuss all things in heaven & earth', she wrote much later. 'Here, however, one had to go through a long ritual first of talking about the weather, the children, whether they went to school and the like, but even after that people weren't ready without taking a special resolution, to talk about a scientific or generally important subject'.

'I believe', Tanya added in parentheses, 'that they are so unaccustomed to it, that they simply do not have thoughts concerning this, and also that they are not accustomed to finding a clear formulation for them in case they do have them'.[11]

She yearned for these conversations about 'all the things in the heavens and on earth'. Her homesickness for Russia must also have been intensified by reports that women in her native country now worked in offices, factories and workshops, as well as actively participating in associations, societies and at conferences. New legislation even forbade the dismissal of pregnant working women, decreed four months of maternity leave and double pay in the month the child was born.[12] At the same time, the Dutch law dismissing married women from government employment had resulted in nine out of ten married female physicists and mathematicians ending up at the kitchen table, or perhaps in the living room, but definitely confined to the home (Kirejczyk 1993).

Tanya was hardly a convinced socialist. In her earlier, idealistic description of the 'new economic order' she would like to live in, 'partial socialisation' had been her starting point. The state should only control the companies, raw materials and agricultural land necessary to provide the basic necessities of life, she argued, noting that 'the definition of basic necessities and products is of course subject to change'. In any case, she thought it 'better to socialise less rather than more companies' because 'excessive socialisation shall – in a different fashion to capitalism – result in the same or perhaps even worse slavery and shall prove to be just as unstable'.[13]

Tanya's ideal was to give everyone, young and old, equal opportunities as much as possible. In notes for a lecture about her ideas in 1924, she emphasised this once more. 'In the current, unorganised situation, impecuniousness in childhood results in individuals' inability to develop themselves to the full'. This led her to supplement her earlier ideas by proposing a basic income. 'Suppose everyone was given the essential minimum their entire lives. This would mean that everyone from their birth would receive a minimum allowance from a state insurance corporation which he would have to pay back to the state when an adult. Old-age insurance [pension schemes] already exists; people would greatly benefit if they insured their children's lives'.[14] Did Tanya, in anticipation of this, want to try to increase all children's opportunities to develop themselves by organising good education in Russia?

In late January 1926, not long after arriving in Moscow, she was granted a permit to live in Ulitsa Volkhonka.[15] Her work was far from lucrative; her flat was cold; food was frugal; paper was scarce and sometimes there was no running water or electricity. But she immersed herself in the Russian language and renewed old contacts. Her old colleague and friend Benjamin Kagan was now a geometry professor at the Moscow State University. And she also regularly visited Lydia Solomonovna and Leonid Isaakovich Mandelstam[16], who had been appointed a theoretical physics professor at the same university—a position the university administrators incidentally would have preferred to give to Paul (Pechenkin 2019, 90).

None of them were convinced Bolsheviks. Sometimes Mandelstam said jokingly that he immediately suffered from a headache when having to converse with a Bolshevik (Pechenkin 2019, 87). But their optimism and idealism outweighed this. In these years, they could move about relatively freely as academics, and Russian Jews were no longer subject to restrictive measures and quota. This meant Kagan finally had a good position, as did Mandelstam and for example Boris Hessen, who became the director of the Physics Institute of the Moscow State University in 1926. Added to which, the Bolsheviks seemed, for the time being, to recognise the crucial societal role of science and research, along with the importance of good education.

Tanya, too, turned her mind optimistically and energetically to plans and ideas for educating all the children and young people who would have remained illiterate and innumerate in the old system. That things were more complicated in practice, as Paul had predicted years before, was clear. But she endured, returning to Moscow after spending the summer with Paul and the children. This time, she organised a seminar on thermodynamics at the Communist Academy—from 1 October to 1 June for 50 roubles a month.[17] Among all the socially relevant subjects given by a hundred researchers at the new academy, her theoretical course was a little out of place. But 'my wife is staying with much pleasure and success in Moscow', Paul wrote in a postscript in a letter to Bohr shortly before Christmas.[18]

Yet it did Paul no good that Tanya was away for so long. Perhaps only love can counterbalance the aversion to other people and the world, he had written long before. But their love had grown a bit stale. Tanya had become a little insistent, some people thought, although this might have partially been because she felt restricted in the Netherlands. Franck sometimes found her 'a little bit difficult to come along with, because she never stopped talking about the things she was doing' (Franck and Sponer 1962). Or was his vision coloured by prevailing mores, which frowned on women talking about work? In any case, she persisted with her plans to travel to Russia. And her perseverance, which had previously been an anchor for Paul, now resulted in him being left behind on his own.

Could this have contributed to Paul's unmooring? Even in the preceding years, his mental state had progressively taken hold of him. 'Pavel Sigizmundovich has lately been tremendously strained', Krutkov wrote in 1923. 'He is writing an article about Bohr for a popular German periodical [*Naturwissenschaften*] and is having difficulty writing it. I tease him a little about it and then he quietens down' (Frenkel 1977, Ch. 5).[19] That same year, Fokker noted with some surprise how Paul bid farewell to Lorentz after his final lecture in Leiden. 'Then Ehrenfest said: "Mr Lorentz, I wish you a pleasant journey", he clasped his hand, kissed it at length and tearfully slipped away through the side door'.[20] It seemed as if Paul had suddenly realised that he would deeply miss Lorentz, the man who in the background had guided so many of his plans in the right direction and who had held such high expectations for him.

Paul's complaints about his lack of original work had increased steadily since then, as had his grumbling about everything that he thought impeded this process. And when Tanya was no longer there to vent his frustration, such self-pity increasingly found its way into letters to friends. Sometimes Paul risked alienating himself even from Einstein. His criticism with Uhlenbeck of Einstein's—yet again pioneering—work on quantum statistics was something Einstein could take in his stride. He could appreciate when Paul and Uhlenbeck sent a critique of his work in the form of a satirical paper that they gave a motto borrowed from Schiller: 'When a sovereign builds, canal cleaners also get work', or, when great physicists such as Einstein think up something, ordinary physicists must do the dirty work.[21] But when Einstein tried to cheer up Paul after yet another litany of complaints about his lack of zest for work and life, his words sometimes seemed to convey slight irritation.

For Tanya, too, it must have been exhausting from time to time. Paul's grousing about the heating, their want of money, lazy students, examinations, Tanya's Russian connections, Tanitchka's serious disposition and Baba Sonya's sophisticated fickleness. His brother Hugo had long appreciated how much Paul needed balanced and calm Tanya. In 1924, well before Tanya's departure, he had urged him to emphasise more strongly to Tanya that 'all these Russians who have now been living abroad for years ... might not be such welcome guests when they return to Russia'. People in the Soviet Union, Hugo thought, 'shall resent outsiders for interfering with the Russia they have built at the cost of much happiness and many lives, and which in the truest sense of the word is "their" Russia'. With this argument, he hoped Paul would keep Tanya with him.[22] But his constant complaining seemed to have quite the opposite effect.

Paul was now often alone in their beautiful house in Leiden. And while 'outside' in physics developments seemed to follow in rapid succession, he began to feel increasingly 'like an asthmatic sausage dog' chasing a tram full of young physicists.[23] Einstein won the great debate, which had dominated physics in 1925. The old law of conservation of energy prevailed and nobody any longer doubted the existence of quanta. But thinking on these quanta was now determined by a new, much younger generation of physicists. They were so young that they were dubbed the *Knaben* (German for 'lads'; Weyl 1946, 216). Even Coster and Kramers, who had returned to the Netherlands from Copenhagen, seemed too old and old-fashioned to be part of this group of up-and-coming talent.[24]

The big questions in quantum theory were still largely the same as the ones he had racked his brains over in 1913. In Bohr's atomic model, electrons could jump back and forth between energy levels, like between rungs of an irregular ladder, but which mechanism determined the position of these rungs? And how did electrons 'know' so exactly where the rungs were when jumping? But whereas Einstein and Bohr kept trying to work in a more 'anschaulich' manner, bringing concepts to live and trying to visualise them, one acutely and the other seeking words, the Knaben dealt with these problems mathematically. Werner Heisenberg, trained by Sommerfeld in Munich and then further educated by Born in Göttingen, was one of them. Another was Pauli, who had followed the same course of study, and together with Heisenberg, met Bohr in Göttingen in 1922.

About three years after this first meeting with Bohr, Heisenberg, a hay fever sufferer, decided while on the relatively pollen-free German island Heligoland to disregard the issue of which forces and principles were acting upon atoms. In other words, he would no longer think about atoms as entities you could play with in a kind of theatre of the mind, as Paul would do. In the spirit of physicist and philosopher Mach, and in a long Prussian tradition, he intended to work exclusively with what could be measured and counted, with observables. From a range of spectroscopic measurements, Heisenberg then induced rules establishing how and when electrons could jump.

Young physicist Pascual Jordan and likewise very mathematically inclined Born, once one of Hilbert's '*Wunderkinder*', subsequently translated these rules into mathematical 'vectors' and 'matrices' in Göttingen. This approach was in the air because introverted British genius Paul Dirac presented the same idea almost simultaneously—with even more elegant quantum algebra. In the space of a year, the three young physicists on both sides of the North Sea and Born transformed the old quantum theory into brand new quantum mechanics, with clear rules of behaviour for atoms.

Paul and many colleagues of his generation were still concerned about which physical concepts lay behind these mathematical matrices. For them, it was a relief when not much later French aristocrat Louis Victor de Broglie revealed in his dissertation a beautiful symmetry in nature. Light waves can be conceived not only as particles but conversely particles can be described as waves, de Broglie demonstrated in 1924 (de Broglie 1924). And in early 1926, the same year as Dirac published his elegant quantum algebra, 38-year-old Erwin Schrödinger elaborated de Broglie's idea.

His starting point was that if electrons were also waves then they travelled like waves around the atomic nucleus. For this wave—and therefore the electron—not to be unstable, the head of the wave must connect to its tail. In other words, the energy levels in atoms occur exactly where a whole number of electron wavelengths fits into the path around the atomic nucleus. This allowed Schrödinger to explain the structure of atoms—and more—with a relatively simple wave equation in such a way that physicists such as Paul could visualise them once again (Schrödinger 1926).

All in all, it was no wonder Paul felt dazzled by the torrent of new publications.[25] And his sense of being overtaken by developments was strengthened by the obsolescence of his adiabatic principle. In 1923, he had published an article emphasising the most important achievement of the principle: giving Bohr a guiding principle with which to link the quantum world to the 'ordinary' classical world (Ehrenfest 1923a). But new quantum mechanics no longer required Paul's adiabatic principle—and Bohr's correspondence principle also lost quite some of its attraction.

Things looked different to the outside world, where these were still golden years for Paul.[26] Fermi and Bohr's student Oskar Klein spoke warmly about the time they spent in Leiden between 1924 and 1926.[27] Uhlenbeck and Goudsmit basked in their success there. For tutorials and marking, Paul had amiable Arend Rutgers' assistance. After completing his chemistry studies in Amsterdam, he moved to Leiden in 1926 to take his doctoral degree under Paul on a subject which came to him during a Wednesday evening colloquium by future Nobel laureate and Harvard professor Percy Bridgman—another of Paul's well-chosen guest speakers (Rutgers 1930). And there was Ioffe's gifted protégé Lev Shubnikov, who lodged on the other side of the street from 1926, and learned the intricacies of cryogenics from Wander de Haas at Onnes' laboratory. Shubnikov's partner, physicist Olga Trapeznikova, who joined him in 1927, later recalled how everything around Paul seemed to light up when with shining eyes he made his entrance (Reinders 2018, 48).

In Göttingen, where he gave summer lectures on Einstein's ideas and on quantum statistics and had fun with Franck and other colleagues, students adored him too (see Figure 31.1). 'From Born we mostly learned essentially mathematical physics', Maria Goeppert-Mayer later said, who attended Paul's lectures in the late 1920s and became the second woman after Marie Curie to win a Nobel Prize for physics in 1963 (Goeppert-Mayer 1962). But 'Ehrenfest spent a great deal of time with the students in Göttingen ... and we learned physics from him – in contrast to mathematics. And he would insist: "say first in words what you are doing, otherwise you don't understand it"'. Slightly younger Victor Weisskopf wrote that Paul taught him to 'distrust the complicated mathematics and formalism so popular in Göttingen at the time', and never to fear posing a 'stupid question' (Klein 1981, 11). And yet others recalled how Paul teased Born, who persevered in his mathematical way of working in quantum mechanics. 'Max, why do you keep stacking up all these matrices? Write down Schrödinger's equation and everything shall be clear in two minutes', Paul would say for instance just as Born wanted to start a lecture (Ryutova-Klemoklidze 1995, 126).

Amid all this, he still found time to befriend his fellow Viennese Schrödinger and Pauli. That Paul and much younger Pauli were both sharp tongued became evident at their first meeting. 'Your papers please me better than you yourself', Paul is reputed to have said. To which Pauli rejoined: 'For me it's exactly the other way

268   THE DELIGHT OF THINKING

Figure 31.1 Paul with students in Göttingen. Right with plaits and with her back to the camera, his daughter Galinka. Left in the foreground his colleague James Franck.—*Photograph Ehrenfest Family Archive.*

around' (Casimir 1983, 85). And alongside the pleasure such an exchange gave him, Paul appreciated that Pauli didn't seem to regard physics as a competition and didn't rush to publish every idea at once. Paul too strove to put ideas in their context and to search calmly for synthesis and broad understanding. '*Klugscheisser*', or big mouths, is what Paul called physicists—of whom there were far too many in his eyes—who wanted to be first with hurried publications using mathematical techniques that they didn't seem to really grasp themselves. 'Always so clever they were! And nobody understood anything' (Uhlenbeck 1962, session I).

All the more, he enjoyed the more sedate discussions of the dramatic developments in quantum mechanics during the Solvay Conference on electrons and photons in Brussels in October 1927. The mathematical ins and outs weren't at the forefront but rather the meaning of these formulas. Were the two descriptions of the quantum world—the matrices and the wave equations – indeed equivalent, as most physicists tacitly assumed? How should this quantum mechanical description of the world at the smallest scales be interpreted? What did it mean that solid electrons could also be described as waves, and light waves as particles?[28]

In long reports to Tanya and his students, Paul gave his heroes Einstein and Bohr the leading role in the debates. The central question in their discussions was whether 'God plays dice with the universe', as Einstein put it.[29] Einstein preferred to hold on to the idea that a chain of 'cause and effect' set nature in motion and maintained this motion. On this occasion, Paul—who alternated between messenger, translator and catalyser in the discussions—took Bohr's side, who interpreted the wave function of particles as a probability distribution, allowing randomness to creep into nature and making it impossible to determine simultaneously certain properties of particles with great precision.[30] He enjoyed the discussions which stretched late into the night and resumed at breakfast. 'It was like a game of chess. Bohr kept shooting out of his philosophical clouds of smoke to make short work of one example after the other. Einstein was like some kind of jack-in-the-box: every morning he sprung up with renewed energy. Oh, it was delightful'.[31]

Paul played a connecting role at the conference, with his obvious pleasure in the questions, with the photographs he took and with the time he spent sparring with Bohr, who invariably knocked at his door around midnight for 'a few words' and then stayed for hours. His colleagues also enjoyed Paul's clear explanations. That summer, Planck had already sent Paul a letter full of praise for his clear explanation of complex quantum algebra. Such a 'neat, illustrative form' was essential for someone like himself who was accustomed 'to think physically instead of geometrically', Planck had written. And just before the Conference, Paul had published a paper with truly original work that, even if the paper itself was slightly wooden, received attention and recognition. Building on Heisenberg and Schrödinger's work it asked the question how it was possible that an electron, at least according to Schrödinger, spread out like a wave in space, while it travelled along a path like a small bullet in classical physics theories?

In addressing this question, he brought together many of the ideas from his earlier work. With his adiabatic principle, Paul had tried to find a guideline that could indicate which parameters in the classical description of a system would prove to be quantised at the smallest scales. Building on this work, Bohr had then developed his correspondence principle to analyse at what scales, and whether, this refined quantum description of nature morphs into the classical description of nature. And now Paul asked how is it that nature seems so fuzzy at the smallest scales, with Schrödinger's wave function describing the whereabouts of particles only in terms of probabilities, while in the world around us, classical mechanics can tell us exactly where a bullet or planet is at every moment of its trajectory.

His earlier studies, with Tanya, of 'ensembles' in statistical mechanics helped him find an answer. An ensemble can be thought of as a large number of virtual copies of a system—such as a moving ball or a quantum system—all measured at exactly the same time. But whereas in classical physics all these measurements would give the same result, in the quantum world, there is a huge variety of results.

Paul's solution was to work with the weighted average, or 'expectation value', of all these possible outcomes.

He then went on to show that such a quantum expectation value of the position of a particle obeys a relationship that (apart from being completely defined in terms of expectation values) is exactly like Newton's second law of classical mechanics. And he also showed that the more spiked the wave function of a particle is, resulting in a very sharply defined position expectation value, the more accurate the Newtonian description becomes. With his graceful demonstration that Newtonian physics is contained in the more refined and comprehensive theory of quantum mechanics, Paul had thus once again managed to bridge the classical and quantum worlds. And by deriving the relations in the context of both Heisenberg's and Schrödinger's work, he had even created a link between Heisenberg's more formal and Schrödinger's more illustrative and conceptual version of this quantum mechanics (Ehrenfest 1927e; Neuenschwander 2013-2014).[32]

It was also Paul's last real original work. He was too formed by this tradition of 'anschaulich'—visualised and conceptualised—working to escape from it. And he seemed too wedded to the old quantum theory to make an original contribution to new quantum mechanics, as Einstein did. In these golden years, the gap grew between his reputation in the outside world and what he himself believed he would still be able to achieve in physics. As did Paul's estrangement from Tanya. 'Nowhere', Tanya is said to have remarked to Einstein, 'does the transition from life to death take place so imperceptibly as in Leiden' (Reve 2019, 325).[33] And Paul could no longer enthral her enough to make her stay with him in the Dutch town.

Quite the reverse. Heavy steam locomotives carried Tanya far beyond Moscow in early autumn 1927. In Simferopol in Crimea, she was appointed professor of didactics in mathematics and physics at the Frunze Pedagogic Institute.[34] 'I am hopeful we shall manage to get mathematics education at schools onto a sound footing', she wrote to Paul in early 1928. 'When this takes shape more clearly, I shall write about it in more detail. It is possible to work with the people here. There is good will here. ... Today I am going to put together the programme for next year with M.L. and N.V. Ogloblin. This is quite a problem. Choosing what future teachers need and what they can grasp. In addition, there are many official meetings—I am in two specialist committees, one for mathematics and one for pedagogics, meaning I have a meeting almost every week. Alongside these committees in which students have representatives, they also have their own committees. They are very intent when it comes to choosing the chair and the board, less so with what comes afterwards. To my surprise, I was brought into this too. I was elected chair of the trade union committee. What this involves, I shall say when I know myself'.[35]

32

Escapism

Radios and travelling

Huddled in his jacket, Paul stood on 9 February 1928 in front of a mass of people around Lorentz's grave at the cemetery on Kleverlaan in Haarlem. Urged by the Leiden administrators, Lorentz's widow Aletta had asked him as a 'friend and successor' to speak, after her husband had died following a short sick bed.[1] And while blasts of wind tugged his jacket, he depicted his predecessor as a master in a sunny workshop with large windows. This master worked there 'preferably with simple instruments' but also with 'unusually fine precision instruments', which he had from other masters such as Christiaan Huygens, Fresnel or Maxwell. And that he had 'invented and constructed' 'many of his fine, ingenious instruments himself' wasn't something 'the master in his workshop filled with sun' said in all his modesty. In an emotional speech, Paul used these metaphors to sketch his predecessor as a harmonious, modest, loyal, dutiful, loving and great scholar. No one was in the same league as Lorentz, whose 'sharp, critical spirit' in physics 'always clearly [separated] the complete from the as yet incomplete' and emphasised 'what is still mysterious' (Ehrenfest 1928).[2]

It was an impressive gathering. An extra train had been arranged to transport Lorentz's colleagues and other interested people from Leiden to Haarlem, and even the State Telegraph Service closed for three minutes to commemorate this man of science who had served his country.[3] Lorentz had represented a version of the Netherlands Dutch people liked to see: a country that fought water rather than its neighbours, calmly held firm and functioned as a mediator. That Rutherford, Langevin and Einstein, representing British, French and German science, also said some words at the grand funeral ceremony further underlined how well respected he was outside of the Netherlands too (Kox and Schatz 2021, Ch. 12).

Paul was distraught. Never had he spoken directly to Lorentz about failing to meet his high expectations. Lorentz's aloof comment that 'everyone must do what they can' had been revealing enough (Kox and Schatz 2021). But even if he had towered demoralisingly high above him, Lorentz had also been a refuge for Paul. How often had Lorentz posted a 'greeting from home to home?' Invited Paul and Tanya to dinners in Haarlem? Asked Paul to elucidate recent publications?[4] Just recently, he had given him carte blanche to bring promising young Russian

Igor Tamm to Leiden. Lorentz arranged for the Lorentz Fund to cover his travel and accommodation expenses, which he even 'calculated liberally' (Kox 2018, doc. 289).

Such support, encouragement and friendship had now disappeared. And while Lorentz's son-in-law De Haas and rather stern Willem Keesom had shared responsibility for the cryogenic laboratory since Onnes' death in 1926, responsibility for theoretical physics in Leiden now rested on Paul's shoulders alone. It must have been an unsettling thought, which together with his feelings of grief had a paralysing effect. Professor Ehrenfest was 'indisposed', the *Leids Dagblad* reported in the ensuing days, and couldn't deliver lectures. In reality, he sunk into a pitch-black hole.[5]

Even Tamm's arrival and Tanitchka's obvious pleasure at this hardly cheered him up over the following weeks. For his eldest daughter, it was a treat that Witte Rozenstraat had become so Russian, with the Shubnikovs across the road, and now with Tamm too, as well as regular visits by other Russian scientists such as Lev Landau, Ivan Obreimov and Pyotr Kapitsa. It recalled warm memories of her childhood years in St Petersburg. But having so often worried about his closed daughter, been annoyed by her surliness and been concerned about her crush on family friend Gijs van Aardenne—seventeen years her senior—Paul now gave little mind to her. More than a year earlier, he had been glad when she put Van Aardenne out of her head. 'It might please you to hear that Tanitchka has coped very well with the end of her affair', he wrote to Bohr at the end of 1926.[6] 'Now, for example, I am being disturbed by cheerful laughter and yelling'. But after Lorentz's death and Tanya's departure to Russia, Paul was preoccupied with his own problems and the radios on which he listened to Bach concerts and crackling radio broadcasts from Russia.

'So, we are alone at home with baba Katya and the radio', he wrote to Ioffe in early April, with Tanitchka having gone cycling with Tamm and the Shubnikovs, and Pawlik and Galinka staying with the Kohnstamms in Ermelo.[7] He preceded to strongly advise Ioffe against sending Tamm, together with Mandelstam, to Leningrad to make the city's physics laboratory a large central top institute. 'This enormous concentration [of talent] inspires deep fear in me', he warned. 'Everything was concentrated in Paris after the French Revolution and now compare the situation in France with that in Germany.'[8] French physicists weren't in the vanguard of the quantum revolution, which Paul attributed to the vast divide between Paris, where all talent was concentrated, and everything outside the city. 'And one of the most pernicious effects of such a place [like Paris] is that a handful of extremely talented old men dominate things. ... Life for young people in a place so full of talent is hell. Just have a word with Langevin and Madame Curie', Paul advised Ioffe. Young people flourish far more when they 'can develop their own activities in well-organised, decentralised places'.[9] Would Ioffe in Leningrad have suspected that, despite the fiery tone, Paul was deep in the doldrums?

A few weeks later, another letter made this abundantly clear. First, Paul discussed his participation at a conference on a steamship in the Volga at the end of 1928, but then he bluntly mentioned another issue. 'At some point when we meet, we must discuss very carefully how I can utilise the probably really very small remnants of my vitality somehow in a meaningful AND FOR ME BEARABLE WAY'. Because 'for me everything is reduced to an ever-smaller circle: first of all, I feel my dearest wish is to be able to die painlessly and calmly'.[10]

'The children and particularly Wassik' of course needed him 'as a source of income for some time', he explained. For another four or five years, he could also make himself useful organising education somewhere. He could even serve science a while as 'company' to 'scientific friends I personally care for a great deal'. But most of all, he longed with wistful melancholy to return to 'serene, warmly hued beauty'. Instead, he found himself in 'drab, ugly haste. How hideous is—just by itself—the mathematical blight which pervades the entirety of physics literature'.[11]

For Ioffe, who was naturally familiar with his friend's mood swings, it must have been a relief when Paul seemed to put his state of mind a little in perspective. 'Well, Ioffe, you shall now have had enough of my mewing'.[12]

Paul's students were also familiar with his occasionally extreme reactions. They all knew the story of Walter Elsasser, who came from Göttingen to Leiden to be Paul's assistant in 1927. Gifted, broadly interested and by his own account slightly nervous and sensitive, Elsasser instantly threw himself into current physics problems, armed with the latest mathematics. Should he have conferred better with Paul? Did Paul feel passed over and, on top of this, annoyed by his assistant's mathematical approach? In any case, years later, Elsasser was still perplexed by Paul's behaviour: when he unsuspectingly entered the institute after visiting the hairdresser, Paul exploded. He ordered him to leave a few days later. Elsasser, whose hair had been pomaded by the hairdresser, was apparently unaware of what all of Paul's students knew by then: Paul couldn't stand lotion or pomade, except Ioffe's, which were of 'superior quality'. Paul's students also knew that this was part of a regimen at the Ehrenfest house: no alcohol, no tobacco (except Einstein), no meat… But while these rules served as personal guidelines for Tanya, increasingly unstable Paul seemed to impose them on others ever more fanatically. Or had he used them this time chiefly as a convenient argument against Elsasser, who evidently grated on his nerves? Returning to his parents in Berlin and then moving to Pauli's research group, it took Elsasser a year to recover his crushed self-confidence (Elsasser 1978, 85–95).

Even so, Paul's students in Leiden found him an inspiring scholar. Tinbergen, Paul's former assistant, returned to take his doctoral degree under Paul in 1928. One of the first conscientious objectors in the Netherlands, he had done part of his civilian alternative for military service at Statistics Netherlands, where he cut his teeth in economics.[13] In Leiden, Paul gave him the space to continue working on the idea of applying physics and mathematics models to economics.

Amiable doctoral student Rutgers was now Paul's assistant, together with Roelf Krans, who had also exchanged Amsterdam for Leiden. To Paul's great pleasure, Rommert Casimir's son Hendrik Casimir also came to study in Leiden. He wasn't only the youngest but the cleverest in the group. Rutgers later clearly recalled how Paul walked, saluting and bowing forwards, backwards and forwards again, when Casimir fathomed a difficult problem involving matrices with unexpected rapidity.[14]

With verve, he imbued the group with what he considered important in physics: synthesis and broad understanding. He insisted it wasn't about stealing a march on others. Or as he wrote to Bohr a few years later: 'I despise those erudite yet impenetrable dissertations, purported to be the pinnacle of the author's work. I would rather dissertations were CHEERFUL, encouraging (!!!!) reading material for all young people. That they proceed with other authors' original work instead of wanting to REPLACE them'.[15]

This broad perspective and his deep interest in his undergraduates and doctoral students also attracted young physicists from abroad. In summer 1928, Wolfgang Pauli and Hungarian Eugene Wigner visited, along with Walter Heitler and Fritz London from Germany, who had worked with Schrödinger and applied quantum ideas to chemistry. Moreover, Robert Oppenheimer travelled back to Europe from the United States especially for Paul at the end of 1928. Born, who had been Oppenheimer's doctoral supervisor, thought Paul could teach Oppenheimer to see the bigger picture instead of scoring with hastily written articles.

The year 1929 then became at least as busy at Witte Rozenstraat. Planck stayed there when he delivered a lecture in Leiden in March; Franck and Maria Goeppert from Göttingen were guests that summer and Debye spent a night in September: four (future) Nobel laureates in succession. Other guests in the large white house that year included Paul's former Russian assistant George Breit and his sons.[16] Pauli stayed when he received a Lorentz medal, presented by Paul, and of course Einstein paid his annual visit to Leiden (see Figure 32.1 and 32.2). For Einstein, Paul was still an important sparring partner, as he was for Bohr, and likewise, he was also still an 'interpreter' between these two scholars who sometimes so misunderstood one another.

Yet he found it increasingly difficult to keep up with developments in the field. Burgers, Kramer or Coster, who had answered Paul's insistent and sometimes outright importunate questions in earlier years, could no longer help him. They had gone their own ways and were no longer well informed of the latest, mostly highly mathematical work by Dirac or Heisenberg, De Broglie or Jordan. For this reason, Paul sent Oppenheimer on to Pauli in Vienna in March 1929, just a few months after arriving in Leiden. 'Oppie' was well liked by the Leiden group, yet 'the poor devil has already spent about a month and a half in Leiden in rain, rain beneath a grey sky, and what is worse, under pressure from pedantry', Paul explained in a letter to Pauli.[17] 'He always has quite ingenious ideas ... But then of course the

Figure 32.1 Einstein with Tanitchka in the dining room, filling his pipe, c. 1930.—*Photograph Ehrenfest Family Archive.*

great misery starts, that I cannot understand, as you know, that which cannot be visualised and conceptualised' (Pauli 1979, doc. 211).

This sense of impotence about being overtaken by developments in his field took an ever-stronger hold of Paul. He started turning away guests, but without guests he couldn't stand it at home. As a result, he increasingly fled Leiden, and it was occasionally almost impossible for others to keep track of him. Sometimes he was in Copenhagen, sometimes in Göttingen or at a congress in the United Kingdom. He roved with an acquaintance around Germany, took a trip with Pawlik and Herglotz to Arosa in Switzerland, was vexed by his brothers in Vienna and vice versa or suddenly appeared on the coast of northern France.

Nor did he need to stay in Leiden for his daughters. Mild-mannered and slightly withdrawn Tanitchka moved to Göttingen in November 1928, nine months after finishing her studies with high marks, aged just twenty-two. She attended lectures by Bohr and Born and just six months after arriving, Born employed her to copyedit the book he had written with Jordan about quantum mechanics. It earned her 150 marks a month, as Paul proudly wrote to friends.[18] But Tanitchka then turned down Born's offer to do her dissertation under him.[19] Her true love was pure mathematics, and while travelling back and forth between

Figure 32.2 Einstein with Tanitchka in Leiden, c. 1930.—*Photograph Ehrenfest Family Archive.*

Leiden and Göttingen, she started her dissertation under her father's mathematical colleague, Adriaan van der Woude.

Galinka, who went her own way dancing and singing, left Leiden with a diploma too in 1928. Having finished her training programme with pre-school educationalist Cornelia Philippi, she went to work at the Trüper Institute in Jena. She was near Wassik and lived with a warm Jewish family Paul had met during one of his visits to Wassik. But work in the large asylum was taxing for her, she contracted scarlet fever and looked, according to Paul, who visited her in early 1929, like a 'whipped dog'.[20] It was a relief when family friend Gorter offered her a job at his paediatric clinic at the university hospital in Leiden. Here, she distributed milk and food, helped recovering children catch up academically and created engaging teaching methods, such as geography lessons building on the postcards which arrived at Witte Rozenstraat from all corners of the world over the course of time.[21] The evenings she again spent making music together with her father cheered him up.

Of the whole family, Pawlik was most often in Leiden, where he spent his days largely at home. Education at Dutch schools was still out of the question for Paul and Tanya. As Miss Hellema, who had taught Pawlik's sisters, had married to become Mrs Kits van Heijningen, Pawlik instead received a few hours of lessons a week from students such as Tinbergen and Rutgers.[22] He had 'a great sense for physics', Paul wrote to friends in 1929, was 'very dextrous', was handy with a lathe, was going to learn to blow glass at Onnes' cryogenic laboratory, and 'everyone is always very fond of him'. But their home-schooling system hadn't been successful in all respects:—Pawlik's spelling and handwriting were 'eerily bad'.[23] Moreover, outsiders such as Rommert Casimir sometimes cautiously remarked that the adolescent saw rather too few people of his own age, and his education might benefit from a little more discipline.[24] And the independent boy, who sometimes must inevitably have felt alone between increasingly deaf baba Katya and more and more peevish Paul, was ostensibly glad when his mother spent summer and autumn 1930 and spring 1931 in the Netherlands.

Tanya's plan to teach her students at the Mathematics Institute in Simferopol the didactics of mathematics, after completing third- and fourth-year classes on differential and integral calculus, had foundered in spring 1930. Before she could start, the students received their diplomas early due to a shortage of teachers.[25] This had made it even easier for her to return to Leiden and stay with her son, while his father again travelled around the United States. Elated, fourteen-year-old Pawlik told him in a letter that Tanya had decided 'TO PAINT THE OUTSIDE OF THE WHOLE HOUSE!!! ... ALSO TO COMPLETELY PAINT THE DINING ROOM (doors, walls, windows and seeling) also to whitwash the entrans hall AND TO PAINT THE GATE'. Under the heading 'department of finance', he then gave Paul the estimated costs and, as he often did during his father's travels, he also summed up all recent expenditures—the gardener, clothing, coal and the Trüper Institute— and he listed the slightly negative balances on various bank accounts. 'Pan! How you would grumble!!!' he typed on Paul's typewriter a few days later.[26] 'Nowadays we eat in YOR STUDY!!!!!! This is because the dining room is been peinted!!' It was a 'much nicer place to eat', Pawlik thought, adding that the culinary odours pervading the house would have horrified Paul.[27]

Tanya also used this calm period with her mother and Pawlik to organise the publication of *Übungensammlung*, a book with almost 200 concrete, simple and playful questions about space (Ehrenfest-Afanassjewa 1931). On quiet nights at her rickety desk in Moscow, at her old bureau in Simferopol and in the study in Leiden, she had developed them and ordered them beneath nineteen headings, such as 'angles', 'lines' and 'symmetry'. Like the geometry laboratory, she had experimented with in St Petersburg twenty years earlier, the questions aimed to stimulate children to measure, draw, reason and above all to observe. 'Today is the first springlike day. Before breakfast I went to Noordwijk to bring my German manuscript with geometry exercises to Mrs Jonker', she wrote to Paul in Pasadena in early April.[28] Her proofreader Mrs Jonker, who was a primary school teacher, had her pupils in the highest classes try some of the exercises in the meantime. 'She has promised to look at it. She has her own little house there, not far from

278 THE DELIGHT OF THINKING

the lighthouse. We went on a fantastic walk. ... It was the first time this year that I heard larks'.[29]

By this time, Paul had traversed the United States from the northwest to the southeast. After visiting Niagara Falls with Hugo and his family, he continued to Ann Arbor in Michigan, where introducing the students to the principles of quantum mechanics during a loosely structured summer course gave him much pleasure. The table at the cafeteria in Ann Arbor where he lunched and dined grew longer and longer as ever more young Americans joined the discussions (see Figure 32.3). The owner of the establishment is even said to have used a special sign to reserve the table for him: 'For Prof. Ehrenfest and his family'.[30]

This was apt because Paul put all his energy and knowledge into the company, which also included part of his Dutch 'club'. He brought along Rutgers and Dieke, who after peregrinations in California and Copenhagen had become a professor in Baltimore, was there too. As of course were Goudsmit and Uhlenbeck, who had become research fellows in Ann Arbor in 1927 thanks to Paul's mediation after their joint defence of their dissertations. Earlier, when he had written to them

Figure 32.3 The table after lunch in Ann Arbor, photographed by Paul.—*Photograph Ehrenfest Family Archive.*

that he had started suffering from 'a kind of persecution mania' because he knew 'absolutely nothing anymore' about modern physics, they had impressed on him that the finer details of quantum mechanics were far less important for this summer school than his infectious enthusiasm and the inspirational atmosphere he could create (Klein 1989, 41–42).[31] And in this inspirational and sometimes rather lofty atmosphere, they endlessly discussed quantum mechanics in Ann Arbor (see Figure 32.4).

At the request of his old colleague and friend Tolman, Paul gave another loosely structured introduction to quantum mechanics in Pasadena. They wrote two publications together; Paul also published a short article with Oppenheimer and—perhaps most importantly to Paul—he spent three weeks with Galinka in Pasadena (Tolman and Ehrenfest 1930; Ehrenfest and Oppenheimer 1931; Tolman et al. 1931). A few months earlier, Gorter, enthusiastic about her remedial classes and her dedication, had offered her a job in charge of a new yet-to-be-established Montessori school for young tuberculosis patients. But Galinka baulked at the idea of committing herself to something at the age of twenty in Leiden, seizing with both hands the invitation her father and her uncle Hugo extended her to go to the United States (Horn 2019). To Pawlik's horror, she borrowed 570 guilders from their neighbour Eli Wiersma, who worked closely with Shubnikov in Onnes' former cryogenic laboratory, and she sailed to New York with the Berengaria.[32]

Figure 32.4 In Ann Arbor with Paul at the centre and from left to right: Paul's doctoral student Hendrik Casimir, Laura Fermi, Paul's assistant Arend Rutgers, George Uhlenbeck, Sam Goudsmit, Jaantje Goudsmit and Enrico Fermi.—*Photograph Ehrenfest Family Archive.*

Following in her father's footsteps, Galinka first stayed with Arthur Schwab and Teddy Schwab-Bryner in New York in September, where she so enjoyed herself among the artists and writers that she stayed much longer than planned.[33] 'She has the same character as you. She is so sociable. Tanitchka isn't like that', baba Sonya had once written to Paul.[34] Galinka easily made contact, imbibed all impressions as enthusiastically as her father and, according to others, was 'as electrifying'.[35] After staying with her uncle Hugo in St Louis, she travelled to her father in California, finding him bronzed, happy and in possession of a Ford Model T. He can 'already drive quite well, even in the busiest places in Los Angeles', Galinka wrote to Leiden.[36] Paul was pleased his daughter was expanding her horizons and called these weeks with her 'perhaps the happiest' in his life.[37]

While Paul then started on a second tour across America, which included MIT, Harvard and Cornell, Galinka stayed in Pasadena with the Model T. She learned to drive, hiked with physicists and astronomers such as Horace Gilbert and Fritz Zwicky and earned money picking peaches, looking after children, doing housekeeping and—her favourite job—as a member of a puppet theatre troop. When Einstein visited Pasadena in 1931, where he received a hero's welcome, the group even made a special play with an Einstein puppet. In the days afterwards, Einstein played the violin in sun-drenched California with the 'sweet bird', as he called Galinka in a scribble on an envelope to Paul.[38]

Galinka didn't make impossible demands of herself, like her father, and wasn't consumed by unbridled ambition, like him. More than anything else, she lived exuberantly. 'You don't have to worry about the Spanish boys. Now I have a new one, Eugenio from Mexico, a real proletarian!' she wrote cheerfully to her mother, who was with Pawlik and baba Katya in Leiden.[39]

Paul, having left the oppressive atmosphere of Leiden behind him, also made a high-spirited impression on his hosts, certainly during the first half of his trip. American students hung on his every word when he jested: 'I am as convinced of the existence of atoms as of the existence of the sun; but whether the sun exists— that is a well-known philosophical question'.[40] Or 'Of non-existence I am unafraid' while also quoting Einstein: 'Life is interesting, death is comfortable'.[41] Or when for example he coyly poked fun at himself when the suitcase containing all his clothes, books and sheet music didn't arrive ('I'm walking around trunkless, without my elegance, my science and my soul').[42] Yet at the same time, this trip didn't have the same freshness as his first trip around America six years earlier. With his uninvited criticism and sometimes thoroughly sarcastic comments he even antagonised Hugo, his only brother still loyal to him, as well as Hugo's children.[43] And while his tour of various institutes extended his network, including Abraham Flexner, the founder of the Institute for Advanced Study in Princeton, neither did meetings or dinners always go as hosts had hoped.[44]

Friendly Rutgers in Ann Arbor evidently hadn't wanted to remark on how Paul's comments sometimes carried a sting. Such as when he said to him: 'Uhlenbeck is an elegant person... you are a simple person'. Or 'you don't miss the opportunity to make a mistake'. Or 'you mustn't forget that physics is in fact too difficult for you'.[45] But hosts didn't always appreciate the sharpness of Paul's comments.

Did Paul's deep, black moods and his sharp tongue also increase Tanya's aversion to Leiden? As well as steadfast and—sometimes rather too—tenacious, colleagues and friends also found Tanya amiable, capable of seeing things in their proper perspective and an excellent listener. 'Finally, back in Europe again. I would like to tell you so much about America', Goudsmit wrote from Copenhagen in 1929 on a postcard also signed by Heitler, Gamow and Jordan. And for Tinbergen she was, alongside Paul, a stimulating person in a crucial phase of his life. When he finished his dissertation *Minimum Problems in Physics and the Economy*, in the preface he thanked 'Prof. and Mrs EHRENFEST', who had advised him 'to apply myself to mathematical economy and statistics'. The 'gripping lectures, the really animated discussions in the colloquium and beyond—by which I chiefly refer to Mrs and Prof. EHRENFEST's hospitable home—[and] the international atmosphere' had an enduring influence on him (Tinbergen 1929).

But when a tanned Paul returned, and they had made a few trips, including one to Göttingen, Tanya soon departed for Russia again. This time to Moscow, where the 'narrow streets and tall houses' made it almost as dark as Leningrad in the winter.[46] She had been 'ordered by the People's Commissariat for Education to assemble a 'brigade' of a small number of mathematicians to formulate a mathematics programme for the teaching training course'.[47] And this time as well, everything was shambolic. 'Time passes quickly, but oh, how slowly things happen. I haven't yet been appointed to my institute [the Mathematics Institute]; changes are taking place at the top', her 'sweethearts' read at the end of November. 'But I am working on the organisation of the training plan, for which I receive help; what does it matter who reaps the benefits'.[48]

She was indefatigable. Already at the Bestuzhev she had argued for realising change step by step. A little later she put aside her pacifist ideals at the Pedagogic Museum of the Military Academy to help design modern educational formats. Evidently, she had hoped that people who learned to think well for themselves would automatically reach sensible decisions about organising their lives and their country. Did she still hope so? Without seeing that the first Five Year Plan was becoming a fiasco, and Stalin was accumulating ever more power? Perhaps it was more important for her not to abandon her friends, family and homeland, particularly in difficult times. And in any case, her life in Russia seemed preferable to her life with Paul in Leiden.

Paul had raised the matter of their 'long-distance marriage' in a letter in spring 1929, more than half a year before planning to travel to Russia to celebrate their twenty-fifth wedding anniversary in Kharkiv.[49] 'That together we brought our three eldest children into the world and what you developed in me and the scientific bits and pieces you learned from me and my friends—that must surely justify our marriage', he wrote to Tanya who was in Simferopol at the time.[50] A little cautiously he mentioned that many of his friends had difficulty with their marriage, and that he didn't like playing the hypocrite. He asked himself whether Tanya as she had been in Göttingen would have found Paul as he was now attractive. He touched upon men's 'natural' desire for younger women and the sexual deprivation which vitiates life.[51] But Tanya hardly responded to this in her reply on thin greyish sheets in pencil. Only once socialism was attained, she wrote, would it be possible to see 'what the problem of sexuality in and of itself is. Perhaps then some marriages shall become more stable. But perhaps children, men and woman shall live in separate houses? All families or only some?'[52] Then she turned to practical matters. In a short postscript, she merely added that she failed to understand what 'you are "bumbling" about us there. I am very contented with you!'[53]

But Paul was unhappy, and after returning from the United States, he was even unhappier than before. He didn't feel like visits from colleagues, and his final crop of doctoral students was now setting out into the world. After the complex discussions on quantum mechanics in Ann Arbor, Rutgers lost all hope of becoming a theoretical physicist and leapt at the opportunity to become a senior assistant at Michel's laboratory in Amsterdam (Hollestelle 2011, 175). Krans, less brilliant than the others, became a teacher in Arnhem after his dissertation. Casimir defended his dissertation in Leiden in November 1931. He had written most of it in Copenhagen, where Paul had placed him 'as a young cuckoo' in 'Bohr's nest' in 1930 (Casimir 1931). Then he went, once more thanks to Paul, to Pauli in Zurich, where he had every chance to develop himself.

Hardly any new students applied for the theoretical physics programme in Leiden after this. Even Fokker, Paul's amicable fellow professor who like Lorentz gave lectures once a week, was dispirited by Paul's grumbling about himself, physics and the world. Relax, Hugo wrote in a letter for Paul's fiftieth birthday in 1930. 'If only you wouldn't take your own and other people's lives so damned seriously!'[54] He called him a fanatic because his younger brother refused to reconcile himself with the fallibility of the world, people and himself, and thus was constantly disappointed. Get some rest, he advised, 'You have done your share of heavy thinking during the past 30 years!'[55]

Slight despair was also betrayed in a letter from Rutgers, after a wonderful Wednesday evening colloquium in 1931 in which Paul clarified a nagging problem and yet stood there 'long faced'. 'When you allow yourself to be forced down this path then you ruin the colloquium'.[56] But Paul could no longer pull himself out of the morass.

And Tanya? After all these years, she could no longer suggest 'anything better either [to] my darling Paulie' than taking at life more philosophically. 'It is surely clear that you and I cannot unravel the point of life, so it is unscientific to argue about what ought to be and what you would like. As long as you are alive you must seize the day'.[57]

33
Love is no longer enough
Nelly, Russia and the Nazis

'It seems we have achieved the greatest density of houses and trams. Now it isn't just difficult getting onto a tram but getting off again at the right moment too.' In early 1932, Tanya was staying in a room in her friend Vera Fedyaevskaya's Moscow flat. She had brought the lamp on the bureau with her from Leiden. Paul had complained that she paid too much for it. Now she described daily life in the Soviet capital to him.[1] 'Today, it was so crowded that an entire group of people couldn't get off, and while they struggled towards the door, the conductor allowed the tram to resume its journey, reluctantly, further increasing the crowdedness. And several stops have already been scrapped this year—for some or other higher consideration—so you must walk quite some distance if you miss your stop.'

Tanitchka also experienced how uncomfortable life in Moscow often was. A month after defending her dissertation in the field of geometry and curved space on 8 December 1931, she joined her mother (Ehrenfest 1931). The *Leids Dagblad* made special mention of her being the first woman in Leiden to take a doctoral degree in mathematics.[2] Now she too would work in Russia, in the country of Tamm, the Shubnikovs, her youth and her mother; a country where there was so much to do. Yet, Tanitchka was unemployed at first, going from one temporary address to another.

For her great-aunt Sonya, it was all too dreadful. Dissatisfied with Leiden and exasperated by Tanya's Russian travels—and, incidentally, also by Paul—she had decided in 1928 to move to Auteuil near Paris, where she had nieces and nephews living with their children. Even if for these children, she felt 'not such affection as for your dear children whom I hold so dear.'[3]

That Paul and Tanya now also let their eldest daughter go to Russia was beyond her comprehension. This 'grave, grave mistake' made her 'afraid and most distressed'. Who knew what the Bolsheviks were planning to do? Had Paul and Tanya considered the season? 'In January, February and March it can freeze severely' and she 'is so sensitive to the cold, her hands freeze even in the Netherlands.'[4] And didn't they see how turbulent things were in the Soviet Union? 'And then you are going to send such a sweet, innocent girl to that hornets' nest; you can surely study in England or France; Russia is in chaos; the journey there is long, and she could be robbed, mugged or worse. ... My dear, precious Tanitchka; what fear I shall endure; I am already in tears'.[5]

In her letter, addressed exclusively to Tanya, baba Sonya preceded to reproach Paul in particular. She wouldn't be going to Leiden any time soon, she wrote, because Paul had sent her a 'stupid, vicious letter'. She scoffed at Russia, a 'paradise', but one 'people try to escape from'; she railed against Ioffe, 'a real Bolshevik' and she was still deeply upset that Paul had said something positive about the 'USSR' in 'their *Pravda*' in late 1929. 'You really must have some nerve to besmirch your reputation by associating yourself with such filth'. But she would leave it at that, she lamented, because 'you believe the fine stories of Bolsheviks such as Ioffe, Kagan and their kind anyhow and won't hear anything bad about them'.[6]

With disquiet, baba Sonya must have read in ensuing letters that after a short stay at the Tamms, Tanitchka moved to the 'dear Mandelstams', who could only host her for a while too. Then she went to stay with a widow who screened off a sleeping area in her room with planks. The job Tanitchka had been promised as a teacher still hadn't materialised. Instead, she was appointed an assistant because of retrenchments. It was 'the same work, except [a teacher's] salary is one-and-a-half times as much', she wrote to her father.[7]

Tanya thought everything would turn out well. Her students were 'very pleasant, do their best and I like being with them, regardless of their progress', she had written previously to Paul. And he was still optimistic about Russia too. Before celebrating his 25th wedding anniversary with Tanya in Kharkov in 1929, he had travelled to Leningrad and Moscow, where he had felt welcome and loved (see Figure 33.1). His hosts were 'enlivened' and 'invigorated' by his coming, and his stay worked 'like doping', Lydia Mandelstam wrote afterwards.[8]

Perhaps Paul and Tanya felt somewhat protected by their Dutch passports on their travels around the Soviet Union. In any case, they didn't appear to take the increasing censorship and danger too seriously. When Tanya wrote in January 1932 that she was in a work group that would examine the physics curriculum of the Moscow State University from a dialectical materialist perspective, Paul was concerned that she would behave as the Bolsheviks expected of her. But the commentary he scribbled in the margin of her letter before forwarding it to Galinka in the United States surely wasn't meant literally: 'Mama shall fight so hard that they shall hang her from the highest gallows'.[9] Yet less than a year earlier, when Tanya was still in Leiden, she had written about the arrest of their friend and colleague Petr Lazarev who was exiled to Sverdlovsk (Yekaterinburg)[10] merely for pointing out that philosopher and revolutionary socialist Friedrich Engels made a trivial mathematical error in *Dialectics of Nature*. Many professors and students in Kharkov were also arrested, Tanya added in the same letter to Paul, who was travelling.[11]

Yet even this didn't seem to be reason for Paul and Tanya to avoid Russia. Indeed, in March 1932 Tanya went—on Paul's advice—to view the designs for the

Figure 33.1 Paul in the study with the typewriter that his brother Hugo gave him to mark his twenty-fifth wedding anniversary. Photographs on the wall include images of the former summer house in Kannuka.—*Photograph Ehrenfest Family Archive.*

Palace of the Soviets, which was planned on the site of the Cathedral of Christ the Saviour near the Kremlin. 'Corbusier wasn't one of the prize winners and—forgive me—quite rightly so', she wrote to him.[12] She wasn't at all taken with the idea of an enormous skyscraper-cum-conference centre, calling it 'a barbaric concept to bring so many people together physically in one place. This is not where exalted solidarity is expressed; this is not where the idea of socialism is faithfully rendered'. Did she still believe that a mild and better form of socialism could be achieved through gradualism?

In Leiden, Paul's letters were becoming extended laments. As he couldn't endlessly bother his colleagues, increasingly he turned to their wives, such as Margrethe Bohr and Hedi Born, who perhaps against their better judgement sent comforting replies.[13] Margrethe Bohr sounded relieved when Tanya, travelling back from the Soviet Union, accompanied Paul in 1932 to the annual conference on quantum mechanics at the Niels Bohr Institute in Copenhagen and returned to Leiden with him. 'I can imagine how happy Paul must be to have you back at home', she wrote to Tanya, who had sent flower bulbs to thank her for hosting them in Copenhagen.[14] Lise Meitner wrote to Paul that she had 'greatly enjoyed spending a few stimulating hours with your wife' in Copenhagen.[15]

What they couldn't have known when Tanya was with Paul in Copenhagen in her 'threadbare' coat,[16] and what Tanya learned just before rushing helter-skelter to Copenhagen, was that Paul wasn't only regularly despondent but also in love—and perhaps they were partially the same to him. In any case, for some time he had been in a relationship with Nelly Posthumus Meyjes, a friend of the Fokkers who was the daughter of a long line of clergymen and who wore a pearl necklace and a skirt.[17] Nearly a year before, he had been deeply impressed by a lecture she gave on painting and hadn't been able to restrain his infatuation. 'Through increasingly gritted teeth, I managed to shield my marriage from "entangling ACTIONS" for twenty-five years. But not anymore', he wrote later to Goudsmit in Ann Arbor.[18] 'Who knows, who knows whether I would have been capable if my wife had been ten years younger than me'.

Nelly was ten years younger than Paul. Her parents had preordained her for marriage and studying had been out of the question.[19] But self-willed Nelly had rejected her 'match', and with various courses she had worked her way up to a voice of some authority in the art world. She was independent, 'a very hard-working woman, fascinated by her work', Paul wrote to Goudsmit.[20] He loved her 'books, reproductions, slides and orderliness'. Her world, which she had built with such 'assiduous labour', was a 'precious gem' to him.

Conversely, Nelly was impressed by Paul, a learned man who hosted Einstein, stayed with Bohr in Copenhagen, with Schrödinger in Berlin and travelled around the world to meet even more learned colleagues. It must have been balsam for her soul that such a man had fallen for her and indeed took her work seriously, while her family dismissed it as a curious pastime. Their relationship blossomed during their encounters at Jan Luijkenstraat, where Nelly lived in a well-heeled Amsterdam neighbourhood, and subsequently in Leiden too, where Paul was terribly embarrassed when they came across acquaintances, who, according to him, raised their eyebrows.

Paul wasn't the only middle-aged physicist with a girlfriend. His friend Herglotz had been in an affair with Hedi Born for years. Einstein regularly went 'sailing' with ladies on the lake near his country house in Caputh. And Schrödinger's great discovery in 1927 would, decades later, be directly linked by the outside world to a sexual escapade with an unknown Viennese mistress: his idea for wave mechanics was alleged to have come to him a year earlier in Arosa 'with a pearl in each ear to keep out distracting sounds and with a woman in bed for inspiration'.[21] Paul wrote to Goudsmit in 1933 that he knew 'fifteen (excellent) colleagues (between 35 and 55 years old)' who were having affairs.[22]

Most of them kept such relationships concealed. Only Schrödinger, who wanted to live rather like a Bohemian, had an open marriage with intrepid Anny Bertel. Paul instead chose an ill-considered middle course. Unhappy with furtive glances when walking around Leiden with Nelly, he decided to receive her at home at Witte Rozenstraat, where the thick walls and lush garden kept curious eyes at bay. But he hadn't informed Tanya, who was far from amused when she heard that Nelly was spending time in the house where Pawlik and baba Katya lived. Her nature, ideas and love for Paul prevented her from directly claiming him for herself. But 'that a few old women on the square secretly wink is truly a trifle compared to the distress you have thoughtlessly caused us'.[23]

Yet the two women approached each other cordially when they met in Leiden in spring 1932. Paul's former assistant Rutgers later described the 'attitude' of Paul's housemates as 'absolutely fair'.[24] It was more Paul who seemed unable to deal with the situation. Tanya mustn't repeat the error of unnecessarily cutting short a journey to Russia, he wrote to her shortly after the Copenhagen trip in an unfriendly letter. He suggested that she receive a set amount of money from him regularly in Leiden, instead of indicating every two weeks how much money she needed from him for all sorts of things. It 'bored' him that she kept repeating the 'refrain' that 'never has a professor in Leiden had such a cheap, undemanding and parsimonious wife as me'.[25] At the same time, he wrote letters to Nelly in which he bounded from writer Musil to Schrödinger and from Bohr and Einstein to Picasso, he studied the expressionists and cubists with her and took her on a trip to Paris.

With a 'grateful heart', Nelly thanked Tanya later that summer for the weeks she had been 'allowed' to spend with Paul in July and in which she and Paul had worked together so much.[26] On a postcard from Paris, she wrote that she had attended a Russian-Orthodox mass with Paul in a church Tanya had recommended, so that 'we would also have a strong Russian atmosphere during those days and so feel you in our midst'. 'I now understand the contrast with Bach very well', she added about the church music.[27]

But sharing their lives and loves amicably like the Schrödingers and their lovers wasn't to be after the false start. In early August 1932, Tanya filed for divorce and left for Spa in Belgium, a complete mess for the first time in their lives together. A few days later, after a visit from Paul, she revoked the divorce filing. 'This is an astonishingly suitable place for me: quiet, green, the beautiful contours of the hills, the wonderful view from my window and magnificent and steeply undulating roads in all different directions', she wrote to her 'dear mummy' in Leiden.[28]

It was Einstein who subsequently took on what for him was the rather improbable role of marriage counsellor. In early August, he arrived in Leiden for a ten-day stay, and Paul's condition shocked him. Immediately, he wrote a pressing letter to the Board of Trustees of the university appealing for the appointment of a second professor of theoretical physics as soon as possible: Paul was overburdened (Lunteren 2003, 1). Then he sent Paul to Tanya in Spa. And finally, less than a week after Paul's short trip to Spa, he travelled to the resort town himself, where Tanya and Pawlik awaited him. 'He is as always terribly kind', Tanya wrote to Paul after Einstein, who had forgotten his passport and first had had to arrange all kinds of paperwork, finally arrived.[29]

In the ensuing days, Tanya, Einstein and Pawlik went for walks in the hills and pine forests.[30] In the mornings, Pawlik and Einstein drunk mineral water 'in the downhill part of the resort', which was 'colder and cheaper', Tanya wrote a few days later.[31] And 'of course E. has been discovered. The landlady is quite moved, but otherwise people do their best not to bother him. They are surprised that he is so much simpler than most professors'. In her letters, she also tried to explain to Paul that their son Pawlik suffered from him constantly berating his spelling mistakes and imperfect knowledge of languages. 'Paps [Pawlik] is truly a fine person, as I want them to be', she wrote.[32] 'I find this no small matter, and it makes life worth living. I would like you to focus more on this'. And ten days after Einstein's coming, with matters clearer, Tanya addressed their relational issues more extensively.

Nelly wasn't the problem in Tanya's analysis. The problem, for the children too, was that Paul was so negative towards Tanya.[33] As before, with their disputes about religious affiliation and Herglotz's militaristic views, now too, Tanya was clear. 'Do you still intend to force Nel upon the children as a member of the family? Then all I can say is that for me this is only possible on the condition of a divorce'. Einstein, 'with whom I can speak completely without reservations, completely agrees with me that you must consort with her outside our home'. Paul and Nel had 'a duty' to shield them from this 'honest, hyper-exposure to your relationship'.[34]

As long ago, she subsequently softened these stern words. Tanya was devoted to Paul. She had only filed for divorce to 'meet you halfway', she wrote.[35] No matter what she would stand firm and make clear to the children that she had maintained her interest for the outside world and her work. And she did. Hiking beneath trees, past rocks and brooks, through fields and over roads, she discussed relativity theory, quanta, thermodynamics, geometry and her *Übungensammlung* with Einstein. As the great physicist, who loved geometry, didn't know her book, Tanya asked Paul to send it to Spa.[36] And Einstein, who extended his stay to the end of September, must have been pleased that peace slowly seemed to return. On the final days of the month, Paul himself even returned, and the three of them went rambling together in the hilly woodlands.

Not much later and according to plan, Tanya travelling to the Agro-Industrial Pedagogic Gorski Institute in Ordzhonikidze, near Georgia (see Figure 33.2). She had landed an interesting job in one of the most beautiful spots in the Caucasus, Paul wrote to friends. He rented a room in central Leiden where he met Nel away from the house[37] and prepared a lecture with her about 'Style Development' in both painting and the 'Scientific Worldview'. Meanwhile Tanya sent her mother reassuring letters. 'I am in top form; I have very pleasant pupils and there is much to do'.[38] She had found time to visit the Georgian village where, from behind a pinkish rock, you could see Mount Kazbek. 'And behind the rock, if you walk up one of the hills a little, Mount Syat is visible (and not mount Elbrus as some anthologies claim) with which Kazbek had its dispute', she wrote, referring to Mikhail Lermontov's famous poem 'The Dispute'.[39] Afterwards, far from her acquaintances in

Figure 33.2 Tanya giving a lesson to students in Ordzhonikidze, early 1933.—*Photograph Ehrenfest Family Archive.*

Moscow and Leningrad, and more importantly, far from wavering Paul in Leiden, she had ensconced herself in 'a room with two windows and a view of the mountains', whose 'edges were outlined against the blue sky'.[40]

Ioffe tried to offer Paul a solution in Russia, which might bring Paul and Tanya closer together again. This time he arranged for Paul to come to discuss a professorship in Kharkov in winter 1932–1933.[41] But Tanya didn't visit him there. And Paul mostly wrote long letters to Nelly during his stay, about the 'sea of electric lights' in the city he could see from high in his hotel room, and that 'people young and old here say that things immediately get *cheerful* when I take part in the discussions'.[42]

'No "morning" stares at me here, questioning and grinning in the eyes', he wrote to her.[43] Among his Russian colleagues, he felt welcome and useful. But he also realised that the circumstances and nature of the new job weren't right for him.[44] It could no longer escape anyone that things were getting grimmer in the socialist utopia. The emphasis on heavy industry hampered housing construction and production of food and consumption goods, harvests failed and while Stalin accumulated ever more power, people in the countryside died of starvation and urbanites lived in fear (Bezemer and Jansen 2008). So, on Paul's return to Leiden at the start of 1933, circumstances seemed like a grimmer version of those a year before. Paul viewed modern physics with dejection, couldn't break with Nel, but neither did he think he wanted to be with Tanya; and Tanya was far away in an increasingly horrific Soviet Union.

Galinka, who would return from America only in early spring 1933, knew of this solely from letters. Tanitchka decided to turn her back on the whole situation as well as the Soviet Union. In Moscow, she contracted measles.[45] And during the trip, she then made across the Soviet Union, she was aghast at the starvation and poverty. 'People were so hungry that they ate the straw roofs of houses', she said later.[46] Still rather ill and exhausted, she had returned to the Netherlands. She has the sharpest mind of all our children, Paul had once written. Einstein, Born, Van der Woude, Van der Waerden and other mathematicians and physicists praised Tanitchka's capabilities, but she left this world for pastures new. She returned to her first love, Gijs van Aardenne, buoyant, dominant and seventeen years her senior, and withdrew to his magnificent house in Dordrecht. Shortly before Christmas, they married, with Tanya in Ordzhonikidze, Paul making his way to Kharkov and Galinka still touring the east coast of the United States. 'Pawlik was the only one of us there', Paul wrote to Margarethe Bohr.[47]

Pawlik was also the only one who constantly tried to reconcile Paul with 'mummykins', who 'as funny as she is in some ways', was also someone 'who knows how to stand on her own two feet'.[48] But Paul could no longer manage, he told his son honestly. Pawlik had a 'fantastic mother', but still, and even if he didn't understand why, he soon felt 'so indescribably miserable and paralysed in her company: everything everything that has failed in my life—everything everything that I particularly hate and loathe about myself starts weighing me down. Mama is my living conscience (not that she reproaches me for anything! Quite the contrary.)'[49]

Paul felt like a 'worthless shipwreck', unable to cope with the loose ends inherent to every decision.[50] Don't take your life and that of others so damned seriously, his brother Hugo had previously written to him. 'It gives you so many problems, disappointments, anger and sadness—and so little joy by comparison'. But Paul couldn't help it, and everything slipped out of his hands in 1933, even more so because of Europe's rapid transformation into a grim continent.

Paul and Tanya hadn't sent Pawlik to Göttingen in summer 1932 as they had done the previous two years. The lessons Pawlik received from Franck and Herglotz were an important part of the educational programme they had developed for him. But in 1932, they sent Pawlik to their old friend Edmond Bauer and his colleague Pierre Auger in Paris because students in Göttingen strode around with black arm bands in the shadow of the timber-framed houses and marched at night with flaming torches, chanting Nazi slogans.[51]

Half a year later, on 30 January 1933, Hitler became chancellor. With astonishing speed, the Reichstag was sidelined, Jewish judges were dismissed, and the Ministry of Education was granted authority to dismiss Jewish students, sidelining university administrations. Jewish colleagues in Germany feared for their jobs or had already lost them, and while the house at Witte Rozenstraat had been an inspiring haven for so many years, Paul felt powerless.[52] In the United States Goudsmit even wondered whether many in the Netherlands didn't think the same as in Germany, where economic concerns were put above the rights of Jews. 'I believe that only Scandinavia, England, [Iceland ? ?], Switzerland and perhaps part of France can be considered truly civilised. I'm not quite sure about Holland; how great is the danger of Hitlerism there? ? ?'[53]

Yet Paul deployed his extensive network one final time. He wrote long letters urging help for Jewish academics, to which he appended lists of names. He corresponded with Ioffe, who was prepared to invite all the physicists listed. He corresponded with Born, who made contacts in modernising Turkey, where president Atatürk took in 200 academics. He contacted the people behind the Institute for International Education in America, which wanted to help academics come over.[54] He wrote to Schwartz in Switzerland as well as his Russian colleague Kapitsa, who was working in Cambridge with Rutherford, the first chairman of the Academic Assistance Council, which collected funds from British scientists to take care of Jewish academics.[55]

Furthermore, he sized up the situation himself during a long weekend in Berlin in early May. Staying with 'the Schrödingers', he told them the morning after arriving that he planned to visit Stark. Many years previously, it had been Stark and his experiments that had inspired Paul to study for a couple of semesters in Göttingen. Subsequently, he and Lorentz had been the first physicists to express their admiration for his discovery in 1913 of what later would be called the Stark effect—splitting spectral lines in an electric field, instead of in a magnetic field like in the Zeeman effect—and to invite him to give a guest lecture in Leiden.[56] But since the 1920s, Stark had also been one of the most prominent physicists trying to drive a wedge between 'Aryan' and 'Jewish' physics. And nobody was surprised when he joined Hitler's NSDAP and subsequently became the chairman of the prestigious Physikalisch-Technisch Reichsanstalt (Imperial Physical Technical Institute) in Berlin.

It was therefore unsurprising that the Schrödingers 'nearly fainted' from horror at Paul's announcement. He explained that he wanted to see whether Stark would have 'the keenness of now showing me the cold shoulder, since after all I am the same Jew now as I was then'.[57] It was another five hours before he returned to the increasingly anxious Schrödingers. 'Excessively polite Stark' had been very apologetic about the dismissal of Jewish colleagues, which he claimed to have nothing to do with, Paul told them. He had even invited his 'dear colleague' for lunch at Kurfürstendamm, the premier shopping street in Berlin. And decades later, a still-astounded Schrödinger wrote that 'the great Johannes [Stark], leader of the (very few) Nazi physicists in Germany', sat in full view of all the passers-by 'in jolly and intimate conversation with a less well-known person in Berlin, but whose dear features and woolly head left no doubt about his high intelligence as well as of his race'.[58]

But had it done Paul good looking into the eyes of evil? It certainly didn't do him any good that on returning from Berlin he was heckled by youths on the station square in Leiden. They taunted him in mock Yiddish: 'Isaac! Isaac! Why are you in such a hurry?'[59] Silently, some of the hate had also trickled across the border. 'Such street pressures—are rapidly increasing here!' he wrote to Goudsmit.[60] Then he explained how persecution in Germany broke Jewish colleagues' morale, such as by banning them from libraries, laboratories and foreign trips, by no longer scheduling them for classes, by opening their post, by spreading slander about them and sometimes by dismissing them point blank. Sympathetic and helpful non-Jewish colleagues seldom dared to express open criticism, Paul ascertained. He included an up-to-date list of professors who should be taken on temporarily or permanently by foreign universities. This included physicists London, Heitler, Nordheim, Bethe, Gordon, Peierls, Szilard, Reichenbach and Born, as well as chemists such as Polanyi and Freundlich.

His lists were used, and 'it was Ehrenfest who came to Berlin to show that our friends abroad had solidarity with us', theoretical physicist Fritz London, who later fled to the United Kingdom, said. It was also Paul who brought London to stay in Leiden for a few weeks in June 1933, where he could breathe the 'air of a free country' and momentarily escape from the hellish atmosphere in Germany. This put him 'in a position to take decisions about my new life' (Gavroglu 1995). His stay was paid for by the Lorentz Fund, with Fokker and Zeeman releasing some of its funds for temporary visits by Jewish scientists from Germany.

But it was a drop in the ocean, and Paul's sense of superfluity grew when De Haas' proposal to bring experimental physicist and Nobel laureate Franck from Göttingen to Leiden came to nothing. Only Coster in Groningen and Ornstein in Utrecht each managed to arrange a position for an academic. 'Any thinking human has the sense that we are being pushed against an impenetrable wall. To be flattened? Or shall the wall collapse? Or the Devil may know', Paul had written a few weeks earlier to Goudsmit, mentioning in passing that 'the plan is to kill oneself' in four or five months.[61] And he increasingly repeated the mantra that he had to commit suicide so a talented Jewish scientist from Germany could take his place.

Einstein chose a different solution, turning his back on Europe. In March 1933, during a trip to the United States, he let it be known he was bidding farewell to Germany, where 'civil liberty, tolerance and equality of citizens before the law' no longer existed.[62] When he disembarked in Belgium after this journey, he handed in his German passport and resigned as the director of the Kaiser Wilhelm Institute in Berlin. With his Swiss passport, he stayed in Ostend before returning to America via the United Kingdom. 'When you go to England, put in a good word for London as well as for Bethe, Teller, Kalman and Gordon', Paul wrote to him at the end of May.[63] Tanitchka and Gijs van Aardenne also still had some contact with the Einsteins. But another meeting between Paul and Einstein didn't take place.

Halfway through the summer, Wassik returned to the Netherlands from Jena. Officially because he had grown a little too old for the Trüper Institute and his papers had expired. But everyone understood that Germany was no longer a safe place for a boy with Down's syndrome and a Jewish father.[64] While awaiting a more affordable and, according to Galinka, suitable place in an institution in Noordwijk, Wassik stayed at professor Waterink's institute next to Vondelpark in Amsterdam.[65]

This same hot summer, on 15 July, Paul asked Tanya for a divorce.[66] He sent her a letter because she was only sailing home via Batumi with a ship across the Black Sea at the end of July. 'Last year I begged you to withdraw the divorce petition. And in the end, it was only the experiences of this year that made me realise that I couldn't go on living like this'. About a month later, he went with Nel to the Dutch island Schiermonnikoog, played the piano at Clara Asscher-Pinkhoff's summer house and told her that he wished that he had such a cheerful Jewish family (Asscher-Pinkhoff 1966). It was so much to his liking on the wide beaches and in the sunny dunes with Nel that he asked Pawlik in Leiden to send more money for a few extra days of vacation.[67]

But after returning, Paul, having already broken with Nelly three times and gone back to her, sunk back into a state of nervous gloom.[68] Did Tanya's return play a role? Did it throw him off balance that it was now Tanya who objected to the divorce? In any case, his instability also became too much for Nelly. When he arrived unannounced with a suitcase in hand at her house on Jan Luijkenstraat and again started doubting his decisiveness while still in front of the stately building, she couldn't stand it anymore. 'Do what you want; leave me or marry me but stop whining'.[69]

34

Broken

A hopeless deed in a hopeless world

Later nobody remembered where Galinka, Pawlik and Rutgers had been at Witte Rozenstraat on 25 September. Were they in the newly painted dining room? Was Tanya there or had she already gone to Amsterdam? And had the Van Aardenne brothers yet broken the news to Tanitchka in Dordrecht, very gently, according to the instructions in Paul's farewell letter? Thirty years later, while emptying the house in Witte Rozenstraat, Tanitchka found the Browning Paul had obtained a licence for in early September.[1] On the afternoon of the 25th, he had used it to shoot Wassik and then himself.

He had planned everything meticulously. Early in the afternoon, he first visited Rutgers in the leafy Plantagebuurt district of Amsterdam. 'We drank coffee together in a little cafe on Plantage Middenlaan', Rutgers wrote two days later to his colleague Zeeman, 'then talked some more in my library [and] sat on a bench in the public gardens next to your lab, where we still talked about physics and other general matters'.[2] Paul told him about Dirac, who had taken him aside after the most recent quantum meeting in Copenhagen to say that his presence had been so valuable. He asked Rutgers to travel back to Leiden with him later that day and spend the night there. And when Paul took a taxi around half past three 'to the institute of Prof. Waterink, where Wassik was looked after', nothing caught Rutgers' attention. 'Neither had Prof. Waterink noticed anything. It happened at five o'clock'.[3] In a waiting room, staff from the institute found Paul dead and Wassik critically injured.

Someone later neatly copied onto Waterink Institute writing paper the instructions Paul had carried with him, including his characteristic exclamation marks and underlining. 'Dr. Gans in Leiden (Psychiatrist)' must be called immediately— 'Tel. 2640'—so he can 'notify the family *and all close friends*'. Also Dr A. Rutgers— 'Tel. 50856'—must be telephoned. 'I request he go *directly* to my family in Leiden'. It must be broken to Tanitchka 'very gently!!!' by the Van Aardenne brothers or a doctor, but certainly *not* by calling the family residence in Dordrecht directly.[4] '*Please please* do *not* try to save my life in the event that I am not immediately dead. Have mercy and let me *now* die'. He asked newspapers to 'report laconically' and 'the fact that I didn't leave behind my idiot child is something anyone can understand who delves into the matter'. Then he ended the list with a 'heartfelt' request for 'forgiveness for my sins against your institute'.[5]

A year earlier Paul had started writing the farewell letters Tanya and the children discovered in envelopes entitled 'open after my death'. The letter to 'Dearest Galinkerl/Dearest Pap [Pawlik]' was dated 4 October 1932.[6] 'I fear very soon there shall be no other way out for me than to kill Wassik and myself—no matter how harmful the consequences of this might be for both of you ... the real underlying cause is that I can no longer maintain a bearable relationship with my work. *Everything* else *follows* from this'. Subsequent sentences elucidated why he didn't address Tanitchka at the same time. 'Tanyitchkerl, who I love so very much, has (quite justly!!!) already completely washed her hands of me—not because she does not love me but because she rightly wants to spare her own soul as much as possible from the sight of this collapse'.[7] If Galinka had been in the Netherlands in the last years, she would (and should!) have done the same, according to Paul.

Then he gave them advice. In Galinka, he had complete confidence. 'You, Galinka, are so grounded in human life that—apart from a short, disturbed period—you can go your own way peacefully and fascinated'. About Pawlik and his studies, he had more concerns. 'But you, Pawlik, I beg: *work conscientiously*, grit your teeth (take care of your physical health!!!). In confusing or worrying times call on the assistance of my friends—De Haas, Rutgers, Tinbergen, Einstein, Goudsmit, Ioffe, Wiersma, Dieke, Uhlenbeck, Holst, Franck ... they shall all be prepared to help you through your difficulties as long as they know you want to work [study]'. Several times he urged his son to do so *'conscientiously'*, *'systematically'*, *'energetically'*, *'with discipline'* and *'without haste'*.[8]

The advice he then gave his son about women and sexuality seemed more related to himself than Pawlik and shed a different light on his idyllic portrayal of his childhood. 'Because your childhood was so very, very, very different to mine and moreover you are much more like Mama in many ways than me, your sexual urge shall probably bring you into confusion less than was the case for me. In my life, this caused much irreparable damage in my *early* childhood and then I was almost on the verge of suicide between thirteen and sixteen because I subsequently became the worst pupil at school and furthermore almost a pathological liar'.[9] In view of this 'troubled childhood history', it was unsurprising, Paul wrote, that he had 'come to the edge of the abyss several times' before his marriage and now had lost 'one's grip of oneself at the end'.

He concluded by describing Tanya as '*a really, really wonderful human being*'. And 'none of you shall ever know how profoundly warm she was simply as a wife for me'. Paul's desire for younger women 'led me to inflict unrestrained suffering upon her—as if I wanted nothing other than to cause her pain constantly—I alone am to blame for this—it would be an injustice to condemn Nelly ... Do not insult Nelly'. Finally, he requested that they, and through them 'dear, dear Tanyitchkerl', live 'a captivating-working, serving life. Forgive me—think of me as I used to be, not as I am now'.[10]

The message to 'dearest, dearest Tanya and Tanitchka', in another envelope, was shorter. Paul no longer dared to address them directly, he noted in calm handwriting in the letter on 27 January 1933, because they could only look upon him with contempt, regardless of whether they could also love him at the same time. Tanya, 'everything I ever truly achieved of real value from my life is entirely connected to how you tried to protect the genuine within me from the false. Tanya, I never stopped loving you. Forgive me'.[11]

Wassik died of his bullet wound the next afternoon. Five days later, he was buried together with his father at the Oosterbegraafplaats cemetery in Amsterdam.[12] Paul had already written down his wishes for his burial shortly after his colleague Kuenen had died in 1923. They were in line with Jewish traditions: as soon as possible, in a small and cheap grave, in the graveyard which happened to be closest, without flowers, with even only the 'technical personnel' allowed to bring his body to the grave. After which, his family should resume work as soon as possible, Paul wrote, make music and, most of all, laugh—and avoid any 'tear symbolism'.[13]

Over the ensuing days and weeks, Tanya sent the farewell letters Paul had written to his colleagues and students. 'I absolutely do not know anymore how I am to drag on the unbearable burden of my life even just through the coming months', he had typed more than a year earlier to his 'dear friends: Bohr, Einstein, Franck, Herglotz, Ioffe, Kohnstamm and Tolman!' Physics had been the cornerstone of his existence and especially his 'interest in understanding advancing physical insights and the great pleasure of passing on to others what I believed to have understood'. But 'ever more enervated and torn', he had 'given up in DESPERATION' his attempts to 'follow developments [in physics] with understanding'.[14] (see also Pais (1991, 409–410)).

Consequently, he was 'completely weary of life'. And to prevent his professorship in Leiden sinking into the mire, he wanted to relinquish his position in Leiden 'at the very latest in autumn 1933'. To minimise the economic and moral damage to his children as much as possible, he decided not to burden them with Wassik's care. If he hadn't found a post in Russia or elsewhere by autumn 1933, he would be left with 'no other practical possibility than suicide and that after having first killed Wassik'. Then he thanked them at length for their friendship, which had kept him going in past times and asked them to support Tanya with advice and to help his children, when necessary, without favouritism. 'May you and those dear to you be well!!'[15]

A comparable letter was addressed to his former students. 'You know how intensely I always joined in your delight at each of your successes. You therefore understand how deeply I wish you and your loved ones all the best for the future!!! Forgive me for being such a weak person!'[16] (Pais 1991, 409–410).

But above all, Paul was loved by many people. In a flood of telegrams in various languages, colleagues and friends expressed their 'shock', 'dejection', 'dismay',

Figure 34.1 Preliminary study for Harm Kamerlingh Onnes' portrait of Paul Ehrenfest painted in 1933, which is now in the Stedelijk Museum in Amsterdam.—*Ehrenfest Family Archive.*

'sad[ness]' and 'deep grief' at *'wachu utratu'* (your loss). Tanya and the children were inundated with cards from neighbours, professors, acquaintances, family members and friends praising his sense of humour. 'Human', 'a good person' and 'vivacious' they called him. Three former women students wrote how grateful they were to Paul for 'tuition given with such love'. Nel Nordström van Leeuwen, Gunnar Nordström's widow, wrote in Finland about the 'many good things' Paul and Tanya had 'brought about together:' 'I ... try as much as possible to give forth your word and spirit'. Famous Leiden historian Huizinga, who had once instituted his own colloquia following in Paul's footsteps, like so many people was dismayed: 'I hardly knew that such sombre spirits had affected all this vitality and brightness'.[17]

Some tried to fathom this darkness. 'Better than anyone among us, he realised which difficulties we face', Langevin said a few weeks later at the opening of the seventh Solvay Conference in Brussels. 'We shall see him no more. The crisis in our discipline was embodied by him and the tragedy which overran a great heart and a great spirit' (Langevin 1934).

De Sitter wrote in the *Leids Dagblad* about Paul's experiences 'in Vienna, later in Göttingen and in Russia, where as a Jew he was constantly exposed to treatment that wounded his sensitive nature, and which is probably partially responsible for the pessimism and lack of self-confidence which, completely unjustly, never left him'.[18]

Born and Schrödinger saw Nazism as the most important reason for Paul and Wassik's death. The last time the three of them were together was during one of Paul's visits to Schrödinger in spring 1933, 'to talk about the situation', as Born wrote decades later. 'It appeared hopeless, and I decided there and then to leave Germany at once. Ehrenfest chose another way out. ... But his decision was probably right. It is appalling to imagine what he might have suffered when the Germans occupied the Netherlands'.[19]

In Moscow, dazed Benjamin Kagan couldn't put pen to paper for days. Of course, he and Leonid Isaakovich Mandelstam had heard Paul complain 'about the waning of his strength and productivity', Kagan finally wrote on 7 October. But 'Leonid Is. [Mandelstam] always thought it was because he set himself excessively high standards. With hindsight we think there could have been a pathological condition. ... I also think that this state of mind could only have developed in the specific conditions in Dutch life. Here, where everyone feels that he is indispensable, that demands are constantly made on him, something like this couldn't have happened, I think'.[20]

Paul's brother Hugo saw things differently. 'He no longer wanted to live & took the boy with him—... a tragedy with the characteristic elements of the inevitable!' With all his sorrow, Hugo also felt 'that this poor, agitated, depressed, truly unhappy and inconsolable Paul now truly for the first time in many years has found peace'.[21] A cheque for 1000 dollars—from the money Paul had earned during his last trip to America—was enclosed in Hugo's letter.

Einstein was in the United Kingdom when Paul died. He was on his way to Princeton, where he would live and work for the rest of his life. Before embarking on the ship to America, he delivered his final European speech at the Royal Albert Hall during a fundraiser for the Academic Assistance Council, the organisation Paul had given the names of so many Jewish academics to. In Princeton, Einstein read the farewell letter Tanya had forwarded. 'If we had managed to take away some of his responsibilities then he could have continued living as a happy person', he wrote back, with his earlier request to the board of trustees in Leiden in mind. 'But we failed. The chaps are always prepared to deliver magnificent speeches, but heaven help you if you want something from them, however modest and justified'.[22]

'You were right to oppose the divorce, even though a woman acting in this fashion plays a curious role', Einstein continued, 'because [Paul] didn't actually want [the divorce] at all, but he was gripped by an unwarranted bourgeois sense of duty'. Paul couldn't alleviate conflicts with frivolity, hypocrisy or irony, Einstein observed. He took everything as absolutely as a mathematical proposition, 'exactly as you are inclined to do. If I took my affairs as seriously, I would be long dead and buried. But I have as much Sancho Panza as Don Quixote in my corpulent body, so everything ends or continues mostly with a good-natured chuckle together'.[23]

Elsa Einstein wrote in her own letter about their feelings of guilt because her husband hadn't taken leave of Paul before departing for the United States. 'Albert and I don't talk about it, but we both feel ... that the fact that Albert didn't go to him shall remain a tormenting thought for the rest of his life'.[24]

A year later Einstein wrote an obituary for the *Almanac of the Leiden Student Fraternity*, the society Paul had so detested. 'I came to know him 22 years ago. ... After a few hours we were real friends—people who through their dreams and ambitions were made for each other' (Einstein 1934, 1976). It was a magnificent piece in which Einstein put his friend's death down to a crisis of conscience. 'Those who knew him well, as was vouchsafed to me, know that this unblemished personality in the main fell victim to a conflict of conscience that in some form or other is spared no university teacher who has passed, say, his fiftieth year' (Einstein 1934, 1976).

Einstein called Paul the 'best teacher in our discipline I ever met', and who, for all this felt 'unhappier than anyone else who was close to me'. Paul 'didn't feel equal to the lofty task that confronted him'. And although this feeling was 'objectively unjustified', it was still 'often robbing him of the peace of mind necessary for tranquil research. So greatly did he suffer that he was compelled to seek solace in distraction'. Einstein mentioned the 'frequent, aimless travels', Paul's 'preoccupation with the radio' and 'many other features of his unsettled life' (Einstein 1934, 1976).

Was he thinking of himself too when he wrote that the 'strangely turbulent development' in theoretical physics had worsened Paul's situation? 'To learn and to teach things that one cannot fully accept in one's heart is always a difficult matter, doubly difficult for a mind of fanatical honesty, a mind to which clarity means everything. Added to this was the increasingly difficult adaptation to new thoughts which always confronts the man past fifty' (Einstein 1934, 1976).

He concluded his piece with a paragraph about Tanya, 'The strongest relationship in his life was that to his wife and fellow worker, an unusually strong and steadfast personality and his intellectual equal. Perhaps her mind was not quite as agile, versatile, and sensitive as his own, but her poise, her independence of others, her steadfastness in the face of all hardships, her integrity in thought, feeling, and action—all these were a blessing to him and he repaid her with a veneration and love such as I have not often witnessed in my life' (Einstein 1934, 1976). In the later English translation of the piece, Einstein added another sentence about the vicissitudes: 'A fateful partial estrangement from her was a frightful experience for him, one with which his already wounded soul was unable to cope' (Einstein 1976). And in this way, he expressed the problems between Paul and Tanya and brought them back together.

In the interim, Tanya answered letters of condolence. She thanked 'Mr Hulsebos', 'Chief Constable of Leidscheplein Police Station, A-dam' for his efforts to 'deter correspondents from sensational elements in their newspaper reports'.[25] She was granted a widow's pension.[26] She would give Russian lessons and perhaps some lectures at the university—the law permitted this now she was no longer married— and she would rent out rooms, as Paul had previously suggested. As Paul had requested, she sent Nelly 500 guilders, but they had no further contact.[27] Fokker knew 'Miss X', as he continued to call her discreetly, quite well.[28] Her face, which had been taut and closed, relaxed once more after Paul's death, he said later.[29] It must also have been a 'relief' that the maelstrom of feelings and the fear Paul would do something to himself were behind her.

On 28 October 1933, Tanya received a handwritten letter from neurologist Cor Ariëns Kappers, who had been the first director of the Netherlands Central Institute for Brain Research (*Nederlands Instituut voor Hersenonderzoek*) in Amsterdam since 1908. He informed her that in the Wilhelmina Gasthuis hospital, the 'precious asset' had been extracted with 'the greatest of care', and after complete fixation it would be transferred to his institute.[30] One of the wishes in Paul's 1926 will had been for his brain to be donated to science.[31]

PART 4

35

Loss after loss

If only we could just be eyes

'You know not only was Pavel Sigizmundovich my dearest and even my only friend, but you and your children are very dear to me too. I have tried to write to you many times but did not manage because of the confusion in my mind. I intended to come to you. But this is impossible owing to formal difficulties related to visas and tickets for the return journey that I have already received in Leningrad etc. Which is why I am writing to you anyway'.

At the end of October 1933, after the Solvay Conference in Brussels on 'the structure and properties of the atomic nucleus', Ioffe, the hesitant letter writer, mustered the courage to write to Tanya.[1] 'Can I help you, Galya or Pawlik in some way?' For Pawlik, Ioffe immediately made a proposal which he had discussed with the 'French physicists'. Pawlik could work in a laboratory in Paris, while at the same time he 'could prepare for an examination especially designed for foreigners; mathematics, physics and a little history, language and geography'. This would allow him to go to university, and a study in Paris would benefit him both 'scientifically and humanly'. Did Ioffe mean it would be better for the boy to make a fresh start beyond Leiden and even the Netherlands?

There was also a drawback: 'it shall take a long time and cost a great deal'. About a cheaper study in 'the USSR (Leningrad, Moscow or Kharkov)' Ioffe sounded reserved. 'You are of course very well aware of the positive and negative aspects of our life here'. 'I would be very grateful to you, Tatiana Alexeyevna, if you would write, even if it were just a few lines, and say what I could do for you. All the physicists in Europe are here'.

Few days later, Pawlik and Ioffe met in Antwerp.[2] Money was available from a fund Hugo had set and managed on behalf of Paul especially for this purpose. Paul had perhaps hoped Pawlik would continue the educational programme he and Tanya had so carefully devised, with glass blowing in the laboratory, Rutgers' mathematics and physics lessons in Leiden, English lessons with the Atwell family in their Norman house and Herglotz and Franck's tuition in Göttingen in the summers. But structure was more important now, Germany was no longer an option, and Auger and Bauer—both with fond memories of Leiden, Pawlik, Paul and Tanya—were keen for him to come to Paris.[3]

The Delight of Thinking. Margriet van der Heijden, Oxford University Press. © Margriet van der Heijden (2021).
DOI: 10.1093/9780198927112.003.0035

In Leiden, Tanya subsequently watched Paul's old network disintegrate. In Göttingen, almost nothing remained of the Mecca of mathematics and physics. 'Mathematics in Göttingen? There is really none anymore', Hilbert is said to have replied when the new German Minister of Education asked in 1934 about the state of mathematics in Göttingen, it having been freed of Jewish influence (Reid 1986, 205).

Connections with the Soviet Union were severed too. The to-and-fro between universities in Leiden and Russia was a thing of the past. Tanya, too, was appalled in the winter of 1932–1933 by the starvation in the countryside caused by Stalin's disruptive agricultural policy and the brutal suppression of even the slightest criticism. After returning in summer 1933, she spent much time convincing Burgers, Paul's communist former student, to cancel his membership of the Communist Party of the Netherlands.[4] Never again would she set foot on her native soil.

Did she ever think of the letter in which Paul had written so long ago that nothing would ever come of all the plans in Russia anyway? Did she regret not having listened? 'A child builds a little house and cries a great deal when it breaks', she wrote on 18 August 1920, Wassik's second birthday, 'may God preserve him'. 'A few years later he no longer has the slightest idea that he ever built houses, but what he learned—manual skill and broadening his spatial insight—shall play an essential role in his life ahead'. The same went for an adult who unsuccessfully pursued a goal for happiness and yet gained wisdom from these failures, she subsequently argued in a notebook.[5]

But do you also gain wisdom when life deals one blow after the other? Pawlik was killed in an avalanche on the mountain slopes above Valloire in France on 7 January 1939. Like his sisters, he had reshaped his life in the preceding years. Tanitchka had had three daughters and a son in Dordrecht. Galinka enrolled in Paul Citroen's Amsterdamse Teekenschool, an art school where she made artistic friends and started a relationship with Jaap Kloot, an artist six years her junior.[6] Pawlik flourished among contemporaries in Paris.

He felt at home there, had almost completed his studies and combined his love of physics with his love of the mountains by alternating work in Auger's lab with visits to the slopes of the Jungfrau Joch in the Alps. Auger had installed an array of Geiger counters along a ridge on these slopes, spaced at intervals of decimetres to roughly a hundred metres. Just a few months earlier, he had been able to interpret the sometimes unexpectedly coincident clicks of these Geiger counters as the 'fingerprint' left by showers of subatomic particles raining down on the Alpine meadows after extremely energetic particles from space set off a chain of collisions high up in the atmosphere. Pawlik had participated in preparations for these experiments. With his dexterous hands, fast mind an amiable character, he had a promising future in the laboratory (see Figure 35.1).

Now Tanya and Galinka buried him in Valloire, without fanfare, like Paul five years before. Then they went to Paris, where they attended the funeral of one

Figure 35.1 Pawlik, about fourteen, tinkering with radios in the study at the Witte Rozenstraat.—*Photograph Ehrenfest Family Archive.*

of Pawlik's fellow travellers, spoke to his fellow students and the girl who had fallen in love with Pawlik—later she would write many letters to Tanya. 'Pauli [Pawlik] was found in the starting position, as taught for such danger, so ready to glide down the slope', Galinka wrote to her boyfriend Jaap in the Netherlands, 'but the avalanche was faster'. And 'mummy just talks and talks continuously with the people in Paris'.[7]

Tanya's son had loved the mountains so much. At only fifteen, he had already taken the train on his own to Arosa to zip down the slopes with the Busch family and afterwards in Davos with Carel de Ridder and sometimes Gijs van Aardenne. He was an experienced skier, the weather had been good and there had been seven of them, accompanied by a guide. But only one member of the group escaped the mass of snow. When Tanya returned to Leiden, a postcard awaited her. 'Hello, I'm back in the mountains again, unfortunately only for five days, seeing as I have to speak at de Broglie's Colloquium on Tuesday! ... Kiss. Paul'. 'The last, 1939', she scribbled in pencil on it.[8]

A deluge of letters of condolence followed once again. 'Of all the young people I know, I always had a special fondness for Paul [Pawlik], as I always felt drawn to all of you', wrote Harm Kamerlingh Onnes, the physicist's painter nephew. 'He had a great life as you and your husband made it for your children, always in contact with the finest minds in various fields'.[9] Pawlik 'was a harmonious person', wrote Paul's former student Dirk Coster, full of plans. Pieter Zeeman remembered Pawlik as 'always very amiable, is that perhaps a Russian heirloom?'[10] Margrethe Bohr invited Tanya for a visit in spring[11] and Niels Bohr wrote: 'we all loved him so much'.[12] Some people also thought of the dark times awaiting Europe: 'who knows, perhaps he was spared suffering, a suffering all of us shall have to contend with'.[13]

Once more, a letter came from Princeton, where Einstein felt lonely too. In the years after Paul's death, Einstein had lost his stepdaughter Ilse, his wife Elsa and his six-year-old grandson, his eldest son Hans Albert's son.[14] 'It appears that fate has deliberately set its sights on the strongest and most steadfast. I know how much it hurts because I lost one of my two grandchildren a few weeks ago and recent years have brought so much grief'. It was out of character for Einstein to be so personal. 'But you are close to my heart due to a sort of inner kinship. It is a somewhat detached attitude towards personal life and a state of fulfilment in objective matters, also here at a distance from contemporaries—yet this is how we both are. I would love to be with you now to give you some company as so often in years gone by. But now we are far away from each other, and I am stuck here in a loneliness which I have learned to love'.[15]

The world at large reflected their small world, he thought. 'Only objective affairs have retained their sparkling lustre for me. They appear in the same glory as the first geometry book I received as a child. It is undoubtedly so for you too'. He enclosed a recent publication for Tanya to read and recommended she read Shakespeare's history plays, which he had devoted much time to. Human fate is marvellous when described objectively by such a wonderful man, he wrote, 'only one shouldn't have to experience it oneself but just have to be eyes'.[16]

But Tanya did what Paul would have wanted and what she always did. She set to work. After Paul's death, she was approached by educationalist Kees Boeke, who restarted the debate about education in the Netherlands and founded the organisation 'Working Community for the Renewal of Upbringing and Education' (known by its Dutch acronym, WVO).[17] He ensured she became part of the WVO Mathematics Study Group, alongside old acquaintances such as Kohnstamm, Minnaert, Struik and Dijksterhuis, and like twenty years before, they regularly met at Witte Rozenstraat. This time her ideas were received more warmly. She was even appointed leader of a summer project week in which the Mathematics Study Group aimed to design guidelines for a modern geometry course. The scheme circulated beforehand for such a course contained nearly all the ideas she had expounded in her essay in 1924. Indeed, in the evenings, she had the study group members in the charming little houses on the Veluwe heathland do small and playful ('*anschauliche*') exercises from her *Übungensammlung*, published in 1931, to explore symmetries and other properties of space 'as if they were schoolchildren'.[18]

Nearly a year after Pawlik's death Tanya extended the long thread running through her work by joining two Mathematics Study Group commissions. With Galinka as her interlocutor, she also worked on *Relevia*, a pamphlet on 'a new economic system, an order, under which I would also like to live'

(Ehrenfest-Afanassjewa 1946). Like in 1920, she fused the best of socialism, such as a basic income, accessible healthcare and sufficient basic products for everyone, with the best 'capitalist' ideas, such as freedom to produce all kinds of products and to pursue all kinds of occupations and—most especially—complete freedom of education. 'Nothing must stand in the way of the freest and most refined possible development of the human mind!'

Galinka shared these ideas with her mother. When she was still taking drawing lessons at the Chouinard School of Art in Los Angeles in 1932, Tanya had written a birthday letter filled with advice to her 'dear Galyunetchka', who had once 'lain on the chest of drawers in Kanuka and cackled like a chicken'. 'Keep behaving well, develop yourself further ... find a like-minded gentleman and remember in time to adopt an objective enough attitude in these affairs, i.e. that you don't lose your opportunities to work under any circumstances! ... The most important is to have a purpose broader than art and to serve this purpose with art, *then* art has value. Illustrations for children, placards to elevate the people, which is tremendous!'[19] And Galinka did indeed, together with Jaap, start publishing colourful and educational children's books.

They married in 1941, around a year after the German invasion of the Netherlands, with well-known Dutch writer Godfried Bomans as one of the witnesses, and theycollected a circle of artists, writers and illustrators around them (Metz 2019).[20] The games and especially the dreamy and cheerful children's books Galinka thought up, wrote and illustrated in consultation with Jaap—or had others illustrate, write and colour—were a great success between 1941 and 1943. *Messy Jan*, *Flora Flower Stall*, *The 1001 Nights Conjuring Book* and especially *Come into El Pintor's House*, written by Galinka, sold tens of thousands copies in the Netherlands, and *The 1001 Nights Conjuring Book* was very popular in Germany too.[21] Their Corunda Trading Company—a 'company' because anti-Jewish laws forbade them from founding a publishing house—flourished (see e.g. Horn, (2019)). But Corunda was also a front, financing illegal activities, such as organising safehouses, accompanying family and acquaintances to such addresses, arranging ration cards for people in hiding and altering identity cards.

Did Jaap and Galinka understand Tanya's acceptance in this period of an appointment as an unsalaried university lecturer 'to teach the foundations and methods of elementary mathematics in physics' at the mathematics and physics faculty in Leiden?[22] And what would Paul have thought? He had been furious in 1933 when Casimir seriously considered going to work for Heisenberg in Göttingen, while Born and other Jewish colleagues had just been dismissed (Hollestelle 2011, 277). Would he have prevented her from taking this step? Or would he have understood her deeper motives? 'H. Kramers drew this up', is written at the top of a draft letter from November 1940 in which 'Mrs Ehrenfest-Afanassjewa' applied for permission from the 'acting head of the Department of

Education, Arts and Sciences'.[23] Had she wanted to protect Galinka and Jaap by taking this job, which showed apparent obedience to authority? She didn't commit her motives to paper any more than she had for her travels year in year out to the Soviet Union. On 11 May 1942, two weeks before the deportations of Jews began, she resigned.[24]

In spring 1943, a Dutch policeman identified Jaap in the street as a Jew and arrested him. Tanya had expected him at Witte Rozenstraat and had been surprised he 'didn't arrive as agreed', she later wrote in a typed account.[25] 'Then I was called one evening by an unknown person, who didn't want to give his name. The unknown person informed me that my son-in-law had been arrested, and he warned me that because my daughter was half-Jewish by ancestry, she was also considered a Jew since her marriage to J. Kloot and would be arrested'. The phone call came two days after an incident in the garden at Witte Rozenstraat, 'where I was working, two detectives appeared and inquired about my daughter'. When visiting a potential place of hiding a few weeks later, Galinka was indeed arrested as a 'full Jewess' and transferred to a prison in Den Bosch.[26] That the *Hauptsturmführer* of the prison prompted her to tell a cover story about Paul Ehrenfest not being her real father and Jaap Kloot not being the father of her unborn child saved her life. He helped her get the story straight by slightly raising his eyebrows during the interrogations, she later told her mother.[27]

Galinka is a 'magnificent person', Paul once wrote to Ioffe (see Figure 35.2). 'She knows how to make something beautiful out of anything'.[28] And to Margrethe Bohr: 'if she only knows two people somewhere, she immediately swims farther by herself!!!'[29] She was released. Shortly afterwards she had a stillbirth.[30] Lawyer Daisy Schaeffer, who had overseen Tanya's divorce procedure, arranged a declaration of Aryan origin for her.[31]

Years before, in 1922, Tanya revisited her earlier text about happiness in her Russian notebook. She proposed a thought experiment. 'Imagine someone in whose life the elementary content always stays the same: he is surrounded by the same things; he does the same work; he spends his free time in the same fashion. But what a different value all of this has depending on his personal fate. If he imagines that someone thinks of him and belongs with him then he shall experience the same work and the same pleasures with joy. But if there is no such person then nothing has any value'.[32] In short, love and solidarity are what makes people happy.

For people who had grown spiritually, this love could extend to more than one person, she added, making 'someone able to share in everyone's sadness and joy'. Did this thought now comfort her? 'An intermediate stage is love for people you consider good. This is followed by love for the good in every human. All of this requires an aptitude for joy: *der Wille zur Freude* [in German]. This is joy in the Lord, best expressed in starets Zosima [the wise old monk in Dostoevsky's *The*

Figure 35.2 Sketch of Galinka as a girl in a tree in the garden of the house on Witte Rozenstraat, by Harm Kamerlingh Onnes, c. 1926.—*Ehrenfest Family Archive.*

Brothers Karamazov]. One lacking this shall, by broadening their own 'I', broaden the scope of their sorrow as this is always amply present. And yet this too is probably easier to bear than one's own sorrow'.[33]

In April 1945, shortly before the end of the war, sixteen-year-old Sari Langedijk cycled from The Hague on a bicycle with wooden tyres to her brother. He had been rounded up in 1942, taken to Herzogenbusch concentration camp in the Netherlands, released again, and had now been living in Leiden for more than two years. As he had managed to obtain a work permit, he was exempt from forced labour in Germany. With his wife Tine, he lived upstairs at 'Mrs Ehrenfest's', and Sari came to stay for the last days of the war.

Later she recalled the hustle and bustle in the house. Baba Katya sat on the terrace in the sun and only spoke Russian and French. Galinka went around with her hair in a red kerchief. And next to Dick and Tine lived Tine's brother Wim, Mrs Brunt from The Hague, Mrs Woltjers and her two sons who likewise wanted to escape forced labour in Germany and Mrs Pinto, a Jewish woman who hoped— in vain—that her husband would return from the camps. They ate small pancakes made of unsifted 'rough' flour. Galinka occasionally managed to get hold of lettuce, and one day Dick came home with a 'delicious tulip-bulb tart'. Sari thought herself foolish, she said later, but 'Mrs Ehrenfest paid honest attention to me'. Tanya showed her the plants, bushes and trees in the garden, and 'you felt intuitively that she had erudition and knowledge, but she was attentive and treated you as an equal. I thought her kind, which is why I remembered her name'.[34]

Had more people lived and hidden temporarily in the house over the preceding years? Galinka and Tanya never spoke about the past, according to Sari. As a reminder of Witte Rozenstraat, she always called one of the plants Tanya had shown her, a saxifrage, the Ehrenfest plant and grew it at every house she lived in.

36

Finding joy in less

Writing and thinking to the end

'I was so pleased with your letter', Einstein wrote on 22 October 1945.[1] 'You seldom come across such a robust and steadfast person. Your interest in the foundations of physics has also remained fresh, as if you hadn't experienced those difficult and threatening years'. Not much later, the company Mimosa Food Products[2] in New York delivered a food parcel to Witte Rozenstraat. Then another, yet another and 'now once again something delicious from you'.[3] Even after Tanya asked Einstein whether the parcels could go to a hungry friend in Prague, they kept arriving in Leiden. He simply added the Czech acquaintance to his list.[4] Only in 1947 did he somewhat reluctantly agree to Tanya's request to send 'her' parcels to Bauer in Paris.[5] 'Are you planning to feed me until my death?'

In early spring 1946, Tanya—'my beloved friend Tanechka'—also received a letter from Vera Fedyaevskaya in Moscow.[6] The war had severely affected Russia, especially Leningrad, where more than a million people had died, including Vera's husband. 'We worried about you the whole time because we fearfully wondered whether you and your family had managed to survive the terrible life under German rule. I was glad to hear someone in Leningrad had received a letter from you. So, you are still alive, thank God'.

Tanya also received a postcard from the Jewish National Fund. Mau Schaap and Bep Schaap-Bedak had arranged for a tree to be planted for Tanya in Palestine, in Westerweel Wood, which commemorated resistance fighter Joop Westerweel.[7] 'We are delighted to inform you that your Jewish acquaintances thought your name should be connected to part of the wood', they explained.[8]

The university of Leiden sent a short letter asking whether she was prepared to resume the post she had resigned in 1942 as an unsalaried university lecturer. Apparently, vacancies were the result of 'the repression [measures against collaborators in the war] now also extending to unsalaried university lecturers'. But her relationship with the university would never truly be cordial: she gave another lecture series on thermodynamics[9] and some Russian lessons but ended her year-old contract in 1950 as assistant lecturer in Russian because her salary was much lower than usual. 'It seems to me that resigning myself to this situation would harm my own prestige too much but also that of the subject I teach', she wrote to the Board of Trustees.[10]

The Delight of Thinking. Margriet van der Heijden, Oxford University Press. © Margriet van der Heijden (2021).
DOI: 10.1093/9780198927112.003.0036

Galinka heard only weeks after the end of the war that Jaap had been transported to Westerbork concentration camp in the Netherlands in late June 1943 and then to Sobibor where he was murdered on 2 July 1943. A year after receiving this message, she travelled to Switzerland, where among others she saw astronomer Zwicky, who had so impressed her in Pasadena in California in the early 1930s. There she was again—but how could Zwicky understand what Galinka had been through?[11] More sometimes restless journeys followed. She travelled to Germany in 1948, where she slept in bunkers and air-raid shelters which served as cheap hotels. 'Left the air-raid shelter as quickly as possible for fresh morning air among the grey ruins as waking people gradually emerged', she wrote to her mother.[12] She visited former *Hauptsturmführer* Benno Samel, the SS officer who had 'saved' her life and probably that of others too[13] and who was trying to continue life under a pseudonym. In Göttingen, she called on Paul's childhood friend Herglotz, who thanked Tanya through her for sending food and other things. 'He cried with joy—apparently, I look a great deal like father after all. He lives in part of a beautiful big house and has nine dogs!'[14]

In the Netherlands, she didn't manage to relaunch Corunda Publishing House. And all her friends and acquaintances were glad when in 1949 she married athletic mathematics teacher Henk van Bommel—who was five years her junior. He would read the newspaper with his long legs crossed on the lawn at Witte Rozenstraat and, as a former member of the resistance, was almost as traumatised as Galinka. The same year they had a daughter, Tamara.

Later, when Tamara was a little older, Tanya regularly travelled to Gronsveld in Limburg, in the far south of the Netherlands, where Galinka had designed a house and had it built. It was much smaller than the house on Witte Rozenstraat, but it too had white, plastered walls. With painted yellow eaves, it was like a little dacha in a garden filled with bramble bushes, apple trees and flowers.

When Tanya arrived in a taxi from the station, she always brought a bouquet of flowers from her garden wrapped in paper, her granddaughter Tamara recalled. And presents. Tanya loved the countryside in Limburg. She appreciated the tree-lined roads and wooded hills so much more than the bare windy polders around Leiden. From Limburg, she sent cards to Witte Rozenstraat, where the female students renting rooms on the upper floors had the house to themselves (see Figure 36.1). 'I hope that on these longest of the year evenings you and your boyfriend want to make use of my lawns', she wrote to one of them.[15]

Her Dutch was a bit shaky to the very end, while she spoke extremely good German, the students found. Tanya found the Dutch language 'uninteresting', they thought. But the card also shows how uninhibited Tanya was in the prudish 1950s, when married women in the Netherlands were still legally incapable, and sex was a subject only for furtive discussion and tittering. 'I still remember walking up the stairs with a boy and Mrs Ehrenfest came into the corridor. "Is this your lover?" she called after me. It resounded through the house', one of them

Figure 36.1 Tanya, or Mrs E as the students called her, with the students who rented rooms from her on the veranda at the back of the house, late 1950s. Sitting in front: Tanya Ehrenfest and on her right Mrs Van Gendt (the housekeeper who came every day). On the balustrade behind them, from left to right: Elly de Meijer, Riejet Wijnbergen, Fransje de Ronde, Tanya Angélique Hajenius, librarian Ida Swellengrebel, Carool Kloos. Standing on the right: Tanya's granddaughter Eudia van Aardenne with her husband Henk Ritsema.—*Photograph Ehrenfest Family Archive.*

later said.[16] Apart from that 'Mrs Ehrenfest' didn't interfere with anything. Sometimes she came upstairs with a bunch of flowers from the garden, and she always encouraged the students to pick flowers but otherwise she 'didn't put the slightest obstacle in their way. Only if we had a party, we would warn her beforehand'. There was just one thing the students weren't allowed to forget: graciousness. And she hated it when people were indifferent to knowledge. 'When you came down the stairs, she sometimes rushed into the corridor with a question. When one of our boyfriends teared past her in the corridor in great haste and shouted what was in hindsight the wrong answer to her question about the translation of an English word, he was in her bad books. Doing things attentively was particularly important to her'.[17]

Tanya was so old and frail by then. When she planted something in the garden with a big spade, it seemed almost bigger than her, according to the students. Of the past, she spoke little. She followed agricultural politics in Cuba with great interest, but she didn't adhere to socialism or any other ideology. For that, she was 'too independent and free spirited. I had the impression she hated the Soviet Union', one of the students observed. Tanya also dismissed her own work in her country of birth. 'I had to teach without books there', she is supposed to have said jokingly.

'All I taught them was to think'.[18] Given her ideals for geometry education, this was of course a wonderful outcome already. But what did the students at the Witte Rozenstraat know about her work in this field?

They sometimes saw Tinbergen come for Russian lessons, as did legal scholar Witkam's daughter, translator and university librarian Wils Huisman and several other young pupils. Did they know that Paul and Tanya had once inspired Tinbergen? That Tanya had published physics articles, initiated mathematics debates, published *Relevia* and corresponded with friends and academics across almost the entire world? Did they suspect that Tanya exchanged letters with Einstein until shortly before his death?

'Do send me the manuscript', Einstein urged in March 1947, 'then I'll try to get a university press interested after reading it myself'.[19] The manuscript was for a textbook about thermodynamics that Tanya had been working on since the 1930s and which she had told him about in previous letters. It included the ideas from her 1925 publication but also a subtle elaboration of them in which she had thought about the states of equilibrium in gases and liquids. In the 1930s, before Pawlik's accident, she recorded thoughts on the subject in long French-language manuscripts,[20] after which she wrote a German-language manuscript, *Die Grundlagen der Thermodynamik* (The Basic Principles of Thermodynamics).

Needless to say, Einstein was already partly familiar with her work on thermodynamics. When baba Katya died in spring 1946, he had also written to Tanya about it. 'For us it is also actually about time ... you really must arrange for your *Thermodynamik* to be translated beforehand'.[21] And referring to their private joke about her analytical way of thinking, he continued: 'I'm convinced that your logical scourer has brought about something good by clarifying the slightly hazy conceptual basis [of thermodynamics]'.[22]

'You are truly a real friend that you immediately set to work reading my manuscript', Tanya wrote enthusiastically one-and-a-half years later, in late summer 1947.[23] Finally, she had managed to get her manuscript across the ocean, 'and I'm thoroughly pleased with the clear and transparent conceptual structure', Einstein wrote in response two weeks later.[24] But as he advanced through the book his enthusiasm cooled. Tanya had wanted to drive off the haze, which made the basic assumptions of thermodynamics a bit vague for incisive thinkers. But she had wanted to convey her subtle insights so precisely that she created an impenetrable thicket of definitions and arguments. Einstein had difficulty distilling the essence of this. 'It is a bit like the performance of a practised conjuror in which there are so many nice details to see that one doesn't notice when your unforgettable P.E.'s frog jumps into the water', he wrote, referring to Paul's favourite expression about discerning this essence.[25]

Fundamentally, Tanya was right, Einstein thought, but her arguments should be shorter and clearer. 'I have the impression you have been possessed by your logical scourer, and this has affected the transparency and comprehensibility of your book'.²⁶ And as Einstein didn't know anyone to whom he dared to entrust the translation, he returned the manuscript 'from which I have learned a great deal' and 'with respect'. To Tanya's disappointment, who remarked ironically: 'that you hastened to wash your hands of my manuscript is very very unfortunate. I had hoped to be able to die in peace. What am I to do now ... in Europe there is too little paper ... and even fewer bigwigs able to recommend my work to a publisher'.²⁷

She was 71 years old. She no longer had the energy to rewrite the manuscript. Instead, she kept looking for translators through Tinbergen, Burgers and others, now German was no longer the lingua franca in physics. It would take a while before she resumed correspondence with Einstein who perhaps hadn't quite caught her irony. But they started writing again and discussed other subjects in their letters: the theory of relativity and quantum mechanics, the world indirectly and sometimes Einstein also read Tanitchka's mathematical reflections, enclosed by Tanya. By this time Tanitchka had six children—Tanja, Eudia, Gijs, Marietje, Carel and Anneke—and Einstein thought it quite right that, despite these 'innumerable children', she continued to do mathematics, so 'this source of intelligence doesn't get blocked'.²⁸

Insomnia was apparently all that indicated Tanya had been through a great deal. Tamara, who often stayed with her grandmother as a young girl, sometimes for long periods, recalled how she would have to wait until eleven o'clock to enter the sunroom, where baba Tanya slept. On a chair next to the bed, in this room with no other furniture than a wardrobe and a mannequin, she would then listen to Tanya reading a large Russian fairytale book, illustrated by Bilibin, or *Alice in Wonderland*. At night, according to Tamara and the students, Tanya would read the 'green' Penguin detective novels, kept on a console table in the dining room, or she would do some work. 'Here you have a contemplation which I wrote on one of my wakeful nights', she wrote to her son-in-law Henk van Bommel in 1950.²⁹

'Why can a group of people so often not bring about substantial improvement in a certain sector of human activity (social, political, economic, intellectual ...). I think: due to a lack of certain positive desires. Most participants in such a constituency know merely that they want to replace something else but come together without a positive plan. *Hate* of what exists, not *love* for what must be'. Tanya applied this to reform of mathematical education: 'Most ... hate the current material; they *do not love* any particular alternative. This is why they are defeated by those who do see something beautiful in what exists'.³⁰

Yet after the war, mathematicians *were* interested in Tanya's alternative ideas about mathematics didactics. The Mathematics Study Group had taken up the matter, and again debate evenings were organised at Witte Rozenstraat. Do you think visually or linguistically, Tanya once asked the participants as soon as they entered.³¹ Small talk was still wasted on her.

In 1951, with her brochure 'Can mathematics education contribute to the development of thinking skills?' she also once again launched a debate with a grandee of mathematics didactics, as she had done with Dijksterhuis in 1924.[32] This time, her sparring partner was Hans Freudenthal, a professor of pure and applied mathematics in Utrecht. In essence, he had adopted most of Tanya's ideas, but this debate focused on an important difference in their thinking. Tanya found good and innovative mathematics education important because mathematics helped 'develop thinking skills'. Freudenthal thought, more pragmatically, that 'one is left on unstable ground when one wants to justify the hours that one or other school subject claims by appealing to the mental exercises on which this time would be spent' (Moor 1999, 327–330).

The students in her house were scarcely aware of such activities either. Tanya would sometimes suddenly ask a somewhat scientific question, when they came to pay their rent, they remembered. 'Such as, do you know why the moon seems to move along with you when you walk outside at night? And when I still didn't know the answer after a few days, she explained it with drawings on the blackboard.'[33] And now and then Tanya's ideas were reflected in practical issues. She shook her head once when a student's room was being painted and all the student's things were moved downstairs. 'Oh you, unhappy possessor of too many things.'[34] And she once wrote to her son-in-law Henk van Bommel: 'I hate having much because I find joy in having less'.[35]

She sent Tamara cards with carefully chosen images such as a Chinese watercolour with a lake and boats. 'Congratulations on your birthday. Are you happy with the package of flower bulbs I sent? ... and have success learning from all the arts and sciences!—"the beautiful art of gardening" as the philosopher Schopenhauer called it, included!' On another card, with two chickens, she asked her granddaughter to '1) draw a line and 2) a right angle and then 3) a triangle. ... Do you find my exercise unpleasant? Too simple? Too difficult? When you get the triangle, you shall certainly like it'.

When she heard, a year before her death, that Tamara 'wants to be disloyal to the ancient Greeks and wants to transfer to the HBS [another kind of secondary school, without ancient Greek]', she offered to bring her granddaughter's mathematics up to standard. 'I would like an overview of what we would have to do beforehand. Do you want to? I have always thought it a shame that Russian girls' schools didn't teach ancient languages. Later I learned Latin grammar on my own, but I never had time for Greek. Is Mama eating enough?'[36] Mostly indirectly, through Tamara and Van Bommel, Tanya was eternally worried about 'Galinka, my darling', who she thought was too thin and unable to cope with the humid Dutch climate.

She continued to exchange letters with Einstein until shortly before his death in 1955, sometimes enclosing articles or calculations, and often after long intervals.

Figure 36.2 Tanya, 'baba Tanya' to her grandchildren, in her son-in-law Henk van Bommel's garden in Utrecht.—*Photograph Ehrenfest Family Archive.*

'You are right', he wrote in one of his last letters to her, 'not to burden your brain with the dates of people's birthdays'.[37] Dutch newspapers had printed photographs of Einstein on his birthday, and Tanya realised she had completely forgotten it. She didn't recall exactly when her birthday was, she had cheerfully explained in a belated letter. She only remembered knew the date of her baptism.[38] But 'I'm just as delighted by wishes from you on any day of the year that you choose', Einstein replied.[39] Tanya still had many plans, for the garden, her work ... 'My daughters are now doing well again, but this winter was marked by sickness. I was ill too and had a real childhood disease: whooping cough. Isn't that a sign of youthfulness?'[40]

She still complained about the Dutch climate. 'I am waiting', she wrote in the hot summer of 1959 to her son-in-law Van Bommel, 'for the weather to change shortly, so one can walk without getting dizzy ... This is how it is: either too cold or at once unbearably hot'.[41] And when one of her renters, Ida, turned down the heating to be economical, Tanya immediately left a letter on the stairs for the students: 'Please, don't ever do that again. I am dying of cold'.[42]

In 1960, she went deaf, as had happened to her mother. Henceforth, she needed a hearing aid. This was 'a surprise which was indeed predictable', she wrote to

Van Bommel.[43] 'Tomorrow the seller of these aids is coming to let me try. ... look forward to finding out. In other respects, I am doing fine, only I am lazier every day'.

Gradually, more and more people disappeared from her life. Tanya's former neighbour on the other side of Witte Rozenstraat, Eli Wiersma, who had worked so much with Shubnikov and Trapeznikova and so often travelled to Russia, died in 1944. Tanya also survived Paul's former student Coster, who, as a professor in Groningen, had smuggled Lise Meitner out of Nazi Germany in 1938 and died shortly after the war (Lewin Sime 1996, Ch. 8). And she outlived Hans Kramers, who died in 1952.

She saw former communist Jan Burgers exchange his position at Delft University of Technology for a professorship at the Institute for Fluid Dynamics and Applied Mathematics at the University of Maryland in the United States in 1955 (Sengers and Ooms 2017). In this same country, Dirk Struik became a successful mathematician, while remaining a dyed-in-the-wool communist, and published what became a classic on the history of mathematics (Powell and Frankenstein 1999). Sam Goudsmit and George Uhlenbeck, who had stayed in America all these years, were both much lauded. Only Hendrik Casimir, who became the director of Philips NatLab in 1946, still lived in the Netherlands. And she still had regular contact with Arend Rutgers, who was a professor of physical chemistry in Ghent in Belgium, where, like Paul, he set up a reading room and, like Onnes, a glassworks.

She contacted her surviving Russian friends and colleagues at most with great caution, fearing they would get in trouble. Leonid Mandelstam died in Moscow in 1944 and Anna Föhringer died in Paris in 1945. Benjamin Kagan taught geometry at Moscow State University until his death in 1953. Yuri Krutkov, who had written such a long letter in 1920, was condemned to ten years imprisonment in 1937 during Stalin's Great Terror and died of a heart condition four years after his release. Boris Hessen, director of the Physics Institute in Moscow and Tamm's best friend, was executed in 1936, and cheerful Lev Shubnikov, who had so often walked and cycled through Witte Rozenstraat, befell the same fate a year later (Trapeznikova 2019). Igor Tamm, who had a narrow escape in the terror, worked with Gustav Hertz and his student Andrei Sakharov on the Russian atomic project and won the Nobel Prize for physics in 1958 for the discovery of Cherenkov radiation. Abram Ioffe, who hadn't wanted to lead the Russian atomic project and had been sidelined by Stalin because of his Jewish descent, died in 1960.

Tanya saw how even the most formidable female mathematicians and physicists were consigned to the margins of the history of science. Lise Meitner, who ended up in the United Kingdom via the Netherlands and Sweden after 1938, had conducted crucial experiments and correctly interpreted them as proof of nuclear fission after leaving Berlin. But the Nobel Prize for this work was awarded in

1944 solely to Otto Hahn, with whom she had worked closely for all those years and who carried on with the final steps of their research after she had fled the Nazis. Awe-inspiring Emmy Noether, whose work was respected far and wide by mathematicians, could only find a position at Bryn Mawr College for women after fleeing Göttingen in 1933 and died two years after reaching America. Only Marie Skłodowska Curie, who died before the war, still enjoyed fame. And Tanya just managed to see how German-American Maria Goeppert-Mayer, who once stayed at Witte Rozenstraat, won the Nobel Prize for physics in 1963, the second in this discipline ever to be awarded to a woman.

The stubborn resistance to women in the field contrasted in hindsight with the progressiveness of her Paul. 'Wouldn't it be amusing if all the publications of my wife and myself could someday be printed together ... in chronological order. ... It matters very much to me that my wife's essays are published as far as possible in whole', he wrote after Onnes' death in 1926 in one of his letters with 'a few wishes' for after his own death.[44] And while this wish wasn't fulfilled, Tanya didn't simply let her work gather dust in archives. After much back-and-forth, she published *Die Grundlagen der Thermodynamik* a year after Einstein's death at Uitgeverij Brill publishing house at her own expense (Ehrenfest-Afanassjewa 1956). Colleagues she sent it to mostly thought it too difficult to use as a textbook.[45] Her meticulous and well-considered reasoning appealed more to mathematically minded philosophers of science than to physicists, she once noted. But at least her subtle ideas, which were before their time, were on paper (Uffink 2000; Uffink and Valente 2020). And her essays on mathematics education were published in Dutch as a compilation, with a foreword by mathematician Bruno Ernst (Ehrenfest-Afanassjewa 1961).

She succeeded in getting a small portion of her other work translated into English. An article about the use of probability in physics, which she had published in Russia in 1911Ra, was out of date by then (Ehrenfest-Afanassjewa 1911, 1956). But the English translation of the overview of thermodynamics she had written with Paul for the *Encyclopaedia* became a classic (Ehrenfest and Ehrenfest 1959). In the preface, she referred to her own latest insights into thermodynamics and commemorated Paul and his work. Tinbergen would refer to *Relevia* in one of his essays in 1979 (Tinbergen 1979, 78). And in that same year Galinka translated the innovative idea of a basic income to the '*Ooievaarsregeling*', or 'Stork Scheme', a social insurance scheme that paid a basic income to mothers who worked less or not at all because they were raising a child, or paid the money to another carer if the mother did work (G. Ehrenfest 1979).

Yet, especially in the Netherlands, Tanya's name is most remembered for her ideas on geometry education. Doubtless, this is partly because many people associate women more easily with education than with thermodynamics. She regarded the belated praise for her didactic ideas and for her *Übungssammlung* with gentle mockery. 'Yesterday the *Mededelingen* [Announcements of the Mathematics

Study Group] issue dedicated to my 85th birthday arrived.[46] Henk [van Bommel] shall probably also receive it and bring it to you', she wrote to Galinka in 1961. 'Very moving, even more so because none of them could understand me! But it is correct that before the publication of my first article in their (*two!*) education journals, there had been no discussion on how come and why mathematics must be taught at school, and everyone started talking about it afterwards. But why did you give them that ghastly photo? ... Mrs Van Gendt [her beloved domestic help, who came daily] thought I looked more like 100 than 85 in that photo. On top of which, I don't even have my eyes open'.[47]

Three years later, Tanya had a stroke. The students took leave of her, at some distance, halfway through the doorway of the dark room in which she lay in bed. The flowers one of them had picked in the garden had to go, Tanya motioned with her arms. She still wouldn't allow flowers in her bedroom, even if she could no longer say so. It went on for a week. 'Can you still think?' Galinka asked her silent mother on one of the last days. The smile on Tanya's face had a touch of irony. Yes, she blinked with her eyes.[48]

Epilogue

A series of letters went back and forth between St Louis and Leiden at the end of the 1930s. In the American city, Hugo Ehrenfest was deeply concerned about his brothers and their families in Nazi-ruled Vienna. His eldest brother Arthur had died in 1931, but his widow Regine was stranded in the Austrian capital, while his son Fritz was trying to escape the Nazis through Greece. Hugo's elder brother Emil and his younger brother Otto had both lost their flourishing companies and fortunes and lived in poverty and fear.

Hugo hoped that old and infirm Emil's life would soon be over. For Otto, his wife Fanny and their two sons, as well as for Arthur's son Fritz, he had arranged American visas. But he disagreed with Tanya Ehrenfest-Afanassjewa's proposal in one of her letters from Leiden. Tanya also wanted to get Regine out of Vienna, as well as Pepperl, Hugo's father Sigmund's widow. She wanted to use the money she and Paul had set aside for Pawlik's studies[1]. But Hugo didn't like the idea. It would be better for Tanya to use this money, for example, to move to Switzerland with Galinka, who she was so worried about. Regine and Pepperl were too old simply to be 'transplanted'. Moreover, Hugo couldn't act as a guarantor for yet more family members in the United States. 'Tanya, I have great reverence towards you as a person, but you are a Russian—and I am an American. You are and always were prepared to give away EVERYTHING you have. I feel forced to think of the future of my wife, children and grandchildren. [...] Tanya, try not to mitigate the suffering of everyone, that is an impossible and disconsolate desire'.[2]

Did Hugo write this because he understood his sister-in-law? When Tanya came into his younger brother Paul's life, Hugo was already living in the United States. And while Paul visited him twice, Tanya focused for a long time mainly on her native country. Consequently, they knew each other almost exclusively from letters and stories, and the American perspective on the socialist USSR might have heavily influenced Hugo's outlook.[3] Yet a mutual appreciation is unmistakable in the letters.

It is, of course, unquestionable that Tanya Ehrenfest-Afanassjewa lived in a terrible age and that she also lived with Paul Ehrenfest in trying and sometimes awful times. Each engaged with this in their own way. For a long time, they withstood the hardships together successfully—both driven by the desire to leave behind the sometimes-stifling environment they had grown up in, to choose for their own ideals and to follow their own principles in life.

Effervescent and curious Paul, as the last-born child, looked up to his elder and clever brothers, who were climbing the Viennese social ladder. In his parents' flourishing 'supermarket', he learned to mix easily with people from different backgrounds. At the same time, it was constantly made clear to him in the streets and at school that as a Jewish boy in Vienna he would never belong.

Amiable and extremely clever Tanya was brought up by her uncle Pyotr and aunt Sonya in affluence but also with many rules and restrictions. It was only possible for her to study at a women's higher education institution once her authoritarian uncle had died. Her mother, a widow and the daughter of a freed serf, played hardly any role in her upbringing. Women, Tanya must have learned from an early age, have little say in life.

With Paul, she could escape the strictures of her youth, and he found the same freedom with her. The impetus of this flight was their shared love for mathematics and physics, alongside—it shouldn't be forgotten—a shared love for music and literature. Their desire to mean something in mathematics and physics and to have lasting influence, as Paul once described it, took them from Göttingen to Vienna, back to Göttingen and St Petersburg and finally to Leiden. On this journey, they— and certainly Paul—befriended just about everybody who counted in physics, as well as many people—and this certainly applies with Tanya—who wanted to raise mathematics didactics to a higher standard.

They were never a conventional couple. Neither in Vienna, where Paul was the odd man out with his Russian peasant's smock and Tolstoyan values and Tanya was an outsider as a woman in the lecture halls. Nor in St Petersburg, where, as a Viennese Jew with a Russian beard, Paul was intensively involved with rather stuffy scientific institutions, and where Tanya—raised Russian Orthodox— would rather discuss probability with friends at the blackboard than frequent well-heeled salons. Even if they weren't dangerous revolutionaries or nihilists, many in the established order found their idiosyncratic and critical attitude somewhat threatening or in any case disturbing. Paul's search for employment or Tanya sitting an examination at an 'ordinary' university (but exclusively reserved for men for centuries) was—also because of this—far from matter of course. Yet their independence also made them an example to a new generation of physicists: participants in the two-weekly discussions at the blackboard in their living room included Alexander Friedmann—the man who exposed the possibility of an expanding universe—famous Abram Ioffe and a succession of other luminaries in Russian (and later Soviet) physics. Later they would call Paul a founder of Russian theoretical physics.

Paul and Tanya played a comparable stimulating role from 1912 in Leiden, a small provincial Dutch city, where their brand-new and comparably vast 'Russian house' was completely out of place. It had plastered instead of brick walls. Neoclassical lines and pediments instead of a step gable or a pitched roof. The letterbox was in the wall of the study, the front door was at the other end of the house, the bell was at the garden gate, and suppliers delivered their goods through a large shutter in the kitchen—all of which was entirely alien to Dutch domestic logic. Alongside the flower bulbs, Russian poppies flowered in the garden every year.

For young Dutch students such as Jan Burgers, Struik, Coster and to a certain extent Kramers and Minnaert, and in later years Uhlenbeck, Goudsmit, Tinbergen, Casimir, Rutgers and Dieke, it was an oasis. It was where they kept up to date with the latest developments in the field. Where just about any subject could be discussed. Where Paul played the piano and the children hung their drawings on the walls and doors. Where Einstein came to recover when he had completed his general theory of relativity, amid the appalling World War I which made it clear in many other parts of Europe 'to what deplorable breed of brutes we belong'.[4]

That Einstein was soon like a member of the family at Witte Rozenstraat—in as much as the distance between Leiden and Berlin allowed—was unsurprising. 'You have that rare ability of smelling unusual personalities & to quickly approach them [...] to establish valuable friendships', Paul's brother Hugo once wrote about him.[5] And this is how it was: almost as soon as the war ended and travel resumed, Niels Bohr came to stay, and Paul and Tanya opened their house to 'people with ideas' from around the world over the ensuing years. On the wall of the guest room, high in the attic, sixteen Nobel laureates' names can still be found among the signatures, as well as that of Lise Meitner, who more than deserved to also have won a Nobel Prize.

Bringing people and ideas together was also a key aspect of Paul and Tanya's role in science. Their own work in physics consisted of intelligent and well-considered comments on often very subtle and complex issues. If, for example, Paul's work in 1912 on the role of quanta in the interaction between light and material had received more attention, then the development of quantum theory might have progressed a little faster. The 'adiabatic principle' he developed in the ensuing years inspired Bohr's correspondence principle and thus certainly moved quantum theory forward. Paul's article in 1909 on an extremely fast rotating disc elegantly illustrated the limitations of Einstein's special theory of relativity. And in his later work too, Paul produced acute commentary on developments in quantum mechanics. But this work never stood on its own, like Einstein's theories or Schrödinger's wave equation. It wasn't formative and didn't produce formulas and concepts that would be widely discussed in lecture halls—or their future equivalent—a century later.

Paul's qualities were elsewhere. He was a 'curator' of science at the highest level. Unerringly, he recognised the quality—or lack thereof—in others' work; he could place it in exactly the right context, inspired others' enthusiasm in it and with such enthusiasm, he subsequently brought physicists across the world in contact with each other. Accordingly, and above all, he was a peerless teacher and mentor. Even decades later, his former students still lauded his crystal-clear explanations and the consummate skill with which he summarised complex issues simply by writing the essential formulas on a blackboard—in an unprecedentedly clear manner. They praised his humour, how he played with language, the charm he used to break the ice and his sometimes intimidatingly sharp but always elucidating Socratic dialogues. But the best proof of his intensive engagement with his students is that their careers were almost without exception successful.

Burgers ran two research groups with great energy, first in the Netherlands and then in the United States. Struik had a long and productive career as a mathematician in America. Coster was the co-discoverer of the element hafnium and a professor at the University of Groningen. Kramers became Bohr's right-hand man in Copenhagen, was then somewhat crushed by Bohr, but—supported by Paul—made a name for himself again in Utrecht. Goudsmit and Uhlenbeck were showered with prizes in the United States. Casimir became the director of Philips NatLab. As a professor in Ghent, Rutgers followed in Pauls footsteps by establishing a reading room and a colloquium series. And Tinbergen, who Paul had given the freedom to apply physics models to economics, was later awarded the Nobel Prize as one of the most important founders of econometrics.

Paul also had lasting influence on the teaching of statistical mechanics. In Leiden and several other Dutch universities, this topic was taught for a long time from Paul's great mentor Boltzmann's perspective.

Long forgotten, overlooked or simply ignored in the Netherlands were the role and influence of Tanya Ehrenfest-Afanassjewa. She had been employed in St Petersburg as a mathematics educationalist. Together with Paul, she had written a still-renowned overview of Boltzmann's work and she had published her own work in the field of probability. But it was a cruel irony that she ended up in Leiden in 1912, where women were firmly confined to the domestic sphere. Worse, the unwritten rule that married women didn't work, except in need, was formalised in Dutch law in 1924. Even if Tanya had designed and largely paid for her own house, and even if she had a string of publications to her name, from then on she was still considered a 'legally incapable spouse', unable to open a bank account, to withdraw money, to decide on large purchases and to have a job—which she found especially distressing.

Did this eventually play a role in driving Paul and Tanya apart? Tanya worked at home. She organised debate evenings on mathematics. She wrote scientific articles, some of which were published and a few of which were even presented, through Onnes and Lorentz, at the Royal Netherlands Academy of Arts and Sciences. Yet, she still contributed from the sidelines, unpaid and counting for little. Even when she initiated a large debate about geometry education in 1924, resulting in the founding of a journal for mathematics didactics and a commission for education reform, she was left empty handed. She wasn't invited to join the commission or the editorial board; the pamphlet didn't earn her a position.

'Wouldn't it be amusing if all the publications of my wife and myself could someday be printed together [...] in chronological order. [...] It matters very much to me that my wife's essays are published as far as possible in whole', Paul had written in 1926.[6] In reality, Tanya was again restricted in Leiden, as she had been as a student in St Petersburg. Again, it was as if engaging in science was at most a leisure activity for women, an elevated form of embroidery, to be performed at a modest level when the home was in order.

Whether things could have gone differently is never possible to answer with certainty. Paul and Tanya were confronted by many other problems too. The revolution and civil war in Russia unsettled Tanya. They rendered her railway shares worthless, sweeping away their financial security—and causing panic, especially for Paul. Tanya must have looked on with increasing dejection how Paul ascribed his lack of creative work to a want of tranquillity and a feeling of continually having to earn extra money with lectures and examination jobs. Conversely, Paul watched sadly how Tanya neglected her scientific work to care for Wassik, their youngest son with Down's syndrome, and how this jeopardised her last refuge in provincial Leiden. It was an important reason to send the boy to the Trüper Institute in Jena.

Later, many attributed Paul's demise in 1933—he killed Wassik and then himself—in hindsight to the quantum mechanics revolution in physics. Paul could no longer keep abreast of rapid developments. He was perturbed by the complex and abstract mathematics used to express quantum mechanics. And he, the man who seemed to come up with metaphors so effortlessly, was driven to despair by the thought that this theory could no longer be illustrated by analogies and examples from the sensory world. Quantum mechanics wasn't truly comprehensible to anyone.

Yet this is as reductive as the regular portrayal of Tanya in the past. A cold and rational woman, who—with a stern white bow around her neck—wanted to give her opinion at all colloquia, forbade smoking and drinking and preferred to help develop education in the Soviet Union between 1926 and 1933 than to be with her family and husband in Leiden.[7] The spirit of the age and the position of women in Tanya's years and many ensuing decades made it easy to assume that Paul had inspired her ideas about a more intuitive, exploratory, and gradual introduction to geometry in mathematics education; to dismiss her role in writing their article about statistical mechanics as that of a sounding board and assistant; to trivialise her later work on thermodynamics as fiddling in the margins. It is true that her work wasn't original in the sense of bold and pioneering, but it seems almost impossible to produce such work in a discipline in which many actors, whether explicitly or not, expected women to fail, and in which a woman therefore continuously had to prove and justify herself. And paradoxically, Tanya's very clear, strict and indeed sometimes somewhat single-minded reasoning did make her work original in a different manner. It gave her insights into the reversibility and recurrence paradox in statistical mechanics and led to conclusions about negative temperatures and about the ambiguity of states of equilibrium in thermodynamics which were far ahead of their time, as were her ideas about mathematics didactics.

Likewise, her influence on Tinbergen, with whom she corresponded for the rest of her life, reached farther than her contemporaries estimated.

It was just as easy for decades to contrast autonomous Tanya, with her head full of physics and mathematics, with the prevailing ideal of a sympathetic, well-groomed, charming and obliging wife, and then to pity Paul a little. Yet it seems reasonable to suppose that Tanya's stable, graceful and indeed ostensibly somewhat distant adherence to their shared ideals and guidelines kept Paul on track as he lurched between electrifying cheerfulness and exhausted lethargy. They had given up their religion for each other—marriage between a Jew and a Russian-Orthodox woman was forbidden in Vienna. Together they bade farewell to the bourgeois life of a good cigar, a fine glass of cognac, an education for the children at a gymnasium and then a good start to social life through a student fraternity. Inspired by 'Tanya's' Tolstoy, they chose homeschooling for the children, vegetarianism and eschewed alcohol and tobacco. But Paul was the one who at times loudly applied these guidelines, defended them rigidly or imposed them on others—perhaps because he sometimes needed something to hold on to the most.

Tanya's travels to Russia (as she kept calling the country) for extended periods from 1926 were undoubtedly prompted by homesickness and a desire to do something useful. For one way or another, she didn't feel very useful in the Netherlands as a recreational scientist or a housewife who had never learned how to cook. Moreover, it must have been increasingly difficult for her to live with a man who complained more and more about himself and their shared life during his ever longer deeply gloomy periods. With all their setbacks, Tanya concentrated even more determinedly on the work she could do and no longer tried to meet Paul's expectations. He could no longer tolerate her presence well, he wrote to former students and his son. The large amount of correspondence between family members illustrates how unyieldingly Tanya loved Paul, and he—at a distance—still loved her. But his annoyance must have affected her and perhaps—having learned as a girl to keep her feelings to herself—contributed to her withdrawing to the Soviet Union.

Paul might be understood better in light of Lise Meitner's carefully chosen words about Boltzmann, who had been both Paul and Meitner's tutor and committed suicide in 1906. Meitner described him as exceptionally sensitive and as someone who 'may have been wounded by many things a more robust person would have hardly noticed'. Yet this 'uncommon humanity' and sensitivity made Boltzmann such a formidable teacher, she thought (Lewin Sime 1996, 15). And her words seem almost directly applicable to Paul, who was battered by life.

At the front of his notebook XXIX, which started on 24 December 1924, he copied a quotation, in 1925, from an article by botanist Albert Hitchcock about success in the sciences. 'When is a scientist old? So long as his [sic] mind is directed forward, so long as his past achievements are to him incidental, so long as he is thinking of what he is going to do instead of what he has done, he is young. When he begins to emphasise those things that have been accomplished, when his mind is facing the past instead of the future, he is getting old. Some men get old at an early age' (Hitchcock 1925). Paul felt himself getting old, but his situation was slightly different to that in this quotation. Paul hadn't produced any great, monumental work to look back at, no formula which would endure the ravages of time. And it must have been alarming to realise that he was highly unlikely to still do so.

He was still surrounded by the giants of the discipline. It gave him a zest for living and peace to find himself with Bohr and Einstein. With them, he felt safe 'from all the ugliness of the world, like in the snow high in the mountains'. Yet it became increasingly clear to him that the high mountains could be cold, bleak and inhospitable for somebody better equipped for life on the slopes just beneath the peaks. It also became ever more difficult for him to compensate his resultant agitation and dissatisfaction with the pleasure gained from his students. Conversely, far from every student could withstand Paul's more and more frequently cutting comments, and not everyone simply wanted to meet the high standards he demanded. Some, such as Jan Burgers' brother Willy, preferred to go their own way and were even somewhat derisive of what they considered Paul's high-flown approach to the discipline.[8]

Nevertheless, until 1928, Paul almost always managed to still gain considerable energy from the enthusiasm and almost unconditional devotion of the students who did come and stay—in Baltimore, Pasadena and Göttingen as much as in Leiden. Until 1926, amiable and listening Tanya also still was a reference point between the extremes of the mountain tops and the new students at the foot of these mountains. Just as Paul's predecessor Lorentz was to a certain extent. Paul mightn't have met his high expectations, but his consoling conversations and friendly notes always calmed Paul. It was Lorentz's death in 1928 that irrevocably brought Paul to the painful realisation that he would never become 'one of the first in the discipline', as Lorentz had once hoped. And after this, even his students' enthusiasm and admiration could no longer really keep him on his feet. While his friend Einstein had become a science idol almost in spite of himself, Paul couldn't content himself with the more auxiliary but still crucial role he played: that of teacher, mentor, critic, sparring partner and populariser who keeps science lively and alive; who connects people, results and insights; and in doing so strengthens, refines and improves the tissue of knowledge. Indeed, his need for attention and pleas of desperation occasionally started to exhaust his students and colleagues.[9]

It is also true that they lived in dreadful times. 'If only you wouldn't take your own and other people's lives so damned seriously! How much trouble, disappointment, anger, sadness this causes you—and how little joy by comparison. ... Paul, you know that I disapprove of much in this world, in the US, in my own home, in my own life—but why expect that all that must be changed at once, quickly—why get so depressed about the evident fact that things, as a rule, change slowly, and not even always for the better', Hugo wrote to his brother in 1931.[10] But Paul simply couldn't take life and the world more light-heartedly. And he was even less capable of doing so two years later when the world rapidly changed for the worse. Night after night, he sat at his typewriter, writing letters to arrange help for Jewish colleagues in Nazi-Germany and fretting about the future. Simultaneously, he was overwhelmed by guilt towards Tanya and himself for his relationship with Nelly. An 'unwarranted bourgeois sense of duty' is how Einstein described the desire of his friend, who had previously so opposed bourgeois conventionalism, to marry Nelly.[11] In the end, Paul didn't dare to choose. He met with disaster, and that he killed Wassik, a boy with a Jewish father and Down's syndrome, too 'is something anyone can understand who delves into the matter', he wrote.[12]

Inevitably, such a death colours an entire life, and all that preceded it is seen through the lens of Paul's suicide. Yet Paul and Tanya brought joy to St Petersburg with their occasionally slightly insistent but always good-natured enthusiasm for physics. Paul and Tanya gave their colleagues in St Petersburg, Petrograd and Leningrad renewed energy and vigour in confusing and gruelling times and—especially Paul—set many of them on a new course. Paul and Tanya were ahead of their time with their progressive ideas on vegetarianism, pacifism, education and economics—take Tanya's pamphlet *Relevia*, in which she argued for a universal basic income or her arguments for people 'to really learn to think' and to attune education better to children's development. And Paul and Tanya opened their home in Leiden to people and ideas.[13]

Finally, it was also Tanya, who was so often portrayed as somewhat harsh and rigid, who after 1933—without her once-so-electrifying Paul, without Wassik, soon without her son Pawlik either, without her network which was destroyed by Nazism and Stalinism and with her grief-stricken daughters—continued to promote these ideas with determination, grace and dignity.

Acknowledgements

In retrospect, it seems rather presumptuous and reckless: starting a double biography about two idiosyncratic natural scientists who lived against the backdrop of two world wars, a revolution and a civil war. Not to mention the revolutions in natural science itself, those of the theory of relativity and quantum mechanics. The fact that the project was completed is thanks to a long line of people.

At the front is Tamara van Bommel, granddaughter of Paul and Tanya Ehrenfest-Afanassjewa, who not only gave me access to the large family archive, but also her unconditional trust. Carel, Anneke and Marietje van Aardenne have coloured the image of their grandparents for me with stories and memories, as did Eudia Ritzema-Van Aardenne, who unfortunately did not live to see the book published.

Where would I have been without Slavist Hans Driessen? He translated many handwritten and sometimes typed letters and manuscripts from Russian for me, lent me books on Russian history, read several versions of the manuscript very carefully and brightened up the days with his dry humour.

Anne Kox, who as professor emeritus and former senior editor of the Einstein Papers Project knows almost everything about physics around the turn of the century, was indispensable as a sounding board and, later, as a reader. Henriette Schatz also read along and made valuable suggestions as a linguist. Their joint input was a great stimulus for the project.

It was wonderful to talk to Jos Uffink, Professor-emeritus of Philosophy of Physics in Minnesota and one of the few who have studied the physics work of Tatiana Afanassjewa for years. His careful commentary improved the book. I also want to thank Frans van Lunteren, professor of History of Science in Amsterdam and Leiden, for reading the Dutch manuscript.

I am also, of course, indebted to a number of other people who have previously looked into the work and lives of Ehrenfest and Afanassjewa. Mathematician Ed de Moor analysed Afanassjewa's contributions to mathematics didactics in articles in Dutch mathematics journals and in his dissertation—it is a great pity that he is no longer with us to open the book. Science historian Marijn Hollestelle had already largely gone through the Ehrenfest archive in Leiden for his dissertation, and Pim Huijnen previously wrote a thesis on Paul Ehrenfest's journey through German universities. Above all, I am greatly indebted to someone who absolutely cannot be left out is the late American historian of science Martin J. Klein, but whose Ehrenfest biography, published in 1970 (and which runs until 1921), I am building on and responding to, and whose wonderful essays on Ehrenfest were also a source of inspiration.

A book like this requires a lot of rummaging through archives: I gained respect for the archivists at the North Holland Archives in Haarlem, at the International Institute of Social History in Amsterdam and at the University of Göttingen; Diana Kormos-Buchwald, Robert M. Abbey, Professor of History at Caltech and director of the Einstein Papers Project, helped me locate letters from Afanassjewa in the digital Einstein archives in Pasadena; Dr Fons van Alkemade guided me through the Burgers Archief in Delft, and Dirk van Delft, former director of Rijksmuseum Boerhaave in Leiden, gave me the digitised Martin Klein Archief before it was added—largely—to the Ehrenfest Archive in this museum. I would especially like to thank Dalila Walle, archivist at Rijksmuseum Boerhaave, without whom (the entire archive was being reclassified during my work) I would have been lost.

In addition to all the paper sources, there were also 'living' sources. It was amazing how often people came up with a personal memory after a lecture or newspaper article about 'Mrs Ehrenfest' and the house. I remember with great pleasure the afternoon with Carola Kloos and Fransje Teulings, who lived on the Witte Rozenstraat as students in the fifties; the afternoon with Sari Langedijk, who stayed at the Witte Rozenstraat at the very end of the war and who received me together with her sister Hanna; the morning with Maria Wiebenga-von Winning, whose father had known Mrs Ehrenfest, and the visit to Herman Burgers and his wife, who received me at home to share memories of Burgers' father Jan and his relationship with Paul Ehrenfest—and who unfortunately did not live to see the book published. Then there was Simon Burgers, grandson of Jan, who gave me relevant documents, Anne Dirk Renting, who sent me a transcription of the names on the signature wall, Carlo Beenakker, who pointed me to a lecture dictation by Tanya Ehrenfest-Afanassjewa, Hette Weijma, who found the old telephone number of the house on Witte Rozenstraat and many others who added colour to the image of Paul and Tanya Ehrenfest-Afanassjewa with smaller or larger memories.

I would like to add that, despite my completely poor Russian and an unannounced visit, I was warmly shown around the building of the Polytechnic University and that of the former Higher Courses for Women in St Petersburg in 2018. I would like to thank the 'Stichting Physica' for making this trip possible as well as part of the translations from Russian to Dutch. It goes without saying that the Dutch version of the book would not have been published without the publisher Prometheus and my wonderful editor Marieke van Oostrom, who guided the—long-term—project. The English translation was made possible by a grant from the Dutch Literature Fund, and another financial contribution from the 'Stichting Physica'—for which I am deeply grateful.

That the translated version has become a 2.0 version is also due to Brendan Monaghan, with whom it was fantastic to work together, and who made the book better with suggestions and questions. And I warmly thank Oxford University Press and Sonke Adlung for their faith in the project.

Lastly, I have benefited enormously from the experience I have gained over the last few years from participating in the WiHQP project (Women in the History of Quantum Physics). I want to particularly mention Patrick Charbonneau and Daniela Monaldi, Michelle Frank, and certainly also Michel Janssen without whose enthusiasm the present translation would not have been made. Likewise, I am indebted to Ruud Tromp and Paul Halpern of the Forum on the History and Philosophy of Physics of the American Physical Society. As with the work on the Dutch edition, my friends were indispensable in this phase of the project as well and I want to mention Anneke, Ellen, Loes and Jochum, Lot and Alain in particular.

Finally, I would like to thank my friends Loes en Jochum, and Lot and Alain, for their encouragement, and to acknowledge my late father, whose wide-ranging interests and love of mathematics have inspired me throughout my work.

Notes

Chapter 1

1. Unless stated otherwise, all anecdotes and family data in this chapter are taken from: Ehrenfest, Paul. Childhood Memories. 1932. Folder 21. EA-MBL. In a note dated 27 January 1933 on a letter to his children dated 4 October 1932 (EFA), Paul Ehrenfest stated that 'At Nelly's insistence, I wrote down my childhood memories in great detail (up to the age of 8)'. (*'Auf Nelly's Andrang habe ich sehr ausführlich meine Kindheits- Erinnerungen nieder geschrieben (bis zum Alter von 8 Jahren)'.*)
2. 'Favoriten'. *Wien Geschichte Wiki.* 3. Dez. 2024, 14:22 UTC. 6 January 2025, 17:33 https://www.geschichtewiki.wien.gv.at/index.php?title=Favoriten&oldid=972660.
3. '- *es that schrecklich weh - ich hatte offenbar für unmöglich gehalten, dass Mama die Thüre schließen würde, wann da noch meine Finger wären*'.
4. '*Die Luft, die man dann einatmet, ist dann intensiv süß!*'
5. Ehrenfest, Paul. Grade lists and school reports. Folder 93. EA-MBL.
6. See also Klein (1970, 28).
7. In 1910, just over 2 per cent of Favoriten's inhabitants were Jewish.
8. '- *deshalb ging ich immer so schrecklich gerne zur alten Frau Schumann zu Besuch, die in diesem Haus gerade im vierten Stockwerk wohnte - da war man schon ganz nahe beim Himmel und sie gab mir immer so gutes Backwerk zum Naschen*'.
9. '*Wie sehr wünschte ich ein Crucifix zu kaufen [aus schwarzem Holz und mit der Christusfigur aus Elfenbein darauf]*'.
10. '*Das trug ich Tag und Nacht unter dem Hemd*'.
11. '*Mutters Stolz war Arthur, Mutters Liebling Hugo*'.
12. '*Alles "irgendwie" um mehr und mehr "Sommerfrische" Kleinbürgerfamilien Geld bringend in diese kleinfenstrigen, dunkeln, comfortlosen, feuchten "Sommerwohnungen" weg stopfen zu können*'.
13. '*Im Garten - heißer Sommernachmittag - ein wackliger Gartentisch ... neben mir Mama in all Ihrer Strenge*'.
14. '*Es waren, scheint mir, alle sehr gütige, hilfsbereite Frauen, aber mangels entsprechender Erziehung und Bildung ungeschickt in der Technik ruhiger Nachbar-Beziehungen*'.
15. Obituary Johanna Ehrenfest-Jellinek (1840–1892). *Freie Neue Presse*, 3 May 1892.
16. Ehrenfest, Paul. 6 August 1912. Family journal. EFA. '*Die Kranken sind - wenn man die Sache ganz bis Ende denkt—schließlich besser davon, wenn sie irgendwo im Süden in der Sonne ruhig verschwinden als wenn sie im Norden, "im Schoß der Familie," sich den Rest ihres Lebens selber vergällen, indem sie ihren Kindern zur erdrückenden Last werden. Es ist absolut unzulässig dass die Jugend und die frohe persönliche Entwicklung der Kinder durch die Krankheit der Eltern erstickt oder auch nur wesentlich geschädigt wird*'.
17. Trauungsbuch der Israelitische Kultusgemeinde Wien. 15 June 1892.
18. Trauungsbuch der Israelitische Kultusgemeinde Wien. 1 July 1894.
19. '*dass auch hier alles mit Schein und Lüge durchsetzt war ... nun wollte ich diese Liebe zerstören - ich zerstörte sie rasch und leidenschaftlich, hassend Durch Durchdenken, Fragen, Lesen, Streiten. Und mit 12-13 Jahren machte es mir eine ganz besondere Freude jedem, den ich nur irgendwie anpacken konnte, mit schärfster Härte alle logische Unsinnigkeit des populären Glaubens zu Bewusstsein zu bringen*'.
20. Ehrenfest, Paul. 5 August 1912. Family journal. EFA; T. Ehrenfest-Afanassjewa to M.J. Klein. 1957. MKEAD.
21. School reports and diplomas of Paul Ehrenfest. Folder 96. EA-MBL.
22. Obituary Sigmund Salomon Ehrenfest (1838–1896). 1896. *Freie Neue Presse*, 10 November.
23. T. Ehrenfest to M.J. Klein. Undated but around 195. MKEAD.
24. P. Ehrenfest (nephew) to M.J. Klein. 24 June 1957. MKEAD.

Chapter 2

1. At the time, the city of Kyiv was part of the Russian Empire and was officially known as Киев (Kiyev), from which the English name 'Kiev' is derived. We use this name here.
2. The Pale of Settlement covered nearly 1.2 million square kilometres and either large parts or all modern Poland, Ukraine, Belarus, Lithuania and Moldova.
3. Personal communication T. van Bommel.
4. H. van Bommel. Notes from conversations with Afanassjewa, c. 1950. EFA.
5. Passport Jekaterina Ivanova (1860–1946). Folder 545. EA-MBL.
6. As note 3.
7. Vladimir Stoyunin (1826–1888), Vladimir Yakovlevich Stoyunin was a Russian educator, educational theorist, essayist and publicist, born in St Petersburg. An influential thinker, Stoyunin was a pioneer in the development of women's education in Russia.
8. '*Maar het onderwijs was zó, als ik het overal zou willen zien. We werden aangemoedigd zelf te denken en nooit hoefden we iets uit ons hoofd te leren*'.
9. Gymnasium diploma T. Afanasyeva, May 1893. Folder 463. EA-MBL. Translated from Russian by HD-BM.
10. Personal communication T. van Bommel.
11. Lyalya Nevzorova to Afanasyeva, May 1893. Folder 463. EFA. Translated from Russian by HD-BM.
12. Undated note. EFA. Translated from Russian by HD-BM.
13. Gymnasium diploma T. Afanasyeva, May 1893. Folder 463. EA-MBL. Translated from Russian by HD-BM.
14. '*Es erschienen Damen mit Pferdeschwänzen und kurzgeschorenen Haaren, Garibaldi-Anhä ngerinnen und andere mehr; und alle hörten atemlos zu*'.
15. A famous example being Vera Bogdanovskaya, who obtained a doctorate in chemistry in Switzerland and had the sad honour of being the first woman chemist killed by an explosion during a chemistry experiment.
16. Nevertheless, the upper classes, which made up 2.5 per cent of the population, produced 80 per cent of the students. It should be noted that class and wealth didn't coincide: the upper classes distinguished themselves by practising an intellectual profession, but there was a great deal of poor nobility and poor clergy. See e.g. Bezemer and Jansen (2008, 150).
17. H. van Bommel. Notes from conversations with Afanassjewa, c. 1950. EFA.
18. Personal communication T. van Bommel.

Chapter 3

1. Course lists. Folders 94 and 95. EA-MBL.
2. Also available on: http://www.gutenberg-e.org/rentetzi/chapter03.html.
3. Student number 486, Meldungsbuch Chemische Fachschule, Technische Hochschule, Wien. 17 October 1899. Folder 94. EA-MBL.
4. Hans Hahn (1879–1934) would later become a professor of mathematics in Czernowitz (Chernivtsi), Bonn and Vienna. He was the PhD thesis supervisor of Kurt Gödel.
5. Heinrich Tietze (1880–1964) would later become a professor of mathematics in Brünn (Brno), Erlangen and Munich.
6. '*Wenn ich wie einst Solon um den Glücklichsten der Sterblichen gefragt würde, ich würde ohne Zagen Kolumbus nennen. ... Aber das Glück ist mitbedingt durch die sinnliche Wirkung und die muss bei Kolumbus am höchsten gewesen sein. Ich kann nie in Amerika landen, ohne ein gewisses Gefühl des Neides gegen ihn oder besser vielleicht der Beseligung, dass ich einen kleinen Teil seiner Freude mitempfinden kann*'.
7. Karoline (Lina) Friedmann-Jellinek, the younger sister of Paul's mother Johanna Ehrenfest-Jellinek, died in 1892, two months before Johanna died.

Chapter 4

1. Miloradovitch to Afanasyeva, 31 October 1899. EFA. Translated from Russian by HD-BM.
2. See e.g. Schiff, Vera Iosifovna—wiki7.org
3. Miloradovitch to Afanasyeva, 18 January 1900. EFA. Translated from Russian by HD-BM.
4. T. Afanasyeva, drafts of the appeal, written in pencil. EFA. Translated from Russian by HD-BM.
5. T. Afanasyeva, draft of the letters. EFA. Translated from Russian by HD-BM. 'If you ... do show solidarity with us, please write the following on a separate sheet of paper: I cast my vote against boycotting the courses and then your name'.

6. Tanya is probably referring to the student community, the 'studentchestvo', of which all students were automatically members and as such each other's comrades.
7. Diplomas T. Afanasyeva. Folder 464, EA-MBL.
8. Schiff to Afanasyeva, 26 June/9 July 1901, EFA. Translated from Russian by HD-BM. Schiff wrote from Friedrichoda, not far from Göttingen.
9. Girls' schools were part of the independent Department of Empress Maria, named after Maria Feodorovna (1752–1828). This included the first girls' gymnasiums, founded in the1860s. These grammar schools were thorough but, despite Schiff's praise, didn't offer the same university-preparatory curriculum as the boys' gymnasiums.

Chapter 5

1. Especially in North America, 'neurasthenia' or weakness of the nerves became a popular diagnosis in the late nineteenth and early twentieth centuries.
2. Based on similar experiments, Giglielmo Marconi was granted the British patent on radio communication a year later; he shared the Nobel Prize in Physics in 1909 for his contributions to the development of wireless telegraphy.
3. X-rays were first seen in experiments by German physicist Wilhelm Röntgen in 1895 and by his French colleague Henri Becquerel in 1896.
4. Johannes Stark (1874–1954) would be awarded the Nobel Prize in Physics in 1919 for his discovery of the so-called Stark-effect.
5. Course list Ehrenfest. Folder 94. EA-MBL.
6. Gauss to Schilling von Canstadt, 11 September 1835 (Mittler and Glitsch 2005, 317). '*Welche Vorzüge auch große Orte in Rücksicht auf andere Genüsse haben mögen, so können Sie doch nirgends eine größere Wärme für diejenigen Bestrebungen antreffen, die darauf gerichtet sind, der Natur ihre Geheimnisse abzulauschen*'.
7. '*Klein dagegen war mehr wie ein Prinz, der seinen Bewunderern die Grösse seines Reichs zeigen wollte, indem er sie auf endlos gewundenen Pfaden durch offensichtlich undurchdringliches Gelände führte und auf jedem kleinen Hügel anhielt, um einen Überblick über die zurückgelegte Wegstrecke zu geben*'.
8. https://sammlungen.uni-goettingen.de/sammlung/slg_1017
9. Thiele cites Klein's former student Constantin Carathéodory: '*Auf der Treppe, die zu Kleins Studierzimmer führte, hatte ich eine Abguss des Kopfes des Capitolinischen Jupiter ... aufgestellt. ... Aber ein Rest von der Distanz, die zwischen einem gewöhnlichen Sterblichen und einem Gott besteht, ist zwischen uns immer geblieben*'.
10. Ehrenfest, December 1902. Notebook I. Folder 256. EA-MBL. '*Glaube mir und lerne jetzt, trachte vorwärts zu kommen, zeichne auf Wo du nicht verstehst, wo du Mängel entdeckst und gehe weiter; sonst kannst du auf einen Stoff geraten, der leicht dein ganzes Leben verschlingt*'. Also cited in: Klein (1970, 43).
11. Mathematician R. Courant.

Chapter 6

1. Strictly speaking, Maria Gaetani Agnesi was the first woman professor of mathematics at a modern university. She was appointed to a position at the University of Bologna in 1750, but never actively took it up.
2. H. van Bommel. Notes on conversations with Ehrenfest-Afanassjewa. c. 1950. EFA.
3. Ehrenfest-Afanassjewa to M.J. Klein. 1957. MKEAD; See also Klein (1970, 41).
4. T. Ehrenfest-Afanassjewa. Biographical notes. Folder 2. EA-MBL; See also Klein (1970, 42).
5. H. van Bommel. Notes on conversations with Ehrenfest-Afanassjewa. Around 1950. EFA.
6. Correspondence between Ehrenfest and Afanasyeva. 1902–1903. EFA.
7. Ludwig Prandtl to Gertrud Föppl. 9 May 1909, Vogel-Prändtl 2004, 43.
8. Emilie Föppl to Ludwig Prandtl. 10 May 1909, ibid.
9. Notebook I. Folder 256. EA-MBL.

Chapter 7

1. '*Per aspera, sed nun ad astra. Frl. T.A. zur freundlichen Erinnerung an die Vorbereitungstage*'. Folder 386. EA-MBL.
2. '*Zur baldigsten Nachahmung, in besserer Auflage!*' Folder 386. EA-MBL.
3. Ehrenfest. Travel diary I. Folder 267. EA-MBL.

4. Lorentz to Ritz and Ehrenfest. 19 April 1903. Folder 149. EA-MBL 149. *'Ich werde morgen am Nachmittag wohl nicht zu Hause sein können und wäre es mir daher lieb wenn Sie mit Herrn Ehrenfest vormittags halb zehn kommen könnten'.*
5. This so-called 'Lorentz force law' is regarded as the first modern formulation of this effect that was previously also found by Oliver Heaviside, see Darigol 2000, 429–430.
6. In fact, the splitting became clearer in later, more precise measurements. Zeeman saw a 'broadening' of the peak in the spectrum, see Kox (1997).
7. In retrospect, the full, correct explanation of the Zeeman effect requires quantum-mechanical treatment.
8. Lorentz himself began to adapt the term 'electron' around 1897; Irishman Johnstone Stoney minted the term in 1891, see Kragh (2023, 75–76).
9. Much later, Paul Ehrenfest described this study and Lorentz's style of receiving guests in a long letter: P. Ehrenfest to A.D. Fokker, 2 February 1912, see also Klein (1970, 215).
10. Afanasyeva. 3 March 1903. In P. Ehrenfest, notebook I. Folder 256. EA-MBL. *'Du wirst dir höchstens Zeit am Anfang dieses Sommersemesters ein Klavier verschaffen—und nicht um zu verstauben—wie Ohren hat, mag hören. ... Brav sein!'*
11. As note 10. *'Möglichst bald lesen: Leo Tolstoi: Krieg und Friede, Anna Karenina, Das Familienglück, Das Kaffeehaus von Surath, Erzählungen von Sebastopolscher Kriegen'.* On 17 March she added to that list: *'Gogol. Lermontoff -ein Held unserer Zeit. Grigorowitsch—Die Fischer, Anton der Unglückliche. Garschin -Vier Tage auf dem Schlagfeld und alles Uebrige'.*

Chapter 8

1. *'Er bleek een groot verschil te zijn tussen dat wat de professoren in Petersburg ons bijbrachten en dat wat in Göttingen (Klein, Hilbert, Minkowski) aan de orde was. Daar kwamen mensen die hun officiële studie op andere plaatsen inde wereld voltooid hadden. Zij waren meestal jonger dan ik en wisten meer'.*
2. *'Toch had ik de moed om in de jaren 1902-1904 twee keer in het seminarie van Klein, Hilbert en Minkowski lezingen te houden. De gesprekken met Paul, en Walter Ritz, met wie ik dikwijls samen was, droegen veel bij tot mijn verdere oriëntering'.*
3. Travel diary I. Folder 267. EA-MBL.
4. For a detailed analysis, see e.g. Badino (2009) and references therein.
5. Correspondence Ehrenfest-Afanasyeva.
6. Ehrenfest to Lorentz, 19 May 1912, NHA. See also Kox (2018, doc. 114). *'Ja selbst — so lächerlich es klingen mag - bei einem mehrwöchentlichen Aufenthalt unter der Fischerbevölkerung von Schiermonnikoog[20] fühlte ich mich sehr rasch mehr heimisch, als ich es in Wien könnte'.*
7. After a short stop in Göttingen; see also Klein (1970, 47).
8. Travel diary II. Folder 267. EA-MBL. *'Herrgott, wenn ich als Kind auch nur einen kleinen Bruchtheil (sic!) dieser Dinge gesehen hätte'.*
9. As note 8. *'Die Perlenschnur der Laterne am Arno! ... Man sollte seine Kinder zur Erzieh. Nach Florenz senden oder mitnehmen'.*
10. Travel diary III. Folder 267. EA-MBL.
11. Hugo to Paul Ehrenfest. 9 April 1924. Folder 64, EA-MBL. In English.
12. Travel diary IV. Folder 267. EA-MBL *'Dass die unsägl.[ich starke] Widerwillens-Empfind.[ung], die ich gegen jeden Menschen empfinde, mit dem ich auch nur für kurze Zeit zusammengekoppelt bin'.*
13. Travel diary IV. Folder 267. EA-MBL. *'Wellen auf dem See - Arbeite Du! - Bangigkeit'.*
14. Travel diary IV. Folder 267. EA-MBL. *'Ist das Gefühl des nervösen Widerwillens, das mir jetzt so geläufig ist nicht lediql. eine andere Form derselben psych. Disposition die mir früher als Sonntags-Langweil. etc. geläufig war. - Wie heilen?'*
15. Afanasyeva to Ehrenfest, undated but based on what is described: summer 1903. EFA.
16. Travel diary IV. Folder 267. EA-MBL. *'...nur durch Liebe'.*

Chapter 9

1. Postcard Afanasyeva (Petersburg) to Ehrenfest (Vienna), undated but based on its content autumn 1904. *'1) Du kennst mich noch immer gar nicht, wenn Du glaubst dass ich irgendeine andere Confession annehmen kann, wenn ich schon einmal meine Kindheitsconfession aufgegeben habe! 2) Die Möglichkeit von Deiner Seite für einen Augenblick anzunehmen dass ich confessionslos werde, indem Du Jude bleibst, empfinde ich als eine Verhöhnung!!!'*

2. As note 1. 'Schon fünf!!! Tage keine Wörtchen ... meinen Lieben, Schlechten, meinen einzigen Pauljutsch ... dass die ganze Welt auf einen Punkt reduziert ist'.
3. As note 1. 'Hast Du abgeschälgt [sic] welche Folgerungen das für Deine Universitätsangelegenheiten und die ganze Tätigkeit haben wird?'
4. As note 1. 'Confessionsl. = nicht Religionslos'.
5. As note 1. 'Ich bin gänzlich unfähig mir weiter irgendwelche Sorgen zu machen—ich warte nur ungeduldig bis wir endgültig beisammen sind. Kindjutsch, schau mich an! Ja?!'
6. Ehrenfest-Afanassjewa to M.J. Klein, 1957, MKEAD. In English.
7. EFA.
8. The family archive contains many pictures of the interior and the maid. EFA.
9. Correspondence between the Ehrenfests 1904/5. EA-MBL.
10. Permission slip University of Vienna, January 1905. Folder 465. EA-MBL.
11. Tanya to Paul Ehrenfest. 17 August 1905. EFA.
12. Paul Ehrenfest. 24 December 1904/6 January 1905. Family journal. EFA. 'Ich verspreche Dir, mit Dir nach Russland zu übersiedeln und dort zu leben, sobald es vernünftigerweise mir überhaupt möglich ist und Du es dann noch verlangst'.
13. As note 12. 'Ich verspreche Dir, lesen, schreiben und schließlich auch sprechen zu lernen'.
14. As note 12. 'Ich verspreche Dir, unsere Kinder in der Religion Deiner Kindheit taufen und erziehen zu lassen; sie mögen hierin im Gegensatz zu mir eine Heimat finden!'
15. Tanya Ehrenfest. 24 December 1904/6 January 1905. Family journal. EFA. 'Ich verspreche Dir jede Trauer energisch zu überwinden—nie mich Deiner Hilfe zu entziehen'.
16. https://www.marxists.org/history/ussr/events/1905/workers-petition.htm; see also: Gapon 1906, 275–277.
17. See e.g. Sharp (2010, Ch. 15).
18. Tanya Ehrenfest. 1905. Under heading 'Events from this year'. Family journal. EFA. 'Mein lieber Mann hat sich als der aller denkbar beste barmherzige Bruder erwiesen'.
19. Paul Ehrenfest. 1905. Under heading 'Events from this year'. Family journal. EFA. 'Ich danke Gott, dass er diese größte Gefahr so bald von uns genommen hat. Schütze er unser Glück'.
20. Herglotz habilitated in Göttingen in 1904 and worked as a Privatdozent at the University of Göttingen; Hahn habilitated in Vienna in 1905 and worked as a Privatdozent at the University of Vienna; Tietze habilitated in Vienna in 1904, where he continued to teach at the University and he worked towards his habilitation in 1908.
21. CPAE 1, doc. 171, 173, 131.
22. CPAE 1, doc. 131.
23. It was an application letter to Professor Adolf Hurwitz.
24. 'Die Hypothese (2), die in der gegenwärtigen Form offenbar nur formal gemeint ist, bedarf dann noch einer weiteren Reduktion'.
25. For a more elaborate discussion, see e.g. Duncan and Janssen (2019, Ch. 2 and 3).
26. As note 19. 'Sternennächte ... über Baumwipfeln und See. ... Steiler Niedermarsch, gar sorgfältig gestützt'.
27. As note 26. '1001 Theoriën von Ritz'; 'Spaziergang ... in den Abendhimmel über den Weiden' '(Tanterl in Besorgnis)'.
28. Paul Ehrenfest. 21 December 1905. Family journal. EFA. 'Nun hat uns Gott schon ein Jutscherl' geschenkt und was für eines!!! Sie wurde geboren am 28/15 October 1905 3/4 6h morgens. Sie heißt Tatiana'.
29. As note 28. 'die Pilonki [nappies]-Camillenthee- Klistirchiks [lavements]—"Nje-nadas [neverminds]"—Triapotschki [clothes] flogen nur so in der Luft herum. ... Ende 4 Wochen kam dann Mama. Und Jutscherl bequietscht nun eine zweite Methode. Und Jutscherl wird Grund haben später einmal Mama dankbar zu sein für die unendliche Arbeit dieser Wochen. Aber ob's Jutscherl maß ist, das verstehen sie alle zusammen nicht!'

Chapter 10

1. Tanya to Paul Ehrenfest. 19 March 1906. EFA. Translated from Russian by HD-BM.
2. Between 1900 and 1910, 71 per cent of bankers, 65 per cent of lawyers, 63 per cent of industrialists, 59 per cent of doctors and more than half of journalists were Jewish. In the private sector, they could avoid the problems they would encounter in government service.
3. Paul Ehrenfest to Casimir. 27 November 1930. Folder 74. EA_MBL. 'Im Grunde begreife ich nicht bei welcher Gelegenheit ein Jude das Woertchen "wir" wirklich ueberzeugt gebrauchen kann, es sei denn in der Verbindung "wir Juden" einfach in Rassensinn. Denn mir scheint es, dass sobald ein

Jude das Wort "wir" in irgendeinen anderen Sinn gebraucht, sehe ich immer, wie sich die anderen Leute, die er so naiv mit "wir" umarmte, sehr hastig "zu bekreuzigen" beginnen'.

4. Remark Paul Ehrenfest. 1905. Under heading 'Events from this year'. Family journal. EFA.
5. Remark Paul Ehrenfest. 1905. Under heading 'Events from this year'. Family journal. EFA. *Arbeit mit Meitner über H-Theorem und nun H-Theorem – Zeit - (vorher Jutscherl: Dimensionsarbeit) - Gespräche und lustiges Geplänkel in den Caffeehäusem und Milchhallen (Stadtpark, Rathauspark)'* This suggests that the conversations took place from 1905 on, likely also with Tanya sometimes present. The discussions then continued in 1906 (Lewin Sime 1996, 18).
6. For example, Paul Ehrenfest drew Meitner's attention to the work of Lord Rayleigh (Lewin Sime 1996, 18).
7. Meitner to M.J. Klein. 12 February 1958. MKEAD.
8. Paul to Galinka and Pawlik Ehrenfest. 4 October 1932. EFA.
9. Remark Paul Ehrenfest. 1905. Under heading 'Events from this year'. Family journal. EFA. '*Arbeit mit Meitner über H-Theorem und nun H-Theorem - Zeit*'.
10. It was read during a session of the Akademie on 22 February 1906.
11. For further discussion of these ideas, see e.g. Regt (1999).
12. Paul Ehrenfest. Undated. Family journal. EFA. '*Im Mai nach Weesen: Früh-Frühling. ... Schnee auf Bergen - Wasser auf Wiese ... Tania' im Wagen und Betterl. ... Arbeiten: Falsche Arbeit zu H-Theorem ... Planck, Jeans*'.
13. '*Eine im Planckschen Modell eingeschlossene Strahlung mag im Laufe der Zeit beliebig ungeordneter werden - sicher wird sie nicht schwärzer*'. Paul had already mentioned this effect in his 1905 paper on Planck's theory. In fact, in 1906, Planck referred to Paul's earlier work at the end of his *Vorlesungen*, in which he once again elaborated on his theory of thermal radiation (Planck 1906). And vice versa, Paul referred to Planck's *Vorlesungen* in his 1906 paper.
14. Meitner followed this schedule for a year and then moved to Berlin to work with Planck, living on an allowance provided by her father.
15. Tanya to Paul Ehrenfest, 19 March 1906. EFA. Translated from Russian by HD-BM.
16. Rental termination form. 12 May 1906. Folder 100. EA-MBL; Notes Ehrenfest. Family journal. EFA. '*September nach Goettingen*'.

Chapter 11

1. Ehrenfest Paul. Registration office, City Archives Göttingen. 1 October 1906. The date of arrival in Göttingen, according to the same file, was 2 September 1906.
2. In the months before his death on 5 September 1905, Boltzmann had suffered from what was then often called neurasthenia—see e.g. Cercignani (1998).
3. Ehrenfest-Afanassjewa to M.J. Klein, c. 1957. MKEAD.
4. Paul Ehrenfest. 17-1/9-2 1906. Notebook 7. Folder 256. EA-MBL. '*Zur größtmöglichen Anregung zur Mathematik, damit die dumme B. möglichst unrecht behält*'.
5. Paul Ehrenfest. Notes dated as '*September nach Goettingen*'. Family journal. EFA. '*Mit Tania oft auf Arm zu Hergl. (Hund-Hühner) - zu Gast bei Hergl. im Zimmer & in Garten*'.
6. Tanya Ehrenfest. Notes on Tanyitchka's first years. Undated but sometime in 1907. Family journal. EFA. Translated from Russian by HD-BM.
7. As note 6.
8. '*Denn früher war Einstein ein richtiger Faulpelz. Um die Mathematik hat er sich überhaupt nicht gekümmert*' (Seelig 1956, 28).
9. For background see Janssen (2002).
10. '*Das Relativitätsprinzip oder - genauer ausgedrückt - das Relativitätsprinzip zusammen mit dem Prinzip von der Konstanz der Lichtgeschwindigkeit ist nicht als ein "abgeschlossenes System," ja überhaupt nicht als System aufzufassen, sondern lediglich als ein heuristisches Prinzip, welches für sich allein betrachtet nur Aussagen über starre Körper, Uhren und Lichtsignale enthält*'.
11. 'The Relativity Principle'—lecture given at the meeting of the Göttinger Mathematical Society on 5 November 1907 (Minkowski 1915). After Minkowski died of appendicitis in 1909, Sommerfeld arranged for the publication of this lecture.
12. '*So genial sie sind, so scheint mir doch in dieser unkonstruierbaren und anschauungslosen Dogmatik fast etwas Ungesundes zu liegen. Ein Engländer hätte schwerlich diese Theorie gegeben; vielleicht spricht sich hierin [, ähnlich wie bei Cohn,] die abstrakt-begriffliche Art des Semiten aus*'.
13. Klein to Paul Ehrenfest. 16 November 1906. Folder 147. EA-MBL and MKEAD. '*Lieber Doctor, zu Ihrer Colloquiumvorlesung anknüpfend, die mir und uns allen besondere Freude machte, habe ich Ihnen heute eine große Frage, oder Bitte, vorzulegen: würden Sie, ev. mit Ihrer Frau-Gemahlin*

zusammen, bereit sein, für Band 4 der Enzyclopädie die Schlussartikel über statistische Mechanik zu schreiben ?'
14. As note 13. 'Jedenfalls möchte ich Sie und Ihre Frau-Gemahlin bitten, vielleicht morgen (Sonnabend) um 6 Uhr zu mir in meine Wohnung zu kommen, sonst wir die ganze Sache durchsprechen können'.
15. Ehrenfest-Afanassjewa to M.J. Klein. Ca. 1957. MKEAD. In English.
16. Zermelo was inspired by Poincaré. See Strien (2013).
17. Much later, Afanassjewa turned out to be not entirely happy with their result: 'I must say that I am not completely happy with this article, but at the time we could not attain more', she wrote to Martin Klein in 1957. Ehrenfest-Afanassjewa to M.J. Klein. 1957. MKEAD. In English.
18. Paul Ehrenfest. Notes 1907. Family journal. EFA. 'Juni 1907. Über Neuenahr nach Berck-Plage. ... Gernet in Göttingen ... In Neuenahr Schiff'.
19. Tanya to Paul Ehrenfest. Undated, but based on the content late in 1904. EFA. 'Mein grenzenlos Lieber! Hättest Du eine Ahnung wie absolut eindeutig mir meine Zukunft vorkommt? Der Verzweigungspunkt ist ja schon überschritten - wahrscheinlich in dem Moment wo ich von der Hochschule die Befreiung erhielt - denn eigentlich war es nur ja zwei Möglichkeiten entweder jetzt schon zu Dir gehen, oder später. Wusstest Du das nicht?!'
20. Since the Tsar shared power with the Council of State, half of whose members were elected by the Tsar and half by the large landowners, it didn't actually have much power, especially since the Tsar retained the right of veto in all cases.
21. Paul to Tanya Ehrenfest. 28 June 1907. Folder 7. EA-MBL. 'Im Coupé eines Waggons der Directen bis fast nach Tübingen allein mit einem jungen Mann, Typus Apotheker ... Er ist ... übel rauchig. ... Du brauchst keine Angst zu haben, dass ich mich in München binde und Dein Petersburger Träumli nicht in Erfüllung geht. Ich wuerde nur versuchen mich dort eine eventuelle Unterkunft zu verschaffen'.
22. Happel to Paul Ehrenfest. Undated, but based on the content early June 1907. Postcard was also signed by mathematician Alexander Brill, physicist Richard Gans, mineralogist Ernst Sommerfeldt, and Gustav Hertz, who studied physics in München but was visiting Tübingen.
23. Paul to Tanya Ehrenfest. 29 June 1907. Folder 7. EA-MBL. 'Im großen heiße, spärlich beleuchteten Esszimmer ... Ein schläfriger Kellner. Es tröpfelt ein Gespräch. ... Happel führt uns durch Winkel und Treppen und Schwebebögen immer höher auf der Zentrale Hügel der Stadt. ... Alte Klöster, Kirchen, Speicher, das alte Universitätsgebäude. Alles war durcheinander geschoben. ... Darüber der Mond. ... Alles fremd entseelt. ... Wenn alles so ist, wie es mir jetzt vorkommt und in Happel und Hertz entgegentritt, so möchte ich lieber gestorben sein als jemals so oder ähnlich zu leben'.
24. Paul to Tanya Ehrenfest. Undated, but based on the content 30 June. Folder 7. EA-MBL. 'Es sind das sehr böse Dinge - Antisemitismus, Experimente verlangt - überflussung. [Also vielleicht] nur Österreich und Petersburg. [Und wenn wir selbst dort wohnen dürfen -] eine Anstellung ist für mich dort sich nur außer der Universität möglich'.
25. As note 24. 'und ohne viel gesehen zu haben, habe ich auf dieser kleinen Ausflug nach Tübingen einmal knapp vor der Nase die Grenzen vorgefunden, auf die ich eingeengt bin, wo ich noch kurz vorher die Illusion hatte, dass nur die ganze Welt offen steht. In Tübingen habe ich erst bemerkt wie anspruchsvoll ich bin. ... Ich hänge sehr, sehr an dir, Tania!!'
26. Paul to Tanya Ehrenfest. 15 August 1907. Folder 7, EA-MBL.
27. Tanya to Paul Ehrenfest. 13 August 1907. Folder 7, EA-MBL. 'Gestern hatte ich starkes Kopfweh und war überhaupt übel dran - und es war mir so, als solltest Du kommen und meinen Kopf in die Hände nehmen'.
28. Paul to Tanya Ehrenfest. 14 August 1907. Folder 7, EA-MBL. 'dass ich ein Drama von Ibsen, Jugendbund, gelesen habe ... mit minderem Genuss'.
29. Paul to Tanya Ehrenfest. 15 August 1907. Folder 7, EA-MBL. Bertha's son was a boy with 'eine enorm lebhafte Intelligenz durch ein trostloses Milieu verschmutzt'.
30. As note 29. Translation of the Russian fragments by HD-BM.
31. As note 29. '[Ich bin mit vielem Vergnügen ein paar Tage hier -] es ist mir lieb sie alle wieder zu sehen. Aber wenn es mir je an Sicherheit gefehlt haben sollte ob es gut ist in Deine Heimat zu gehen - habe ich erst meine "Heimat" vor der Nase so bin ich plötzlich ganz sicher'.

Chapter 12

1. Translated from Russian by HD-BM.
2. Paul Ehrenfest to Hedwig Born. 1929. MKEAD. 'Es ist fast grotesk: Alle behandeln mich dort als einen ganz nahen, vertrauten Verwandten. Nicht nur meine Freunde oder die Verwandten.

Auch die Kinder auf der Straße, die Arbeiter und ihre Frauen, zwischen die ich auf einer Festversammlung geriet. Und das, obwohl sie an meiner Sprache natürlich sofort bemerken, dass ich Ausländer bin'.
3. V.O. 11th Linie 24, App. 19. Correspondence EFA; Frenkel (1977, Ch. 3).
4. Koyalovich (1867–1941) ultimately became better known as a chess player than a mathematician.
5. Koyalovich to Tanya Ehrenfest. 13 September 1907. EFA. Translated from Russian by HD-BM.
6. Tanya Ehrenfest to Koyalovich. 31 December 1907. EFA. Translated from Russian by HD-BM.
7. Tanya Ehrenfest to Gernet. Undated but based on the content at the end of 1907. EFA. Translated from Russian by HD-BM.
8. Koyalovich to Tanya Ehrenfest. 3 January 1908. EFA. Translated from Russian by HD-BM.
9. Tanya Ehrenfest to Koyalovich. Undated but based on the entire correspondence (shortly) after 14 January 1908. EFA. Translated from Russian by HD-BM.
10. Koyalovich to Tanya Ehrenfest. 10 April 1908. EFA. Translated from Russian by HD-BM.
11. These ideas are discussed in more detail in her essay in (Ehrenfest-Afanassjewa 1961) 43–77.
12. This was the basis for the collection of exercises she would publish much later in life: (Ehrenfest-Afanassjewa 1931).
13. *'[Voor deze lerares was mijn idee geheel nieuw en het lokte bij haar altijd weer] heftig protest [uit. Desalniettemin werd het bedoelde onderwijs voortgezet. (onder heftig hoofdschudden, ook van de zijde van sommige ouders:] "Zij spreekt over een bol, terwijl leerlingen de definitie van een rechte lijn nog niet eens geleerd hebben!")'.*
14. Paul and Tanya Ehrenfest. Summer 1908. Family journal. EFA.
15. Tanya Ehrenfest. Summer 1908. Family journal. EFA. Translated from Russian by HD-BM.
16. Tanya Ehrenfest. 1 April 1909. Family journal. EFA. Translated from Russian by HD-BM.
17. Paul Ehrenfest. Summer 1908. Family journal. EFA.
18. The physics section was published as a separate edition from 1907, alongside a second separate edition, *Problems in Physics*.
19. Nowadays the Ulitsa Akademika I.P. Pavlova.
20. Stolypin survived but was assassinated a few months later.
21. The museum was founded in 1864. From 1872, it offered popular science lectures for the general public and from 1885, meetings where teachers discussed their subject by discipline. The Mathematics Department was also founded in 1885.
22. Introduction to the *Proceedings of the First All-Russian Congress of Mathematics Teachers*, which took place in Petersburg from 27 December 1911 to 3 January 1912. Translated from Russian by HD-BM.
23. Maksheyev to Tanya Ehrenfest. 1 April 1908. EFA. Translated from Russian by HD-BM.
24. As note 23. This committee organised a series of lectures on contemporary trends in mathematics education and the principles that should guide the mathematics teacher in the academic years 1909/1910 and 1910/1911, following Henri Fehr's guidelines in the journal *L'Enseignement Mathématique*, see also Ch. 13 in this book.
25. Paul Ehrenfest. Autumn 1908. Family journal. EFA. According to Tropp et al. (1993, 36), she simply called them 'the thin one' and 'the fat one' or sometimes also 'little spider' and 'little elephant'.

Chapter 13

1. Ioffe to Paul Ehrenfest. Undated but based on the content 1908. Folder 145. EA-MBL. *'In Beantwortung Ihrer Anfrage Theile [sic] ich Ihnen mit, 1. Dass Ihre Vorlesung stattfinden wird im....... 2. Dass der nächste physikalische Kruschok stattfindet am Freitag .... 3. Dass es mir leider nicht möglich ist für die Nacht von Freitag am Samstag Ihnen ein Nachtquartier zu verschaffen. Seihen [sic] Sie herzlich von uns allen gegrüßt.... was sagen Sie dazu, dass ich Ihnen so prompt geantwortet habe?!!'*
2. Paul Ehrenfest. 21 October 1910. Notes and drawing. Family journal. EFA.
3. Rayleigh to Paul Ehrenfest. 20 October 1908. Folder 153. EA-MBL.
4. Ioffe to Paul Ehrenfest. 11 July 1909. Folder 145. EA-MBL. Translated from Russian by HD.
5. In two later short articles in the *Physikalische Zeitschrift* (Ehrenfest 1910b; Ehrenfest 1911a), Ehrenfest hit the Russian Waldemar von Ignatowski over the head with this, after the latter had dared to criticise the paradox. He later regretted this high tone.
6. Paul Ehrenfest. 1909. Family journal. EFA.
7. Tanya Ehrenfest. 4 August 1909. Family journal. EFA. Translated from Russian by HD.
8. Tanya Ehrenfest to Selivanov. 7 June 1909. EFA. Translated from Russian by HD.
9. Tanya Ehrenfest to Selivanov. 20 June 1909. EFA. Translated from Russian by HD.

10. Tanya Ehrenfest. 22 December 1909. Family journal. EFA. Translated from Russian by HD.
11. Paul Ehrenfest. Undated but based on the content and the heading 'Moskau' end of 1909. Family journal. EFA. *'Tania Influenza - Ecyklopädie fertig -kaum noch Lust zu fahren –0– Fahren abends los –Lopuchinskaja – Frost – Schnee - kalt, kalt... Bahn: Ioffe, Shaposhnik [?] - Gernett [sic], Biron, Bulgakoff, Rosing [- Die Russian-Dame: Gespräche über ihre pädag. Erfahr. Mit Biron und Chemik-Officier & Bulgakow]. Einfahrt in Moskau – Omnibus – Vorstadt - Frost! – Sonne – Straße - Kreml. Thor – Hotel - Kreml. bei scharfe Frost. Der große Platz - Kloster roth, weiße Facaden Linien, grüner Aufbau - violetten Schatten - auf Dach schnee -rechts blaue Kuppeln - blauer Himmel - ganz links alte Kirche mit feinem Kuppeln - Vorn großer Thurm mit schwarz-gold Borte. Über Fluss - Häusermeer mit 1000 lustigen Rauchwolkerln'.*
12. Information about the Russian discussions is taken from the Introduction to the *Proceedings of the First All-Russian Congress of Mathematics Teachers,* held in St Petersburg from 27 December 1911 to 3 January 1912. Translation from Russian by HD. Ehrenfest-Afanassjewa's own observations, for example (Ehrenfest-Afanassjewa 1961, Introduction), show that similar discussions were also held during the 'Twelfth Congress' in 1909–1910. Such earlier discussions in fact led to the organisation of separate Congresses of Mathematics Teachers in later years.
13. In 1899, together with Charles-Ange Laisant, Henri Fehr founded the magazine *L'enseignement mathématique,* which quickly became a success. In 1905, it was awarded a gold medal at the International Exhibition in Brussels. Three years later, in 1908, the General Assembly of the International Congress of Mathematicians in Rome recognised the importance of a thorough examination of the methods of teaching mathematics in the secondary schools of different nations and appointed Felix Klein, George Greenhill and Henri Fehr to form an International Commission to 'study these questions and report to the next Congress'. This *Commission Internationale de L'Enseignement Mathématique,* also known as the International Commission on Mathematical Instruction (ICMI), still exists today. See e.g. https://mathshistory.st-andrews.ac.uk/Biographies/Fehr/
14. These insights are derived from the reports of Tanya Ehrenfest's lectures as recorded in the Proceedings of the First All-Russian Congress of Teachers of Mathematics, held in St Petersburg from 27 December 1911 to 3 January 1912 (and as translated and summarised by H.D.) in combination with her own observations in, for example, (Ehrenfest-Afanassjewa 1961, Introduction).
15. Paul Ehrenfest. Undated, about the 'Moskau' Congress, hence end of 1909. Family journal. EFA.
16. Paul Ehrenfest. Undated, but based on the content after the summer of 1909. Family journal. EFA. *'Habe jetzt oft die Empfindung dass ich niemals in Russland werde festen Fuß fassen können. Allerdings ebenso wenig in Göttingen'.*

Chapter 14

1. Paul Ehrenfest. 11 July 1910. Family journal. EFA. *'In der Nacht van 9–10 VII 3/4 4h kam Galja zu uns. Abends war Tania noch ganz munter. Ich konnte oben (Tania' bei mir) bis 3/4 3h nicht einschlafen. 3/4 4h wurde ich aufgeweckt. Kam - [Lampe brennt - Höre schon Gequieke -] Tania leuchtet und lächelt -[bald brach der Morgen an - der Artz kam noch Nacht anlegen.]/ Morgens war Tania' sehr froh, strahlt über ganzes Gesichterl. Guckt furchtbar neugierig auf ihr neugebackenes Schwesterl./ G. liegt wie ein Häuflein Wäsche auf dem Wickeltisch und schneidet unendlich komische Grimassen'.*
2. Tanya Ehrenfest. 21 July 1910. Family journal. EFA. Translated from Russian by HD-BM.
3. As note 2. A similar anecdote is on a postcard she sent to Paul Ehrenfest around 21 July, Folder 8. EA-MBL.
4. Paul to Tanya Ehrenfest. 27 July 1910. Folder 7. EA-MBL. *'Tante von (Russisch sprech.) Artz Dr. Simon untersucht in nette Pension ("Schweizer") Größe sauberes Südzimmer'.*
5. Herglotz to Paul Ehrenfest. 28 April 1909. Folder 144. EA-MBL.
6. Herglotz to Paul Ehrenfest. 6 June 1910. Folder 144. EA-MBL. *'Ich bin äußerlich ganz wohl und gesund, aber psychisch furchtbar müde, unfähig etwas aufzugreifen und länger drüber nachzudenken. Hoffe dass es nach einer Kur in Kissingen im Sommer wieder besser geht'.* See also (Huijnen 2003, 45).
7. Paul to Tanya Ehrenfest. 11 August 1910. Folder 7. EA-MBL. *'absolut glattrasiert !!!!!!!! ohne Bart u. Schnurrbart - goldener Zwicker - fettglänzendes Gesicht'.*
8. Paul to Tanya Ehrenfest. 3 August 1910 (Julian calendar). Folder 7. EA-MBL. *'¼ 11h einschlafen. Fundamentalperiode erschöpft'.*
9. E.g. Paul to Tanya Ehrenfest. 11 August 1910. Folder 7. EA-MBL. *'sehr enttäuscht von Dir keine Nachricht zu finden. Das ist sehr schade!!!'*
10. Tanya to Paul Ehrenfest. 29 July 1910 (Julian calendar). Folder 7. EA-MBL.

11. Tanya to Paul Ehrenfest. 27 July 1910 (Julian calendar). Folder 7. EA-MBL. *'Bin schrecklich gespannt, wie ihr gereisst seit'.*
12. Ehrenfest-Afanassjewa to Ehrenfest. 26 July 1910 (Julian calendar). Folder 7. EA-MBL. *'Gott gebe, dass wir bald zusammenkommen alle, und zwar in Russland'.*
13. Paul to Tanya Ehrenfest. 17/18 Aug 1910. Folder 7. EA-MBL. *'- da musst Du es schon in den Kauf nehmen, dass Du vieles Unerfreuliche zu hören bekommst. Mich beherrscht, seit meinem Zusammentreffen mit Herglotz ununterbrochen folgendes Gefühl: in Russland ist für mich irgend eine normalere Entwicklung (Arbeit und Lebensunterhalt) außerordentlich unwahrscheinlich; anderseits habe ich jeden seelischen und materiellen Anschluss in Deutschland u. Oesterreich verloren. Hätten wir keine Kinder, so würde mir das ... nicht mit Sorgen erfüllen, wie es jetzt der Fall ist'.*
14. As note 13. *'und dass ich nachprüfen kann, ob die Erinnerungsvorstellungen mit denen ich operiere, richtig sind. Eine große Zahl von ihnen erweist sich als richtig: ... Auch meine Erinnerung an den Durchschnittsmenschen auf der Straße, in Eisenbahnwaggon und in Restaurant ... Auch die Erinnerung wie die Sauberkeit der Straßen und Häuser etc. angenehm und unangenehm wirken kann ("sonntagnachmittags Stimmung"!)'.*
15. As note 13. *'Überall kannst Du da Dinge sehen, die man ohne jedes "Aber" ansehen kann, die einfach wirklich sehr schön sind, wo speciell[sic] auch keinerlei historischer, sinnloser Herumspielerei stört, wie sie uns in Göttingen so häufig zum Lachen herausforderte'.*
16. As note 13. *'Wenn Du willst, kann ich Dir ihn so schildern, dass Du mit mitleidigem Lächeln die Ochsen über ihn zuckst. ... weißer Weste, gelbe Schuhe, grauer Hut, alles ungemein bequem und teuer ... ganz gute Zigarre—langsame leidenschaftslose Bewegungen, eine lässige Humorvolle, nicht geradezu witzelnde Redeweise. ... Er habe sich durchaus überzeugen müssen, dass nur in Zeiten wirklichen Kämpfe bei denen Buchstäblich das Leben vielen Individuen auf dem Spiel steht, großem individuelle Leistungen auftreten; dass kampflose Zeit immer zu kränklichem greisenhaftem Verfall überall, überall führe'.*
17. As note 13. *'Du leidest an einer Krankheit, die bei mir nicht so tief saß - von der ich mich aber jedenfalls schon vollkommen geheilt habe: siehst in alle Menschen und Dingen vor allem ihre Beziehungen auf Dich - nicht die Beziehungen, in denen sie untereinander stehen - dies nimmt Dir alle Ruhe und Kraft - mir ging es auch so. ... Ich habe keinerlei Bedürfnis auf irgendwelche Menschen zu wirken'.*
18. As note 13. *'Durch nichts wird er in meinen Augen überzeugender gerechtjertigt als durch seine Arbeiten die gleichmäßig und in steigender Entwicklung fortschreiten. ... Was wird aus uns in Russland werden? Vielleicht noch unsere Kinder können so halb in die Proletarienschicht abrutschen?'*
19. As note 13. *'Du könntest ruhig Deine Wahrscheinlichkeits-Kruschok machen aussondern könnten wir es vielleicht irgendwie so einrichten, dass Du abends weggehst, so oft Du Lust hast aber, dass im Übrigen ich absolut isoliert wäre - gewissermaßen offiziell als krank erklärt. Leider erwarte ich aber gerade das Gegentheil, schon deshalb, weil Du ja jetzt kein anderes Empfangszimmer zur Verfügung hast außer meinem Zimmer'.*
20. Tanya to Paul Ehrenfest. Undated. Folder 7. EA-MBL. *'Für so dumm habe ich Dich nie im Leben gehalten: Dein lieber Herglotz umgibt sich mit äußerstem Sorgfalt mit allen denkbaren Bedingungen, die nur eine äußerster Friedenszeit bieten kann ... und ergötzt sich an der vergangenen "Individualisten" (die ihn nicht stören sollen) welche durch Krieg und Revolution geschafft wurden - und Du lässt Dich durch so eine Philosophie irremachen?'*
21. As note 20. *'Man kann ja auch zu Grunde gehen auf eine vernünftigere Weise, als dass man eine stupide "chair a canon" aus sich vorstellt - zu Gunsten jener Speculanten die einen Krieg zu Stande bringen'.*
22. As note 20. *'Du begehst wieder den Fehler, den der Herglotz selber Dir abräth: bei einem anderen durchaus etwas für sich lernen zu wollen, wo Dein eigenes Temperament und Habitus, Gott sei Dank, ganz anders ist! ... Du wirst sehen, dass man auch ohne rasiert zu sein, ein volles, mit Zweck erfülltes Leben führen kann!'*
23. As note 20. *'Für meine Person, steht mir das Leben durchaus nicht wie ein Loch da! Im Gegentheil, ich sehe vor mir mehr Aufgaben, (die ich meinem Verständnisse und Interesse nach erfüllen könnte) als ich Zeit und Kraft habe - was brauche ich also zu suchen? Du machst aber niemals dass, was vor Deiner Nase wartet, und glaubst dass man zuerst die Griechen studieren muss um dem Herglotz ähnlich zu werden'.*
24. As note 20. *'In dieser Zeit wirst Du also Deinen Magister zu Ende bringen. Und Du wirst sehen, wie das regelmäßige Aufstehen und Spazieren Dir dazu verhelfen werden. Spazieren kannst Du mit Lieberl!!!!!!!!!!!!!!!!!!!!!'*

25. It seems likely that this was inspired by a similar survey carried out in 1908 by the Swiss mathematician and educational reformer Henri Fehr (Fehr 1908).
26. Maksyeyev to Tanya Ehrenfest. 4 May 1910. EFA. Translated from Russian by HD-BM.
27. Tanya to Paul Ehrenfest. 13 August 1910 (Julian calendar). Folder 8. EA-MBL. '*Heute habe ich endlich die Enquete zum Absenden gestellt und fühle mich sehr froh u. erleichtert. Kann wieder an weitere Arbeit denken. Jetzt hoffe ich auch die Wahrscheinlichkeitsgeschichte einmal zu Ende zu bringen*'.
28. It would be published in a Russian academic journal in May 1911 (Ehrenfest (Tatiana) 1911Ra).
29. Paul to Tanya Ehrenfest. 26 August 1910. Folder 7. EA-MBL. '*Drei Jahre lang habe ich kaum einmal gelacht. Jetzt aber lache ich mit Herglotz oft halbe Stunde lang*'.
30. Paul to Tanya Ehrenfest. 29 August 1910. Folder 7. EA-MBL. '*Heute Morgen ging ich in die russische Kirche hier ... Abends ging ich die Zionisten hören*'.
31. Paul to Tanya Ehrenfest. 1 September 1910. Folder 7. EA-MBL. '*In Norden Damascus in Süden Ägypten machen eine rapide wirthschaftliche Entwicklung durch. Die Bahn Damascus-Mekka die durch Palästina geht wird bald eine enorme Bedeutung erwerben, denn jetzt beginnt der Bau der Bagdad Bahn die Europa mit der Inneren von Asien verbindet (über Damaskus) und in Aegypten soll dann eine Bahn gebaut werden die Suez mit Südafrika in Verbindung setzt. ... Vorlauf Ziel der Colonisation dieses: in Palästina ein Gebiet zu schaffen ... der Art dass alle Juden und Nichtjuden, die hinkommen sich auf jüdische Boden fühlen*'.
32. Paul to Tanya Ehrenfest. 29 August 1910. Folder 7. EA-MBL. '*Dass die Juden körperlich degeneriert seien, schwerer Arbeit unfähig: Er erzählt darüber wie die jungen Einwanderer zu Ackerbauern würden. Selbst ein ganz flüchtiger Besuch zeige dass jene Annahme falsch ist. ... In summa: ... es gibt eine 13 Millionen Juden 2/3 von ihnen sitzen in Galizien und Russisch-Polen ... Gesichts dieser Thatsache ist es selbstverständlich, dass alle Culturstaaten strengere und strengere Sparmaßnahmen treffen und das überall auf den Liberalismus der Antisemitismus folgen muss. Das Schlagwort von der Assimilation ist für jene Massen absolut lehr und wertlos. Dieses Problem kann in hundert Jahren ungeahnt furchtbar werden*'.
33. Paul Ehrenfest, Notebook I. December 1902, Folder 256. EA-MBL. '*Onze geest kan wereldburger zijn, ons hart niet*'.
34. Paul to Tanya Ehrenfest. 26 August 1910. Folder 7. EA-MBL. '*Hier sind enorm viele Juden - deutsche, polnische (!) und russische. Die polnische sind weitaus weniger fatal als die deutschen und russischen - die benehmen sich so ungemein unangenehm, dass ich ganz deprimiert bin: kommt ein Russe hierher, so muss er sich leider nur zu leicht sagen können: na Gott sei Dank, dass unsere barbarischen Ausnahmegesetze uns diese Leute von Halse schaffen*'.
35. As note 34. '*Sehr lehrreich wird mir sein gerade mit Joffe zusammen nach Wien. U. Boskowitz zu fahren - denn er wollte nie die Antisemitismus-Schilderungen und vor allem die Schilderungen diejeniger Zustände, die ihn rechtfertigen glauben. Nun werd ich ihm einmal alles ad oculus demonstrieren ... Ich bin dah sehr froh, das wenigstens für mich persönlich die Sache die Lösung gefunden hat, dass meine Kinder Nichtjuden sind*'.
36. As note 34. '*Man müsste die Kinder taufen lassen (orthodox, wenn Du es wünschtest - oder protestantisch, wenn ich wählen sollte) ... Ich brauche wohl nicht die Gründe auseinanderzusetzen, die Herglotz und Joffe vorbrachten. Ich bitte Dich, über diesen Punkt nachzudenken ... Ich glaube fast, wir haben ja da eine Sünde begangen, die wir rasch gutmachen sollten*'.
37. Tanya to Paul Ehrenfest. Late August 1910. Folder 8. EA-MBL. '*Was Deine Anpassungsrücksichten betrifft bezeuglich unserer Kindli's, so darfst Du nicht vergessen, dass man von ihnen bei jeder Entscheidung einen Einwand zu befürchten hat—so bleibt nichts übrig, als dasjenige zu wählen, was uns selber nach unserem besten Wissen als wünschenswert vorkommt. Und den Unbequemlichkeiten kann man am besten dadurch helfen, dass man möglichst viel arbeitet um die Kindli's dadurch mat. Unabh. Zu machen. Und dann wird es mit der Zeit in dieser Beziehung immer besser und nicht schlechter—die Leute werden doch vernünftiger, trotzdem dass selbst Onkel Gustl sich zu rasieren für nötig befunden hat*'.
38. Paul Ehrenfest. 14 November 1910. Family journal. EFA. '*Galja sitzt jetzt und sieht dabei wie ein armseliges Chimpanserl aus (ist aber sehr zufrieden dabei!)*'.
39. Paul Ehrenfest. 14 November 1910. Family journal. EFA. '*Tolstoi starb. ... Tante liest Reden vor. Tania' hört zu*'.
40. Paul Ehrenfest, 9 March 1912. Family journal. EFA. '*In letzten Tagen viel Zeit auf Abfassung der Moskau-Adresse vertrödelt*'.
41. Paul Ehrenfest. 9 March 1912. Family journal. EFA. '*Ich schlief auf Tragbett in Speisezimmer*'.
42. Paul Ehrenfest. 3 February 1912. Family journal. EFA. '*Pest in Silezien – Studentstreick - ... - Wir müssen wegfahren*'.

Chapter 15

1. Einstein to Besso. 21 October 1911. CPAE 5, doc. 305.
2. Hugo to Paul Ehrenfest. 23 September 1910. Folder 5. EA-MBL. 'Ich kann mir vorstellen dass mit einer "constanten" Arbeitslust Dein ganzes Leben, Deine Erfolge und damit auch Gemütsstimmung und (auf dem Wege eines Circulus viciosus) schliesslich die Arbeitslust selbst Dich vollkommen befriedigen würden. ... Wenn ich auch vollkommen durchdenken bin [ich] von der Überzeugung, dass Du Russland verlassen sollst—bald als möglich. Schweiz, Holland, beide Lander scheinen mir aus äußerlichen Gründen viel besser geeignet. [...] Du hast Dir ein neues Vaterland erwählt & es will Dir nicht—Deine Wahl war unglücklich'. Partly cited also in Huijnen and Kox (2007, 195).
3. Herglotz to Ehrenfest. 20 December 1911. Folder 145. EA-MBL. 'Ich fange nun wirklich an zu glauben, dass Du die Grenze des Pathologischen überschritten hast und - bei der "inertia foetida" gelandet bist. ... [Es ist doch wirklich unerhört! Die Sache ist ja doch wohl fertig:] handelt sich also bloss ums Niederschreiben. Du hast gar keine Entschuldigung mit einer Nervendesorganisation oder sonst was, wenn Du andere Sachen arbeiten kannst, wie es der Fall ist. Das ist bloss Oesterreichisch-Slavische Schlamperei'.
4. Ehrenfest to Ehrenfest-Afanassjewa. 7 January 1912. Folder 8. EA-MBL.
5. Ehrenfest to Ehrenfest-Afanassjewa. 9 January 1912. Folder 8. EA-MBL. 'Er hat sehr fein geschnittenes Gesicht überraschend jung aussehend. ... Fein sitzende "intellectuelle" Brille. ... [Planck] war sehr überrascht und sehr interessiert, dass ich bewiesen habe, dass die Wiensche Formel ebenfalls auf Energiestufen beruht. [Speciell war ihm ganz neu alle Resultate die ich in der Richtung hatte: Welche Züge der Planckschen Theorie sind unentbehrlich, welche entbehrlich. Das Gespräch war so, dass ich deutlich sah,] wie erstaunt und wie erfreut er darüber ist wie ungemein genau ich in allen Nuancen verstanden habe und zu welchen Gesichtspunkten ich in genauer Fortführung seiner Ideen gelangte. ... Er sagte er hätte niemals genau gewusst, welche Züge seiner Theorie dies und jenes leisten - jetzt sei das sehr klar'. See also Huijnen (2003, 77).
6. For Meitner in Berlin see: Lewin Sime 1996, 23–45.
7. Ehrenfest to Ehrenfest-Afanassjewa. 11 January 1912. Folder 8. EA-MBL. 'Planck müsste Initiative auf sich nehmen, dann würde er [Rubens] keinerlei Widerstand machen'.
8. Ehrenfest to Ehrenfest-Afanassjewa. 12 and 13 January 1912. Folder 8. EA-MBL. See also: Huijnen (2003, 79–80).
9. Ehrenfest to Ehrenfest-Afanassjewa. 18 January 1912. Folder 8. EA-MBL. 'Zwei Stunden lang hat mir Frau Epstein Tschaikowsky, Rachmaninow, Brahms, Beethoven vorgespielt - besondere die russische Sache sehr schön!'
10. Ehrenfest to Ehrenfest-Afanassjewa. 19 January 1912. Folder 8. EA-MBL. 'Vielleicht müsste ich einen Tag lang ganz ruhig liegen und lange schlafen. Vielleicht tuhe ich das morgen (Sonntag) ... Ich ernähre mich ganz regelmäßig - esse mit großem Appetit mittags und meist auch abends einen riesigen (eier-)Pfannkuchen mit Compott. Sonst noch 2 Töpfe Jogurt (Jeden Tag!), Eier, Käse, Brot, Butter. Gelegentlich sonst noch etwas Süßes. Mein Magen ist ganz in ordnung'.
11. Ibid. 'Sieh doch wie die Sache liegt: Ein Doctor der Universität A sucht den Doctor der Universität B, um sich von der Universität C ----- als Privatdozent habilitieren zu machen. Man muss wirklich noch sehr jung sein um das ruhig machen zu können. Das ist wirklich sehr demüthigend'.
12. Ibid. 'Das aber ist total unmöglich: Fabelhafter Kohlendunst, ewig trüb und traurig, eine unbeschreiblich trostlose Stadt, wirklich unbeschreiblich, und zu guter Letzt noch sehr theuer! Du kannst Dir gar nicht vorstellen, wie übel Leipzig ist. ... Alles in allem wäre es also geradezu Unsinn jetzt hier den Doctor. zu machen. Das ist mich wirklich Evident. Und dass Du jetzt plötzlich umgedacht hast, [thut mir sehr leid und] ist eigentlich nur dazu gut, mich traurig zu machen'.
13. Ehrenfest to Ehrenfest-Afanassjewa. 1 February 1912. Folder 8. EA-MBL. 'So liegt hier jetzt hoher, schöner Schnee. Ach ... wenn du nur sehen könntest wie blühend hier alle Kinder aussehen!!!!!;' 'Sag T'erl dass hier die Hunde und Pferde nur Deutsch verstehen und niemand Russisch'.
14. Ehrenfest to Ehrenfest-Afanassjewa. 5 February 1912. Folder 8. EA-MBL. 'Habe an Poincaré Strahl-Arbeit und Encyklop. Artikel und Begleitbrief gesendet. Ohne Brief dasselbe an Jeans, Rayleigh, Lorentz, Langevin'.
15. Ehrenfest to Ehrenfest-Afanassjewa. 6 February 1912. Folder 8. EA-MBL. 'Heute (sehr gut) in Laue-seminar über Braun-Lechatelier vorgetragen.... (Sommerfeld lässt mich jetzt schon ersichtlich ungern gehen)'.

16. Ehrenfest to Ehrenfest-Afanassjewa. 5 February 1912. Folder 8. EA-MBL. '*Einstein soll ganz bestimmt nach Zürich gehen, dann thue ich ganz bestimmt dasselbe. Dann ist ja Einstein u. Debye dort. Du machst dann dort dein Doctorat*'.
17. Ehrenfest to Ehrenfest-Afanassjewa. 6 February 1912. Folder 8. EA-MBL. '*Damit ist für mich jeder Zweifel verschwunden, wohin ich will. Ich will ganz bestimmt nach Zürich - und selbst dann wenn ich dort kein Docentur erhalten könnte. ... Morgenmittag fahre ich nach Zürich und werde dort sofort und energisch Habilitation am Polytechnikum anbahnen*'.
18. Ehrenfest to Ehrenfest-Afanassjewa. 8 February 1912. Folder 8. EA-MBL. '*Da ich im Augenblick noch auf Zürich Hoffnung setze - bin ich wieder sehr froh - das heißt, ich bin wieder in jener Stimmung wo ich sehr bedauern würde sterben zu müssen. ... Und wenn nur die Zürich Sache herauskäme, ja dann würde eben Zürich für die Theor. Physik rasch das werden, was Goettingen für die Mathematik war (und nicht mehr ist!!!)*'.
19. Ehrenfest to Ehrenfest-Afanassjewa. 9 February 1912. Folder 8. EA-MBL. '*Wäre ich auf dem "Quanten Congress" gewesen, so hätte ich ihnen alles das schon dort erzählt. Na, hol's eben der Teufel*'.
20. As note 19. '*Das Polytechnikum-Universität-Vierteil liegt vortrefflich. ... weil Straßen rapid steigen so überall Licht und Sonne*'.
21. Ehrenfest to Ehrenfest-Afanassjewa. 10 February 1912. Folder 8. EA-MBL. '*Er ist jetzt schon Ordinarius und wird trotzdem wahrscheinlich eine Berufung nach Utrecht annehmen*'.
22. As note 21. '*Weiss ist ein hagerer, sehr zarter, ziemlich jugendlicher nordisch-blonder Franzose. Spricht Deutsch mit accent. ... Wenn auch er nicht helfen wird, ist es dann schon, dass er eben nicht helfen kann*'.
23. As note 21. '*Inzwischen mit Weiss gesprochen. Aussichten für Züricher Polytechnikum fast exakt null eben gerade, weil Einstein ins Polytechnikum kommt. Da bleibt neben ihm kein Platz übrig. ... Sobald Einstein da ist reichen Sie jedenfalls Gesuch ein ... ich [Weiss] werde dann Ihr Gesuch sehr lebhaft unterstützen. ... Zürich ist sehr schön. Ich war heute mehrmals sehr traurig. Wie nötig wirst Du mir heute!!! Och, Lieberl Lieberl*'.
24. Ehrenfest to Ehrenfest-Afanassjewa. 12 February 1912. Folder 8. EA-MBL. '*Debye sagte mir, dass er mit Prof. Kleiner gesprochen habe, dass dieser aber durchaus ablehnt, es sei zu voll. Herr Gott, es ist wahrhaft unglaublich - wovor fürchtet sich dieser alte Kerl*'.
25. CPAE 5, doc. 382 and 447.
26. Ehrenfest to Ehrenfest-Afanassjewa. 12 February 1912. Folder 8. EA-MBL. '*Es ist ganz außerordentlich schön hier und ich möchte doch ganz gerne hier bleiben. Du könntest da so schön Doctor machen und für die Kinder wäre es ideal*'.

Chapter 16

1. Paul to Tanya Ehrenfest. 13 February 1912. Folder 8. EA-MBL. '*(Morgens, hoch über Wolken Berge!! Erinnerst Du Dich, Lieberl?) ... Die Schweiz die ist doch sehr schön. Und wenn Kinder hier aufwachsen so können sie hier "Heimath" fühlen—selbst in Zürich*'.
2. As note 1. '*Du solltest nebeneinander die Schweizer und Oesterr. Zollbeamten sehen. Einfache, liebe, jugendliche Bauern oder alte Bauern dort. Hier ausgemergelte, arrogant-kriecherische Hohlköpfe unbestimmbaren altern—"unappetitlich" etwa im Sexuellen und Ethnischen Sinne*'.
3. Tanya to Paul Ehrenfest. Undated but sent around 13 February 1912. Folder 8. EA-MBL. '*Grüß die Wiener von mir, aber bleibe mit ihnen nicht zu lange!*'
4. Paul to Tanya Ehrenfest. 18 February 1912. Folder 8. EA-MBL. '*Morgens fuhr ich mit Emil und Tietze im Automobil auf alle Plätze in Wien, die für mich mit Kindheit Erinnerungen verknüpft sind. Eigentlich ganz, ganz trostlose, hässliche Plätze*'.
5. Paul to Tanya Ehrenfest. 19 February 1912. Folder 8. EA-MBL. '*Am Sonntag gab's schon ein Zusammenstoß mit Arthur. Habe nun schon über über genug von Wien!*'
6. Tanya to Paul Ehrenfest. Various postcards, date illegible, but before 12 February 1912. Folder 8. EA-MBL.
7. Paul to Tanya Ehrenfest. 11 February 1912. Folder 8. EA-MBL. See also CPAE 5, doc. 342 and 357.
8. Paul to Tanya Ehrenfest. 13 February 1912. Folder 8. EA-MBL. '*Ich müsste das natürlich unbedingt annehmen. Aber eigentlich wäre es doch ganz übel*'.
9. Paul to Tanya Ehrenfest. 15 February 1912. Folder 8. EA-MBL. See also CPAE 5, doc. 342 and 357. '*In einer Universität wie z. B. Innsbruck, muss man schon ein Physiker von sehr großer Begabung sein, um nicht ganz zu versumpfen. Es ist ganz übel: kein Physikal. Verkehr, kein Seminar, keine wissenschaftl. Gesellschaft, sehr wenige Zeitschriften u.s.w—denke da z. B. an Prag*'.

10. Paul to Tanya Ehrenfest. 23 February 1912. Folder 8. EA-MBL. '*Seine Frau ist sehr hübsch und lieb ... singt ganz gut. ... Im Gegensatz zu früheren Wochen, in ganzer letzter Woche nichts Neues gelernt!*'
11. As note 10. '*Dafür kommt ja jetzt Prague!! Also, in wenigen Stunden ein Knick in meiner Welt Linie: Begegnung mit Einstein [der doch ganz unzweifelhaft mein Freund werden wird]. Überall nur ganz begeisterte Urteile über ihn als Mensch gehört—special auch von einem Antisemiten*'.
12. Paul to Tanya Ehrenfest. ½ 2 at night 23/24 February 1912. Folder 8. EA-MBL. '*Kam 2^{50} in Prag an. Vor Ausgang Einstein mit seiner Frau. Erkannte Ihn nach Photographie. Er mich "weil er mich gerade so vorstellte." Seine Frau Serbin*'.
13. As note 12. '*seine Bücher (sehr wenige!!) ... Fühle es doch als "wunderbares" Erlebnis, dass ich nun so ganz mir nichts dir nichts in Einsteins Zimmer sitze. Bin mir denn doch scharf bewusst dass sein Name noch genannt werden wird, wenn ich längst vergessen sein werde*'.
14. As note 12. '*Er: breit, einfach, Cigarre im Mund, wunderbar schöne braune Augen ... sofort über Physik und dieses ganz ohne Unterbrechung. ... Nicht einen einzigen unangenehmen Zug in seinem Wesen*'.
15. As note 12. '*über alles was mit Strahltheorie & kinet. Gastheorie zusammen hängt*'.
16. Einstein and Marić lived in a newly built apartment, 7 Třebizského, nowadays Lesnická 7, on the left bank of the Vltava.
17. As note 12. '*Er sagt: dass mir einer sagt "ah das glaube ich Ihnen nicht"—ja, das geschieht mir häufig—aber dass er es auch begründet, das ist äußerst selten*'.
18. Paul to Tanya Ehrenfest. 25 February 1912. Folder 8. EA-MBL. '*Heute (Sonntag) waren wir den ganzen Tag beisammen und der heutige Tag gehört mit zu den schönsten und genussreichsten meines Lebens*'.
19. As note 18. '*Wenn noch irgendein Zweifel bestehen könnte, dass wir ganz nahe Freunde werden müssen, so verschwand er in diesem Moment*'.
20. As note 18. '*Immer munter, fröhlich, unermüdlich. Ein großer, lieber Jungen*'.
21. As note 18. '*Er sagte: Taufen hätte ich mich nicht lassen. In dem Schoss Abraham's zurückzukehren—das war ja gar nichts. Ein unterschriebenes Papier—Sie müssen dasselbe thun. ... Du kannst Dich denken, wie erbittert ich ihm tadelte, und nachgegeben hätte*'.
22. As note 18. '*Er sagte: wenn Sie nur diese sündhaft-dumme Confessionslosigkeit aufgeben, so will ich Ihnen sogar gestatten, sich ausschließlich von Gemüse zu nähren*'.
23. Paul to Tanya Ehrenfest. Undated, but based on the content 1 March 1912. '*Gestern sammelten Lampa u. Einstein alle alten Pistolen (von Mach—Schallgeschwind Experiment) setzen mir die an die Brust: Ich sollte schwären, dass ich der Prager Professur annehme—über dies aber redeten Sie mir sehr lieb zu*'.
24. As note 23. '*Einstein sagte: "Aus Religiosität frisst er kein Fleisch und dem Vieh zu liebe ist er Confess.los—ein abscheulich verrücktes Lüder. [Dabei sagt er aber nach jedem zweiten Wort: Sie sind] wie meiner liebster Freund Besso."*'.
25. As note 18. '*Es kann kein Zweifel darüber bestehen, dass ich nun ohne die geringste Schwankung diesen Weg einschlagen muss*'.
26. As note 23. '*Eines ist ganz unglaublich. Wie viel Lebensfreude und Muth die Berührung mit Einstein mir gibt. Ich möchte jetzt ganz gewiss nicht sterben. Du wirst dich sehr darüber freuen wie Einstein musiciert*'.
27. Paul to Tanya Ehrenfest. Postcard from the train. 1 March 1912. Folder 8. EA-MBL. '*Ach, wenn doch jetzt Ioffe von Röntgen als Debye's Nachfolger nach Zürich gesetzt würde!*'
28. Herglotz to Paul Ehrenfest. 26 May 1912. Folder 145. EA-MBL. '*Ob es gleich wohl das richtige war Prag fallen zu lassen—und ob nicht höhere Rücksichten maßgebend gewesen wären ist eine andere Sache: z. B. dass Du durch das erhöhte Einkommen Deinen Kindern bedeutende Vortheile verschaffen kannst—ich meine nicht materielle als etwa Ausbildung—Reisen—bei Krankheit oder Kränklichkeit alles Günstige [...] u.s.w*'.
29. Tanya to Paul Ehrenfest. Folder 8. Undated. EA-MBL. Translated from Russian by HD-BM.
30. Paul to Tanya Ehrenfest. 2 March 1912. '*Es ist doch gar kein Zweifel, dass mir das Slavische Wesen ungeheuer viel näher liegt als das "Wienerische" ... Die Schmoluchowskys wollen uns ganz bestimmt in einem Jahre in Zürich besuchen*'.
31. Paul to Tanya Ehrenfest. Undated but based on the content early March 1912. '*Cernowitz ist ein ganz unmögliches Nest*'. And: '*Sehr viele Interessante Mathematische Gespräche hatte ich mit Hahn. ... Habe mir Themata notiert um Dir darüber erzählen zu können*'.
32. As note 23. '*Oijoijoi Lieberl. Du wirst jetzt schimpfen über mich. Oijoijoi!!!*'

33. Paul to Tanya Ehrenfest. 7 March 1912. Folder 8. EA-MBL. '*In der kleinen Stadt sind eben die Universitätsleute sehr enge aufeinander angewiesen. Für dich kann es aus diesem Verkehr allerdings nur dann wirklicher Vortheil anwachsen, wenn du es verstehst dir die uninteressanten Frauen der Professoren tüchtig vom Leibe zu halten. Eine Frau wie die Frau von Smoluchowsky würde Dich allerdings nie langweilen können ... Ich hoffe, dass dies auch für die Frau von Prof. Weiss der Fall sein wird. Leider nur in sehr geringer Maße für die Frau von Einstein*'.
34. Tanya to Paul Ehrenfest. Undated but based on the content early March 1912. Folder 8. EA-MBL. '*Gestern Abend war das Ehepaar Joffe da. Ich habe alles von dir erzählt. Joffe glaubt, dass nun Zürich wirklich das Beste wäre, was du dir wünschen könntest. Er hat ganz fein bemerkt: wenn er nicht durchaus Professor werden wollte, so könnte er noch ruhig hier sitzen: Geld verdienen bis 1000 Rub. könnte er auch in Petersburg in allerlei "Curzen." Aber er ist einmal ein Mensch, der es nötig hat Professor zu sein. ... Ich muss hinzufügen, Lieberl, dass ich mit dem ironischen Verhalten von Joffe nicht einverstanden bin, mir gefällt es gar nicht Geld zu verdienen durch eine Arbeit, an die man nur halbwegs glaubt*'.
35. Tanya to Paul Ehrenfest. Undated but based on the content early March 1912. Folder 8 EA-MBL. '*Aber doctorieren hätte für mich kaum einen Sinn. Eine Docentur hätte ich doch nicht gekriegt*'.
36. Paul to Tanya Ehrenfest. Undated but based in the content 1 March 1912. Folder 8. EA-MBL '*Es passt mir jetzt natürlich sehr in den Kram, dass dich die Pedagog. Museumsleute durch Ihre Verständnislosigkeit Abstoßen*'.
37. Tanya to Paul Ehrenfest. Undated but based on the content late February/March 1912. Folder 8 EA-MBL. '*[Wir sprachen] über Entropie, Wahrscheinlichkeiten, Mechanik ohne <Russian word> auf leeren Raum ... Ich bin überzeugt dass ich bei ihm die Math. besser gelernt hätte, falls er zu meiner Studentenzeit in Petersburg gewesen wäre. Er glaubt, dass es ganz vernünftig wäre, falls ich Magister machen thüte [sic]!*'
38. Tanya to Paul Ehrenfest. Undated but based on the content second week of March 1912. Folder 8 EA-MBL.
39. Paul to Tanya Ehrenfest. March 1912. Folder 8. EA-MBL '*Hier müsste im Frühjahr sehr schön sein. Jetzt ist es grau, dreckig und hässlich*'.
40. Paul to Tanya Ehrenfest. March 1912. Folder 8. EA-MBL '*Durch alle Straßen strömen in Cascaden Bäche von Schmutz. Wasser. Bodenloser Schmutz. Abgesehen von einigen sehr schönen, sehr alten und ganz neuen Bauten alle anderen Häuser ganz abgesackte langweilige Ziegelbauten*'.
41. Paul to Tanya Ehrenfest. March 1912. Folder 8. EA-MBL '*Hässlicher als alles was ich auf dieser Reise bisher sah. Wirr, schmutzig, poesielos, geschmacklos, unfreundlich. ... schon jetzt sehe ich: Es würde mich total zugrunde richten, wenn wir in einer russischen Provinzial-Universität leben würden*'.
42. Paul to Tanya Ehrenfest. March 1912. Folder 8. EA-MBL '*Hier zu leben wäre mir ganz, ganz unmöglich. Erstens gerade die groteske Hässlichkeit der Stadt, zweitens Mangel einer Wissenschaftl. Atmosphäre*'.
43. Paul to Tanya Ehrenfest. At night. 3/4 March Eastern calendar 1912. Folder 8. EA-MBL.

Chapter 17

1. Ehrenfest. 20 April–2 May 1912. Folder 268. EA-MBL.
2. Ehrenfest-Afanassjewa to Ehrenfest. Summer 1910. Folder 7. EA-MBL. '*Vergesst nicht, dass der Mensch, der Russland gesehen und die besten Menschen kennen gelernt hat, gar nicht in derselben Lage ist, wie ein Deutscher für den wirklich das bisschen idealisierte Griechenthum das äusserste Mass von Leben, Empfindung und Schönheit darstellen muss*'.
3. Paul to Tanya Ehrenfest. 18 January 1912. Folder 8. EA-MBL. '*Lieberl, liebes, - verzeihe mir, dass ich nicht verstehe mein Schicksal mit fester Hand zu lenken. Ich sollte es, ich kann es nicht. ... Ich wundere mich geradezu wie leicht es mir nun mehr und mehr fällt alles zu unterlassen, was ich unterlassen will und wie unmöglich es mir wird das zu thun, was ich thun soll*'.
4. Tanya to Paul Ehrenfest. Undated, based on content February 1912. Folder 8. EA-MBL.
5. Tanya to Paul Ehrenfest. Undated, based on content February 1912. Folder 8. EA-MBL. '*Bist du einverstanden? Du bist ja der Meinung, dass wir nicht das unterrichten sollen, was wir nicht selbst verstehen, u. ausserdem dachte ich, dass Lieberl dadurch besser zur Idee der systematischen Arbeit erzogen wird*'.
6. Einstein to Paul Ehrenfest. 12 April 1912. CPAE 5, doc. 384.
7. As note 6.
8. Tanya to Paul Ehrenfest. Undated, to Vienna, Mid-February 1912. Folder 8. EA-MBL.
9. These questions are taken from the completed draft of the survey in EFA. Translated from Russian by HD-BM.

10. As note 9.
11. Tanya to Paul Ehrenfest. Undated, to Vienna, Mid-February 1912. Folder 8. EA-MBL. 'Sollte der Solvey helfen!!!'
12. Tanya to Paul Ehrenfest. Undated, to Vienna, Late-January 1912. Folder 8. EA-MBL. 'ein ganz entscheiden begabter u. energischer Mensch, u. kann sicher nicht das alles Wissenschaftliches leisten, was er suchte wegen Geldfragen'.
13. Wir sind doch ganz ausserordentlich glücklich, Lieberl. Und natürlich dürfen wir einige, als jeder andere, Compromisse machen'.
14. Paul Ehrenfest. Diary 13. May 1912. Folder 268. EA-MBL.
15. Paul Ehrenfest. Diary 13. May 1912. Folder 268. EA-MBL.
16. Lorentz to Paul Ehrenfest. 13 May 1912. Folder 149. EA-MBL; see also: Kox (2018, doc. 113).
17. In fact, already in February Sommerfeld confirmed Lorentz' suspicion that Lebedev was informed by Ehrenfest, via Sommerfeld, and he shared Ehrenfest's address in St Petersburg. See: Sommerfeld to Lorentz, 25 February 1912. See: Kox (2008, doc. 239).
18. Lorentz to Paul Ehrenfest. 11 April 1912. Folder 149. EA-MBL; See also: Kox (2018, doc. 111). 'und da ich vermute, Sie haben ihn zu diesem Schreiben veranlasst, nachdem Sie von Prof. Sommerfeld etwas von der neuen Stiftung gehört hatten'.
19. As note 18. 'Ich benutze diese Gelegenheit um Ihnen für die freundliche Zusendung Ihres schönen und gründlichen Encyklopädie-Artikels ... meinen besten Dank auszusprechen. ... Nachdem ich vor vielen Jahren das Vergnügen hatte, Sie mit Ihrem Freunde Ritz auf kurze Zeit hier zu sehen haben unsereWege sich nicht wieder gekreuzt ... so wäre es mir sehr lieb ... auch etwas von Ihrem weiteren Lebenslauf und von Ihrer jetzigen Stellung zu vernehmen; ich habe damals nicht einmal gewusst, Sie seien ein Russe'.
20. Paul Ehrenfest to Lorentz. 24 April 1912. Folder 149. EA-MBL; See also: Kox (2018, doc. 112).
21. Lorentz to Van der Waals. 10 February 1912. Lorentz Archive. NHA; See also: Kox (2018, doc. 109).
22. Einstein to Lorentz. 18 February 1912. CPAE 5, doc. 360.
23. Einstein to Ehrenfest. 20–24 December 1912. CPAE 5, doc. 425.
24. Sommerfeld to Lorentz. 24 April 1912. NHA; See also: Kox (2008, doc. 244). 'Er trägt meisterhaft vor. Ich habe noch kaum einen Menschen so fesselnd und glänzend reden hören. Prägnante Wortbildungen, witzige Pointen, Dialektik steht ihm in ungewöhnlicher Weise zur Verfügung. Charakteristisch ist seine Art, die Tafel zu behandeln. Die ganze Disposition seines Vortrags steht auf das anschaulichste für den Hörer auf der Tafel vermerkt. Er versteht es die schwierigsten Sachen anschaulich und konkret zu machen. Die mathematischen Überlegungen übersetzt er in fassliche Bilder'.
25. As note 24. 'Aus dem persönlichen Verkehr hatte ich, mehr wie aus seinen Arbeiten, den Eindruck, dass es ihm um die physikalischen Tatsachen zu tun ist. In seinen Arbeiten ist er wohl mehr Logiker und Dialektiker. Die Mathematik ist ihm, wie es sein soll, nicht Selbstzweck. Im persönlichen Verkehr giebt er sich vielseitiger wie in seinen Abhandlungen. Die experimentellen Ergebnisse verfolgt er, soweit sie principiell sind'.
26. As note 24. 'Planck schätzt ihn als sehr scharfsinnig, wie er mir in Brüssel sagte'.
27. As note 24. 'Er geht mit der Absicht um, sich in Berlin, München oder Zürich zu habilitieren. Ich würde ihn sehr gern hier haben, nachdem ich durch seinen Besuch belehrt bin, dass er ... eine starke physikalische Ader hat'.
28. Einstein to the College of Professors of the Philosophical Faculty in Prague. Before 23 May. CPAE 5, doc. 400.
29. Hasenöhrl to Lorentz. 1 June 1912. Lorentz archive NHA. 'sehr viel, wie ich glaube, fast zu viel Idealismus'.
30. As note 24. 'Von Klein hörte ich gelegentlich, dass er durch Stimmungen und Verstimmungen in der Arbeit gelegentlich gehemmt würde'.
31. Lorentz to the Board of Trustees of Leiden University. 20 June 1912. Curators Archive. Leiden University. '... tot de eersten gerekend zal worden'.
32. As note 16.
33. Paul Ehrenfest to Lorentz. 19 May 1912. Kox (2018, doc. 114). 'Für meine Erlebnisse in den letzten 10 Jahren ist vor allem meine ungewollte Heimatlosigkeit charakteristisch Ich bin seit jeher überzeugt, dass - abgesehen von Fällen einer exceptionellen Begabung - nur dann eine Entfaltung der Kräfte möglich ist, wenn man die Menschen mit denen man normal zu thun hat nicht als fremd empfindet. - Im jetzigen Wien fühlte und fühle ich mich so fremd wie kaum irgendwo sonst. Ungleich mehr fühlte ich mich im Kreise meiner Göttinger Kameraden "zuhause." Ebenso später in der deutschen Schweiz. Ja selbst - so lächerlich es klingen mag - bei einem mehrwöchentlichen

Aufenthalt unter der Fischerbevölkerung von Schiermonnikoog fühlte ich mich sehr rasch mehr heimisch, als ich es in Wien könnte (trotz höchst mangelhafter Kenntnis der holländischen Sprache)'.

34. As note 33. '*Unzweifelhaft aber wäre Russland meine Heimat im ernstesten Sinne des Wortes geworden, wenn man mich hier nur irgendwie zur regulären Unterrichtsthätigkeit zulassen wollte: trotz meiner mangelhaften Beherrschung der Sprache fühle ich mich im Verkehr mit allen Bevölkerungsschichten (außer den politischen Beamten!) absolut heimisch und merkwürdigerweise nehmen auch mich die Russen aller möglichen Bevölkerungsschichten nicht als Fremden'.*
35. As note 33.
36. As note 33. '*Ohne eine solche verantwortliche Thätigkeit fürchtete ich aber ganz zu stagnieren: ich fühlte die zunehmende Desorganisation meiner Beschäftigungen; auch ist für mich "arbeiten" und mündliche Mitteilung meiner Überlegungen untrennbar'.*
37. As note 33.
38. Lorentz to Paul Ehrenfest. 30 May 1912. Kox (2018, doc. 115).
39. Lorentz to Paul Ehrenfest. 13 May 1912. Kox (2018, doc. 113). '*... dass unsere Fakultät sich am Ende für einen jungen Holländer entscheidet'.*
40. Ehrenfest-Afanassjewa. 5 January 1912. Family journal. EFA. Translated from Russian by HD-BM.
41. Paul Ehrenfest. Summer 1912. Diary 13. Folder 268. EA-MBL. See also Klein (1970, 191).
42. Einstein to Paul Ehrenfest. Before 20 June 1912. CPAE 5, doc. 409.
43. Telegram Lorentz to Ehrenfest. 29 September 1912. Kox (2018, doc. 125). '*Ehrenfest professor leiden ernant herzliche glueckwusche [sic] brief folgt Lorentz'.*
44. Paul Ehrenfest to Lorentz. 5 October 1912. Kox (2018, doc. 127). '*Wir hoffen zwischen dem 15. und 20.X in Leiden eintreffen zu können'.*
45. Herglotz to Paul Ehrenfest. 6 August 1912. Folder 145. EA-MBL. '*Ich hab' doch wirklich über Dich lachen müssen: schreibt er: Diese letzten Jahre waren so schön, dass es gar nicht schöner kommen kann'.*

Chapter 18

1. *Leidsch Dagblad.* 4 December 1912. LNA online.
2. Lorentz to the Ehrenfests. 12 November 1912. Folder 149. EA-MBL. '*Ik vergat gisteren te vragen of mevrouw, zoo zij het niet te druk heeft met de inrichting van het huis, morgenavond het colloquium zou willen bijwonen'.*
3. Lorentz to the Ehrenfests. 14 November 1912. Kox (2018, doc. 131).
4. Paul Ehrenfest. November 1912. Family journal. EFA.
5. Paul Ehrenfest to Lorentz. Draft 17 November 1912. Folder 149, EA-MBL. Original: Kox (2018, doc. 132).
6. The Physical Laboratory at Leiden (Holland). *Nature* 54, 345–347 (1896). https://doi.org/10.1038/054345a0
7. Paul Ehrenfest. December 1912. Family journal. EFA.
8. In Berlin: Chemists Johannes van 't Hoff, Emil Fischer and Eduard Buchner. In Göttingen: Chemist Otto Wallach. In München (and nearby Würzburg): chemist Adolf von Bayer and physicists Wilhelm Röntgen and Wilhelm Wien.
9. Paul Ehrenfest to Lorentz. 17 November 1912. Kox (2018, doc. 132). '*Übrigens bin ich momentan noch nicht in jenen Zustand gerathen (er wird noch kommen!) wo ich mich jede Nacht im Traume antrittsredend sehe'.*
10. '*Gestatten Sie mir über eine Krise zu sprechen, die gegenwärtig eine fundamentale Hypothese der Physik – die Ätherhypothese – schwer bedroht'.*
11. For context and further discussion, see e.g. Martínez (2004), and Hollestelle (2016).
12. '*Nehmen wir also an, wir hätten eine riesige Hohlkugel vor uns. Viel größer als die Erde, viel größer als die Erdbahn. ... Genau im Mittelpunkt der Hohlkugel sitzt ein Experimentator'.*
13. '*Es wäre unbescheiden, wenn ich die besondere Methode, durch die Herr Lorentz diese Aufgabe bewältigt hat, durch irgend ein Epitheton bewerten wollte'.*
14. Paul Ehrenfest. December 1912. Notebook 9/Diary 15. Folders 257 and 268. EA-MBL.
15. Lorentz to Paul Ehrenfest. 6 December 1912. Kox (2018, doc. 134). Translation A. Kox.

Chapter 19

1. Paul to Tanya Ehrenfest. From the train from Leipzig to München, January 1912. Folder 8. EA-MBL. *'Ich will dieses: So leben, dass ich nicht den Contact mit der fortschreitenden Physik verliere und dass ich anderseits selber was mache und weiter mich versichern für den Fall wir Russland verlassen müssen - dann noch ein wenig Geld - und alles so dass Du nicht Dein Leben verlierst'.*
2. Personal communication T. van Bommel.
3. This thermal noise is nowadays called Johnson-Nyquist noise after John Johnson, who first measured it at Bell Laboratories in 1926, and Harry Nyquist, who subsequently explained the results.
4. Ehrenfests. Late 1912 and early 1913. Family journal. EFA.
5. Ehrenfest-Afanassjewa. Sketches and construction drawings. 1913. EFA.
6. The house has since been designated a national monument ('rijksmonument') with ID: 516,100 and described in more detail in document 20,031,222. See: https://rijksmonumenten.nl/
7. Advertisement in *Leidsch Dagblad*. 6 August and 13 September 1913. LNA online.
8. Ehrenfest to Lorentz. 5 October 1912. Kox (2018, doc. 127). *'ich will wirklich alle meine Kräfte anspannen um wenigstens einigermaßen das Vertrauen derjenigen Männer zu rechtfertigen, die es für möglich fanden, mich auf Ihre Lehrkanzel zu berufen'.*
9. *'De belangrijkste formules verschenen op het bord bijna als afzonderlijke esthetische scheppingen en niet alleen als schakels in een samenhangende redenering. Hij vermeed lange berekeningen en besteedde vaak weinig aandacht aan numerieke factoren. "4π" kon bijna alles betekenen'.*
10. *Leidsch Dagblad*. 20 May 1913. LNA online.
11. Ehrenfest to Lorentz. 23 December 1912. Kox (2018, doc. 135).
12. Ehrenfest to Ehrenfest-Afanassjewa. 19 April 1913. Folder 9. EA-MBL. *'Ist doch ein lieber Fleck auf Erden - für uns beide'.*
13. Ehrenfest to Ehrenfest-Afanassjewa. 21 April 1913. Folder 9. EA-MBL. *'Du musst unbedingt herkommen - sofort!'*
14. Ehrenfest to Ehrenfest-Afanassjewa. 22 April 1913. Folder 9. EA-MBL. *'Sehr Schade dass Du nicht da bist - noch nicht da (?)'.*
15. Lorentz to Ehrenfest. 2 July 1913. See Kox (2018, doc. 145).
16. Ehrenfest 1913. Diary 15. Folder 268. EA-MBL. This visit lasted from 18 June until 1 July. A whole group of mathematicians and physicists were present, among them also the Dutch astronomer Betty Biegel, the physics students Eva Bruins and Catherina Franekamp and chemist Marie Scanavy-Grigorieff. See: Dane and Verhoef (2024).
17. In the paper, she briefly analyses the methods of Th. Svedberg for calculating these 'Svedberg variations', which led to some debate in which Lorentz and Schmoluchowsky took part. See also Ch. 20.
18. Ehrenfest-Afanassjewa to Schmoluchowsky. Undated but before 20 March 1913. Korespondencja Mariana Schmoluchowskiejo 1892–1917. T.2 Daniel-Exner. Jagellonian Digital Library. *'Wir sind seit drei Tagen in Zürich und schwelgen in der neuesten Relativitätstheorie, Quanten und sonstigen schöne Sachen'.*
19. Ehrenfest to Lorentz. 19 May 1912. See Kox (2018, doc. 114). *'Meiner Frau bot sich Gelegenheit zu einer sehr wesentlichen Antheilnahme an der Reformbewegung betreffs des mathemat. Unterrichtes in Russland'.*
20. Ehrenfest to Ehrenfest-Afanassjewa. 9 January 1914. Folder 9. EA-MBL. *'Bier-Idee'.*
21. Ehrenfest to Ehrenfest-Afanassjewa. 5 January 1914. Folder 9. EA-MBL. *'Augenblicklich bist Du also schön tief in Russland.... Um den Schnee beneide ich Dich. Hier Patent - Dreck Wetter - Kindlis sehr brav. Alles in bester Ordnung (auch innerlich)'.*
22. Ehrenfest to Ehrenfest-Afanassjewa. More postcards 4 and 5 January 1914. Folder 9. EA-MBL. *'Sie haben in T' Cabineterl einen Riesenversammlung von Pupperln und Thiereln zusammen getragen'. 'T' war es sehr interessant so früh in der Finsternis mit mir durch Leiden zu gehen. [...] Galja behauptet dass ich die Nägel schön mache. Sie sitzt auf deinem Platz'.*
23. As note 22. *'Alle 3 Giebel-Dreiecke provisorisch gefärbt. Die gelbe Borte contrastiert zu wenig - studiere genau die Farbe in Petersburg!!'*
24. As note 22. *'Ich habe ihm mehrmals wiederholt er möge nur ja alle diese Treppen exactest nach Deiner Vorschrift machen'.*
25. Ehrenfest to Ehrenfest-Afanassjewa. 9 January 1914. Folder 9. EA-MBL. *'Er besah es sehr aufmerksam von alle Seiten ... und fand es sehr gut! Dann zeigte ich ihn auch alles im Inneren. Selbst viel ihm auf: die Flachkeit der Treppe und die Abrundung aller Holzprofile die er sehr sympathisch findet'.*

26. Ehrenfest to Ehrenfest-Afanassjewa. 5 January 1914. Folder 9. EA-MBL. 'Regen-Wind-kalt- So sehr dass wir T' doch nicht zur Lehrerin schickten'.
27. Ehrenfest to Ehrenfest-Afanassjewa. 6 January 1914. Folder 9. EA-MBL. 'Wetter graulich. Zerrinnender Regen. Schnee. Rechts ist Wasser, links ist Wasser u. dazwischen noch viel Wasser'.
28. Ehrenfest to Ehrenfest-Afanassjewa. 11 January 1914. Folder 9. EA-MBL. 'Aus Deinen Briefen schliesse ich, dass Du von Ioffe nichts gehört hast. Selbstverständlich thut mir das sehr Leid, aber er ärgert mich riesigh ... Mache Dir irgendwie täglich einige Notizen um mir dann alles gut erzählen zu können. Es ist sehr gut das Du Dich "austobst"'.
29. Ehrenfest to Ehrenfest-Afanassjewa. 11 January 1914. Folder 9. EA-MBL. 'Formuliere in Deinem Vortrag sehr prägnant Deine Endthesen! Die Provinz will Anweisungen zum Handeln haben!'
30. Proceedings of the Second All-Russian Congress of Teachers of Mathematics held from 27 December 1913 to 3 January 1914 (Julian (Russian) Calendar). Translated from Russian by HD. Tanya presented her results on 2 January (15 January in Western-Europe). Two days earlier she had been honorary chair of the session of Section B.
31. Ehrenfest to Ehrenfest-Afanassjewa. 17 January 1914. Folder 9. EA-MBL. 'Heute oder gestern hast Du also Deinen Vortrag gehalten. Na ich hoffe, dass Du Dich nicht allzu sehr herumzankst - Vergesst nicht, dass die Leute jedenfalls viel mehr praktische Erfahrung als Du und auch dieses: dass Sie Dir lieb entgegenkommen und streiten nichts hilft falls es zu Verärgerung führt'.
32. 'Verkühle Dich nicht - Es richtig!!' 'Streite nicht unnöhtig. Reagiere nicht auf Boshafte Redenationen'.
33. Ehrenfest to Ehrenfest-Afanassjewa. 4 January 1914. Folder 9. EA-MBL 'Bitte, übernehme von keiner Seite die Verpflichtung irgend etwas bestimmtes zu schreiben'.
34. Ehrenfest to Ehrenfest-Afanassjewa. 5 January 1914. Folder 9. EA-MBL. 'Ich vermuthe dass Du so unvorsichtig sein wirst Dich darauf einzulassen. Es ist durchaus notwendig dass wir im neuen Haus vor allem etwas gemeinsam arbeiten zur Orientierung über dem Gebrauch der Zimmer und Schaffung'.
35. Ehrenfest to Ehrenfest-Afanassjewa. 12 January 1914. Folder 9. EA-MBL. 'Du verstehst, dass diese Herumlauferei und Fahrerei für mich ermüdend und deprimierend ist'.
36. Ehrenfest to Ehrenfest-Afanassjewa. 11 January 1914. Folder 9. EA-MBL. 'Ich schäme mich vor Dir, dass ich Dir so gar nichts Interessantes schreiben kann. Erklärung: ich arbeite gar nichts und denke auch über nichts und Dich an jammern jetzt wo Du froh sein sollst (ja, Lieberl, das wünsche ich sehr!) hat doch keine Sinne'.
37. Ehrenfest to Ehrenfest-Afanassjewa. 14 January 1914. Folder 9. EA-MBL. 'Alles in allem bin ich schadhafte unzufrieden und mache ich absolut nichts. Aber ... du weisst bei mir geht's ja auch bald vorüber - in dem Moment wo Du die Karte liest, bin ich vielleicht schon wieder ja froh'.
38. Ehrenfest to Ehrenfest-Afanassjewa. 15 January 1914. Folder 9. EA-MBL. 'Hat sehr froh über die Stockholmer Zeit erzählt. Er hat ganze grosse Haufen von Briefen bekommen. Es ist wirklich unglaublich wie viel.... Er ist in diese Zeit sehr froh und lieb'.
39. Ehrenfest to Ehrenfest-Afanassjewa. 18 January 1914. Folder 9. EA-MBL. 'Alle Leute laufen auf die Canäle Schlittschuhe.... Backer mit Coseyn laufen bis Hoek van Holland. De Sitter mit älterem Jungen bis Haag'.
40. As note 3. 'Wohl ein paar ältere Physiker. Aber gerade die Jungen benützen das Lesezimmer so gut wie gar nicht. Überhaupt arbeiten jetzt alle sehr schlect. Ewige Examenjagt, oder glatte Nichtsthunerei. Kein Mensch wohnt in Leiden etc'.
41. Ehrenfest to Ehrenfest-Afanassjewa. 9 January 1914. Folder 9. EA-MBL. 'Du fühlst dass ich fürchte, dass Dich das Congress-Fieber ganz gegen das Leben hier aufreißt'.
42. As note 41. 'Und denke nicht zu hässlich über Deine westeuropäische Verbannung - bedenke 3 Dinge: 1) Dass es natürlich 'viel dringende Arbeit in Russland zu thun gibt. Aber warum? Weil niemand Sie niemanden thun lässt. ... 2. Auch Du und in besonderer Du könntest dort so gut wie nichts thun 3. Wenn du nur nur die Sache vom richtigen Zipfel anfangen wolltest, könntest Du auch hier vieles gutes machen an dem Du Dich entwickelst und dauernde principielle Wirkung hast. In besonderer auch durch Ausrüstung Deiner Kinder und durch Schreiben und Denken'.

Chapter 20

1. The *Hogere Burgerschool* (HBS) was a relatively new type of secondary school (established 1863) with a strong focus on mathematics and the sciences. It was meant to prepare boys for a vocational profession in the field of engineering or—with an additional state examination in Latin and Greek—for further academic studies in these fields.
2. Ehrenfest-Afanassjewa. 7 July 1914. Family journal. EFA. Translated from Russian by HD-BM.

3. Photographs in the family archive; personal communication T. van Bommel. Details, such as the lamp on the ceiling and the blackboard, were still present in the house during a visit in 2015.
4. Ehrenfest. July 1914. Family journal. EFA.
5. Ehrenfest. 2 September 1914. Family journal. EFA.
6. *Leidsch Dagblad*. 5 September 1914. LNA online.
7. As note 5. '*Man möchte wissen was jetzt in zwei Jahren sein wird. Es ist aber vielleicht gut, dass wir es nicht wissen*'.
8. Ehrenfest. 9 August 1914. Family journal. EFA. '*Erst Geld-Warenpanik. Dann hier mehr und mehr Militair. Niemand mag mehr still sitzen*'.
9. Yearbooks of the historical society of old Leiden 1914, 1915, 1916.
10. Ehrenfest. Childhood Memories. Folder 21. EA-MBL. '*So könnte mir Wien und Deutsch-Oesterreich nie nie etwas anderes als "feindliche Fremden" bleiben*'. On second page under the heading: '*Wie wir bis zu meinem 8sten Lebensjahr wohnten*'.
11. As note 8. '*Endlose Hetzereien der Zeitungen in Deutschl., England und Belgien*'.
12. De la Court to Ehrenfest-Afanassjewa. 10 December 1914; 18 October 1915. Folder 9. EA-MBL. In the first letter, he announces the arrival of the first Belgian family; in the last, he returns the key with a list of items that two Belgian families were still borrowing for use elsewhere.
13. See also the correspondence (1914–1918) in Kox (2008).
14. Ehrenfest. 25 May 1916. Family journal. '*T' und Galinka lieben schrecklich "Vergaderingen" (Phys. Coll., Math. Pädag. Coll, Damesdebating)—stellen Stühle, "halten Vorträge", lauern ausgezogen auf Treppe, gucken in's Zimmer*'.
15. Ehrenfest-Afanassjewa. 30 August 1914. Family journal. Translated from Russian HD-BM.
16. Ehrenfest-Afanassjewa. 15 October 1914. Family journal. Translated from Russian HD-BM.
17. Rommert Casimir was the father of Hendrik Casimir, who later became one of Ehrenfest's students, scientific director of Philips NatLab and book author.
18. It was the first 'Lyceum' in the Netherlands, with a curriculum that was the same for all students for the first two years, after which they could choose to follow either the HBS curriculum, with a strong focus on mathematics and science, or the Gymnasium, with a stronger focus on the humanities and with Latin and Greek.
19. '*De rol der axioma's en bewijzen in de wiskunde*' in *Weekblad voor Voorbereidend en Hooger Onderwijs*.
20. Proceedings of the First All-Russian Congress of Teachers of Mathematics, general session, 2 January 1912, discussions on the presentations of S.A. Bogolomov, P.A. Dolgushin and R.A. Kulisher: contribution T.A. Ehrenfest-Afanasyeva (S-Petersb). EFA. Translated from Russian HD-BM.
21. As note 20.
22. Others said similar things, see e.g. Hollestelle (2011, 55).

Chapter 21

1. Analysis in leading Dutch newspaper *Nieuwe Rotterdamsche Courant*, 9 October 1914. Delpher Newspaper Archive online.
2. Ehrenfest to Smoluchowski. 18 October 1915. Korespondencja Mariana Schmoluchowskiejo 1892–1917. T.2 Daniel-Exner. Jagellonian Digital Library. '*Übermorgen würde meine Frau im Colloquium über Ihre letzten Publicationen berichten*'.
3. Ehrenfest-Afanassjewa to Schmoluchowsky. Undated but before 20 March 1913. Korespondencja Mariana Schmoluchowskiejo 1892–1917. T.2 Daniel-Exner. Jagellonian Digital Library. '*Dass Sie meine Ausführungen für irgendwie beachtenswert halten, freut mich gewaltig!*'
4. As note 3. '*Der ganze Witz liegt darin, dass ich immer unsicher war, ob ich es drucken sollte, [schliesslich nahm ich meinen zweiten Correcturbogen nicht mit auf die Reise]. Sie werden mir übrigens besser wie andere nachfühlen können, da Sie so viele wirklich schöne Dinge, viel später als Sie es könnten, veröffentlichen!*'
5. The focus specifically on kinetic energy came from her, he wrote in a letter to Ioffe (20 February 1913). See also Klein (1970, 263): '*By the way, I essentially owe the discovery of the correct generalisation to arguments with Tanya*'.
6. Einstein to Ehrenfest. Before 7 November 1913. CPAE 5, doc. 481.
7. The then known set of spectral lines of light emitted by hydrogen atoms was described by the so-called Rydberg formula. A more refined picture of the atom was developed in consecutive years using newly discovered spectral lines in the ultraviolet or infrared.
8. He made a similar remark in a letter to Lorentz, August 1913, stating that Bohr's work had driven him to despair; see: Klein (1970, 278).

9. Ehrenfest to Lorentz. 4 September 1914. Folder 150. EA-MBL. See also: Kox (2018, doc. 161). *'Ich zweifle doch keinen Augenblick, dass ein großer Theil der naturwissenschaftlichen Gelehrten in allen diesen Ländern das charakteristische Element des Häkkelschen Schrittes (Behandlung eines Cambridger Ehrendoctorates als ob es ein militärisches Ehrenzeichen wäre) als total unsinnig und verkehrt erkennen'.*
10. For the English translation: https://www.jstor.org/stable/pdf/25122278.pdf?acceptTC=true
11. Ehrenfest to Ehrenfest-Afanassjewa. 8/21 January 1912. Folder 8. EA-MBL. See Ch. 15.
12. Ehrenfest to Lorentz. After 16 December 1914. Kox (2018, doc. 167). *'überhaupt in dem Chaos das jetzt auf Erden geschieht muss man schon ein ganz besonderes Temperament und besondere Reife haben um jedesmal wirklich sicher zu wissen, was einem das Gewissen befiehlt'.*
13. Einstein to Ehrenfest. Second half of November 1913. CPAE 5, doc. 484.
14. Einstein to Elsa Löwenthal. 19 July 1913. CPAE 5, doc. 453.
15. Einstein to Elsa Löwenthal. After 19 and before 24 July 1913. CPAE 5, doc. 454.
16. Einstein to Ehrenfest. After 2 December 1913. CPAE 5, doc. 489.
17. Einstein to Ehrenfest. 19 August 1914. CPAE 8a, doc. 34. *'In solcher Zeit sieht man, welcher traurigen Viehgattung man angehört. Ich döse ruhig in meinen friedlichen Grübeleien und empfinde nun reine Mischung von Mitleid und Abscheu'.*
18. The others included David Hilbert, Michele Besso and Heinrich Zangger.
19. See e.g. Burgers (1962, 59) and the correspondence in CPAE 8a.
20. Einstein also collaborated with other Leiden physicists: in Zurich he had worked with Adriaan Fokker. See e.g. Pais (1982, 236–237). Later, he would also work with Nordström in Leiden. See: Renn and Janssen (2015).
21. As note 19.
22. Ehrenfest. Retrospective list. 12 January 1916. Family journal. EFA.
23. Ehrenfest-Afanassjewa. 18 July 1915. Family journal. EFA. Translated from Russian by HD-BM.
24. Ehrenfest-Afanassjewa. 19 July 1915. Family journal. EFA. Translated from Russian by HD-BM.
25. Ehrenfest. 18 May 1915. Family journal. EFA. Under the heading *'Lieblingslieder der Kindlis'.*
26. Ehrenfest-Afanassjewa. February 1914. Family journal. EFA. Translated from Russian by HD-BM.
27. Ehrenfest. 11 May 1915. Family journal. EFA. *'Bis gestern arbeitete Lieberkerl noch jeden Tag paar Stunden im Garten ... Prachtvoller Mei [sic] – Apfel – Erdbeer - Ombrasia [sic] blühen. ... Haus ganz in Sonne gebadet. ... Gestern Abend war ich bei De Sitter - als ich nach Hause kam plauderten wir bis 1/2 1^h. Dann aber wird Tania ängstlich'.*
28. As note 27. *'Pawlik kommt (als Wanotschka oder Lenotschka erwartet!) ... Kräftige Schreierei. / (Galinka war gefallen - T' später auch ...)'.*
29. Ehrenfest-Afanassjewa. 2 July 1915. Family journal. EFA. Translated from Russian by HD-BM.
30. Ehrenfest-Afanassjewa. July 1915. Family journal. EFA. Translated from Russian by HD-BM.

Chapter 22

1. Einstein to Lorentz. 16 January 1916. CPAE 8a, doc. 177. *'Eure lockenden Einladungen machen es mir schwierig, hier hocken zu bleiben'.*
2. Einstein to Lorentz. 23 January 1915. CPAE 8a, doc. 47. *'Von niemandem als von Ihnen wünsche ich sehnlicher, dass er die in heissem Ringen gefundenen Überlegungen über die allgemeine Relativität genau nacherlebe'.*
3. See Einstein to Ehrenfest. 3 January 1916. CPAE 8a, doc. 179.
4. See e.g. also Janssen (2005), Kox (2008), and the correspondence in CPAE
5. Ehrenfest. October 196. Family journal. EFA.
6. As note 5.
7. Einstein to Ehrenfest. 24 October 1916. CPAE 8a, doc. 269.
8. Einstein to Ehrenfest. 18 October 1916. CPAE 8a, doc. 268.
9. As note 8.
10. Einstein to Besso. 21 July 1916. CPAE 8a, doc. 238.
11. For a more detailed discussion of this 'dimensionality debate', see Jalloh (2024).
12. Ehrenfest. Childhood Memories. 1932. Folder 21. EA-MBL.
13. As a result, these examinations were called the 'Kwatta-exams', see Hollestelle (2011, 48).
14. By this time, Sommerfeld had extended Bohr's model, introducing elliptical electron orbits and quantising the z-component of angular moment, i.e. introducing an additional type of quantisation. The latter explains his (limited) interest in Ehrenfest's work as well as Ehrenfest's (much greater) interest in his. The new Bohr-Sommerfeld-model was an attempt to explain the fine structure in observed spectra, as caused by the Zeeman and Stark effect, and Sommerfeld took

experimental results as guidance. For more details and nuance, see e.g. Duncan and Janssen (2019).
15. Ehrenfest to Ioffe. 20 February 1913. Folder 146. EA-MBL. His publications indeed show a variety of topics, see 1913b, 1913c, 1915a, 1915b, 1915c, and 1916a.
16. Schrödinger to M.J. Klein. 17 April 1957. MKEAD. In English.
17. Ehrenfest to Ehrenfest-Afanassjewa. 8 April 1929. Folder 30. EA-MBL. '*Kramers sieht aus, als ob die ganze Welt ihm gehört, er aber viel zu faul sei, sie in Besitz zu nehmen*'.
18. Ehrenfest. After 15 and before 22 October 1916. Family journal. EFA. '*Ich telefoniere. Sie kommen ... aus Garten sieht T' ... durch Thüre (Glas) baba Sonja in Speisezimmer. Sie rührt sich nicht aus Angst "Traum zu zerstören*"'.

Chapter 23

1. Ehrenfest to Einstein. 8 May 1918. CPAE 8b, doc. 534.
2. Einstein to Ehrenfest. 3 June 1917. CPAE 8a, doc. 350.
3. Ehrenfest-Afanassjewa. Various notes 1916–1917. Family journal. EFA. Translated from Russian by HD-BM.
4. Recurrent theme in notebooks and correspondence in EA-MBL.
5. Ehrenfest to Mr and Mrs Burgers. 14 August 1917. BA-DUT.
6. Ehrenfest to Burgers. August 1917. BA-DUT.
7. Ehrenfest to Burgers. 29 August 1917. BA-DUT.
8. Ehrenfest to Burgers. 27 August 1917. BA-DUT.
9. The Physics Laboratory was funded by the Pieter Teyler Foundation in Haarlem, which also funded two learned societies: the First and Second Teyler's Society.
10. *Algemeen Handelsblad*. 19 September 1917. Delpher newspaper archive online. The newspaper cites British correspondent Harold Williams from the *Daily Chronicle*.
11. He made the remark, probably in the 1920s, to his student Jan Tinbergen.
12. They weren't the only ones affected. In 1917, Dutch investors had invested an estimated one to two billion Dutch guilders in Russian government bonds and railway loans guaranteed by the Russian state. They never recuperated any of this, and it was partly because of this that it took until 1942 before the Netherlands recognised the Soviet Union as a sovereign state.
13. Finland declared its independence as early as 1917, and Ukraine in 1918. Russia reincorporated Armenia, Azerbaijan, Georgia and the eastern part of Ukraine in 1921.
14. Personal communication T. van Bommel. The texts on Galinka's sheets of paper may have been slightly different.
15. As note 1.
16. Ehrenfest to Einstein. 27 March 1918. CPAE 8a, doc. 494. See also note 5 of document 494.
17. Einstein to Hilbert. 12 April 1918. CPAE 8b, doc. 503, In this letter, he described what he would more or less being to Paul: '*Liebes Paulchen, stelle Dir vor, du wärest vor ein paar Jahren nach Göttingen statt nach Leyden berufen worden und hättest im Frühjahr 1918 den Kollegen Keesom* [from Leiden] *freundlich eingeladen, und kriegtest von ihm dafür eins an den Kopp!*'
18. Einstein to Ehrenfest. 1 May 1918. CPAE 8b, doc. 528.
19. As note 1.
20. Hilbert to Einstein. 27 April 1918. CPAE 8a, doc. 524.
21. Einstein to Ehrenfest. 3 June 1917. CPAE 8a, doc. 350.
22. Ehrenfest to Einstein. 27 March 1918. CPAE 8a, doc. 494.
23. Ehrenfest to Ehrenfest-Afanassjewa. 25 July 1918. Folder 15. EA-MBL. '*so dass sie einigen Eindruck davon haben was höhere Mathematik ist und wie ihre Anwend, auf Mechanik etc. aussehen*'.
24. Ehrenfest to Ehrenfest-Afanassjewa. 23 July 1918. Folder 15. EA-MBL.
25. Ehrenfest to Ehrenfest-Afanassjewa. 18 July 1918. Folder 15. EA-MBL.
26. Ehrenfest to Ehrenfest-Afanassjewa. 21 July 1918. Folder 15. EA-MBL. '*Er liebt alles was alt und schön ist—speziell stöbert er gerne in alten Gebäuden auf "Solders" und im Keller rund*'.
27. Ehrenfest to Ehrenfest-Afanassjewa. 23 July 1918. Folder 15. EA-MBL. (As note 24.) '*Mir sehr leid, dass Du in Leiden sitzt mit soviel Arbeit und Mühe und ich es so herrlich habe. Aber nun wäre es doch verkehrt nicht voll auszunützen was mir nun zu Gebote steht*'.
28. Ehrenfest-Afanassjewa to Ehrenfest. 18 July 1918. Folder 15. EA-MBL.
29. Ehrenfest to Ehrenfest-Afanassjewa. 17 July 1918. Folder 15. EA-MBL. '*Ich hielt Sie für 4 Tage jünger!*'

NOTES   353

30. Ehrenfest to Ehrenfest-Afanassjewa. 22 July 1918. Folder 15. EA-MBL. '*Bitte Dich um Folgendes: sende T' (für v. Aardenne !!) nach Leidsche Dagblad mit folgender annonce (spoed !!!): Twee of drie rustige, frische kamers gevraagd voor 2 heeren bij voorkeur aan de Zuid-Westkant der Stad*'.
31. As note 29. '*Bitte an Dich: lass Sie beiden bei uns übernachten. Sie werden Bettwäsche mitbringen. Versorge Sie bitte den übrigen*'.
32. Ehrenfest to Ehrenfest-Afanassjewa. 26 July 1918. Folder 15. EA-MBL. '*Dank für die Sendung (Hemd, Rubaschka, Strümpfe, Hose) aber nun habe ich dringend nöthig: Brotkarte, Vettbon (recommandiert senden!!) und vielleicht 2–3 überzähl. Brotbons als Geschenk für mich, falls es euch nicht zu knapp wird*'.
33. Ehrenfest-Afanassjewa to Ehrenfest. 30 July 1918. Folder 15. EA-MBL.
34. 'Stadsnieuws, een onhoudbare toestand' in *Leidsch Dagblad*. 12 July 1918. LNA online.
35. Personal communication Maria Wiebenga (daughter of Carl Gustav Julius von Winning), 14 April 1915. Her father, from a noble family and born in Wiesbaden in 1887, had been a private physician in Leiden since 1917.
36. Einstein to Ehrenfest, 4 September 1918. CPAE 8b, doc. 608.
37. Ehrenfest to Burgers. 24 Oct 1918. Folder 598. EA-MBL. '*Nur um alles in der Welt, komm nicht in Bolschewik-Uniform*'.
38. Ehrenfest to Roosenschoon. 21 August 1918. Folder 598. EA-MBL. '*Vielleicht ist es ein Irrthum— aber Sie wissen dass meine Frau und ich überzeugt sind, dass die Hauptbedingung einer dauernden Ehefreundschaft ist: gemeinsame Liebe an etwas Drittem. Gewöhnlich sind dieses Dritte die Kinder, aber bei einem Mann der so sehr wie Jan die intellectuelle Arbeit als Glücksquelle liebt, ist es sehr, sehr gut, wie mir scheint, dass seine Frau eine waschechte Physikerin ist! Ein Grund mehr für mich zu wünschen, dass Sie Ihr Studium kräftig durchführen*'.

Chapter 24

1. Ehrenfest to Burgers. 2 October 1918. Folder 139. EA-MBL.
2. Paul asked Burgers to help out with Krutkov's paper, as he didn't fully grasp the mathematics, see Hollestelle (2011, 125).
3. Bohr to Ehrenfest. 18 May 1918. Folder 136. EA-MBL. In English.
4. Bohr to Rutherford. 27 December 1917.
5. The prize went to Burgers instead (see also Ch. 23). Burgers would feel a bit guilty about that the rest of his life. Personal communication Herman Burgers, conversation in The Hague, 16 April 2015.
6. These matters are discussed in: Bohr to Ehrenfest. 25 January 1919; Ehrenfest to Bohr. 12 March 1919. Folder 136. EA-MBL.
7. *Leidsch Dagblad*. 25 April 1919. LNA Online.
8. Ehrenfest. Notes of the speech for Bohr. Folder 136. EA-MBL. '*total unverständlich ... diese gegenwärtig so mühevolle und zeitraubende Seereise zu unternehmen ... Wir allen kennen mehr oder weniger genau seine bahnbrechenden Arbeiten über die Structur der Atome u. Moleküle. Aber wir sind überzeugt, dass die persönliche Begegnung mit Prof. Bohr zu einer tieferen Einfühlung in seinen Ideen kreist*'.
9. Ehrenfest. 29 April–1 May. Research notebook XXV. Folder 269. EA-MBL.
10. Bohr to Ehrenfest. 1 June 1919. Folder 136. EA-MBL. In English. The same letter is cited in Klein (2010, 313).
11. Ehrenfest to Bohr. 4 June 1919. A copy/draft of this letter is in folder 316. EA-MBL.
12. Einstein to Lorentz. 21 September 1919. CPAE 9, doc. 108.
13. Ehrenfest to Einstein. 21 September 1919. CPAE 9, doc. 109.
14. This news came to Lorentz through astronomer Balthasar van der Pol who had attended the meeting of the British Association for the Advancement of Science in Bournemouth, where Eddington had shown his preliminary results.
15. Lorentz to Einstein. 10:40 am, 22 September 1919. See Kox (2008, doc. 339). '*prof einstein huberlandstrasze 5 berlin / eddington fand sternverscheidung [sternverschiebung] am sonnenrand vorlaeufig grusse [grösse] zwischen neun zehntel sekunde und doppeltem / lorentz*'.
16. Expression used by Paul in various letters, see e.g. CPAE 9, doc. 98.
17. Einstein to Ehrenfest. 8 November 1919. CPAE 9, doc. 160.
18. As note 17.
19. Ehrenfest to Einstein. 24 November 1919. CPAE 9, doc. 175.
20. Einstein to Ehrenfest. 4 December 1919. CPAE 9, doc. 189.
21. Ehrenfest-Afanassjewa to Ehrenfest. 28 March 1932. EFA. Translated from Russian by HD-BM.
22. Oppenheimer. Typed note, undated. Folder. EA-MBL.

23. O. Klein to M.J. Klein. 30 August 1957. MKEAD. EA-MBL. English.
24. Ehrenfest to Einstein. 2 September 1919. CPAE 9, doc. 98.
25. As note 24.
26. As note 24.
27. Einstein to Lorentz. 21 September 1919. CPAE 9, doc. 108.
28. Einstein to Ehrenfest. 12 September 1919. CPAE 9, doc. 103.
29. Lorentz to Einstein. 21 December 1919. CPAE 9, doc. 229.
30. Ehrenfest to Einstein. 2 September 1919. CPAE 9, doc. 98. See also: Klein (1970, 310–311).
31. Ehrenfest to Einstein. 9 December 1919. CPAE 9, doc. 203.
32. Ehrenfest to Einstein. 24 November 1919. CPAE 9, doc. 175.
33. Ehrenfest-Afanassjewa. Late 1919. Second Family journal. EFA. Translated from Russian by HD-BM.
34. Ehrenfest-Afanassjewa. Letter about the 'Abolition of night and Sunday work in the Leiden bakery'. 24 March 1919. *Leidsch Dagblad*. LNA online.
35. Einstein to Lorentz. 19 January 1920. CPAE 9, doc. 265.
36. Baumgart to Ehrenfest. 10 August 1920. Folder 136. EA-MBL. '*Lieber Freund, Ihren Brief von 4. Juli habe ich Dienstag 3 Aug. enthalten. Ich kann Ihnen nicht sagen, wie sehr es mich erfreut hat und nicht nur mich, es war ein Hallo am ganzen Physikalische Institut und am Polytechnicum. Der Brief wurde in optischer und in Röntgen-Institut vorgelesen. Mir scheinte es förmlich, ich höre Ihre Stimme. Mit einem Wort die Freude war sehr groß. Jetzt beeile ich mich auf alle Ihre 45 Frage schnell Antwort zu geben*'.
37. As note 36. '*1. Die Mutter Ihrer Frau wöhnt in Moskau, ist gesund… ist von Frl. Föhringer von Ihrem Brief benachrichtet 2. Joffe, Professor in Polytechnikum, Director des Röntgen-Institut, jetzt mitglied der Akademie des Wissenschaften, schreibt Ihnen selbst. 3. Frl. Föhringer, bibliotekar … assistent an der Universität (Hohere Kurse sind jetzt mit der Universität zusammengeschmolzen), hält in Pädagogischen Institut an der Universität vorlesungen über Geschichte der Physik. / Krutkow macht Große Fortschritte, arbeitet über Quantenhypothese (Formelle Fragen), und Adiabat. Hypothese, hat Ihnen seine Arbeit gesandt, haben Sie dieselben erhalten? / Rozhdestvensky, Professor an der Universität, Director des Optischen Institut - arbeitet sehr viel. … / Bursian arbeitet fleissig … Arbeit über Lechatelier-Braun Prinzip, Röntgenstrahlen, Zeeman-effect, Stark-effect. / Tymenschenko ist jetzt im Ausland und hoffentlich bei Ihnen / Weichart tot, ertrunken - näheres schreibt Ihnen Friedmann / … / Chvolson schreibt selbst, hat eine Arbeit über die Struktur der Atomkerne veröffentlicht. / … / Perlitz, irgendwo im Ausland, wir wissen nicht wo / Fränkl- Dozent in Süden Russlands Fühlung mit ihm verloren / Markov und Stekloff sehr gealtert, besonderes Markoff / Kriloff, akademiker, arbeitet in Optisch Institut. / Friedmann und Tamarkin schreiben Ihnen selbst*'.
38. As note 36. '*Seit Anfang 1917 haben wir überhaupt nur zufällige Nachrichten aus dem Ausland. … Über Sonnenfinsternis und Gravitation wussten* [sic] *wir aus Zeitungen, aber ohne Details*'.
39. As note 36. '*Was ist in der Optik erreicht experimentell u. theoretisch. Arbeit von Paschen und seiner Schule. Lorentz's Arbeit. Alles über quanta. Ihre Arbeiten. Burgers. Neues von Bohr. … Alles was Sommerfeld geleistet hat. Sommerfeld's Buch Atombau u. Spektrallinien, wenn möglich und überhaupt alles, was Quanten Optik … betrefft*'. Note: First published in 1919, Sommerfeld's book and its successive editions, appearing almost annually in the early 1920s, chronicled progress in this field.
40. As note 36. '*Ich freue mich sehr, dass Sie glücklich in Leiden sind und so ein intensives wissenschaftliches Leben führen und Erfolg und Anerkennung von wirklich kompetenten Leuten haben. Wie gern möchte ich bei Ihnen sein*'.

Chapter 25

1. *Svobodnaia Assotsiatsia dlia Razvitiia i Rasprostraneniia Polozhitel'nykh Nauk*—SARRPN.
2. *Nieuwe Rotterdamsche Courant*. 23 January 1920. Delpher newspaper archive online.
3. '*De man zal van zijn troon, als "heer der schepping", moeten afdalen en zich in het vervolg tevreden moeten stellen met een plaatsje naast de vrouw, die als politiek gelijkwaardige, als vriendin, als medewerkster op elk gebied, steeds aan zijn zijde zal staan. … Tronen zijn uit in onze tijd. We hebben de laatste jaren herhaaldelijk gezien, dat de bezitters er vrijwillig afstand van deden*'.
4. '*de politieke vrijmaking der vrouwen… een heilzame invloed [zou] uitoefenen op mannen en hunne karaktervorming*'.
5. See e.g. also: *An icon in the world of modern science*. NobelPrize.org. Nobel Prize Outreach 2025. https://www.nobelprize.org/prizes/physics/1903/marie-curie/article/

6. It wasn't until 1962 when Marguerite Perey, who obtained her PhD-degree at the Radium Institute in Paris, was admitted as a member of the Académie. See e.g. Adloff and Kaufmann (2005).
7. Herman Weyl. 18 April 1935. Funeral Speech. Archiv der sozialen Demokratie, Bonn. https://mathshistory.st-andrews.ac.uk/Extras/Weyl_Noether/
8. Annual Report of the Association of Women Students Leiden 1918/1919. '*Niet het minst te danken aan het 3-tal musici dat met strijkjesmuziek een feestelijk gerucht onderhield, en aan de 10 eerstejaars aan wie het presenteren van thee en koekjes was toevertrouwd*'.
9. In a handwritten postcard, Berta De Haas-Lorentz later wrote of the 'many hours we spent discussing thermodynamics and during which I learned so much from you'. De Haas-Lorentz to Ehrenfest-Afanassjewa. 2 May 1956. File 482. EA-MBL.
10. The Dutch system differed from the Anglo-Saxon one, with students first obtaining their *kandidaats* diploma, usually after two or three years of study, and then after another three years a *doctoraal* diploma, after which some went on to obtain a PhD.
11. Ehrenfest to Einstein. 22 January 1921. CPAE 12, doc. 30.
12. This is evident from documents. Folder 112. EA-MBL.
13. Einstein to Ehrenfest. 13 February 1921. CPAE 12, doc. 48.
14. The event took place on 27 October 1920.
15. Ehrenfest to Einstein. 21 February 1921. CPAE 12, doc. 55.
16. Einstein to Ehrenfest. 1 March 1921. CPAE 12, doc. 71.
17. Ehrenfest to Einstein. 28 February 1921. CPAE 12, doc. 68.
18. Einstein to Ehrenfest. 10 December 1919. CPAE 9, doc. 204.
19. Ehrenfest to Einstein. CPAE 12, doc. 343.
20. As note 19.

Chapter 26

1. Ehrenfest-Afanassjewa to Ehrenfest. 13 September 1922. Folder 437. EA-MBL.
2. The German-Swiss violinist, composer and director Adolf Busch (1891–1952) was best known for his Busch Quartet and its performances of Beethoven, Brahms and Schubert.
3. Ehrenfest to Ehrenfest-Afanassjewa. 4 September 1922. Folder 437. EA-MBL.
4. Ehrenfest-Afanassjewa. 16 December 1918. Second Family Diary. EFA. Translated from Russian by HD.
5. Passport paperwork. Folder 112. EA-MBL.
6. Ehrenfest to Ehrenfest-Afanassjewa. 4 September 1922. Folder 437. EA-MBL
7. Ehrenfest to Einstein. 30 July 1922. CPAE 13, doc. 316.
8. Ehrenfest to Einstein. 29 August 1922. CPAE 13, doc. 333. Paul had not seen his brother Hugo for 23 years.
9. Ehrenfest to Ehrenfest-Afanassjewa. 23 August 1922. Folder 437. EA-MBL. '*Lieberl, liebes, liebes! Komme soeben von Erziehungsheim - war 3 Stunden da - sehr günstigen Eindruck von den Leuten und guten Eindruck von den Einrichtungen die für Wapsili von Wichtigkeit sind*'.
10. As note 7.
11. Ehrenfest to Einstein. 30 July 1922. CPAE 13, doc. 316.
12. Ehrenfest-Afanassjewa to Ehrenfest. 18 September 1922. Folder 437. EA-MBL.
13. Ehrenfest to Ehrenfest-Afanassjewa. 26 September 1922. Folder 437. EA-MBL. '*Bleibe nun ganz ruhig bis 1. November dort um Dich einmal gründlich auszuruhen. Nun spare nicht am Essen!!!!! Und wenn Du erst einmal ausgeruht bist, wäre es gut, dass Du Dich wieder übst herumzuwandern*'.
14. De Vereniging voor Vrouwelijke Studenten Leiden was founded in 1900.
15. As note 13. '*Wozu Du T' nach Berlin begleitest, kann ich nicht capieren. ... Aber warum in aller Welt willst Du nach Berlin mitfahren? Das ist doch nur Sentimentalität. ... Fahre nicht nach Berlin mit Tania'. ... Außerdem ist die Gefahr von Influenza groß. "[Die Gefahr eines Englisch-Türkischen Krieges ist sehr groß ... Falls ich Dir in irgend einem Moment telegraphieren, dass Du kommen sollst oder Du selber siehst, dass es geraten ist,] so fahre nicht via Berlin nach Hause, sondern sofort: Stralsund, Rostock, Hamburg, Bremen, Groningen. Nöthigenfalls mit reduziertem Gepäck*'.
16. As in note 13. '*Gestern war ich auf Professorenkranz. Es ist fürchterlich, dass man dort so systematisch nur über "Koetjes en kalfjes" praat. ... Abscheulichstes Wetter. Es regnet in allen denkbare Arten u. Variationen*'.
17. Ehrenfest to Ehrenfest-Afanassjewa. 25 September 1922. Folder 437. EA-MBL. '*Heute Nacht ist ganz plötzlich Kuenen gestorben*'.
18. Ehrenfest to Ehrenfest-Afanassjewa. 1 October 1922. Folder 437. EA-MBL. '*Wenn Du noch immer zweifeln solltest, dass ich zu sehr recht eine hohe Lebensversicherung abgeschlossen habe, so wird Dir aller Zweifel vergehen, wenn Du sehen wirst in welcher Sorge nun das Haus Kuenen ist*'.

19. Ehrenfest-Afanassjewa to Ehrenfest. 27 September 1922. Folder 437. EA-MBL.
20. Ehrenfest-Afanassjewa to Ehrenfest. 12 October 1922. Folder 437. EA-MBL.
21. Ehrenfest to Ehrenfest-Afanassjewa. 20 October 1922. Folder 437. EA-MBL. '*Versäume in Tübingen ja nicht Paschen kennen zu lernen und sage ihm mit welchem Respect ich über ihn Dir erzählte*'.
22. Later, Tanya would try to organise funding for Yanitsky and two other Russian scientists, biophysicist Boris Rajewsky and physicist-philosopher Semyon Frank, via Einstein and the CIC. CPAE 14, 304.
23. Ehrenfest-Afanassjewa to Ehrenfest. 17 September 1922. Folder 437. EA-MBL.
24. Ehrenfest-Afanassjewa to Ehrenfest. 30 September 1922. Folder 437. EA-MBL.
25. Ehrenfest-Afanassjewa to Ehrenfest. 12 September 1922. Folder 437. EA-MBL.
26. Ehrenfest to Einstein. 22 January 1921. CPAE 12, doc. 30.
27. It resulted in two publications on the application of quantum rules to a simple mechanical system, in relation to Bohr's correspondence principle (Ehrenfest and Breit 1922; Ehrenfest and Breit 1923).
28. *Nature* 54(1398), 347 (1896).
29. See also: https://iifiir.org/en/iir-presentation
30. *Algemeen Handelsblad.* 15 November 1921. '*Leiden als internationaal wetenschapscentrum*'. Delpher online newspaper archive. For further discussion, see Van Besouw and Van Dongen (2013).
31. Delpher online newspaper archive.
32. Einstein to Ehrenfest. 7 April 1920. CPAE 9, doc. 371.
33. Einstein to Ehrenfest. 27 February 1922. CPAE 13, doc. 63
34. Einstein to Curie. 4 July 1922. CPAE 13, doc. 262
35. CPAE 13. Introduction, xl.
36. Einstein to Ehrenfest. 23 May 1922. CPAE 13, doc. 200
37. Einstein to Ehrenfest. On or after 21 August 1922. CPAE 13, doc. 329
38. Ehrenfest to Einstein. 4 February 1922. CPAE 13, doc. 45
39. Einstein to Ehrenfest. 12 February 1922. CPAE 13, doc. 47
40. Ehrenfest to Einstein. 13 February 1922. CPAE 13, doc. 54
41. Ehrenfest to Einstein. 13 July 1922. CPAE 13, doc. 316
42. Einstein to Ehrenfest. On or after 21 August 1922. CPAE 13, doc. 329 (Here '*Indianerleben*' is translated as 'camping experience').
43. Ehrenfest to Ehrenfest-Afanassjewa. 20 October 1922. Folder 437. EA-MBL. '*Sie ist ein sehr, sehr lieber Mensch und bringt ein wohliges Element ins Haus. Auch ist sie selber so froh bei uns zu sein*'.
44. Ehrenfest to Ehrenfest-Afanassjewa. 1 October 1922. Folder 437. EA-MBL.
45. Ehrenfest to Ehrenfest-Afanassjewa. 20 October 1922. Folder 9. EA-MBL. '*ich werde Dir ihm zu lesen geben wenn es gelegentlich nöhtig sein wird*'.
46. As note 45. '*Heute habe ich wieder den ganzen Tag verloren, weil ich die Reparatur des Ofens und In-Betriebstellung der Zentralen-Heizung überwachen musste. Mit Schrecken sehe ich die nächste Woche entgegen. Weißt Du es muss nun endlich wirklich eine solche Anordnung getroffen werden, dass ich nicht fortwährend irgendetwas "muss" das mich von der Erfüllung meiner Amtspflichten und der wissenschaftl. Arbeit wegholt. Dazu gehört auch, dass man endlich auch Dämme gegen das unbegrenzte "helfen-müssen" mit Geld oder Dienstleistungen aufgeworfen werden müssen. Ich will meine Schulden bezahlen, ich will uns kaufen können, was wir haben möchten, ich will nicht fort während Geldsorgen haben (notabene bei der jetzt vorstehende Verminderung des Einkommens von 8–10%) - vielleicht werde ich dann arbeiten können, und dann werde ich für meine Mitmenschen dadurch etwas Spezifisches leisten können. Während ich jetzt durch dieses fortwährende müssen, zu einem inneren Hass gegen alle komme. Wenn ich mich fortwährend quälen muss, und gar nicht mehr zu ruhiger, spielender Arbeit kommen kann, weil ich mich fortwährend vertheidigen muss (etwas was ich seit zwei Jahren immer intensiver fühle), dann kann ich schließlich auch das nicht mehr thun, was ich sonst sicher freudig thun würde: einen Krutkow aufmerksam und mit Liebe entgegensehen. Wenn aber auf meinen Schreibtisch immer mehr Rechnungen und russische "müss-Briefe" aller Art liegen (eben jetzt wieder ein Brief von Hessen) dann kann es gar nicht anders dass ich alle zum Teufel wünsche und mich immer wieder abfrage, ob ich denn wirklich so viel wertloser bin als alle die Briefschreiber, dass meine Entwicklung und meine spezifischen Wirkungsmöglichkeiten für sie aufgeopfert werden müssen. Es ist ja sehr begreiflich dass ?Chochnujka?, Tante, Dr. Babkin, ein junger russ. Jude Weinstein (Brief von Weyl aus Zürich), Kasterin, Boguslawsky, Marussja, die Universität von Simferopol, das Comite von Hessen, Kulischer, Bursian, Besikowitsch, Friedmann, ein Astmaleidender russ. Professor u.s.w. u.s.w. gerade an uns sich*

wenden, aber was das alles zusammen auf mich, der nun einmal nicht geschickt dafür ist für einen abscheulichen Druck ausübt! Ich will, will, will wirklich endlich Ruhe von diesem Hexensabbath haben. Denn wirklich nach einer kurzen Periode von Hoffnung da Wassik unterrichten würdest können und ich hoffen durfte, dass Du nun vielleicht wieder Dich aufrichten würdest können, bin ich nun wieder ganz erdrückt. Ich sehe mehr und mehr: es ist uns beschieden widerstandslos und rettungslos unter diesem russischen Schutt begraben zu werden. Und mehr als einmal dachte ich schon dass das nicht meine Sache ist. Dazu existiere ich nicht'.

Chapter 27

1. The institute is nowadays called the Ioffe Institute.
2. Ehrenfest to Ioffe. 2 June 1928. Folder 152. EA-MBL. 'Denn in seiner [Bohrs] und Einsteins naehe fuehle ich mich immer gegen alle Haesslichkeit der Welt beschuetzt, wie in Hochgebirgsschnee'.
3. Ehrenfest to Ehrenfest-Afanassjewa. 13 August 1923. Folder 438. EA-MBL. 'Es war eine Art Erholung für mich, dass das ein Mensch war, den ich etwas lernen kann, wo ich jetzt fortwährend mit großer Anspannung lernen muss was Bohr, Kramers, Pauli, Sommerfeld, Debye ausdenken'.
4. As note 3. 'Ich plage mich sehr mit der Übertragung des Boltzmannschen microcanon. Beweises auf Quantensysteme. Komme aber absolut nicht vorwärts. Und wenn ich es finde, werde ich's als schief, trivial und desillusionierend empfinden'.
5. Einstein to Ehrenfest. Between 11 and 13 March 1922. CPAE 13, doc. 79
6. *Leidsch Dagblad*. May 1920. LNA online.
7. *Leidsch Dagblad*. 13 November 1923. LNA online. The journalist added that it was quite 'understandable' that Einstein wanted to come to Leiden 'because he needs a quiet environment for his work, which cannot be said of Berlin at the moment, or indeed of Germany as a whole'.
8. *Leidsch Dagblad*. 11 May 1923. LNA online. This was Prince Hendrik who was in Leiden as honorary chairman of the Nautical College.
9. Anecdote Prof C.J. Gorter shared with Leiden colleagues. See also Gorter (1962).
10. *Leidsch Dagblad*. 4 February 1923. LNA online.
11. Ehrenfest to Ehrenfest-Afanassjewa. 23 October 1922. Folder 438. EA-MBL. '[Habe schrecklich gequatscht aber natürlich waren gerade deshalb alle überaus zufrieden.] Es ist das eine schrecklich verderbliche Thätigkeit sowohl für Sprecher als Zuhörer'.
12. In the Netherlands, students first obtained their *kandidaats* diploma and then, if they did well, their *doctoraal* diploma, allowing them, if they wished and were talented, to obtain a doctorate. The *kandidaats* (two years) was a bit shorter than a bachelor's degree (three years) and the *doctoraal* comparable to or a bit more advanced than a master's degree.
13. The letter contained Paul's response to an article on statistical mechanics that Fermi published in the *Physikalische Zeitschrift* in 1923'.
14. See note 12.
15. This education was of a high quality. In 1969, Jan Tinbergen became the first winner of the Nobel Prize in Economics; four years later, in 1973, his younger brother Nikolaas won the Nobel Prize in Physiology (Medicine). Their youngest brother, Luuk, would become a renowned ornithologist.
16. Ehrenfest-Afanassjewa. Undated, but before 20 August 1920. Notebook Afanassjewa. EFA. Translated from Russian by HD.
17. As note 16.
18. As note 16.
19. Ehrenfest to Schumpeter. 3 May 1918. Folder 158. EA-MBl. Also cited in Jolink (2003, 27). Biographical accounts of Tinbergen, e.g. in (Klein 1970; Jolink 2003; Buitenhuis 2016) usually omit Afanassjewa.
20. Some of this language appears in the speech delivered by Hermann Weyl on 18 April 1935 at the funeral of Emmy Noether: https://mathshistory.st-andrews.ac.uk/Extras/Weyl_Noether/
21. 'Ik geloof dat we toen begrepen dat een oorspronkelijke gedachte op het didactisch gebied even belangrijk kan zijn als een wetenschappelijke ontdekking'.
22. Minnaert in the *Mededelingenblad van de Wiskundewerkgroep der WVO*, Jaargang 9 (1961). Archive: Archief Werkgemeenschap voor Vernieuwing van Opvoeding en Onderwijs, IISH, Amsterdam. '[Haar] liefde voor het onderwijs, bevrijd van de neiging tot schoolmeesterij, en bezield door de frisse geest der universitaire wetenschap'.
23. Ehrenfest to Ehrenfest-Afanassjewa. 7 August 1923. Folder 438. EA-MBL. 'Dass Du mit den Kindern spazieren gehst ist ganz unmöglich. Die abstände sind viel zu Groß. Aber vielleicht kannst Du ein Student oder Studentin finden die bereit ist mit ihnen täglich einen weiten Spaziergang zu

machen ... *das würde für die Kinder viel größeren Nützen abwerfen, als wenn Du Dich mit ihnen müde herumschleppst*'.
24. Ehrenfest-Afanassjewa to Ehrenfest. 12 February 1924. Folder 27. EA-MBL. Translated from Russian by HD-BM.
25. Ehrenfest-Afanassjewa to Ehrenfest. 9 March 1924. Folder 27. EA-MBL. Translated from Russian by HD-BM.
26. Ehrenfest-Afanassjewa to Ehrenfest. Undated, but likely early 1912. Folder 27. EA-MBL. Translated from Russian by HD.
27. Leidsch Dagblad. 10 July 1923. LNA online. '*ruime recreatiezalen, twee op elke verdieping, waar de meisjes na iedere les van 45 minuten een kwartier mochten hollen*'.
28. Ehrenfest to Ehrenfest-Afanassjewa. 7 August 1923. Folder 438. EA-MBL. '*Es wäre gut, dass Du mit dem II H.S. [zweitem Hauptsatze] noch ein wenig vorwärtskommst ehe Du nach Russland fährst. Sonst wird diese Arbeit, von der ich mir viel Gutes für Dich verspreche, auf nichts auslaufen. (Mit Doctor-titel hast Du auch ganz andere Chancen in Russland als ohne denselben.)*'.
29. Ehrenfest to Ehrenfest-Afanassjewa. 10 August 1923. Folder 438. EA-MBL. '*Je näher Deine russische Reise kommt, desto mehr Sorge bereitet sie mir. Ich bemerke jetzt, dass ich in Grunde gehofft hatte, dass Du die Erlaubnis doch nicht bekommen würdest. Am meisten fürchte ich, dass Du krank werden wirst (Verkühlung!!!!! + Infections-Wahrscheinlichkeit). Nun ich halte mich nicht für berechtigt, Dich zurück zu hohlen aber wenn Du hier bist müssen wir doch noch einmal sehr ernst alles durchbesprechen. Und verstehe mich recht: er ist wirklich allein die Sorge, ob Du gesund und wohl zurückkehren wirst und absolut nicht Dingen von secundierter Bedeutung die mich nun unsicher machen*'.
30. Ehrenfest-Afanassjewa to Ehrenfest. 13 August 1923. Folder 26. EA-MBL. '*Was soll ich nun mit Dir anfangen? Alles, was Du mir an Zeitungsberichten schickst, erweckt in mir nur mehr den Wunsch irgend, irgendetwas zu thun, um es besser zu machen und sonst nichts*'.
31. As note 30. '*Ob ich hier richtig ausruhen werde, weiß ich nicht - wenn die Kinder irgendwie am Horizont sind, ist es eigentlich unmöglich, u. sicher werde ich am letzten ausruhen, wenn ich ganz allein nach Russland gehe*'.
32. Ehrenfest to Ehrenfest-Afanassjewa. 15 August 1923. Folder 438. EA-MBL. '*Wärest Du hier, so kam ich wohl auch weiter mit der Quanten Interpretation der "Microcanonische" Ableitung des II H.S. so ist sie leider total stecken geblieben*'.
33. Leidsch Dagblad. 17 December 1923. LNA online. '*Prof. P. Ehrenfest is uitgenodigd om aan de universiteit te Pasadena (Californië) lezingen te houden. Ook zal de hoogleraar in enkele andere Amerikaanse steden lezingen houden. Omstreeks Pasen denkt de hoogleraar weer naar Nederland terug te keren*'.

Chapter 28

1. Hugo to Paul Ehrenfest. 10 December 1923. Folder 64. EA-MBL. '*Hoffentlich hast Du gutes Wetter dieses mysteriöse Auftauchen der New York Skyline - dieser mächtigen Gebäude gegen einen blauen Himmel voll genießen zu können*'.
2. Hugo to Paul Ehrenfest. '*Sonntag 4 p.m.*' on stationary with letterhead '*Ellen Ehrenfest*', undated but probably 11 December 1923. Folder 64. EA-MBL. '*Lebhaft empfinde ich im Augenblick meine nunmehr 23 Jahre alten ersten Eindrücke dieser mächtigen Stadt und wie ich entlang Broadway wackelte, immer den einen oder anderen Fuß viel zu kräftig auf das Pflaster setzte*'.
3. Hugo married Sophie Schwab in 1904.
4. As note 1. Georgia O'Keeffe and Alfred Stieglitz were members of this circle. '*Arthur (Schwab) & seine Frau (die wir alle einfach Teddy nennen) bestehen darauf, dass Du bei ihnen wohnst ... in der Mitte einer Gruppe interessanter Menschen (Künstler, Schriftsteller, etc.)*'.
5. Edna Clare (Teddy) Bryner, Arthur Schwab's wife, was a writer and expert in Tibetan Buddhist literature.
6. Ehrenfest to Ehrenfest-Afanassjewa. 17 December 1923. Folder 438. EA-MBL. '*Gestern Sonntag 11h Ankunft Hoboken. Sophys Cousin A. Schwab mich abholend. Mit Boot über Hudson. In seine kleine Wohnung wo ich mit ihm und seiner Frau wohne. ... Eindrücke von gestern und heute sehr, sehr überwältigend. ... Im Augenblick nur dieses: A. Schwab und seine Frau und einige Freunde die ich bei Ihnen fand, gehören zu den feinsten, besten, lebendigsten Menschen, denen ich je in meinem Leben begegnet habe*'.
7. Millikan to Ehrenfest. 22 June 1922. Folder 151. EA-MBL.
8. In 1926, he became a professor at the same university, a position he combined with his work as head of the gynaecology and obstetrics department at the Jewish Hospital in St. Louis.
9. Hugo to Paul Ehrenfest. 25 February 1924. Folder 64. EA-MBL. In English.

10. Hugo to Paul Ehrenfest. 10 December 1924. Folder 64. EA-MBL. *'Jetzt bist Du in Amerika - sei gnädig & zurückhaltend im Deine Urtheile - ich weiß, das Du im Mai besser von uns denken wirst - wir leben in einer anderen Welt'.*
11. Ehrenfest to Ehrenfest-Afanassjewa. 7 January 1924. (Accidentally dated 1923). Folder 438. EA-MBL. *'Gestern Vormittag erst durch Wüste - dann plötzlich endlose Orangen-Plantagen - unendlichen Ebenen mit mannshohen ganz gleichen Orangen-Bäumen in Dreiecks-Netz-Verteilung - Alle überladen mit großen goldgelben Orangen - Die Sonne (nicht Luft!) so heiß, dass trotz Abstellung der Heizung Ventilatoren hart arbeiten müssen, um etwas zu kühlen. Luft schrecklich durchsichtig so, dass große ferne Bergen nahe klein erscheinen'.*
12. As note 11. *'Sonntag frühmorgens -frühling in einem kleinen Dorf irgend in Thüringer-Wald oder in Württemberg - der Herr Doctor, Apotheker, Schullehrer und Post-Director 50–60 Jahre zurück von heute in einem kleinen hellen sonnigen Zimmer vor dem Fenster der Apfelbaum in Blühte oder auch der Kirschenbaum - sie werden gleich wieder ein anderes Quartett von Haydn spielen - jetzt trinken sie ein Glas Apfelwein, Eigenproduct des Schullehrers, und rauchen ein Pfeiflein. Ihre Apfelbäume und Kirschenbäume und Birnbäume geben recht verschiedene Früchte - nicht immer besonderes gut - ein bisschen "wie es Gott gefällt", aber es war "alles beisammen" – Apfel – Haydn – die Bibel – ein Schnäpslein - die Schnupftabak des Schullehrer - vielleicht Newton oder Kant beim Post-director - ein Größe-Misthaufen - Gottergebenheit und ein wenig Typhus'.*
13. As note 11. *'Präpariere dies alles auseinander: Apfelplantage - 1.000.000 Volt Laboratory - Cathedral mit drahtlos-telegrafischer Aussendung der Predigt – Concerthall - - - - - cultiviere das linke vordere Viertel eines Mathematiker-Gehirns - das Herz eines lyrischen Dichters - den Kehlkopf einer Primadonna und die Faust eines Boxers - kultiviere sie "in vitro" (außerhalb des Körpers in passenden Nährflüssigkeiten in Reagenzgläsern) und Du erhältst gigantische Resultate von denen die unscientific oldfashioned Method als alles beisammen nicht träumen konnte - - - - - die California Orangen sind extrem groß, sehr gut und absolut uniform - - - - - nichts kann dagegen auf seinen Beinen stehen bleiben'.*
14. Hugo to Paul Ehrenfest. 2 February 1924. Folder 64. EA-MBL. In English. Hugo thought the letters were 'literary gems'.
15. Ehrenfest to Ehrenfest-Afanassjewa. 9 January 1924. Folder 438. EA-MBL.
16. As note 15. *'Sie verstehen, scheint es, nicht wie einen sensitiven Fremden das Gefühl des heimisch sein zu suggerieren. Wie ganz anders war es mit Arthur und Teddy Schwab in New York'.*
17. Hugo to Paul Ehrenfest. 9 April 1924. Folder 64. EA-MBL. In English.
18. Ehrenfest to Tinbergen (et al.). 10 January 1924. Folder 02, Box 2, Samuel A. Goudsmit papers. Niels Bohr Library & Archives, AIP. *'... alle College-girls gepudert und ein größer Prozent-Satz (~50%) geschminckt + gefärbte Lippen und Augenbrauen ... übrigens hier im Technologie Inst. in Pasadena gibt es keine weiblichen Studenten ... und die Jungen laufen den ganzen Tag in ihren Football-Kleidung herum'.*
19. Various postcards and letters to the Netherlands. Folder 439–440. EA-MBL.
20. Ehrenfest to Ehrenfest-Afanassjewa. Various letters. Folder 440. EA-MBL.
21. Hugo to Paul Ehrenfest. 2 April 1924. Folder 64. EA-MBL. English.
22. Hugo to Paul Ehrenfest. 1 May 1924. Folder 64. EA-MBL. *'Lasse Dir nicht diesen letzten Tagen in Amerika durch unsinnige & unberechtigten Grübeleien verderben. Deine momentanen Geldsorgen sind Halluzinationen'.*
23. Hugo to Paul Ehrenfest. 4 May 1924. Folder 64. EA-MBL. In English.
24. Hugo to Paul Ehrenfest. 2 April. Folder 64. EA-MBL. In English.
25. As note 24.
26. e.g. Ehrenfest to Ehrenfest-Afanassjewa. 7 January 1924. Folder 439. EA-MBL.
27. Hugo to Paul Ehrenfest. 8 May 1924. Folder 8. EA-MBL. In English.
28. For further discussion, see e.g. de Moor and Treffers (1974/1975); Moor (1993); Kemme and de Moor (2012).
29. Personal communication Fransje Teulings and Carola Kloos. 22 March 2019.

Chapter 29

1. Personal communication T. van Bommel.
2. As note 1 and correspondence and photo albums from those years. EFA.
3. Franck (1882–1964) and Gustav Hertz (1887–1975) shared the 1925 Nobel Prize in Physics for discovering the rules that describe the collision of electrons and atoms. Franck was professor of experimental physics in Göttingen from 1920, moved to Baltimore in 1933 and married physicist

Hertha Sponer there in 1946. Hertz, who was also Jewish, survived World War II, thanks to his isotope research at Siemens, and went on to have a career in East Germany.
4. Patrick Blackett (1897–1974) worked on cosmic rays, on cloud chambers to detect them, and was the first to show that radioactivity can cause one chemical element to change into another.
5. Pyotr Kapitsa (1894–1984) graduated from the Petrograd Polytechnic Institute in 1918 and then went to Cambridge University. There he worked for over ten years with Ernest Rutherford at the Cavendish Laboratory and became the first director (1930–1934) of the Mond Laboratory. After a visit to his parents in 1934, the USSR authorities refused to allow him to return to Britain. In the USSR he then turned to low temperature physics—the field Kamerlingh Onnes had also worked in—and in 1937 he discovered superfluidity, an invention which was (one of) the reasons why he was awarded the Nobel Prize in Physics in 1978.
6. See note 3.
7. Robert Millikan (1868–1953) won the 1923 Nobel Prize in Physics for his measurements of elementary charge in 1909–1910—in his famous oil drop experiments—and for his work on the photoelectric effect in 1914.
8. Julia Ivanova Faussek (1863–1942 or 1943) graduated from the Bestuzhev Courses in 1884, studied Montessori education in Italy and opened the first Montessori kindergarten in St Petersburg in 1913.
9. Correspondence Ehrenfest-Ehrenfest-Afanassjewa. Folder 27. EA-MBL.
10. Ehrenfest-Afanassjewa to Ehrenfest. 3 March 1924. Folder 27. EA-MBL. Translated from Russian by HD.
11. Ehrenfest to Ehrenfest-Afanassjewa. 5 January 1924. Folder 41. EA-MBL. '*Sehr wichtig: gerade am letzten Tag meines Aufenthaltes in St. Louis erfuhr ich durch Gespräch mit Hugo und zwei mit ihm befreundeten Kinderärzten..., dass ich in Leiden eine Bemerk. Von Hugo total missverstanden hatte: Wapsiks Defect hat absolut nichts mit einer "erblichen Belastung" meiner oder Deinerseits zu thun und umgekehrt folgt aus Wapsiks Defect nicht das geringste bezüglich der Nachkommenschaft unserer anderen Kinder. Auch die Hypothese, dass dieser (Mongoloid)-Defect auf Erschöpfungs- oder Ermüdungs-zustände von Vater oder Mutter zurückführen ist, ist durch nichts gestützt*'.
12. Petr Lazarev studied medicine, mathematics and physics in Moscow, graduating in physics in 1912 under Pyotr Lebedev. In 1911, Lazarev joined Lebedev in protesting against the university policies of Minister Kasso (Chapter 14 of this book).
13. These students included A.N. Arsenyeva, B.M. Gochberg, V. Tomashevsky, N.S. Ustaya, B. Ya. Pines, N.A. Brilliantov, A.I. Shalnikov and V.S. Gorsky.
14. Ehrenfest to De Ridder. 19 November 1924. T. van Aardenne-Ehrenfest Archive. NHA.
15. Ehrenfest to Ehrenfest-Afanassjewa. 15 August 1923. Folder 438. EA-MBL. '*Nah, wie steht's mit Deiner Thermodynamik?!*'
16. Ehrenfest-Afanassjewa to Ehrenfest. 29 January 1924. Folder 27. EA-MBL. Translated from Russian by HD.
17. As note 16.
18. Her correspondence with Carathéodory shows that her work made him realise that his definition of 'simple thermodynamic systems' was incomplete; he also urged her to be more radical in drawing conclusions about irreversible systems: Carathéodory to Ehrenfest-Afanassjewa, 2 March 1925. File 482. EA-MBL.
19. She sent a short erratum to the journal in September 1925 with a more precise formulation of the entropy axiom. This wasn't included but was published separately in December 1925 (Ehrenfest-Afanassjewa 1925b). For a full discussion, see Uffink (2000) and Uffink and Valente (2020).
20. '*Die Entropie der Welt strebt einem Maximum*', in Clausius (1865).
21. This follows from a letter in which Jan Burgers describes a lecture by Ramsey that he attended at Harvard University: Burgers to Ehrenfest-Afanassjewa. 11 May 1956. File 482. EA-MBL.
22. Carathéodory to Ehrenfest-Afanassjewa. 23 July 1926. Folder 482. EA-MBL. '*Mit Großer Freude sehe ich im Programm der British Association, dass Ihr Mann auch dort sein wird. Hoffentlich kommen Sie auch dorthin. Sie werden wohl sehr böse auf mich sein, dass ich so lange nichts habe von mir hören lassen und auch Ihren letzten Brief nicht beantwortet habe*'.
23. As note 22. '*weil ich in den letzten Monaten wieder reine Mathematik getrieben habe. Wenn das kleine Buch fertig sein wird, würde ich Sie bitten, wenn Sie wollen die Sache zu lesen*'.
24. *Lehrbuch der Thermodynamik*.
25. Kohnstamm to Afanassjewa. 6 September 1926. Folder 482. EA-MBL. '*Waarde mevrouw Ehrenfest, [hartelijk dank voor de toezending van uw manuscript dat ik] met veel belangstelling en bijna doorlopend met instemming [las]*'.

26. Kohnstamm to Hernn Dr A. Meiner in Firma Ambr. Barth. Enclosed in the letter of note 24. Folder 482. EA-MBL. *'Inzwischen hat sich diese Tage aus Besprechungen mit Frau Ehrenfest-Afanassjewa [über die als Anhang geplante Axiomatik der Thermodynamik,] mir gezeigt dass wesentliche Punkte des Manuskriptes schärfer gefasst werden könnten [, und dass man sogar die Phasenregel, und was damit zusammen hängt, vermutlich vor dem ersten Hauptsatz wird abhandeln können]. Um unnötige Extra-korrektur zu vermeiden, möchte ich Sie deshalb bitten, sofort die weitere Drucksetzung des Manuskriptes zu sistieren, und mir zu berichten wie weit man schon mit dem Druck gelangt ist. Danach will ich entscheiden, ob es besser ist, bei dem ursprünglichen Plan zu bleiben, also das Manuskript unverändert zu lassen und die Axiomatik als Anhang zu behandeln, oder das Manuskript umzuändern und die Axiomatik gleich in dem systematischen Aufbau aufzunehmen'.*
27. 'Wichtige Ergänzungen sind ferner T. Ehrenfest-Afanassjewa zu verdanken'.
28. 'Van Mensch en Ding. Interview bij Zeemans afscheid als hoogleraar'. *Algemeen Handelsblad*, 22 September 1923. Delpher newspaper archive online. In Amsterdam, Zeeman had seen three women graduate in physics, compared to fifteen men. *'Een vrouw die een hoogeschool heeft bezocht, zal later inde maatschappij, ook al komt ze, als duizend andere vrouwen, aan het hoofd van een huishouden te staan, een wetenschappelijke sfeer kunnen scheppen.... Zij zal de opvoeding in andere banen leiden en daardoor de wetenschap dienen, ook al werkte ze niet direct mede aan haar bevordering'.*

Chapter 30

1. Ehrenfest to Ehrenfest-Afanassjewa. 4 December 1923. Folder 63. EA-MBL. *'Meine Reise wird Früchte tragen—vielleicht nicht für mich aber für die Jungen in Holland, Deutschland u. Russland'.*
2. As note 1. *'Ich begann erst diplomatisch—aber nach zwei Minuten schimpfte ich alles vergessend über das amerikanische Hochdruck-Stipendium-System, und gerade mein unvorsichtiges Herumschimpfen gewann mir sein Vertrauen. Ich sagte: "man muss doch auch Wälder säen und nicht nur Fabriken bauen. Gebe Sie 50% ihrer Gelder an Leute die träumen".*
3. Hugo to Paul Ehrenfest. 2 February 1924. Folder 64. EA-MBL. In English.
4. This becomes clear from his correspondence in which he mentions all the *plaudern* (chatting).
5. *'De hoofdpersoon te Leiden is echter voor mij nog steeds Prof. Ehrenfest, al is zijn handelwijze tegenover mij, en zijn standpunt tegenover het activisme, onvergefelijk. Maar toch is hij de ideaal-professor en bijna de* **ideaal-mensch***'.*
6. With perhaps one exception: Jan Tresling, whose name hardly ever appears in the record. There even seems to be no trace of him at all after he obtained his PhD degree in 1919 (with the dissertation *'Deformaties en Trillingen in het vaste lichaam bij afwijkingen van de wet van Hooke, ook in verband met de toestandsvergelijking'*).
7. Ehrenfest to Tinbergen et al. 10 January 1924. Folder 2, Box 2, Samuel A. Goudsmit papers. Niels Bohr Library & Archives, AIP. *'Lieber Tinbergen und Ihr alle lieben Jungen drüber so fern fern! Ich denke sicher 5 Mal so oft an jeden einzelnen von Euch als Ihr alle zusammen an mich!'*
8. *'Woensdagavond, bijvoorbeeld, ging ik weer op de trein terug naar Leiden en dan had ik altijd zo'n gevoel dat ik iets moest omschakelen, want de grapjes die je in Amsterdam hoorde en kon vertellen, die kon je in Leiden niet gebruiken. Dat ging niet. Dat was niet netjes genoeg. In Amsterdam was het heel gezellig'.*
9. *'Daar in Italië..., daar kennen ze alleen maar de klassieke fysica'.*
10. *'Je moet maar met hem samenwerken, een beetje, dan kan hij iets van die nieuwe atoomstructuur en al dat spectraalgedoe leren'.*
11. *'En toen was er een klein stukje in "Naturwissenschaffen" en een heel lang artikel in "Archives Néerlandaises des Sciences exactes et naturelles", dat in Holland in het frans gepubliceerd werd om zeker te zijn dat niemand het zou lezen'.*
12. Goudsmit (1971) adds that he doesn't want to be credited in retrospect, since Millikan had new experimental data that proved the formula correct, whereas he himself had only made an educated guess.
13. Ehrenfest to Goudsmit. 14 April 1924. Folder 2, Box 2, Samuel A. Goudsmit papers. Niels Bohr Library & Archives, AIP. *'Sie sind ein Dummerl. Verstehen Sie nicht, dass Ihre Erlebnisse gerade das normale Erlebnis aller gut vorgebildeten Anfänger ist? ... Aber begreifen Sie nicht dass es eine sehr schöne Sache ist, dass was Sie finden zusammen anfällt mit dem was die best geschulten reifen Spezialisten eben in gleichen Zeit finden?'*
14. In a lecture in the Netherlands, translated into English by J.D. Van der Waals, as well as in a related paper, Goudsmit (1971) credits De Laer Kronig, the American physicists Earle Kennard and Harrold Urey, and Compton with having proposed similar ideas earlier, but he also reminds

the audience that Uhlenbeck and Goudsmit could work with new interpretations of the hydrogen spectrum and thus give a far stronger and more substantial underpinning to the idea: https://www.lorentz.leidenuniv.nl/history/spin/goudsmit.html
15. Ehrenfest to Lorentz. 16 October 1925. 'Maandag wenschen Uhlenbeck, Goudsm [sic] en ik U oordeel en raad te hooren in veband [sic] met een zeer geestig idee van Uhlenbeck over spectra dat zeer velijdelijk [sic] maar misschen [sic] fout is'.
16. 'Jullie hebben nog geen reputatie opgebouwd, dus jullie hebben niets te verliezen'.
17. Einstein to Ehrenfest-Afanassjewa. 4 February 1925. Folder 418. EA-MBL. 'Am Samstag werde ich wohl bei Euch eintreffen mit dem Protz-Rüstzeug für das Fest. Alles ist mir recht, wenn ich nur keine Rede halten muss'.
18. Note Ehrenfest on a letter from Bohr to Ehrenfest. 14 October 1925. Folder 137. EA-MBL. 'Der "Bohr-Kramers-Slater-Company" werden hartnäckig die falschen Wechsel unter die Nase gehalten mit denen sie versuchte 'etwas dafür zu kaufen'.
19. After a disastrous flood in 1916, in which about twenty people drowned in the northern Netherlands, and also to create extra agricultural land, Minister Lely had ordered a study to close off the Zuiderzee. The Afsluitdijk was then built for this purpose, with the help of Lorentz's calculations.

Chapter 31

1. Draft of a letter to the highly esteemed Albert Petrovitsch, undated. Ehrenfest-Afanassjewa notebook. EFA. Translated from Russian by HD-BM. It is almost certainly Pinkevich (1883–1937), who in the Soviet Union was, among other things, Rector of the Third Petrograd Teachers' Academy (1918–1920) and of the Second Moscow State University (1926–1930) and who was executed on 25 December 1937 during the Great Terror (and rehabilitated in 1956). The Second University had emerged from the Higher Courses for Women.
2. As note 1.
3. As note 1.
4. At the instigation of Otto Schmidt, whose beard was as impressive as his status and who, besides being an important advisor at the People's Commissariat for Education, was also a mathematician and a passionate polar explorer: Schmidt. 4 December 1925. Folder 473. EA-MBL. Translated from Russian by HD-BM.
5. Copy of appointment decision. Folder 473. EA-MBL. Translated from Russian by HD-BM.
6. Ehrenfest to Bohr. 12 January 1926. Folder 137. EA-MBL. 'Meine Frau ist via Jena Göttingen Berlin weg nach Moskau [um] dort 3-4 Monate zu arbeiten'.
7. Leidsch Dagblad, 17 December 1924. LNA online. A kandidaats and doctoraal roughly correspond to a bachelor's and master's degree.
8. Personal communication T. van Bommel; Various correspondence. Folders 29, 30. EA-MBL.
9. Recurring theme in conversations among Leiden professors and/or their offspring.
10. The mysterious logic was brought forward by grandson Carel van Aardenne, conversation, May 2017.
11. Ehrenfest-Afanassjewa to M.J. Klein. 1957. MKEAD.
12. Just a year after the Bolsheviks took power, the government enacted a new 'Code on Marriage, the Family and Guardianship'. This code notably enshrined marriage as a civil, not religious act. It also mandated the equality of men and women before the law and allowed for divorce at the request of either spouse. Two years later, in 1920, the Soviet Union legalised abortion, giving women unprecedented control over their own reproductive systems.
13. Notes on a new economic order. Notebook Ehrenfest-Afanassjewa. Undated but before 18 August 1920. EFA. Translated from Russian by HD-BM.
14. Brief notes for a lecture on such an economic order. Notebook Ehrenfest-Afanassjewa. 28 October 1924. EFA. Translated from Russian by HD-BM.
15. Extract from the house register, Volchonkastreet 16. 16 February 1926. Folder 473. EA-MBL. Translated from Russian by HD-BM.
16. Mandelstam and Ehrenfest had known each other since 1912, when Mandelstam worked at the University of Strasbourg and Ehrenfest had written to him asking if there were any vacancies there: Ehrenfest to Mandelstam. 27 January 1912. Folder 150. EA-MBL.
17. It appears from a tax declaration of the Communist Department of the Central Executive Committee of the USSR dated 17 May 1927 that T.A. Ehrenfest received a fee of 400 roubles for conducting this seminar from 1 October 1926 to 1 June 1927. An extract from the list of resolutions of the Section of Natural and Exact Sciences of the Communist Academy shows that she received 50 roubles a month for conducting a seminar on thermodynamics in the academic year 1926/1927. Folder 473. EA-MBL. Translated from Russian by HD-BM.

18. Ehrenfest to Bohr. 14 December 1926. Folder 137. EA-MBL. *'Meine Frau arbeitet mit sehr viel Erfolg und Genuss in Moskau'.*
19. Krutkow wrote this in a letter dated 29 April 1923.
20. Fokker to Wander and Berta de Haas. 10 May 1923. Fokker Archive. MBL. *'Toen zei Ehrenfest: "Mijnheer Lorentz, ik wensch u een goede reis", hij greep zijn hand, zoende die lang, en maakte snikkend dat hij door de zijdeur wegkwam'.*
21. CPAE 15, doc. 420. Sent as *'Manuscript f[or] J[ournal of] Wretched Physics Does the Bose-Einstein statistics for ideal gases lead to a degenerative condensation? G. E. Uhlenbeck and P. Ehrenfest. (Leyden)'.*
22. Hugo to Paul Ehrenfest. 25 February 1924. Folder 64. EA-MBL
23. Paul to his brothers Ehrenfest. 28 August 1928. Folder 2. EA-MBL. *'So fuehle ich mich ueberhaupt stets wie ein asthmatisches Dackel, dat einer Tram nachlaeuft. Und ich fuehlte "ich kann nicht mehr" und war sehr verzweifelt'.*
24. In his book, *The Copenhagen Network: The Birth of Quantum Mechanics from a Postdoctoral Perspective* (2020), Alexei Kojevnikov notes that: 'Over 80 authors took part in that brainstorming effort: The majority of them were under 30 years of age and they authored almost 70% of all publications. Some were still working on their dissertations, but more commonly, they were recent PhDs, having obtained their degrees after 1920 and would have been considered postdoctoral students by today's standards' (Kojevnikov 2020, 3).
25. The number of publications doubled every two months in these years, see Kragh (2002, 168).
26. Alexei Kojevnikov describes for example how it was also Paul who in 1924 understood the novel Bose Einstein as describing indistinguishable particles, as he understood the discrepancy between the statistics of independent quanta and that of indistinguishable quanta better than either Einstein or Bose (Kojevnikov 2020, 63).
27. These discussions eventually led to two papers by Ehrenfest and Uhlenbeck, one particularly focused on Klein's work in five dimensions: Ehrenfest and Uhlenbeck (1926a; 1926b).
28. Although articles at the time argued that the two theories were equivalent, the formal equivalence of the two methods wasn't proven until 1932, see e.g. Muller (1997a; 1997b; 1999).
29. The literal quote comes from a letter from Einstein to Born in December 1926: 'the theory produces a good deal but hardly brings us closer to the secret of the Old One. I am at all events convinced that *He* does not play dice'.
30. Such uncertainty was, incidentally, in line with the 'uncertainty relations' that Heisenberg had proposed a year earlier.
31. Ehrenfest to his students. October 1927. See Bohr (1972, 415–418) and for further context also Mehra and Rechenberg (1982–2000, Ch. 2).
32. Earlier in 1927, he had also published a paper on the Pauli principle, followed by an erratum, and the English editions of both papers in *Nature*. Ehrenfest (1927a, 1927b, 1927c, 1927d).
33. *'Nergens heeft de overgang van leven naar dood zo ongemerkt plaats als in Leiden'.*
34. Letter of appointment. 1 September 1927. Folder 473. EA-MBL. Translated from Russian by HD-BM.
35. Ehrenfest-Afanassjewa to Ehrenfest. 3 February 1928. Folder 29. EA-MBL. Translated from Russian by HD-BM.

Chapter 32

1. Lorentz-de Kaiser to Ehrenfest. 5 February 1928. Folder 150. EA-MBL.
2. A handwritten version is in the Leiden archive: Folder 150. EA-MBL. *'Een werkplaats met groote ramen, doorstraald door de opwekkende ochtendzon. ... Het liefst het eenvoudigste werktuig ... Maar ook zeldzaam fijne precisie-werktuigen ... En omdat ons de meester met zoo bijzonder veel liefde laat zien, welk van de werktuigen hij juist van Christiaan Huygens heeft en welk van Fresnel en welk van andere meesters, zoo moeten wij wel gissen, dat hij zoo menig ander van zijn vernuftige werktuigen zelf zal hebben uitgevonden en vervaardigd ... Zijn scherp critische geest, die steeds naar de uiterste eerlijkheid streeft, scheidt klaar en helder het telkens voltooide van het nog onvoltooide en wijst met nadruk op alles, wat nog raadselachtig is'.*
3. *Leidsch Dagblad.* 9 February 1928. LNA online.
4. The Ehrenfest Archive contains many such letters: Folder 150. EA-MBL.
5. *Leidsch Dagblad.* 15 February 1928. LNA online.
6. Ehrenfest to Bohr. 14 December 1926. Folder 136. EA-MBL. *'Es wird Euch vielleicht freuen zu erfahren, dass Tanitschka die Liquidation Ihrer Affaire ganz besonders gut überstanden hat. Zum Beispiel werde ich augenblicklich beim Briefschreiben sehr durch Ihr fröhliches Lachen und Schreien gestört!'*

7. Ehrenfest to Ioffe. 13 April 1928. Folder 152. EA-MBL. 'Wir sind nun allein mit Baba Katja ... und dem Radio'.
8. As note 7. 'Diese enorme Conzentration floesst mir eine tiefe Angst ein. Die Französische Revolution hat alles nach Paris conzentriert und vergleiche jetzt die Situation dort mit der in Deutschland'.
9. As note 7. 'Eine der toedliche Gifte der Conzentration ist diese, dass eine Handvoll alt gewordenen Leute absolut alles beherrschen, im guenstigsten Fall Maenner, die in ihrer Jugend ganz hervorragendes geleistet haben, oft auch nicht diese. Das Leben der Jungen wird eine Hoelle. Du sollst einmal Madame Curie oder Langevin darueber hoeren. [Junge Menschen] müssen die Gelegenheit haben an halbwegs gut versorgten dezentralisierten Plaetzen sich zu eigenen Thaetigkeit zu entfallen'.
10. Ehrenfest to Ioffe. 2 June 1928. Folder 152. EA-MBL. 'Wenn wir zusammenkommen muessen wir doch wirklich sehr sorgfaeltig beraten, wie ich doch den wahrscheinlich wirklich kleinen Rest meiner Lebenskraft noch irgendwie nuetzlich UND FUER MICH ERTRAEGLICH verwerten koennte. Alles reduziert sich fuer mich auf einen stets kleinerer Kreis: ganz primaer fuehle ich, dass fuer mich das liebste waere schmerzlos, ruhig sterben zu koennen'.
11. As note 10. 'Zweitens fuehle ich, dass die Kinder und ganz besonders Wassik noetig haben, dass ich als Quelle von Einkommen noch einige Zeit fortwurstle. Drittens glaube ich ... dass ich noch ... einige organisatorische nuetzliche thun koennte, ich meine Unterrichtsorganisation. ... Viertens glaube ich, dass ich noch rein wissenschaftlich nuetzlich sein koennte als "Gesellschafter" von wissenschaftlichen Freunden'.
12. As note 10. 'So Ioffe, jetzt wirst Du wieder genug von meinem Gemiau haben!'
13. In 1923, the Conscientious Objection Act came into force, under which men who objected to military service (of whatever nature) could perform alternative military service. 'From May 1926 to June 1928, in accordance with the law of 13 July 1932, Sn. No. 357, I was in compulsory state service, first for 15 months at the prison in Rotterdam ... as an assistant scribe, then for 10 months at the Central Bureau of Statistics in The Hague', Tinbergen wrote in the preface to his dissertation.
14. Rutgers to M.J. Klein. 3 May 1957. MKEAD.
15. Ehrenfest to Bohr. 24 January 1930. Folder 137. EA-MBL. 'Ich hasse die hochgelehrten unlesbaren Dissertationen die "eine Kroenung" der Arbeit des betreffende jungen Mensch darstellen muessen. Ich habe gerne, dass seine Dissertation LUSTIGE, ermüthigende (!!!!) Lecture fuer all die juengeren Leute liefert. Dass Sie zu den Originalabhandlungen anderer Autoren hinfuehrt ohne sie ERSETZEN zu wollen'.
16. Breit had moved to the USA. Leiden was his springboard from the USSR to the West.
17. Also in the Leiden archive: Ehrenfest to Pauli. 26 November 1928. Folder 152. EA-MBL. 'Der arme Teufel sitzt nämlich seit etwa anderthalb Monaten bei uns in Leiden i[m] Regen, Regen unter grauem Himmel, und was arger ist, unter dem Druck meiner Schulmeisterlichkeit. Er hat stets sehr witzige Ideen und gibt auch unseren Diskussionen in unserem kleine[n] Kreis fast immer eine besonders interessante Wendung. Aber dann beginnt natürlich die große Misere, daB ich, wie Sie wissen, nichts kapieren kann, was sich nicht "veranschaulichen" lässt'.
18. Ehrenfest to Föhringer. 4 October 1929. Folder 142. EA-MBL. (Anna Föhringer was their old friend from their St Petersburg years, now living in Paris.)
19. Ehrenfest-Afanassjewa to Ehrenfest. 27 December 1929. Folder 152. EA-MBL. Born had suggested she would investigate the Ramsauer effect.
20. Ehrenfest to Ioffe. 11 January 1929. Folder 146. EA-MBL. 'Wie ein gepruegelter Hund'.
21. e.g. Ehrenfest to Ioffe. April 1929. Folder 152. EA-MBL. Postcard anecdote: personal communication T. van Bommel.
22. Rutgers to Ehrenfest. 4, 10, 20 December 1929. Folder 154. EA-MBL.
23. Ehrenfest to Föhringer. 4 October 1929. Folder 142. EA-MBL. 'Besonders im Lesen und der Orthographie. Das ist so gespenstartig schlecht, dass er nun selber begreift, dass das liquidiert werden muss. Aber in einigen Dingen ist er sehr gut. Vor allem was Physik betrifft. ... Auch ist er sehr geschickt mit seinen Fingern. Arbeitet sehr gut an der Drehbank und sollte nun mit Glasblasen beginnen'.
24. See e.g. Casimir to Ehrenfest-Afanassjewa. 9 April 1931. Folder 32. EA-MBL. 'Wil u eens uw gedachten laten gaan over Paul [jr]? Moet hij eventueel nog naar een vierde of vijfde klas HBS?'
25. Ehrenfest-Afanassjewa (in Simferopol) to 'E'. Ehrenfest (baba Katya) in Leiden. 24 March 1930. Translated from Russian by HD-BM.
26. Pawlik to Paul Ehrenfest. 12 July 1930. Folder 6. EA-MBL. 'OM HET HELE HUIS VAN BUITEN TE VERVEN!!! ... VERDER OM DE EETKAMER HELEMAAL TE VERVEN (deuren, muuren, ramen en plafond) verder de wistibuule te WITTE EN HET HEK TE VERVEN!'
27. Pawlik to Paul Ehrenfest. 15 July 1930. Folder 6. EA-MBL. 'Pan! Wat zou je brommen!!! ... we eten tegenwoordig in JOU STUDEERKAMER!!!!!! DAT KOMT OMDAT DE EETKAMER GEVERVD WORDT!! ... een veel leukere plek'.

28. Tanya to Paul Ehrenfest. 8 April 1931. Folder 32. EA-MBL. Translated from Russian by HD-BM.
29. As note 28.
30. Paul to Tanya Ehrenfest. 24 July 1930. Folder 3. EA-MBL.
31. Ehrenfest to Goudsmit, Dieke and Uhlenbeck. 5 May 1930. Folder 30, Box 2. Samuel A. Goudsmit papers, Niels Bohr Library & Archives, AIP. '*Jungens ich weiss absolut nichts nichts von der neueren Physik. ... Es beginnt eine Art Verfolgungswahn für mich zu werden*'.
32. Pawlik to Paul Ehrenfest. 4 September 1930. Folder 6. EA-MBL. '*Pans. Stuur geld! ... Ik vind het erg akelig*'/ 'Dad *Send money! ... I find it very awful*', he wrote.
33. Ehrenfest to Margrethe Bohr. March 1933. Folder 137. EA-MBL.
34. Sonya (in Auteuil) to 'E' (Katya) and P. Ehrenfest. 27 March 1931. EFA. Translated from Russian by HD. (Letters were often forwarded, in this case from Leiden to Paul in the States).
35. Personal communication Herman Burgers (son of Paul's student Jan Burgers), conversation in The Hague, 16 April 2015.
36. Galinka to Tanyitchka Ehrenfest. 17 March 1931. Folder 496. EA-MBL. '*Vaasje ziet er anders erg goed uit en reuze jong! En hij kan al aardig auto "driven" zelfs op de drukste punten in Los Angeles—en dat zegt wat!!*'
37. Ehrenfest to the Atwells. 16 May 1931. Folder 4. EA-MBL. In English.
38. Einstein to Ehrenfest. 1 February 1931. Folder 141. EA-MBL.
39. Galinka to 'E' Ehrenfest. 3 January 1931 [wrongly dated 1930]. EFA. And within this letter a note to Tanya. Translated from Russian HD-BM.
40. Rutgers to M.J. Klein. 3 May 1957. MKEAD. '*Von der Existenz der Atome bin ich ebenso fest überzeugt wie von der Existenz der Sonne; aber ob die Sonne besteht, das ist ja ein berühmtes philosophisches Problem*'.
41. As note 40. '*Vor dem Nicht-sein habe ich keine Angst*'. Quoting Einstein: '*Leben ist interessant—Tod sein ist bequem*'.
42. As note 40. '*Ich hüpfe kofferlos herum, ohne meine Eleganz, meine Wissenschaft und meine Seele*'.
43. Hugo to Paul Ehrenfest. 15 January 1931. Folder 65. EA-MBL.
44. The meeting with Flexner was arranged by P.A.T. Levene from St Petersburg who had emigrated to the US. A dinner at Levene's place, sometime earlier however, didn't go very well. Correspondence in folder 196. EA-MBL.
45. Rutgers to M.J. Klein. 3 May 1957. MKEAD. '*Der Uhlenbeck ist ein eleganter Mensch ... Du bist ein einfacher Mensch*'. '*Die Gelegenheit einen Fehler zu machen, lässt Du Dir nie entnehmen*'. '*Du musst nicht vergessen, dass die Physik eigentlich zu schwierig für Dich ist*'.
46. Ehrenfest-Afanassjewa to Ehrenfest. 17 November 1931. EFA. Translated from Russian by HD.
47. Ehrenfest-Afanassjewa to 'E'. Ehrenfest (baba Katya). 29 November 1931. EFA. Translated from Russian by HD.
48. Ehrenfest-Afanassjewa to all the Ehrenfests in Leiden. 9 December 1931. EFA. Translated from Russian by HD.
49. Ehrenfest described the plan for this celebration in several letters, e.g. to Föhringer, 4 October 1929. Folder 142. EA-MBL.
50. Ehrenfest to Ehrenfest-Afanassjewa. Undated. Folded into the letter of n. 46. '*Dass wir miteinander unsere drei älteren Kinder in die Welt setzten und was Du "in mir entwickelt hast" und wohl auch einiges was Du von mir und meinen Freunden rein wissenschaftlich gelernt hast—das dürfte doch wohl unsere Ehe rechtfertigen*'.
51. As note 50. '*Die Sehnsucht nach dem Sehen nach dem Berühren schöner junger Frauen ist so natürlich in uns gelegt. ... Ich denke gerade: Wenn wir einander morgen in Goettingen begegnen würden. Du so wie Du damals wärst ich so, wie ich jetzt bin. Ich würde Dir sofort hastig nach zulaufen beginnen. Du würdest Dich kopfschüttelnd abwenden*'.
52. Ehrenfest-Afanassjewa to Ehrenfest. 10 April 1929. Folder 30. EA-MBL. '*Man muss wirklich abwarten bis der Sozialismus sich völlig realisiert—dann wird man sehen können, was das Sexuelle Problem an und für sich ist. Vieleicht werden dan manch Ehen gherade stabiler. Aber vielleicht werden Kinder, Mann und Frau in drei verschiedene Häuser wohnen? Alle Familien so oder nur manche Typen?*'
53. As note 52. '*Ich verstehe nichts was Du da immer herum brummst. Ich bin mit Dir sehr zufrieden*'.
54. Hugo to Paul Ehrenfest. 15 January 1931. Folder 65. EA-MBL. In English.
55. As note 54.
56. Rutgers to Ehrenfest. 5 March 1931. Folder 154. EA-MBL. '*Nou hebben we tenminste weer eens een behoorlijk coll[oquium] gehad, en U staat daar met een afgezakt gezicht en laat U inschüchtern. ... als U zich die weg laat opdringen, dan verpest U het colloquium*'.
57. Ehrenfest-Afanassjewa (Leiden) to Ehrenfest (U.S.). 11 April 1931. Folder 32. EA-MBL. Translated from Russian by HD-BM.

Chapter 33

1. Ehrenfest-Afanassjewa to Ehrenfest. 24 January 1932. Folder 33. EA-MBL. '*Ich sitze bei der Lampe, die ich von Franny übernommen habe ... Heute - wegen ja größer Dichtheit war es unmöglich für eine Mehrzahl Menschen vor der betreffenden Haltestelle sich durch den Wagen von der Eingangs- bis zur Ausgangstür durchzuarbeiten - und, wie einige von ihnen angefangen hatten herauszukommen, liess der Kondukteur den Wagen weiter fahren, trotz aller Proteste, die Dichtheit somit weiter erhaltend. Und die Haltestellen sind dieses Jahr aus irgend welchen höheren Erwägungen sehr reduziert, sodass man recht weit zurücklaufen muss, wenn man eine Haltestelle versäumt hat*'.
2. *Leidsch Dagblad*. 8 December 1931. LNA online.
3. Afanassjewa-Maslova to Ehrenfest-Afanassjewa. Undated but based on the content around 1931. EFA, Translated from Russian by HD-BM.
4. As note 3.
5. As note 3.
6. As note 3.
7. Tanyitschka to Paul Ehrenfest. 2 February 1931. Folder 32. EA-MBL. '*Wegen Ersparung werde ich, scheint es, nur Assistenzplatz erhalten und nicht Dozentplatz, obwohl es wahrscheinlich leicht waere es anders einzurichten (die Arbeit ist dieselbe, nur die Bezahlung 1 ½ mal groesser)*'.
8. Mandelstam to Ehrenfest-Afanassjewa. 6 January 1930; Mandelstam to Ehrenfest. 1 February 1930. Folder 150. EA-MBL. Translated from Russian by HD-BM.
9. Scribble on the letter of note 1. '*Mama zal er zo strijden dat men haar er aan hoogste galgen zal hangen*'.
10. Lazarev was arrested on 5 March 1931 and exiled to Sverdlovsk (Yekaterinburg) in September. At Molotov's intercession, he was allowed to return to Moscow in February 1932. During his pre-trial detention, his wife committed suicide.
11. Ehrenfest-Afanassjewa to Ehrenfest. 8 April 1932. Folder 32. Translated from Russian by HD-BM.
12. Ehrenfest-Afanassjewa to Ehrenfest. 21 March 1932. EFA. Translated from Russian by HD-BM.
13. Correspondence in folders 137.
14. Margrethe Bohr to Ehrenfest-Afanassjewa. 29 April 1932. Folder 137. EA-MBL. '*Ich kann mich vorstellen, wie froh Paul ist Ihnen wieder zu Hause zu haben*'.
15. Meitner to Ehrenfest. 24 May 1932. Folder 151. EA-MBL. '*Ich habe mich aufrichtig gefreut über unser Zusammentreffen in Kopenhagen und in besonderes auch über die Anwesenheit Ihrer Frau, der ich ein paar sehr anregende Stunden verdanke*'.
16. Ehrenfest-Afanassjewa to Ehrenfest. 28 March 1932. EFA. Translated from Russian by HD-BM. She wrote that Paul '*would be embarrassed to walk around Berlin with me, my coat is so threadbare!*'
17. Personal communication Nelleke Posthumus Meyjes, conversation, 9 January 2020.
18. Ehrenfest to Goudsmit. 7 March 1933. Folder 42, Box 3, Samuel A. Goudsmit papers. Niels Bohr Library & Archives, AIP. '*25 Jahre lange vermöchte ich meine Ehe mit staerker und staerker zusammengebinnenen Zahnen in wesentlichen von verwirrenden HANDLUNGEN frei zu halten. Aber nicht langer. / Vielleich ielleicht haette ich es vermöcht, wenn meine Frau zehn Jahre juenger waehre als ich*'.
19. As note 17.
20. As note 18. '*[die unbedwingbare Liebe zu] einer (sehr hart und mehr fasziniert arbeitenden) unverheirateten Frau*'.
21. Teresi, Dick (7 January 1990). 'The Lone Ranger of Quantum Mechanics'. *The New York Times*. ISSN 0362-4331.
22. As note 18. '*15 mir naher bekannten (exzellente) Physiker (zwischen 35 und 55 Jahren)*'.
23. Ehrenfest-Afanassjewa to Ehrenfest. 22 August 1932. EFA. '*Dass ein paar Tanten auf dem Broedersplein heimlich winken, ist wirklich eine Kleinigkeit vergleichen mit dem Übel das Sie ohne nachdenken an uns gestiftet hat [und mit etwas darf Sie doch selber auch bezahlen]*'.
24. Rutgers to Zeeman. 27 September 1933. Zeeman Archive. NHA.
25. Ehrenfest to Ehrenfest-Afanassjewa. 27 May 1932. Folder 7. EA-MBL. '*Damit Du Dir (und incidentell auch mir) refrainartig wiederholen kannst, dass nie einer Leidener Professor eine so billigem anspruchslose, extrem ökonomische Frau gehabt hat wie ich. Was mich [(gleichgültig ob es richtig oder falsch ist) jedenfalls] langweilt*'.
26. Posthumus-Meyjes to Ehrenfest-Afanassjewa. 26 July 1932. EFA. '*Ik dank je Tanja, met een dankbaar hart, dat ik deze tijd mag hebben en hoop dat je deze dank van mij hebt willen aanvaarden*'.

27. Posthumus-Meyjes to Ehrenfest-Afanassjewa. Undated, but based on content in mid-July 1932. EFA. '*[Ik wilde zoo graag vanmorgen naar de kerk gaan die jij aangeraden hadt [sic], omdat ik zoozeer wilde dat] wij in deze dagen ook een sterken Russische indruk zouden hebben en daarbij jou in ons midden zouden voelen. ... Heel goed begrijp ik nu het contrast met Bach*'.
28. Ehrenfest-Afanassjewa to 'E'. Ehrenfest (baba Katya). 6 August 1932. EFA. Translated from Russian by HD-BM.
29. Ehrenfest-Afanassjewa to Ehrenfest. 12 August 1932. EFA. '*Er ist wie immer schrecklich lieb*'.
30. Some days earlier, 9 August 1932, Tanya wrote to Paul about how she wanted to take Pawlik and Einstein for a walk in a forest with ferns close to their hotel: '*Es ist hier herrlich. Habe heute endlich einen Fährenwald entdeckt ... Ich bin überzeugt dass es sowohl dem Papserl [Pawlik] als auch Einstein gefallen wird*'.
31. Ehrenfest-Afanassjewa to Ehrenfest. 16 August 1932. EFA. '*Natürlich ist E. schon entdeckt. Die Wirtin ist ganz gerührt, aber sonst tun die Leute ihr Bestes, um ihm nicht lästig zu fallen*'.
32. As note 29. '*Paps [Pawlik] ist wirklich ein feiner Mensch, so, wie ich ihn haben wollte! Ich finde, Lieberl, dass das keine Kleinigkeit ist, und des Lebens wert. Ich möchte, dass Du imstande wärest Deine Aufmerksamkeit auf diese Seite zu richten [und zu erkennen was für glückliche Menschen wir eigentlich sind.]*'.
33. Ehrenfest-Afanassjewa to Ehrenfest. 22 August 1932. EFA. '*Du scheinst noch immer nicht zu begreifen, dass es uns allen gar nicht auf diese eine Nel oder auf siebzehn andere ankommt, sondern auf Dein Verhältnis zu mir!*'
34. As note 33. '*Hast Du noch immer die Absicht die Nel den Kindern als ein Familienmitglied aufzuzwingen? Also, da kann ich nur sagen, dass dieses für mich nicht anders möglich ist, als unter Bedingung unserer Scheidung. ... E., der sehr lieb ist, und mit dem ich ganz rückhaltlos sprechen kann, ist mit mir vollkommen eins, dass Du mit ihr nur außerhalb unseres Hauses verkehren sollst. [Auch ist E. mit mir eins,] dass es Eure Pflicht ist, Euch von den "ehrlichen" hyperexponieren Eurer Verhältnisse zu enthalten, [und zwar: eine Pflicht gegen mich und meine Kinder.]*'.
35. As note 33. '*dass ich an Dir hänge und jeden Augenblick sprungbereit bin Dir freudig entgegenzukommen*'.
36. Ehrenfest-Afanassjewa to Ehrenfest. 15 August 1932. EFA. Translated from Russian by HD-BM.
37. Paul to Galinka and Pawlik Ehrenfest. 24 September 1933. Letter to be opened after his death, in which he asks Galinka, among other things, to stop renting this room.
38. Invitation to the lecture *Style Development in the Natural World View* at the Kennemer Lyceum, where Nel's course 'The Development of Modern Painting' was given. Undated, but probably late 1932 or early 1933. Folder 88. EA-MBL.
39. Ehrenfest-Afanassjewa to 'E'. Ehrenfest (baba Katya). 26 October 1932. EFA. Translated from Russian by HD-BM. An English translation of Mikhail Lermontov's poem 'The Dispute' is available online: https://allpoetry.com/The-Dispute
40. As note 39.
41. Ehrenfest to Ioffe. 23 October 1932, and further correspondence with Ioffe. Folder 152. EA-MBL.
42. Ehrenfest to Posthumus Meyjes. 22 December 1932. EFA. '*Schönes elektrisches Lichtermeer am Abend. ... Die Jungen und Alten Leute sagen, dass es gleich froh wird, wenn ich an den Discussionen teilnehme*'.
43. As note 42. '*Hier starrt mir kein "Morgen", fragend, grinsend in die Augen!*'
44. As note 42.
45. Ehrenfest-Afanassjewa to E. Afanasjeva (baba Katya). EFA. Translated from Russian by HD-BM. Natalya Tamm and Marya Kagan had lovingly taken care of her.
46. Personal communication A. van Aardenne, conversation in Amsterdam, March 2019; T. van Bommel.
47. Ehrenfest to M. Bohr. 2 March 1933. Folder 137. EA-MBL. '*[Die Zeit meiner Abwesenheit benuetzte Tanitschka um frei von der last größen Familienanwesenheit ... sich mit Van Aardenne zu verheiraten.] Pawlik war der einige von uns dabei*'.
48. Pawlik to Paul Ehrenfest. 12 September 1933. Folder 67. EA-MBL. '*Hoe grappig zij in sommige opzichten ook is, is dat toch iemand die weet hoe die op zijn beenen moet staan*'.
49. Paul to Pawlik Ehrenfest. 13 December 1932. Folder 67. EA-MBL. '*Du hast eine wunderbare Mutter - je älter du wirst, desto besser wirst du das verstehen. Und sie können sich immer noch nicht vorstellen, wie vertraut und warmherzig sie mir in der schwierigsten Zeit ihres Lebens geblieben war. Ich kann durch bewusstes Nachdenken und Analysieren nicht begreifen, warum ich nur für kurze Zeit bei Mama sein kann. Dann werde ich so unbeschreiblich traurig und wie gelähmt: Alles, alles, was in meinem Leben schiefläuft, alles, was ich an mir besonders hasse und verachte, beginnt mich niederzudrücken. Mama ist mein lebendiges Gewissen*'.

50. Ehrenfest to Ioffe. 8 November 1931. Folder 152. EA-MBL.
51. Already in 1931, Tanya had written a shocked letter about Nazi propaganda in Germany in a letter to Paul (8 April 1931). Folder 32. EA-MBL. Translated from Russian by HD): '*The Nazi newspapers leave a strong impression, and it is anyway chilling to think that people of all kinds who will stop at nothing to achieve their goals have so much power these days (Al Capone)*'.
52. Between 1933 and 1934 over 1145 Jewish scholars lost their jobs in Germany.
53. Goudsmit to Ehrenfest. 1 May 1933. Folder 42, Box 3, Samuel A. Goudsmit papers. Niels Bohr Library & Archives, AIP. '*Ich glaube dass nur Skandinavien, England, Island?, die Schweiz und vielleicht einen Teil von Frankreich als wirklich zivilisiert gelten können. Von Holland bin ich nicht sehr sicher, wie gross ist die Gefahr für Hitlerismus dort ? ? ?*'
54. Correspondence in folders 235–240. EA-MBL.
55. The Council was founded in late April 1933 by William Beveridge and still exists today, as the Council for At Risk Academics (CARA). From 1933 to 1938, the council raised 100,000 British pounds and saved about 2000 people. Of all the scholars who received aid in these pre-war years, sixteen would win the Nobel Prize.
56. Schrödinger to Klein. 17 April 1957. MKEAD. In English. According to Schrödinger Paul later, in 1933, said: '*he [Stark] always had a foible for the Netherlands and particularly for Leiden because we were the first to invite him as a guest lecturer after his important discovery; indeed the first to express our great admiration*'. In 1919, Stark was awarded the Nobel Prize in physics for this discovery.
57. As note 56. According to Schrödinger, Paul also said: 'I want to see his face when he sees me'.
58. Schrödinger to Klein. 17 April 1957. MKEAD. In English.
59. '*Isak! Isak! Was heb Du o Hast?*'
60. Ehrenfest to Goudsmit. 13 May 1933. Folder 42, Box 3, Samuel A. Goudsmit papers. Niels Bohr Library & Archives, AIP. '*Solche Strassenbelastungen*'.
61. Ehrenfest to Goudsmit.7 March 1933. As note 60. '*Jeder nachdenkliche Mensch hat das Gefuehl, dass wir gegen eine voellig undurchsichtige wand herangeschoben werden. Um zerquetscht zu werden? Oder wird die Wand umfallen? Oder weiss der Teufel was? ... Pläne kann man doch nun wirklich nicht machen bis etwa die nächsten vier-fünf Monate vorbei sind. (Es sei denn der Plan mich zu toeten)*'.
62. Einstein's explanation for his decision to turn his back on Germany is engraved on the Einstein Memorial of the American Academy of Sciences: 'As long as I have any choice in the matter, I shall live in a country where civil liberty, tolerance and equality of all citizens before the law shall prevail'.
63. Ehrenfest to Einstein. 22 May 1933. Folder 141. EA-MBL.
64. Trüpers to Ehrenfests. May 1933. Folder 67. EA-MBL. Correspondence in EFA.
65. Correspondence in Folder 6. EA-MBL.
66. Ehrenfest to Ehrenfest-Afanassjewa. 15 July 1933. EFA. '*Ich bitte Dich jetzt um die Scheidung unserer Ehe. ... Voriges Jahr bat ich Dich die Scheidungsklage zurueck zu nehmen. Und in der that haben mir nur erst die Erfahrungen dieses Jahres mir die Sicherheit gegeben, dass ich umoeglich so weiterzuleben vermag*'.
67. Paul to Pawlik Ehrenfest. August 1933. Folder 67. EA-MBL. Pawlik answered on 1 September that he hadn't sent any money because he expected Paul to come to Leiden to see them all.
68. As note 65.
69. Personal communication Nelleke Posthumus-Meyjes, conversation in Amsterdam, 9 Jan 2020. See also (Casimir 1983, 149) who cites this last remark slightly differently, 'stop dawdling'.

Chapter 34

1. According to T. van Bommel, after Afanassjewa's death in 1964 Tanitchka told Galinka that she had found the Browning revolver in a cupboard. Anne Kox confirmed in private communication that this is consistent with what biographer Martin J. Klein told him: that the police had returned the Browning to Afanassjewa in 1933. The revolver case and licence have been at the archives of National Museum Boerhaave since 2025. The licence for the 9 mm Browning revolver is dated 4 September 1933 and signed by the head of the Leiden police. The manual describes it as a 'Pistolet Automatique Browning, Modèle 1910 Brevetè', delivered by the 'Fabrique Nationale d'Armes de Guerre, Herstal, Liège'. It remains unclear how Paul obtained the Browning.
2. Rutgers to Zeeman. 27 September 1933. Zeeman Archive. NHA. '*We hebben samen koffie gedronken in een cafeetje aan de Plantage Middenlaan, toen nog gepraat in mijn bibliotheek, en*

gezeten op een bankje in het plantsoentje naast uw lab, waar we nog steeds over natuurkunde en ook over meer algemene dingen gepraat hebben'.
3. As note 2. *'naar het instituut van prof. Waterink, waar Wassik verpleegd werd.... Ook prof. Waterink heeft niets gemerkt. Om 5 uur was het geschied'.*
4. Ehrenfest. 25 September 1933. Copied onto stationery of the Pedagogic Institute (the Waterink Institute). EFA.
5. As note 4. *'Versucht bitte, bitte nicht mein Leben zu retten falls ich nicht sofort tot sein sollte. Habt Erbarmen und lasst mich jetzt sterben'.*
6. Paul to Galinka and Pawlik Ehrenfest. 4 October 1932. EFA. *'Ich fürchte, dass es schon sehr sehr bald keinen anderen Ausweg mehr für mich geben wird als Wassik u. mich zu töten—so schädliche Folgen das auch für Euch Beide haben mag—[es ist doch weniger arg als wenn ich mich unbegrenzt weiter zersetze. Die Zersetzung aber aufzuhalten vermag ich nicht. Ihre eigentliche Kern-Ursache ist: ich kann absolut nicht mehr ein erträgliches Verhältnis zur Arbeit finden. Alles andere ist die Folge davon'.*
7. As note 5. *'Tanitschkerl, die ich so sehr liebe hat sich (mit Recht!!!) schon völlig von mir abgewendet—nicht weil sie mich nicht liebt aber weil sie mit Recht ihre eigene Seele von dem Anblick dieser Zersetzung freihalten möchten so gut es geht. / Wärest Du hier Galinker, auch Du würdest es thun (oder thun sollen)'.*
8. As note 5. *'denn Du Galinkerl bist ja so unendlich fest mit den Leben der Menschen verankert, dass Du abgesehen von vielleicht einer kurzen Zeit von Störung, freudig und fasziniert Deinen Weg gehen wirst. Aber Dich, Pawlik, fehl ich an: arbeite sorgfältig, beisse die Zähne zusammen (sorge gut für Deine körperliche Gesundheit!!!) Lass Dir in der Zeit der Verwirrung und Sorge durch meine Freunde beistehen—De Haas, Rutgers, Burgers, Tinbergen, Einstein. Goudsmit, Dieke, Uhlenbeck, Kohnstamm, Herglotz, Bohr, Holst, Franck ... jeder von ihnen wird bereit sein Dir über Schwierigkeiten wegzuhelfen, wenn sie nur wissen, dass Du arbeiten willst'.*
9. As note 5. *'Da Deine Kindheit so ganz ganz ganz anders war als die meine und Du überdies in sehr viele Beziehungen Mama viel ähnlicher bist als mir, so wird wahrscheinlich der Sexualtrieb Dich viel weniger in Verwirrung bringen als das für mein Leben der Fall war. In mein Leben kam dadurch schon in früher Kindheit sehr viel Not und Schaden und zwischen 13 und 16 kam ich schließlich ganz nahe an den Rand des Selbstmordes, weil ich auf der Schule schließlich der schlechteste Schüler wurde und über dies ein fast pathologischer Lügner wurde'.*
10. As note 5. *'Eure Mutter ist ein ganz ganz wunderbarer Mensch—je tiefer sich Euer Leben entwickeln wird, desto mehr werdet Ihr das begreifen. Aber nie werdet Ihr wissen können wie innig-warm die einfach als Frau gegen mich war.... Die Sehnsucht nach Frauen die jünger sind als ich hat bewirkt, dass ich Mama unbändiges Leid zufügte ... Daran bin allein ich schuld. Es wäre ein tiefes Unrecht Nelly zu verurteilen ... Im letzter Instanz fällt alle Schuld auf mich allein. Bitte beleidigt nicht Nelly. /Meine Bitte an Euch Beide und durch Euch auch an das liebe liebe Tanitschkerl ist: Liebt das fasziniert-arbeitende, dienende Leben./ Verzeiht mir, denkt an mich, wie ich war nicht wie ich jetzt bin'.*
11. Paul to Tanya Ehrenfest. 27 January 1933. In an envelope *'Tania! Im Fall meines Todes ... öffnen'.* EFA. *'Tania—alles was ich überhaupt aus meinem Leben herausgeholt habe hängt aufs genaueste damit zusammen wie Du das Echte in mir gegenüber das Unechte zu stützen versucht hast—/Tania—ich habe nie aufgehört Dich zu lieben. Verzeih mir'.*
12. Personal communication T. van Bommel. The cemetery records (of formally Nieuwe Ooster) show that the grave has been cleared.
13. Ehrenfest. Written statement 'na de beëdiging van Sande Bakhuyuzen' in envelope *'Technische Verfügungen betreffs Begräbnis etc'.* 11 January 1933. EFA. *'alleiniger Anwesenheit des Technische [sic] Personales.... Keine Ceremonien, kleine Blumen, keinerlei Grabverzierung irgendwelcher Art ... Arbeit, Musik, Lachen und des Alltagsthätigkeiten beginnen sofort wieder ... Jedenfalls werde jede Trauersymbolik ... bestimmt vermieden'.*
14. Ehrenfest. 14 August 1932. EFA. Copies of this letter are also in the Einstein Archive in Jerusalem and the Niels Bohr Archive in Copenhagen. Pais (1991) suggests in his book the letter was never sent. The correspondence between Einstein and Afanassjewa suggests that she shared the letter with Einstein after Ehrenfest's death. *'Ich weiss absolut nicht mehr, wie ich auch nur die naechsten Monate die unertraeglich gewordene Last meines Lebens weiter schleppen soll. ... Mein Interesse fuer das Begreifen der fortschreitenden physikalischen Erkenntnisse und die grosse Freude an andere weiterzugeben, was ich selber begriffen zu haben glaubte, war, wie Ihr wisst, das eigentliche Rueckgrat meines Lebens./ Immer schwerer wurde es mir in den letzten Jahren*

begreifend der Entwicklung zu folgen. Ich habe es schliesslich nach immer nervoeseren, zerfetzter werdenden Versuchen VERZWEIFELT aufgegeben'.

15. As note 13. '*Aber ebenso sicher weiss ich, dass ich aller spaetestens im Herbst 1933 meine Leidner Stelle frei machen muss. ... habe ich in meinem Gesichtsfeld keine "practische" Moeglichkeit als den Selbstmord, und zwar nach vorhergehender Toetung von Wassik. ... Moege es Euch selber und denen die Euren Herzen besonders nahe stehen wohl eingehen!!*'

16. Ehrenfest to '*Meine Lieben Freunde: Burgers, Casimir, Coster, Dieke, Goudsmit, Kramers, Rutgers, Tinbergen, Uhlenbeck, Wiersma*'. 15 August 1932. EA-MBL 598. '*Ihr wisst wie herzlich ich mich stets mit Euch ueber jeden Eurer Erfolge mitgefreut habe. / Ihr begreift also, wie innig ich Euch und allen Euren Lieben alles Beste fuer die Zukunft zuwuensche!!! / Verzeiht mir, dass ich so ein schwacher Mensch bin! Euer P. Ehrenfest*'. According to S. Burgers, grandson of Jan Burgers, who was one of the recipients and based on the personal archive of his grandfather that also contained the letter, it was sent to those with the underlined names. Personal communication, e-mail, 21 January 2017.

17. Letters, telegrams and cards in EA-MBL 476 and EFA.

18. *Leidsch Dagblad*. 26 September 1933. LNA online. '*in Weenen, later in Göttingen en in Rusland, waar hij als jood voortdurend aan behandelingen was blootgesteld die zijn gevoelige natuur diep krenkten en waarschijnlijk mede verantwoordelijk zijn voor het pessimisme en gebrek aan zelfvertrouwen die hem, geheel ten onrechte, nooit verlaten hebben*'.

19. Born to M.J. Klein. 21 January 1957. MKEAD. In English.

20. Kagan to 'dear Tatyana Alexeyevna'. Undated and 7 October 1933. EFA. Translated from Russian by HD-BM.

21. Hugo Ehrenfest to Ehrenfest-Afanassjewa. 28 September 1933. EFA. '*Er wollte nicht länger leben und nahm den Jungen mit. [Eine mutige Tat. ...] Eine Tragödie mit dem charakteristischen Element der Unabwendbarkeit. [Alles ist so unsäglich traurig und erschütternd und schließlich kann ich doch nur das Eine fühle -] das dieser arme, geschätzte, deprimierte, wirklich unglückliche und trostlose Paul jetzt nun ersten Mal seit vielen Jahre wirklich Frieden gefunden hat*'.

22. Einstein to Ehrenfest-Afanassjewa. Undated. EFA. '*Wenn es gelungen wäre, ihm Verantwortung abzunehmen, so hätte er als froher Mensch weiter leben können. Aber nichts ist gelungen. Die Kerle sind immer bereit, glänzende Reden zu halten, aber wehe, wenn man etwas von ihnen will, und sei es auch noch so bescheiden und berechtigt*'.

23. Ibid. '*Sie hätte bestimmt gut daran gethan, sich der Scheidung zu widersetzen, obwohl eine Frau, die so handelt, eine kuriose Rolle spielt. Denn im Grunde wollte er es gar nicht sondern stand im Banne eines unberechtigten bürgerlichen Pflichtgefühles. Seine Konflikte waren nicht durch Heuchelei oder Leichtsinn oder Ironie gemildert, sondern er nahm alles absolut wie einen mathematischen [Lehrsatz] genau wie Sie es auch zu thun geneigt sein. Wenn ich meinen Sachen so ernst nähme, würde längst das Gras über mir wachsen. Aber ich habe ebenso viel Sancho Panza als Don Quichotte in meinem dicken Leib, sodass alles zumeist mit einem gutmütigen allseitigen Schmunzeln endet oder sich vorsetzt*'.

24. Elsa Einstein to Ehrenfest-Afanassjewa. 30 October 1933. Folder 141. EA-MBL. '*Albert und ich sprechen nicht darüber, aber wir fühlen beide .... dass Albert nicht zu ihm ging, bleibt ein peinigender Gedanke sein Leben lang*'.

25. Ehrenfest-Afanassjewa to Hulsebosch. 12 October 1933. Copy in EFA. '*welke moeite U op U hebt genomen om ... de correspondenten van het sensationeel element in hun artikelen af te houden ... mijn oprechte dank*'.

26. By order of 14 November 1933. Folder 477. EA-MBL. It was 1400 guilders per year.

27. Request in a letter from Paul to Galinka and Pawlik, 24 September 1933, at night. EFA.

28. Kohnstamm to Ehrenfest-Afanassjewa. 1 February 1934. EFA.

29. Notes of a conversation between Fokker and M.J. Klein. 21 November 1953. MKEAD.

30. Ariëns Kappers to Ehrenfest-Afanassjewa. 28 October 1933. Folder 476. EA-MBL. '*Ik hoef u niet te zeggen dat dit met de uiterste zorgvuldigheid is geschied (in het Wilhelminagasthuis). Eerst na volledige fixatie ... zal dit voor mij zo kostbare bezit naar mijn instituut worden overgebracht*'. It wasn't uncommon: the brains of, for example, Gauss and Dirichlet were preserved as well, and later Einstein's would too.

31. As note 12.

Chapter 35

1. Ioffe to Ehrenfest-Afanassjewa. 22 October 1933. EFA. Translated from Russian by HD-BM.
2. Pawlik met Ioffe, who was still in Brussels, in Antwerp on Sunday, 29 October 1933. According to a note from Ioffe to Ehrenfest-Afanassjewa (26 October 1933) and a telegram (draft, undated),

NOTES   371

Ioffe picked up Pawlik at the train station at 11:39 that Sunday. EFA. Both translated from the Russian by HD-BM.
3. This was arranged by Ioffe, but there were already close ties between Leiden and French scientists, strengthened by the Solvay Conferences chaired by Lorentz, especially in the aftermath of World War I, see also Kox (2021). In the thirties, circumstances brought French and Dutch physicists even closer together.
4. Ehrenfest's former student Jan Burgers would remain a member of the Communist Party for many years after these conversations. Personal communication of Jan Burgers' son, Herman Burgers, conversation in The Hague, 16 April 2015.
5. Ehrenfest-Afanassjewa. 18 August 1920. Notebook. EFA. Translated from Russian by HD.
6. Galinka 'paid' for her lessons by helping Lien Citroen, the wife of the headmaster, Paul Citroen, who was in poor health, with housework. Personal communication from T. van Bommel.
7. Galinka Ehrenfest to Kloot. 13 January 1939. EFA. *'Paulie [Pawlik] is gevonden in de "start" houding, zoals geleerd wordt bij dergelijk gevaar, dus klaar om [van] de helling terug te glijden, maar de lawine was sneller. ... Moesje praat en praat alsmaar met mensen in Parijs'.*
8. Pawlik Ehrenfest to Ehrenfest-Afanassjewa. 5 January 1939. EFA.
9. Onnes to Ehrenfest-Afanassjewa. January 1939. EFA. *'Van alle jonge menschen die ik ken had ik altijd een bijzonder voorliefde voor Paul, zoals ik mij steeds aangetrokken voelde tot u allen. Een mooi leven heeft hij wel gehad in een omgeving zoals U en Uw man die maakten, steeds in contact met de fijnste geesten op verschillend gebied'.*
10. Zeeman to Ehrenfest-Afanassjewa. 20 January 1939. EFA. *'Hij was altijd zeer beminnelijk. Is dat misschien een Russisch erfstuk?'*
11. Margrethe Bohr to Ehrenfest-Afanassjewa. 13 January 1939. EFA. German.
12. Bohr to Ehrenfest-Afanassjewa. 31 January 1939. EFA. *'Wir hatten alle Ihren Sohn so lieb'*. Bohr was in Princeton at the time.
13. Gorter-Muller family to Ehrenfest-Afanassjewa. January 1939. EFA. *'Wie weet is hij voor ellende gespaard gebleven, een ellende die wij allen nog te doorworstelen krijgen'.*
14. Ilse Einstein died in July 1934, her mother Elsa in December 1936. Grandson Klaus Einstein died on 5 January 1939. His father, Hans Albert Einstein, worked as professor of Hydraulic Engineering in Berkeley.
15. Einstein to Ehrenfest Afanassjewa. 22 Jan 1939. EFA. *'Es scheint, dass es das Schicksal auf die Starken und Aufrechten besonders abgesehen hat. Ich weiss wie [weh] es thut, zumal ich vor wenigen Wochen das eine von meinen zwei Enkelkindern verloren habe, und viel Schweres brachten die früheren Jahre. ... Solche Worten kommen sonst nicht über meine Lippen. Sie aber stehen mir besonders nahe durch eine Art innerer Verwandtschaft. Es ist so eine gewisse Abseits-Einstellung zum persönlichen Leben und ein Erfülltsein mit objektiven Dingen, auch hier abseits von Zeitgenossen - so sind wir doch beide. Am liebsten wäre ich bei Ihnen, um Ihnen ein wenig Gesellschaft zu leisten wie so oft in früheren Jahren. Aber nun sind wir weit auseinander, und ich sitze hier fest in einer liebgewonnenen Einsamkeit'.*
16. Ibid. *'Nur die objektiven Dinge haben für mich ihre leuchtende Schönheit behalten. Sie erscheinen in derselben Glorie wie das erste Geometrie-Buch, das ich als Kind erhielt. So geht es gewiss auch Ihnen. ... Nur sollte man es selbst nicht erleben müssen sondern nur Auge sein können'.*
17. The 'Werkgemeenschap voor Vernieuwing van Opvoeding en Onderwijs' was de Dutch department of the American New Education Fellowship. See e.g. Moor (1999).
18. Documentation of the Mathematics Working Group of the WVO. 24 May 1938; programme of the working week of the mathematics working group, 22–27 July [1938] at the Veluwe. IISH-1618-79d.
19. Ehrenfest-Afanassjewa to Galinka Ehrenfest. 12 July 1932. EFA. Translated from Russian by HD.
20. Personal communication T. van Bommel.
21. The actual titles of the books in Dutch were: *Rommel Jan, Flora bloemenstalletje, Het toverboek van 1001 nacht,* and *Kom binnen in het huis van El Pintor.*
22. *Leidsch Dagblad.* 13 March 1941. LNA online. Letter of appointment. 10 March 1941. Folder 466. EA-MBL.
23. Draft letter. Undated. Scribble added on 18 November 1940. Folder 466. EA-MBL. *'H. Kramers heeft dit ontworpen. En hij heeft heel het werk gedaan opdat dit tot stand kome. Oostduinlaan, Den Haag'.*
24. Letter of the Ministry of Education confirming her resignation. 15 June 1942. Folder 466. EA-MBL.
25. Ehrenfest-Afanassjewa. Typed account, 30 April 1948 On behalf of EA-MBL. *'Begin juni 1943 werd mijn schoonzoon, Jacob Kloot, ... op straat in Leiden door een hollands [sic] politieagent op*

straat als jood herkend en aangehouden. Verscheidene dagen wist ik niets daarvan en verbaasde me dat hij steeds steeds niet was gekomen zoals het afgesproken was. Toen werd ik op een avond door een onbekende, die zich niet noemen wou, opgebeld. De onbekende man deelde my mee dat myn schoonzoon gearresteerd was en hy waarschuwde me dat myn dochter, die vanwege haar halfjoodsche afkomst na het huwelijk met J. Kloot ook als joodsche gold, zou ook gearresteerd worden [sic] en dat dit de reden was dat hij belde, om haar de mogelijkheid te geven te ontsnappen. Dit telefoongesprek had plaats een dag na dat in in de tuin, waarin ik werkte, twee detectiven verschenen waren om zich over mijn dochter te informeren'. See also Metz (2019). Police registers show that Jacob Kloot, whose falsified identity papers presented him as Klaas Leeuwrik, was arrested on 7 June 1943. (See: www.Dossier071.nl, frame NL-LdnRAL_AR_0552_2684_00204). Five days later, he was taken to the 'Oranjehotel' prison in Scheveningen and from there, via Westerbork concentration camp, to Sobibor (on 29 June).

26. Declaration Hauptsturmführer Samel. 26 June 1943. EFA. It states that Galinka was in prison, in 's Hertogenbosch from 21 to 26 June and that her identity card was temporarily withheld for investigation.
27. Personal communication T. van Bommel. Supported by a letter Galinka wrote to Ehrenfest-Afanassjewa immediately after her release, that was concealing but understandable for her mother. Dated 1943. Further supported by a letter from lawyer Daisy Schaeffer (2 July 1943) to Ehrenfest-Afanassjewa in which she discusses how to prove that Alexander Archangelsky was Galinka's father. EFA.
28. Ehrenfest to Ioffe. 9 July 1933. Folder 146. EA-MBL. 'Galinka ist ein Pracht-Mensch'.
29. Ehrenfest to Margrethe Bohr. 2 March 1933. Folder 137. EA-MBL. '*Hat Galinka nur erst an Ort und Stelle zwei Menschen kennen gelernt so schwimmt sie sofort schon weiter aus eigener Kraft!!!*'
30. Personal communication T. van Bommel.
31. C.J. van Oost, acting head of the national inspection of population registers to Schaeffer. 23 September 1943. Letter indicating that Galinka is no longer 'obliged to register' as someone of Jewish decent. EFA.
32. Ehrenfest-Afanassjewa. 22 April 1922. Notebook. EFA. Translated from Russian by HD-BM.
33. As note 32.
34. Personal communication Sari Tokaya Langendijk and Hanna Langendijk, conversation in Utrecht, 16 December 2015.

Chapter 36

1. Einstein to Ehrenfest-Afanassjewa. 22 October 1945. Folder 425. EA-MBL. '*Ich habe mich so sehr über Ihren Brief gefreut. So eine gefestigte und standhafte Person findet sich nicht so leicht wieder. Auch das Interesse an den logischen Grundlagen der Physik ist bei Ihnen lebendig geblieben, wie wenn Sie nicht die schweren, bedrohenden Jahre durchlebt hätte!*'
2. Card Mimosa Food Products. Folder 425. EA-MBL.
3. Ehrenfest-Afanassjewa to Einstein. 4 April 1946. Folder 425. EA-MBL.
4. Dukas to Ehrenfest-Afanassjewa. 17 April 1946. Folder 425. EA-MBL.
5. Einstein to Ehrenfest-Afanassjewa. 28 March 1947. Folder 425. EA-MBL.
6. Fedyaevskaya to Ehrenfest-Afanassjewa. 10 February 1946. Folder 425. EA-MBL. Translated from Russian by HD-BM.
7. Joop Westerweel (1899–1944) led a resistance group that helped about 300 Jews escape from the Nazis.
8. Joods Nationaal Fonds (JNF). Postcard. April 1946. Folder 453. EA-MBL.
9. Ehrenfest-Afanassjewa. Lecture Notes, c. 1947. Gorlaeus Library. Leiden University.
10. Ehrenfest-Afanassjewa to the Board of Trustees of Leiden University. 18 May 1950. '*Het komt mij voor dat ik door het berusten in deze situatie de prestige [sic] niet alleen van mijn eigen persoon, maar ook van het vak dat ik onderricht, te zeer zou schaden*'.
11. Personal communication T. van Bommel.
12. Galinka Ehrenfest to Ehrenfest-Afanassjewa. 11 August 1948. Folder 418. EA-MBL.
13. Prof. H.J. Pos declares in a letter dated 30 July 1945 that he can provide a favourable testimony about Samel during his period in internment camps Sint Michielsgestel and Haaren; that Samel saved the family of Dr Slager from Eindhoven from arrest and that he later made efforts for Jewish internees in Den Bosch. EFA.
14. As note 12.
15. Personal communication Fransje Teulings and Carola Kloos, conversation in The Hague, 22 March 2019. '*Ik hoop dat je wilt op deze langste van het jaar avonden met je vriend van mijn grasvelden gebruik te maken*'.

16. As note 15.
17. Personal communication Carola Kloos, conversation in The Hague, 22 March 2019 and email, 19 March 2021.
18. As note 15.
19. Einstein to Ehrenfest-Afanassjewa. 28 March 1947. Folder 425. EA-MBL. '*Schicken Sie mir doch einmal das Manuscript, ich werde dann versuchen einen Universitäts-verlag dafür zu interessieren, nachdem ich es selber gelesen habe*'.
20. Ehrenfest-Afanassjewa. '*Le second principe de la thermodynamique et l'accroisement "de entropie"*'. Three handwritten versions. Folder 481. EA-MBL.
21. Einstein to Ehrenfest-Afanassjewa. 5 April 1946. Folder 425. EA-MBL. '*Und für uns ist es eigentlich auch ungefähr an der Zeit, ... Sie sollten wirklich schauen, dass Ihre Thermodynamik vorher übersetzt wird*'.
22. Ibid. '*Ich bin überzeugt, dass Ihre logischen Putzteufel dort etwas gutes hat unterrichten können für die Klärung des etwa nebulosen begrifflichen Basis*'.
23. Ehrenfest-Afanassjewa to Einstein. 17 August 1947. AEA-huj 10-312. '*Sie sind ein wirklich echter Freund, dass Sie sich so unmittelbar an das Lesen meines Manuskriptes setzen!*'
24. Einstein to Ehrenfest-Afanassjewa. 6 August 1947. Folder 482. EA-MBL. '*Ich ... bin ganz entzückt über die klare und durchsichtige Begriffsbildung*'.
25. Einstein to Ehrenfest-Afanassjewa. 12 August 1947. Folder 482. EA-MBL. '*Es ist ein bisschen wie die Vorstellung des routinierten Zauberers, wo es so hübsche Details zu schauen gibt, dass man nicht merkt, wenn der Frosch Ihres unvergesslichen P. E. ins Wasser springt*'.
26. Ibid. '*Ich habe den Eindruck gewonnen, dass Sie ein bisschen von logischen Putzteufeln besessen sind, und dass daran die Übersichtlichkeit des Buches leide*'.
27. Ehrenfest-Afanassjewa to Einstein. 18 August 1947. AEA-huj 10-312. '*Dass Sie sich beeilt haben Ihre Hände von meinem Buch zu waschen, ist sicher sehr traurig: Ich hoffte, dass ich nun ruhig sterben dürfte! Was soll ich nun beginnen? ... In Europa gibt es zu wenig Papier ... und noch minder Bonzen, die mein Buch einem Herausgeber empfehlen könnten!*'
28. As note 23.
29. Ehrenfest-Afanassjewa to Van Bommel. 28 April 1950. '*Hier hebt U een beschouwing, die ik op een van mijn wakkere nachten geschreven heb*'.
30. As note 29. '*Waarom kan een groep mensen zoo vaak geen krachtige verbetering op een bepaald [sic] sector van de menselijke activiteit kweken? (sociaal, politiek, oeconomisch, intellectueel ...) Ik denk: door gebrek aan bepaalde positieve wensen. De meeste deelnemers van zoo een constitutie, weten slechts wat ze willen door iets anders vervangen, maar komen bij elkaar zonder een bepaald positief plan. De haat tegen het bestaande, maar niet de liefde voor hetgeen wat moet zijn. ... de meesten ... haten de stof, die er is; ze houden niet van een bepaald ander stof. Daarom worden ze door diegenen verslagen die wel iets moois in het bestaande vinden*'.
31. Minutes mathematics working group WVO. IISH-1618-79.
32. Many of the ideas from this essay are reproduced in Ehrenfest-Afanassjewa (1961, 14–24).
33. As note 15.
34. As note 15.
35. As note 29. '*Ik haat het te vele [sic] omdat ik van het weinige houd*'.
36. Ehrenfest-Afanassjewa to T. van Bommel. 7 June 1963. EFA. '*Ik zoude graag van tevoren een overzicht hebben van wat we zouden moeten doen. Wil je? Ik heb altijd jammer gevonden, dat op de russische meisjesscholen geen oude talen onderricht werden. Later heb ik zelfstandig de Lateinsche [sic] grammatica geleerd, maar voor 'T grieks [sic] had ik nooit tijd. Eet mama voldoende?*'
37. Einstein to Ehrenfest-Afanassjewa. 18 April 1953. Folder 425. EA-MBL. '*Sie haben ganz recht, wenn Sie ihr Gehirn nicht mit Geburtsdaten belasten*'.
38. Baptism certificate. 8 November 1876. Folder 471. EA-MBL.
39. As note 35. '*Ich empfange eine Gratulation von Ihnen mit Begeisterung an jedem Tage des Jahres, den Sie hiefür [sic] auswählen*'.
40. Ehrenfest-Afanassjewa to Einstein. 23 March 1953. Folder 425. EA-MBL.
41. Ehrenfest-Afanassjewa to Van Bommel. 20 July 1959. EFA. '*Ik wacht dat dezer dagen het weer voldoende veranderd, dat men zal kunnen lopen zonder duizelyk te worden ... Zo is het: of te koud, of direct ondraaglyk warm*'.
42. As note 14.
43. Ehrenfest-Afanassjewa to Van Bommel. 29 January 1960. EFA. '*Een verrassing die trouwens al vooruit te zien was ... Morgen komt de verkoper om my te laten proeven. ... ben benieuwd. In andere opzichten gaat het mij best, alleen ben ik iedere dag meer lui*'.

44. Ehrenfest. 23 February 1926. EFA. 'Es wäre doch lustig, wenn die Publicationen meiner Frau und von mir selber noch einmal vereinigt in chronologischer Folgorde [sic], in möglichst bescheidener Form wiederabgedrückt werden könnten. ... Sehr viel ist mir daran gelegen, dass die Aufsätze meiner Frau möglichst vollständig erscheinen'.
45. Correspondence of Schrödinger, Westerdijk, Casimir, Tinbergen. Folder 482. EA-MBL.
46. Newsletter of the Mathematical Working group of the WVO. Nov 1961. IISH.
47. Ehrenfest-Afanassjewa to G. Ehrenfest. 15 November 1961. EFA. Translated from Russian by HD. 'Erg roerend, te meer zo omdat ze me geen van allen konden begrijpen! Maar het juist, dat er voor het verschijnen van mijn eerste artikel in hun (twee!) onderwijstijdschriften nooit gediscussieerd was van hoezo en waarom er op school wiskunde gegeven moest worden en daarna begon iedereen erover. Maar waarom heb je ze die vreselijke foto ... gegeven. Mevrouw van Gendt vond dat ik er op die foto als 85 maar meer als 100 uitzag. Bovendien heb ik zelfs mijn ogen niet open'.
48. Personal communication T. van Bommel.

Epilogue

1. Folder 65, EA-MBL.
2. Hugo Ehrenfest to Ehrenfest Afanassjewa. 27 December 1938. Folder 65. EA-MBL. 'Tanya ich fuehle sehr grosse Verehrung fuer Dich als Mensch, aber Du bist eine Russin - ich ein Amerikaner. Du bist und warst immer bereit ALLES wegzugeben, was Du hast. Ich fuehle mich gezwungen an die Zukunft meiner Frau, Kinder und Enkeln (wirklich nicht an meine) zu denken. / Tanja, versuche nicht die Leiden aller zu lindern, das ist ein unmoegliches und trostloses Verlangen!'
3. They met once in 1931, after the death of the eldest brother Hugo Ehrenfest.
4. Einstein to Ehrenfest. 19 August 1914. CPAE 8a, doc. 34.
5. Hugo to Paul Ehrenfest. 9 April 1924. Folder 64. EA-MBL. See Ch. 8 and Ch. 28.
6. Ehrenfest. 'A few small wishes for after my death'. 23 February 1926. EFA. See Ch. 36.
7. The stern white bow is what she wears throughout a comic book describing the life of Ehrenfest (Maas and De Heij 2015) although in this book, focussing on Ehrenfest, she only speaks once or twice
8. Anecdote and materials such as the farewell lecture 'Reading and Telling' that Prof. Dr W. (Willy) Burgers gave on 30 June 1967 at the University of Delft—provided by Simon Burgers, grandson of brother Prof. Dr J. (Jan) Burgers, January 2017.
9. It also soured relations with his brothers, as becomes clear from their correspondence.
10. Hugo to Paul Ehrenfest. 15 January 1931. Folder 64. EA-MBL. In English.
11. Einstein to Ehrenfest-Afanassjewa. End of 1933. EFA. See Ch. 34.
12. Ehrenfest. Goodbye Letter. 25 September 1933. EFA. 'Dass ich mein idiotes Kind nicht zurück lassen kann wird jeder begreifen, der sich in die Frage eindenkt'. (In those days 'idiot' was a commonly used word for children with Down Syndrome.)
13. The text 'she opened the house to people and ideas' ('maakte dit huis open voor mensen en ideeën') is also written on one of the two facade stones placed in memory of Paul and Tanya in the facade of the house on Witte Rozenstraat.

Bibliography

Aaserud, Finn and Heilbron, John L. 2013. *Love, Literature, and the Quantum Atom: Niels Bohr's 1913 Trilogy Revisited*. Oxford: Oxford University Press.
Adloff, Jean-Pierre and George B. Kaufmann. 2005. 'Triumph over Prejudice: The Election of Radiochemist Marguerite Perey (1909–1975) to the French Académie des Sciences.' *Chemical Educator* 10: 395–399. https://doi.org/10.1333/s00897050955a
Alberts, Gerard. 1994. 'On Connecting Socialism and Mathematics: Dirk Struik, Jan Burgers, and Jan Tinbergen.' *Historia Mathematica* 21: 280–305. https://doi.org/10.1006/hmat.1994.1026
Alkemade, A. J. Q. 2002. 'Burgers, Johannes Martinus (1895–1981).' In *Biografisch Woordenboek van Nederland*. Den Haag. http://resources.huygens.knaw.nl/bwn1880-2000/lemmata/bwn5/burgers
Asscher-Pinkhoff, Clara. 1966. *De danseres zonder benen*. Den Haag: Leopold.
Aswal, Stephen (Ed.). 1990. *The Founders of Child Neurology*. San Francisco: Norman Publishing.
Badino, Massimiliano. 2009. 'The odd couple: Boltzmann, Planck and the application of statistics to physics 1900–1913).' *Annalen der Physik* 521(2–3): 81–101. https://doi.org/10.1002/andp.200952102-303
Baneke, David. M. and Ad Maas. 2018. 'De Hogere Burgerschool: onderwijs en emancipatie.' *Studium: Tijdschrift voor Wetenschaps- en Universiteitsgeschiedenis* 10(3): 117–122. https://doi.org/10.18352/studium.10155
Bastide van Gemert, Sascha la. 2006. *'Elke positieve actie begint met critiek.' Hans Freudenthal en de didactiek van de wiskunde*. Doctoral dissertation, Utrecht University. Hilversum: Uitgeverij Verloren.
Bauer, Edmond. 1963. Interview of Edmond Bauer by Thomas S. Kuhn and Theo Kahan on 1963 January 8, Niels Bohr Library & Archives, AIP. www.aip.org/history-programs/niels-bohr-library/oral-histories/4498-1
Beller, Steven. 1991. *Vienna and the Jews, 1867–1938: A Cultural History*. Cambridge: Cambridge University Press.
Berends, Frits. 2015. 'The Solvay councils and the physics institute.' *The European Physical Journal Special Topics* 224(10): 2091–2111. https://doi.org/10.1140/epjst/e2015-02525-5
Berends, Frits. 2016. 'Lorentz en Einstein. "U hier als opvolger te mogen begroeten is een wens die ik te lang gekoesterd heb".' *Studium* 9: 6–25. https://doi.org/10.18352/studium.10125
Berends, Frits and Dirk van Delft. 2019. *Lorentz. Gevierd fysicus, geboren verzoener*. Amsterdam: Prometheus.
Berkel, Klaas van. 1996. *Dijksterhuis. Een biografie*. Amsterdam: Bert Bakker.
Berkel, Klaas van. 2000. 'De geboorte van een tijdschrift.' *Euclides* 75(4): 111–116.
Besouw, Jip van and Jeroen van Dongen. 2013. 'The reception of relativity in the Netherlands.' In *Physics as a Calling: Studies in Honour of A. J. Kox*, edited by Ad Maas and Henriëtte Schatz, 1–20. Leiden: Leiden University Press.
Bezemer, J. W., and Mark Jansen. 2008. *Een geschiedenis van Rusland. Van Rurik tot Poetin*. Amsterdam: Van Oorschot.
Blaauboer, Miriam. 2015. 'Hendrika J. van Leeuwen – Portret van de eerste vrouwelijke lector bij Technische Natuurkunde in Delft.' *Nederlands Tijschrift voor Natuurkunde* 81: 4–7.

Blaauboer, Miriam and Margriet van der Heijden. 2025. 'Jo van Leeuwen, the other physicist behind the Bohr-Van Leeuwen theorem.' In *Women in the History of Quantum Physics: Beyond Knabenphysik*, edited by Patrick Charbonneau et al., 56–81. Cambridge: Cambridge University Press.
Blackmore, John T. 1995. 'Boltzmann, his later life and philosophy, 1900–1906.' In *Boston Studies in the Philosophy of Science*. Dordrecht: Springer.
Bohr, Niels. 1913a. On the constitution of atoms and molecules. *The London, Edinburgh, and Dublin Philosophical Magazine and Journal of Science* 26(151): 1–25. https://doi.org/10.1080/14786441308634955
Bohr, Niels. 1913b. On the constitution of atoms and molecules. Part II Systems containing only a single nucleus. *The London, Edinburgh, and Dublin PhilosophicalMagazine and Journal of Science* 26(153): 476–502. https://doi.org/10.1080/14786441308634993
Bohr, Niels. 1976. In *Niels Bohr, Collected Works Vol 3. The Correspondence Principle (1918–1923)*, edited by J. Rud Nielsen, 67–166. Amsterdam: North Holland Publishing Company. https://doi.org/10.1016/S1876-0503(08)70084-7
Bohr, Niels, Kramers, Hans. A., and John C. Slater. 1924. Über die Quantentheorie der Strahlung. *Zeitschrift für Physik* 24: 69–87. https://doi.org/10.1007/BF01327235
Boltzmann, Ludwig. 1905. *Populäre Schriften*. Leipzig: Johann Ambrosius Barth.
Boltzmann, Ludwig. 1974. 'On the development of the methods of theoretical physics in recent times.' In *Theoretical Physics and Philosophical Problems* Vol. 5, edited by Brian McGuinness, 90–101. Dordrecht: Springer.
Brillouin, Marcel. 1926. 'H. A. Lorentz en France et en Belgique. Quelques souvenirs.' *Physica* 6: 30–35.
Brinkman, H. 2013. Coster, Dirk (1889–1950). In *Biografisch Woordenboek van Nederland*. http://resources.huygens.knaw.nl/bwn1880-2000/lemmata/bwn3/coster
Broglie, Louis de. 1924. *Recherches sur la Théorie des Quanta*, Doctoral dissertation. Published as: Broglie, Louis de. 1925. Recherches sur la Théorie des Quanta. *Annales de Physique* 10(3): 22–128. https://doi.org/10.1051/anphys/192510030022
Brower, Daniel R. 1970. 'Review: Reformers and rebels: Education in Tsarist Russia.' *History of Education Quarterly* 10(1): 127–136. https://doi.org/10.2307/367055
Bühler, Walter K. 1981. *Gauss. A Bibliographical Study*. Berlin, Heidelberg, New York: Springer.
Buitenhuis, Manuel. 2016. *Jan Tinbergen's Move from Physics to Economics, 1922–1930*. Master's thesis, Utrecht University.
Burgers, Jan. 1918. *Het atoommodel van Rutherford-Bohr*. Doctoral dissertation, Leiden University. Archives du Musée Teyler, Series III, Vol IV. Haarlem: De Erven Loosjes. https://www.lorentz.leidenuniv.nl/IL-publications/dissertations/sources/Burgers_1918.pdf
Burgers, Jan. 1962. *Autobiographical Notes*. AIP 'Niels Bohr Library & Archives Manuscript Biographies Collection.' Version used here, with a foreword by Jan Sengers and Gijs Ooms: https://ipst.umd.edu/sites/default/files/documents/burgers_autobiography.pdf
Bussato, Vittorio. 2015. *Vier grondleggers van de pedagogiek: Ph. A. Kohnstamm, M. J. Langeveld, H. W. F. Stellwag, S. Strasser*. Amsterdam.
Carathéodory, Constantine. 1909. 'Untersuchungen über die Grundlagen der Thermodynamik.' *Mathematische Annalen* 67: 355–386. https://doi.org/10.1007/BF01450409
Casimir, Hendrik. 1931. *Rotation of a rigid body in quantum mechanics*. Doctoral dissertation, Leiden University. Groningen, Den Haag: J.B. Wolters.
Casimir, Hendrik. 1983. *Haphazard Reality. Half a Century of Science*. In the 2010 edition by Amsterdam University Press.
Cercignani, Carlo. 1998. *Ludwig Boltzmann. The Man Who Trusted Atoms*. Oxford: Oxford University Press.
Clausius, Rudolf. 1865. Pogg. Ann. 125: 400. Quoted in J. Willard Gibbs. 1961. *The Scientific Papers of J. Willard Gibbs* Vol. L, 55. Thermodynamics.
Clements, Barbara Evans. 2012. *A History of Women in Russia*. Indiana Press.

Coster, B. 1926/1927. 'De ontwikkeling van het ruimte-inzicht.' *Bijvoegsel van het Nieuw Tijdschrift voor Wiskunde* 5: 143–154.
Coster, Dirk, Ehrenfest, Paul, and K. Baldus. 1924. 'Optische Dubletts und Röntgendubletts.' *Naturwissenschaften* 12: 724–726. https://doi.org/10.1007/BF01504821
Crull, Elise (2025). Hertha Sponer, Maven of Quantum Spectroscopy. In P. Charbonneau, M. Frank, M. van der Heijden, & D. Monaldi (Eds.), *Women in the History of Quantum Physics: Beyond Knabenphysik*, 82–101. Cambridge: Cambridge University Press.
Dane, J. and Chris Verhoef. 2024. 'Who's that lady? — Applying open source intelligence in a history context.' *Endeavour* 48(4): 1–10. https://doi.org/10.1016/j.endeavour.2024.100967
Darrigol, Olivier. 2000. *Electrodynamics from Ampère to Einstein*. Oxford: Oxford University Press.
Darrigol, Olivier. 2018. *Atoms, Mechanics, and Probability: Ludwig Boltzmann's Statistico-Mechanical Writings – An Exegesis*. Oxford: Oxford University Press.
Dekkers, Geertje. 2024. *Veel, klein en curieus. De wereld van Antoni van Leeuwenhoek (1632–1723)*. Amsterdam: Spectrum.
Delft, Dirk van. 2005. *Heike Kamerlingh Onnes. Een biografie. De man van het absolute nulpunt*. Amsterdam: Bert Bakker.
Delft, Dirk van. 2009. 'Koude drukte. Het laboratorium van Heike Kamerlingh Onnes als Internationaal Centrum van Lage temperaturen onderzoek.' In *Over de grens*, edited by L. Dorsman and P. R. Knegtmans, 31–52. Hilversum: Verloren.
Delft, Dirk van and Peter Kes. 2016. 'Kamerlingh Onnes en Einstein: "Uw hulp zal dus veel goeds tot stand kunnen brengen."' *Studium* 9(1–2): 51–62. https://doi.org/10.18352/studium.10127
Dijksterhuis, Eduard J. 1924/1925a. 'Moet het meetkunde-onderwijs gewijzigd worden? Opmerkingen naar aanleiding van een brochure van Mevrouw EHRENFEST-AFANASSJEEWA.' *Bijvoegsel van het Nieuw Tijdschrift voor Wiskunde* 1: 1–26.
Dijksterhuis, Eduard J. 1924/1925b. 'Antwoord aan mevrouw Ehrenfest-Afanassjewa.' *Bijvoegsel van het Nieuw Tijdschrift voor Wiskunde* 1: 60–68.
Dongen, Jeroen van. 2012. 'Mistaken identity and mirror images: Albert and Carl Einstein, Leiden and Berlin, relativity and revolution.' *Physics in Perspective* 14(2): 126–177. https://doi.org/10.1007/s00016-012-0084-y
Dresden, Max. 1987. *H. A. Kramers, Between Tradition and Revolution*. New York: Springer Verlag.
Driessen, Jozien J. 1996. *Tsaar Peter de Grote en zijn Amsterdamse vrienden*. Amsterdam: Uitgeverij Kosmos.
Driessen, Jozien J. 2006. *De Kunstkamera van Peter de Grote*. Hilversum: Uitgeverij Verloren.
Dumancic, Mirta and Shirin A. Enger. 2024. 'Pioneering women in nuclear and radiation sciences.' *Radiotherapy & Oncology* 197(110374). https://doi.org/10.1016/j.radonc.2024.110374
Duncan, Anthony and Michel Janssen. 2014. The trouble with orbits: The Stark effect in the old and the new quantum theory. *Studies in History and Philosophy of Science Part B* 48: 68–83. https://doi.org/10.1016/j.shpsb.2014.07.008
Duncan, Anthony and Michel Janssen. 2015. 'The Stark effect in the Bohr–Sommerfeld theory and in Schrödinger's wave mechanics.' In *One Hundred Years of the Bohr Atom: Proceeding from a Conference*, edited by Finn Aaserud and Helge Kragh, 217–271. Copenhagen: Royal Danish Academy of Sciences and Letters.
Duncan, Anthony and Michel Janssen. 2019. *Constructing Quantum Mechanics: Volume 1: The Scaffold: 1900–1923* (Oxford, 2019; online edn, Oxford Academic, 24 October 2019). https://doi.org/10.1093/oso/9780198845478.001.0001
Duncan, Anthony and Michel Janssen. 2023. *Constructing Quantum Mechanics Volume 2: The Arch, 1923–1927* (Oxford, 2023, online edn, Oxford Academic, 17 augustus 2023). https://doi.org/10.1093/oso/9780198883906.001.0001

Dunnington, G. Waldo. 1955. *Carl Friedrich Gauss, Titan of Science*. Washington: Mathematical Association of America.
Eckert, Michael. 1993. *Die Atomphysiker: Eine Geschichte der theoretischen Physik am Beispiel der Sommerfeldschule*. Wiesbaden: Vieweg + Teubner Verlag.
Eckert, Micheal. 2015. 'From X-rays to the h-hypothesis.' *The European Physical Journal. Special Topics: The Early Solvay Councils and the Advent of the Quantum Era* 224(10): 2057–2073. https://doi.org/10.1140/epjst/e2015-02523-1
Ehrenfest, Galinka. 1979. *De Ooievaarsregeling, een volksverzekering voor wie het kind verzorgt*. Amersfoort: Werkgroep 2000.
Ehrenfest, Hugo. 1922. *Birth Injuries of the Child*. London, New York: D. Appleton And Company.
Ehrenfest, Paul. 1903. 'Zur Berechnung der Volumkorrektion in der Zustandsgleich von Van der Waals.' *Sitzungsberichte der Kaiserliche Akademie der Wissenschaften in Wien, Mathematisch-Naturwissenschaftliche Klasse* 112: 1107–1115.
Ehrenfest, Paul. 1904. *Die Bewegung starrer Körper in Flüssigkeiten und die Mechanik von Hertz*. Doctoral dissertation, Universität Wien. https://www.lorentz.leidenuniv.nl/IL-publications/dissertations/sources/Ehrenfest_1904.pdf
Ehrenfest, Paul. 1905. 'Über die physikalischen Voraussetzungen der Planck'schen Theorie der irreversiblen Strahlungsvorgänge.' *Sitzungsberichte der Kaiserliche Akademie der Wissenschaften in Wien, Mathematisch-Naturwissenschaftliche Klasse* 114: 1301–1314.
Ehrenfest, Paul. 1906a. 'Bemerkungen zur Abhandlung des Hrn. H. Reissner: "Anwendungen der Statik und Dynamik monozyklischer Systeme auf die Elastizitätstheorie".' *Annalen der Physik* 324: 210–214. https://doi.org/10.1002/andp.19063240113
Ehrenfest, Paul. 1906b. 'Zur Stabilitätsfrage bei den Bucherer-Langevin-Elektronen.' *Physikalische Zeitschrift* 7: 302–303.
Ehrenfest, Paul. 1906c. 'Bemerkung zu einer neuen Ableitung des Wienschen Verschiebungsgesetzes.' *Physikalische Zeitschrift* 7: 527–528.
Ehrenfest, Paul 1906d. 'Zur Planckschen Strahlungstheorie.' *Physikalische Zeitschrift* 7: 528–532.
Ehrenfest, Paul 1906e. 'Bemerkung zu einer neuen Ableitung des Wienschen Verschiebungsgesetzes (Antwort auf Herrn Jeans' Entgegnung).' *Physikalische Zeitschrift* 7: 850–852.
Ehrenfest, Paul. 1906f. 'Ludwig Boltzman.' *Mathematisch-Naturwissenschaftliche Blätter*: 12.
Ehrenfest, Paul. 1907a. 'On the partition of heat energy in the molecules of gases.' *Proceedings of the Royal Society of Edinburgh* 27: 195–202. https://doi.org/10.1017/S0370164600017399
Ehrenfest, Paul. 1907b. 'Die Translation deformierbarer Elektronen und der Flächensatz.' *Annalen der Physik* 328: 204–205. https://doi.org/10.1002/andp.19073280615
Ehrenfest, Paul. 1909a. 'Die Bedeutung des Unterbrechungsfunkens für das Funktionieren elektromagnetischer Stromunterbrecher.' *Mathematisch-Naturwissenschaftliche Blätter*, 6.
Ehrenfest, Paul. 1909b. 'Wie sieht die Curve y=(−1)x aus?' *Mathematisch-Naturwissenschaftliche Blätter*, 6.
Ehrenfest, Paul. 1909c. 'Graphische Veranschaulichung des einfachsten Falles von ungleichförmiger Reihenkonvergenz.' *Mathematisch-Naturwissenschaftliche Blätter*, 6.
Ehrenfest, Paul. 1909d. *Journal of the Physics and Chemistry Society* 4: 347–349. (Russian)
Ehrenfest, Paul. 1909e. 'Gleichförmige Rotation starrer Körper und Relativitätstheorie.' *Physikalische Zeitschrift* 10: 918.
Ehrenfest, Paul. 1910a. 'Ungleichförmige Elektrizitätsbewegungen ohne Magnet- und Strahlungsfeld.' *Physikalische Zeitschrift* 11: 708–709.
Ehrenfest, Paul. 1910b. 'Zu Herrn v. Ignatowskys Behandlung der Bornschen Starrheitsdefinition.' *Physikalische Zeitschrift* 11: 1127–1129.

Ehrenfest, Paul. 1910c. 'Mißt der Aberrationswinkel im Fall einer Dispersion des Äthers die Wellengeschwindigkeit?' *Annalen der Physik* 338: 1571–1576. https://doi.org/10.1002/andp.19103381624

Ehrenfest, Paul. 1911a. 'Zu Herrn v. Ignatowskys Behandlung der Bornschen Starrheitsdefinition II.' *Physikalische Zeitschrift* 12: 412–413.

Ehrenfest, Paul. 1911b. 'Das Prinzip von Le Chatelier-Braun und die Relativitätstheorie der Thermodynamik.' *Zeitschrift für physikalische Chemie* 77: 227–244.

Ehrenfest, Paul. 1911c. 'Welche Züge der Lichtquantenhypothese spielen in der Theorie der Wärmestrahlung eine wesentliche Rolle?' *Annalen der Physik* 341: 91–118. https://doi.org/10.1002/andp.19113411106

Ehrenfest, Paul. 1912. 'Zur Frage nach der Entbehrlichkeit des Lichtäthers.' *Physikalische Zeitschrift* 13: 317–319.

Ehrenfest, Paul. 1913a. 'Zur Krise der Licht Aether-Hypothese. Rede gehalten beim Antritt des Lehramts an der Reichs-Universität zu Leiden.' Berlin: Julius Springer.

Ehrenfest, Paul. 1913b. On Einstein's theory of the stationary gravitation field *KNAW Proceedings* 15: 1187–1191. https://dwc.knaw.nl/DL/publications/PU00013051.pdf

Ehrenfest, Paul. 1913c. 'Bemerkung betreffs der spezifischen Wärme zweiatomiger Gase.' *Verhandlungen der Deutschen physikalischen Gesellschaft* 15: 451–457.

Ehrenfest, Paul. 1914a. 'Zum Boltzmannschen Entropie-Wahrscheinlichkeits-Theorem I.' *Physikalische Zeitschrift* 15: 657–663.

Ehrenfest, Paul. 1914b. 'A mechanical theorem of Boltzmann and its relation to the theory of energy quanta.' *KNAW Proceedings* 16(2): 591–597. https://dwc.knaw.nl/DL/publications/PU00012891.pdf

Ehrenfest, Paul. 1915a. 'On the kinetic interpretation of the osmotic pressure.' *KNAW Proceedings* 17: 1241–1245. https://dwc.knaw.nl/DL/publications/PU00012761.pdf

Ehrenfest, Paul. 1915b. 'Über die kinetische Interpretation des osmotischen Druckes.' *Annalen der Physik* 353: 369–374. https://doi.org/10.1002/andp.19153531905

Ehrenfest, Paul. 1915c. 'Zur Kapillaritätstheorie der Kristallgestalt.' *Annalen der Physik* 353: 360–368. https://doi.org/10.1002/andp.19153531904

Ehrenfest, Paul. 1916a. 'Some remarks on the capillarity theory of the crystalline form.' *KNAW Proceedings* 18: 173–180. https://dwc.knaw.nl/DL/publications/PU00012490.pdf

Ehrenfest, Paul. 1916b. 'On adiabatic changes of a system in connection with the quantum theory.' *KNAW Proceedings* 19: 576–597. https://dwc.knaw.nl/DL/publications/PU00012386.pdf

Ehrenfest, Paul. 1916c. 'Adiabatische Invarianten und Quantentheorie.' *Annalen der Physik* 51: 327–352. https://doi.org/10.1002/andp.19163561905

Ehrenfest, Paul. 1917a. 'On adiabatic changes of a system in connection with the quantum theory.' *KNAW Proceedings* 19(3): 576–597. https://dwc.knaw.nl/DL/publications/PU00012386.pdf

Ehrenfest, Paul. 1917b. 'Adiabatische Invarianten und Quantentheorie.' *Philosophical Magazine* 33: 500–513.

Ehrenfest, Paul. 1918a. 'A paradox in the theory of the Brownian movement.' *KNAW Proceedings* 20: 680–683. https://dwc.knaw.nl/DL/publications/PU00012258.pdf

Ehrenfest, Paul. 1918b. 'In what way does it become manifest in the fundamental laws of physics that space has three dimensions?' *KNAW Proceedings* 20: 200–209. https://dwc.knaw.nl/DL/publications/PU00012213.pdf

Ehrenfest, Paul. 1920. 'Welche Rolle spielt die Dreidimensionalität des Raumes in den Grundgesetzen der Physik?' *Annalen der Physik* 366: 440–446. https://doi.org/10.1002/andp.19203660503

Ehrenfest, Paul. 1921a. 'Le principe de correspondance.' In *Proceedings Solvay Conference 'Atomes et Électrones.'*

Ehrenfest, Paul. 1921b. 'Bemerkung über den Paramagnetismus von festen Körpern.' *Zeitschrift für Physik* 5: 35–38.

Ehrenfest, Paul. 1921c. 'Note on the paramagnetism of solids.' *KNAW Proceedings* 23: 989–992. https://dwc.knaw.nl/DL/publications/PU00014758.pdf

Ehrenfest, Paul. 1922. 'The difference between series spectra of isotopes.' *Nature* 109: 745–746. https://doi.org/10.1038/109745c0

Ehrenfest, Paul 1923a. 'Ein alter Trugschluß betreffs des Wärmegleichgewichtes eines Gases im Schwerefeld.' *Zeitschrift für Physik* 17: 421–422. https://doi.org/10.1007/BF01328698

Ehrenfest, Paul 1923b. 'Kann die Bewegung eines Systems von s Freiheitsgraden mehr als (2s−1)-fach-periodisch sein?' *Zeitschrift für Physik* 19: 242–245.

Ehrenfest, Paul 1923c. 'Das Gleichgewicht zwischen räumlichen Phasen und zweidimensionalen Phasen, die als einmolekulare Adsorptionsschichten kapillaraktiever Stoffe auftreten.' *Recueil des Travaux Chimiques des Pays-Bas* 42: 784–786. https://doi.org/10.1002/recl.19230420911

Ehrenfest, Paul. 1925. 'Energieschwankungen im Strahlungsfeld oder Kristallgitter bei Superposition quantisierter Eigenschwingungen.' *Zeitschrift für Physik* 34: 362–373.

Ehrenfest, Paul. 1927a. 'Besteht ein allgemeiner Zusammenhang zwischen der wechselseitigen Undurchdringlichkeit materieller Teilchen und dem 'Pauli-Verbot'?' *Naturwissenschaften* 15: 161–162. https://doi.org/10.1007/BF01505103

Ehrenfest, Paul. 1927b. 'Erratum: Besteht ein Zusammenhang zwischen der Undurchdringlichkeit materieller Teilchen und dem "Pauli-Verbot"?' *Naturwissenschaften* 15: 268. https://doi.org/10.1007/BF01506067

Ehrenfest, Paul. 1927c. 'Relation between the reciprocal impenetrability of matter and Pauli's exclusion principle.' *Nature* 119: 196. https://doi.org/10.1038/119196a0

Ehrenfest, Paul. 1927d. 'Erratum: Relation between the reciprocal impenetrability of matter and Pauli's exclusion principle.' *Nature* 119: 602. https://doi.org/10.1038/119602c0

Ehrenfest, Paul. 1927e. 'Bemerkung über die angenäherte Gültigkeit der klassischen Mechanik innerhalb der Quantenmechanik.' *Zeitschrift für Physik* 45: 455–457. https://doi.org/10.1007/BF01329203

Ehrenfest, Paul. 1928. 'Grafrede.' *Physica* 8: 101–104.

Ehrenfest, Paul. 1929. 'Bemerkungen über den Diamagnetismus von festem Wismut.' *Zeitschrift für Physik* 58: 719–721.

Ehrenfest, Paul. 1932. 'Einige die Quantenmechanik betreffende Erkundigungsfragen.' *Zeitschrift für Physik* 78: 555–559.

Ehrenfest, Paul. 1933. 'Phasenumwandlungen im üblichen und erweiterten Sinn, classifiziert nach den entsprechenden Singularitäten des thermodynamischen Potentiales.' *KNAW Proceedings* 36: 153–157. https://dwc.knaw.nl/DL/publications/PU00016385.pdf

Ehrenfest, Paul and George Breit. 1922. 'Ein bemerkenswerter Fall von Quantisierung.' *Zeitschrift für Physik* 9: 207–210.

Ehrenfest, Paul, and George Breit. 1923. 'A remarkable case of quantization.' *KNAW Proceedings* 25: 2–5. https://dwc.knaw.nl/DL/publications/PU00014844.pdf

Ehrenfest, Paul, and Tatiana Ehrenfest. 1906a. Über eine Aufgabe aus der Wahrscheinlichkeitsrechnung, die mit der kinetischen Deutung der Entropievermehrung zusammenhängt. *Mathematisch-Naturwissenschaftliche Blätter*, 3.

Ehrenfest, Tatiana, and Paul Ehrenfest. 1906b. 'Bemerkung zur Theorie der Entropiezunahme in der 'Statistischen Mechanik' von W. Gibbs.' *Sitzungsberichte der Kaiserliche Akademie der Wissenschaften in Wien, Mathematisch-Naturwissenschaftliche Klasse* 115: 89–98.

Ehrenfest, Paul, and Tatiana Ehrenfest. 1907. 'Über zwei bekannte Einwände gegen das Boltzmannsche H-Theorem.' *Physikalische Zeitschrift* 8: 311–314.

Ehrenfest, Paul, and Tatiana Ehrenfest. 1911. 'Begriffliche Grundlagen der statistischen Auffassung in der Mechanik.' In *Encyklopädie der mathematischen Wissenschaften IV, 2. Teil*, edited by F. Klein, and C. Müller, 3–90. Leipzig: Teubner. https://doi.org/10.1007/978-3-663-16028-1_11

Ehrenfest, Paul, and Tatiana Ehrenfest. 1959. *The Conceptual Foundations of the Statistical Approach in Mechanics*. Ithaca New York: Cornell University Press.

Ehrenfest, Paul and Paul Epstein. 1927. 'Remarks on the quantum theory of diffraction.' *Proceedings of the National Academy of Sciences USA* 13: 400–408. pnas01831-0052.pdf
Ehrenfest, Paul and J. Robert Oppenheimer. 1931. 'Note on the statistics of nuclei.' *Physical Review* 37: 333–338. https://doi.org/10.1103/PhysRev.37.333
Ehrenfest, Paul and Richard C. Tolman. 1924. 'Weak quantization.' *Physical Review* 24: 287–295. https://doi.org/10.1103/PhysRev.24.287
Ehrenfest, Paul and Arend Joan Rutgers. 1928. 'Bemerkung zur wellenmechanischen Deutung des limitären Ramsauer-Effektes.' *Naturwissenschaften* 16: 184. https://doi.org/10.1007/BF01506449
Ehrenfest, Paul and Arend Joan Rutgers. 1929a. 'Zur Thermodynamik und Kinetik der thermo-elektrischen Erscheinungen in Krystallen, insbesondere des Bridgman-Effektes. I.' *KNAW Proceedings* 32: 698–706. *https://dwc.knaw.nl/DL/publications/PU00015759.pdf*
Ehrenfest, Paul and Arend Joan Rutgers. 1929b. 'Zur Thermodynamik und Kinetik der thermo-elektrischen Erscheinungen in Krystallen, insbesondere des Bridgman-Effektes. II.' *KNAW Proceedings* 32: 883–893. *https://dwc.knaw.nl/DL/publications/PU00015785.pdf*
Ehrenfest, Paul and George E. Uhlenbeck. 1926a. 'Die wellenmechanische Interpretation der Boltzmannschen Statistik neben der der neueren Statistiken.' *Zeitschrift für Physik* 41: 24–26.
Ehrenfest, Paul and George E. Uhlenbeck. 1926b. 'Graphische Veranschaulichung der De Broglieschen Phasenwellen in der fünfdimensionalen Welt von O. Klein.' *Zeitschrift für Physik* 39: 495–498.
Ehrenfest, Paul and George E. Uhlenbeck. 1927. 'Zum Einsteinschen "Mischungsparadoxon".' *Zeitschrift für Physik* 41: 576–582.
Ehrenfest, Tatiana P. 1931. *Oppervlakken met scharen van gesloten geodesische lijnen*. Doctoral dissertation, Leiden. Amsterdam: H.J. Paris.
Ehrenfest, Tatiana. 1913. 'Zur Frage über die Konzentrationsschwankungen in Radioaktiven Lösungen.' *Physikalische Zeitschrift* 14, 675–676.
Ehrenfest-Afanassjewa, Tatiana. 1916. 'On Tolman's principle of similitude.' *Physical Review* 8, 1–7. https://doi.org/10.1103/PhysRev.8.1
Ehrenfest-Afanassjewa, Tatiana. 1924. *Wat kan en moet het Meetkunde-onderwijs aan een niet-wiskundige geven?* Groningen en Den Haag: J.B. Wolters.
Ehrenfest-Afanassjewa, Tatiana. 1924/1925. 'Moet het Meetkunde-onderwijs gewijzigd worden? Een antwoord aan den heer E. J. Dijksterhuis.' *Bijvoegsel van het Nieuw Tijdschrift voor Wiskunde* 7: 47–59.
Ehrenfest-Afanassjewa, Tatiana. 1925a. 'Zur Axiomatisierung des zweiten Hauptsatzes der Thermodynamik.' *Zeitschrift für Physik* 33: 933–945. https://doi.org/10.1007/BF01328381
Ehrenfest-Afanassjewa, Tatiana. 1925b. 'Berichtigung zu der Arbeit: Zur Axiomatisierung des zweiten Hauptsatzes der Thermodynamik.' *Zeitschrift für Physik* 34: 638. https://doi.org/10.1007/BF01328510
Ehrenfest-Afanassjewa, Tatiana. 1931. *Übungensammlung zu einer geometrische Propädeuse*. Dordrecht: Springer. https://doi.org/10.1007/978-94-011-9597-3
Ehrenfest-Afanassjewa, Tatiana. 1946. *Relevia, een nieuw economisch systeem, een orde, waarin ik zelf ook graag zou willen leven*. Den Haag: Boucher.
Ehrenfest-Afanassjewa, Tatiana. 1956. *Die Grundlagen der Thermodynamik*. Den Haag: Brill.
Ehrenfest-Afanassjewa, Tatiana. 1958. 'On the use of the notion "probability".' *American Journal of Physics* 26: 388. https://doi.org/10.1119/1.1996167
Ehrenfest-Afanassjewa, Tatiana. 1961. *Didactische opstellen*. Zutphen: Thieme.
Ehrenfest, Tatiana. 1911Ra. 'On the application of probability theory to regular phenomena.' *Journal of the Natural and Chemical Society, Physics Section*, 5. (Ra).

Ehrenfest, Tatiana. 1911Rb. 'The principle of similarity and its applications.' *Journal of the Natural and Chemical Society, Physics Section* 7 (Rb).
Ehrenfest, Tatiana. 1912R. 'The principle of dimensions.' *Journal of the Natural and Chemical Society, Physics Section* 7 (Russian).
Ehrenfest, Tatiana. 1914R. 'On the principle of corresponding states.' *Journal of the Natural and Chemical Society, Physics Section* 7 (R.).
Ehrenfest-Afanassjewa, Tatiana. 1928Ra. 'Irreversibility, one-sidedness and the second law of thermodynamics.' *Journal of Applied Physics* 3–4 (Ra.).
Ehrenfest-Afanassjewa, Tatiana. 1928Rb. 'Geometric intuition and physical experience.' *Proceedings of the Teachers' Academy M.V. Froenze Volume 2, Section 5* (Russian).
Ehrenfest-Afanassjewa, Tatiana. 1930R. 'How to begin teaching geometry.' *Physics, Chemistry and Engineering in the Soviet School* 5 (R).
Ehrenfest-Afanassjewa, Tatiana. 1930R. *What are Coordinates and Physical Quantities*. Publication of the Teachers' Academy. Simferopol (Russian).
Ehrenfest-Afanassjewa, Tatiana. 1930R. 'On the interpretation of the second law of thermodynamics in the theory of Max Planck.' *Journal of Applied Physics* 1 (R).
Ehrenfest-Afanassjewa, Tatiana. 1931R. 'The results of the kinetic theory.' *Journal for Experimental and Theoretical Physics* 6 (R).
Ehrenfest, Tatiana, and Paul Ehrenfest. 1906a. 'Bemerkung zur Theorie der Entropiezunahme in der 'Statistischen Mechanik' von W. Gibbs.' *Sitzungsberichte der Kaiserliche Akademie der Wissenschaften in Wien, Mathematisch-Naturwissenschaftliche Klasse* 115: 89–98.
Einstein, Albert. 1905a. *A new determination of molecular dimensions*. Doctoral dissertation, Universität Zürich.
Einstein, Albert. 1905b. 'Über einen die Erzeugung und Verwandlung des Lichtes betreffenden heuristischen Gesichtspunkt.' *Annalen der Physik* 17: 132–148. https://doi.org/10.1002/andp.19053220607
Einstein, Albert. 1905c. 'Über die von der molekularkinetischen Theorie der Wärme geforderte Bewegung von in ruhenden Flüssigkeiten suspendierten Teilchen.' *Annalen der Physik* 17(8): 549–560. https://doi.org/10.1002/andp.19053220806
Einstein, Albert. 1905d. 'Zur Elektrodynamik bewegter Körper.' *Annalen der Physik*, 17(10): 891–921. https://doi.org/10.1002/andp.19053221004
Einstein, Albert. 1905e. 'Ist die Trägheit eines Körpers von seinem Energieinhalt abhängig?' *Annalen der Physik* 18(13): 639–641. https://doi.org/10.1002/andp.19053231314
Einstein, Albert. 1907. 'Bemerkungen zu der Notiz von Hrn P. Ehrenfest: Die Translation deformierbarer Elektronen und der Flächensatz.' *Annalen der Physik* 23: 206. https://doi.org/10.1002/andp.19073280616
Einstein, Albert. 1911. 'Zum Ehrenfestschen Paradoxon.' *Zeitschrift für Physik* 12: 509.
Einstein, Albert. 1914. 'Beiträge zur Quantentheorie.' *Verhandlungen der Deutschen Physikalischen Gesellschaft* 16: 820–828.
Einstein, Albert. 1920. 'Äther und Relativitäts-Theorie.' In *Äther und Relativitäts-Theorie*, 1–20. Springer. https://doi.org/10.1007/978-3-642-50746-5_1
Einstein, Albert. 1934. 'Nachruf Paul Ehrenfest.' In *Almanak van het Leidsche Studenten-Corps*, 94–97. Leiden.
Einstein, Albert. 1953. 'H. A. Lorentz als Schopfer und als Personlichkeit.' In *Tentoonstellingscatalogus*, 8.
Einstein, Albert. 1954. *Ideas and Opinions*, edited by Carl Seelig. New York: Crown Publishers.
Einstein, Albert. 1976. 'Paul Ehrenfest in Memoriam.' In *Out of My Later Years*, 236–239.
Einstein, Albert, and Paul Ehrenfest. 1922. 'Quantentheoretische Bemerkungen zum Experiment von Stern und Gerlach.' *Zeitschrift für Physik* 11: 31–34. https://doi.org/10.1007/BF01328398

Einstein, Albert and Paul Ehrenfest. 1923. 'Zur Quantentheorie des Strahlungsgleichgewichts.' *Zeitschrift für Physik* 19: 301–306. https://doi.org/10.1007/BF01327565

Elsasser, Walter. 1978. *Memoirs of a Physicist in the Atomic Age.* New York: Science History (Neal Watson).

Engel, Barbara Alpern. 2004. *Women in Russia 1700–2000.* Cambridge: Cambridge University Press.

Epstein, Alice. 1965. *Paul Epstein. Memoir Recorded with His Wife Alice Epstein.* Pasadena: Archives California Institute of Technology.

Epstein, Paul and Paul Ehrenfest. 1924. 'The quantum theory of the Fraunhofer diffraction.' *Proceedings of the National Academy of Sciences USA* 10: 133–139. pnas01865-0015.pdf

Fehr, Henri. 1908. 'Enquête de l' «Enseignement mathématique» sur la méthode de travail des mathématiciens.' *Revue de Métaphysique et de Morale* 16(5): 10–11.

Fermi, Enrico. 1924. 'Berekeningen over de intensiteiten van spectraallijnen.' *Physica* 7: 340–343.

Frewer, Magdalene. 1979. *Das mathematische Lesezimmer der Universität Goettingen unter der Leitung von Felix Klein: 1886–1922. Hausarbeit z. Prüfung f.d. höheren Bibliotheksdienst.* Cologne: Bibliothekar-Lehrinstitut des Landes Nordrhein-Westfalen.

Flamm, Dieter. 1985. *Hochgeehrter Herr Professor! Innig geliebter Louis! Ludwig Boltzmann, Henriette von Aigentler, Briefwechsel.* Wien: Bohlau Verlag.

Forman, P. 1975. Ritz, Walter. In *Dictionary of Scientific Biographies* 11, 475–481. New York: Charles Scribner's sons.

Franck, James and Hertha Sponer. 1962. Interview of James Franck and Hertha Sponer, session V Franck by Thomas S. Kuhn and Maria Goeppert Mayer on 1962 July 13, Niels Bohr Library & Archives, AIP. www.aip.org/history-programs/niels-bohr-library/oral-histories/4609-5

Frenkel, Viktor Y. 1977. *Paul Ehrenfest.* Moscow: Atomizdat.

Frenkel, Viktor Y. 2001. *Yakov Ilich Frenkel, His Life, His Work, His Letters.* Basel: Birkhäuser Verlag.

Fuchs, Stefan, Janina von Stebut, and Jutta Allmendinger. 2001. 'Gender, science, and scientific organizations in Germany.' *Minerva* 39(2): 175–201. https://doi.org/10.1023/A:1010380510013

Gapon, George. 1906. *The Story of My Life.* London: Chapman & Hall.

Gavroglu, Kostas. 1995. *Fritz London. A Scientific Biography.* Cambridge: Cambridge University Press.

Gay, Peter. 1988. *Freud. A Life for Our Time.* In the 1998 edition by New York: Norton.

Gerhard, Uwe-Jens, Schönberg, A., and B. Blanz. 2008. 'Johannes Trüper – Mittler zwischen Kinderpsychiatrie und Pädagogik.' *Zeitschrift für Kinder- und Jugendpsychiatrie und Psychotherapie* 36, 55–63. https://doi.org/10.1024/1422-4917.36.1.55

Goeppert-Mayer, Maria. 1962. Interview of Maria Goeppert Mayer by Thomas S. Kuhn on 1962 February 20, Niels Bohr Library & Archives, AIP, www.aip.org/history-programs/niels-bohr-library/oral-histories/4770

Goldhammer, Leo. 1927. *Die Jude Wiens.* Vienna: Löwit.

Görs, Britta P., Psarros Nilolaos, and Paul Ziche. 2005. *Wilhelm Ostwald at the Cross Roads Between Chemistry, Philosophy and Media Culture.* Leipzig: Leipziger Universitätsverlag.

Gorter, Cornelis J. 1962. Interview of Cornelis J. Gorter by John L. Heilbron on 1962 November 15, Niels Bohr Library & Archives, AIP. www.aip.org/history-programs/niels-bohr-library/oral-histories/4639

Goudsmit, Samuel. 1921. 'Relativitische Auffassung des Doublets.' *Die Naturwissenschaften* 49: 995. https://doi.org/10.1007/BF01489623

Goudsmit, Samuel. 1922. 'Les doublets dans les spectres visbles.' *Archives Néerlandaises des Sciences exactes et naturelles* IV, 116.

Goudsmit, Samuel. 1925. 'Über die Komplexstruktur der Spektren. *Zeitschrift für Physik* 32: 794–798. https://doi.org/10.1007/BF01331715

Goudsmit, Samuel. 1963. Interview of Samuel A. Goudsmit by Thomas S. Kuhn on December 5, 1963. Niels Bohr Library & Archives, AIP. www.aip.org/history-programs/niels-bohr-library/oral-histories/4640-1

Goudsmit, Samuel. 1971. 'De ontdekking van de electronenrotatie.' *Nederlands Tijdschrift voor Natuurkunde* 37(16): 386–392.

Goudsmit, Samuel and George Uhlenbeck. 1926. 'Over Het Roteerende Electron En de Structuur der Spectra.' *Physica* 6: 273–290.

Gray, Jeremy. 2008. 'Modernism in mathematics.' In *The Oxford Handbook of the History of Mathematics*, edited by E. S. Robson. Oxford: Oxford University Press.

Guichelaar, Jan. 2016. 'De Sitter en Einstein. 'Het lijkt mij dat Einstein hier een vergissing begaan heeft.' *Studium* 9(1-2): 77–95. https://doi.org/10.18352/studium.10129

Halbertsma, Th. 1922. 'Over mongoloïde idiotie, naar aanleiding van een paar gevallen bij tweelingen.' *Nederlands Tijdschrift voor Geneeskunde* 66: 22–32.

Halpern, Paul. 2004. 'Nordström, Ehrenfest and the role of dimensionality in physics.' *Physics in Perspective* 6: 390–400. https://doi.org/10.1007/s00016-004-0221-3

Halpern, Paul. 2007. 'Klein, Einstein, and five-dimensional unification.' *Physics in Perspective* 9: 390–405. https://doi.org/10.1007/s00016-006-0319-x

Hamm, Michael F. 1993. *Kiev. A Portrait 1800–1917*. Princeton: Princeton University Press.

Hans, Nicholas. 1963. *The Russian Tradition in Education*. London: Routledge.

Harman, P. 1995. *The Scientific Papers and Letters of James Clerk Maxwell, Volume II*. Cambridge: Cambridge University Press.

Heijden, M. W. van der. 2016. 'Afanassjewa en Einstein. Wederzijdse waardering.' *Studium* 9(1-2), 63–76. https://doi.org/10.18352/studium.10128

Heijden, Margriet W. van der. 2020. 'Tatjana Ehrenfest Afanassjewa: no talent for subservience.' In *The Legacy of Tatjana Afanassjewa, Philosophical Insights from the Work of an Original Physicist and Mathematician*, edited by Jos Uffink et al., 1–20. Cham: Springer. https://doi.org/10.1007/978-3-030-47971-8_1

Heilbron, John L. 1996. *The Dilemma's of an Upright Man. Max Planck and the Fortunes of German Science*. Berkeley: University of California Press.

Heilbron, John L. 2015. 'British participation in the first Solvay councils on Physics.' *The European Physical Journal. Special Topics: The Early Solvay Councils and the Advent of the Quantum Era* 224(10): 2044–2051. https://doi.org/10.1140/epjst/e2015-02522-5

Heindl, Waltraud E. 1993. *Durch Erkenntnis zu Freiheit und Glück ... Frauen an der Universität Wien (ab 1897)*. Wien: Universitätsverlag.

Hevesy, Georg. 1962. Interview of Georg von Hevesy by Thomas S. Kuhn, Emilio Segrè and John Heilbron on 1962 May 25, Niels Bohr Library & Archives, AIP. www.aip.org/history-programs/niels-bohr-library/oral-histories/4670-1

Hilbert, David and S. Cohn-Vossen. 1952. *Geometry and the Imagination*. New York: Chelsea Publishing Company.

Hitchcock, A. S. 1925. Success in Science. *Science* 62(1598): 141–144. http://www.jstor.org/stable/1650497

Hoffmann, Dieter. 2006. 'Peter Debye (1884–1966) – ein typischer Wissenschaftler in untypischer Zeit.' *Gewina* 29: 141–168. https://dspace.library.uu.nl/handle/1874/251645

Hollestelle, Marijn. 2011. *Paul Ehrenfest. Worstelingen met de moderne wetenschap (1880–1933)*. Doctoral dissertation, Leiden University. Leiden: Leiden University Press.

Hollestelle, Marijn. 2016. Ehrenfest en Einstein. Menselijke katalysator van het heldere denken. *Studium* 9(1-2): 26–50. https://doi.org/10.18352/studium.10126

Horn, Linda. (2019). *Galinka Ehrenfest en El Pintor: 'Vraag Einstein of hij mijn viool meeneemt*. Amsterdam: Uitgeverij De Buitenkant.

Hsu, Jong-Ping and Yuan-Zhong Zhang. 2001. *Lorentz and Poincaré Invariance: 100 Years of Relativity*. Singapore: World Scientific.

Huijnen, Pim. 2003. *Die Grenze des Pathologischen – het leven van fysicus Paul Ehrenfest, 1904–1912*. Master's thesis, Rijksuniversiteit Groningen.
Huijnen, Pim and Anne Kox. 2007. 'Paul Ehrenfest's Rough road to Leiden: A physicist's search for a position, 1904–1912.' *Physics in Perspective* 9: 186–211. https://doi.org/10.1007/s00016-006-0287-1
Icke, Vincent. 2011. *De Sterrewacht straalt weer*. Universiteit Leiden.
Ioffe, Abraham. 1967. *Begegnungen mit Physikern*. Leipzig: Teubner.
Isaacson, Walter. 2007. *Einstein, His Life and Universe*. New York: Simon & Schuster.
Jacobs, Aletta. 1919. 'Openingswoord.' *Maandblad van de Nederlandsche Vereeniging van Staatsburgeressen (vroeger Vereeniging voor Vrouwenkiesrecht)* 10: 1.
Jalloh, Mahmoud. 2024. 'Metaphysics and convention in dimensional analysis, 1914–1917.' *HOPOS: The Journal of the International Society for the History of Philosophy of Science* 14(2): 275–322. https://doi.org/10.1086/731678
James, Ioan. 2002. *Remarkable Mathematicians. From Euler to Neumann*. Cambridge: Cambridge University Press.
Janssen, Michel. 2002. 'Reconsidering a scientific revolution: The case of Einstein versus Lorentz.' *Physics in Perspective* 7: 421–446. https://doi.org/10.1007/s000160200003
Janssen, Michel. 2005. 'Of pots and holes: Einstein's bumpy road to general relativity.' *Annalen der Physik* 517 Series 8, 14(S1): 58–85. https://doi.org/10.1002/andp.2005517S105
Janssen, Michel and Anthony Duncan. 2023. *Constructing Quantum Mechanics Volume Two: The Arch, 1923–1927* (Oxford, 2023; online edn, Oxford Academic, 19 October 2023). https://doi.org/10.1093/oso/9780198883906.001.0001
Janssen, Michel and Christoph Lehner (Eds.). 2014. *The Cambridge Companion to Einstein*. Cambridge: Cambridge University Press.
Janssen, Michel and Matthew Mecklenburg. 2006. 'From classical to relativistic mechanics: Electromagnetic models of the electron.' In *Mathematics, Physics and Philosophy, 1860–1930*, edited by Vincent F. Hendricks et al., 65–134. Dordrecht: Springer.
Jolink, Albert. 2003. *Jan Tinbergen. The Statistical Turn in Economics: 1903–1955*. Rotterdam: Chimes.
Jones, Claire. 2000. 'Grace Chisholm Young: Gender and mathematics around 1900.' *Women's History Review* 9(4): 675–693. https://doi.org/10.1080/09612020000200266
Josephson, Paul R. 1991. *Physics and Politics in Revolutionary Russia*. Berkeley: University of California Press.
Kassow, Samuel D. 1989. *Students, Professors, and the State in Tsarist Russia*. Berkeley: University of California Press.
Kaufmann, Walter. 1906. 'Über die Konstitution des Elektrons.' *Annalen der Physik* 19: 487–553. https://doi.org/10.1002/andp.19063240303
Kemme, Sieb and Ed de Moor. 2012. 'Meetkundeonderwijs op gymnasium en hbs 1900–1968.' *Nieuw Archief voor Wiskunde* 5(13): 102–109. naw5-2012-13-2-102.pdf
Kirejczyk, Marta. 1993. 'Vrouwen kozen exact; studie en beroepsuitoefening rond de eeuwwisseling.' *Gewina* 16: 234–247. https://dspace.library.uu.nl/handle/1874/251237
Klatt, Johanna and Robert Lorenz (Eds.). 2010. *Manifeste: Geschichte und Gegenwart des politischen Appells*. Transcript Verlag. http://www.jstor.org/stable/j.ctv1fxfsm
Klein, Martin J. 1965. 'Einstein, Specific Heats, and the Early Quantum Theory.' *Science, New Series* 148(3667): 173–181. http://doi.org/10.1126/science.148.3667.173
Klein, Martin J. 1970. *Paul Ehrenfest. The Making of a Theoretical Physicist*. Amsterdam, Londen: North-Holland Publishing Company.
Klein, Martin J. 1981. 'Not by discoveries alone: The centennial of Paul Ehrenfest.' *Physica* 106A: 3–14. https://doi.org/10.1016/0378-4371(81)90201-6
Klein, Martin J. 1989. 'Physics in the making in Leiden: Paul Ehrenfest as a teacher.' In *Physics in the Making. Essays on Developments in 20th Century Physics, In Honour of H.B.M.G. Casimir*, 1–20. Haarlem, Amsterdam, Oxford, New York, Tokyo: North-Holland.

Klein, Martin J. 2010. 'Paul Ehrenfest, Niels Bohr and Albert Einstein: Colleagues and friends.' *Physics in Perspective* 12: 307–337. https://doi.org/10.1007/s00016-010-0025-6

Kloek, Els. 2009. *Vrouw des huizes. Een cultuurgeschiedenis van de Nederlandse huisvrouw.* Amsterdam: Uitgeverij Balans.

Koblitz, Ann. 1993. *A Convergence of Lives: Sofia Kovalevskaia – Scientist, Writer, Revolutionary.* New Brunswick: Rutgers University Press.

Koblitz, Ann. 2000. *Science, Women and Revolution in Russia.* New York University Press.

Kojevnikov, Alexei. 2002. 'The Great War, the Russian Civil War, and the invention of big science.' *Science in Context*, 15(2), 239–275. https://doi.org/10.1017/s0269889702000443

Kojevnikov, Alexei. 2004. *Stalin's Great Science.* London: Imperial College Press.

Kojevnikov, Alexei. 2020. *The Copenhagen Network: The Birth of Quantum Mechanics from a Postdoctoral Perspective.*

Kolmogorov, A. Y. and A.P. Yushkevich. 1998. *Mathematics of the 19th Century*, translated by R. Cooke. Basel, Boston, Berlin: Birkhäuser.

Kox, Anne J. 1993. 'Einstein, Lorentz, Leiden and general relativity.' *Classical and Quantum Gravity* 10, S187. https://doi.org/10.1088/0264-9381/10/S/020

Kox, Anne J. 1997. 'The discovery of the electron: II. The Zeeman effect.' *European Journal of Physics* 18: 139–144. https://doi.org/10.1088/0143-0807/18/3/003

Kox, Anne J. 2008. *The Scientific Correspondence of H. A. Lorentz, Volume I.* New York: Springer.

Kox, Anne J. 2018. *The Scientific Correspondence of H. A. Lorentz, Volume II.* Cham: Springer.

Kox, Anne J. and Henriëtte F. Schatz. 2021. *A Living Work of Art: The Life and Science of Hendrik Antoon Lorentz.* Oxford: Oxford Academic. https://doi.org/10.1093/oso/9780198870500.001.0001

Kox, Anne J. and W. Troelstra. 1996. 'Uit het Zeeman-archief: de ontdekking van het Zeeman-effect.' *Gewina* 19: 153–166.

Kragh, Helge. 2002. *Quantum Generations. A History of Physics in the Twentieth Century.* Princeton: Princeton University Press.

Kragh, Helge. 2012. *Niels Bohr and the Quantum Atom.* Oxford: Oxford University Press.

Kragh, Helge. 2023. A terminological history of elementary particle physics. *Archive for History of Exact Sciences* 77: 73–120. https://doi.org/10.1007/s00407-022-00299-2

Kramers, Hendrik A. 1919. *Intensities of spectral lines.* Doctoral dissertation, Leiden University. Kopenhagen: Bianco Lunos Bogtrykkeri. https://www.lorentz.leidenuniv.nl/ILpublications/Kramers.html

Kramers, Hendrik A. 1951. Levensbericht D. Coster. In *Jaarboek, 1951–1952*, 198–201. Amsterdam.

Krutkow, G. (Yuri). 1919. 'Contribution to the theory of adiabatic invariants.' *KNAW Proceedings* 21(8): 1112–1123. https://dwc.knaw.nl/DL/publications/PU00012165.pdf

Lambert, Franklin. 2015. 'Einstein's witches' sabbath in Brussels: The legend and the facts.' *The European Physical Journal. Special Topics: The Early Solvay Councils and the Advent of the Quantum Era* 224: 2023–2040. https://doi.org/10.1140/epjst/e2015-02521-9

Landé, Alfred. 1926. 'Axiomatische Begründung der Thermodynamik durch Carathéodory.' In *Handbuch der Physik 9*.

Langevin, Paul. 1934. *Rapports et Discussions du Septième Conseil de Physique tenu a Bruxelles du 22 au 29 Octobre 1933. Sous les Auspices de l'Institut International de Physique Solvay.* Paris.

Lieven, Dominic. 2015. *Oorlog & Revolutie.* Houten: Spectrum.

Linders, A. 2018. Corry Tendeloo. In *1001 Vrouwen uit de Nederlandse geschiedenis*, edited by Els Kloek, 646. Nijmegen: VanTilt.

Historische Vereniging Oud Leiden. 1914, 1915, 1916. *Korte kroniek van Leiden en Rijnland. Leidsch Jaarboekje.* Leiden.

LeBlanc, Ronald D. 2001. Vegetarianism in Russia: The Tolstoy(an) Legacy. *The Carl Beck Papers in Russian and East European Studies 1507.*
Lewin Sime, Ruth. 1996. *Lise Meitner. A Life in Physics.* Berkeley and Los Angeles: University of California Press.
de Haas-lorentz, Geertruida L. 1912. *Over de theorie van de Brown'schen beweging en daarmede verwante verschijnselen.* Doctoral dissertation, Leiden University. Leiden: Eduard IJdo. https://ilorentz.org/history/proefschriften/sources/deHaasLorentz_1912.pdf
Lorentz, Hendrik A. 1875. *Over de theorie der terugkaatsing en breking van het licht* Doctoral dissertation, Leiden University. Leiden: Universiteit Leiden.
Lorentz, Hendrik A. 1892. La Théorie Électromagnétique de Maxwell et Son Application Aux Corps Mouvants. In: Collected Papers 1936. Dordrecht: Springer. https://doi.org/10.1007/978-94-015-3447-5_4
Lorentz, Hendrik A. 1895. *Versuch einer Theorie der electrischen und optischen Erscheinungen in bewegten Körpern.* In: Collected Papers 1937. Dordrecht: Springer. https://doi.org/10.1007/978-94-015-3445-1_1
Lorentz, Hendrik A. 1904. Electromagnetic phenomena in a system moving with any velocity smaller than that of light. *KNAW Proceedings 6:* 809–831. In: Collected Papers 1937. Dordrecht: Springer. https://doi.org/10.1007/978-94-015-3445-1_5
Lorentz, Hendrik A. 1913. 'De internationale wetenschap bevordert den vreede.' *Vrede door Recht* 14.
Lunteren, Frits H. van. 2003. *Paul Ehrenfest: de Leidse onderzoekschool van een fysicus in diaspora.* Leiden.
Maas, Ad. 2001. *Atomisme en individualisme. De Amsterdamse natuurkunde tussen 1877 en 1940.* Doctoral dissertation, University of Amsterdam. Hilversum: Verloren.
Maas, Ad and Fred de Heij. 2015. *Ehrenfest!* Leiden: Rijksmuseum Boerhaave.
Margolis, Alexandre D. 2004. *Encyclopaedia of St. Petersburg.* Moscow: Rosspen.
Martínez, Alberto A. 2004. 'Ritz, Einstein, and the emission hypothesis.' *Physics in Perspective* 6: 4–28. https://doi.org/10.1007/s00016-003-0195-6
Maxwell, James C. 1965. *Scientific Papers Vol II*, edited by W. D. Niven. New York: Dover Publications.
McMeekin, Sean. 2017. *De Russische revolutie, een nieuwe geschiedenis*, translated by G. van Wardt. Amsterdam: Nieuw-Amsterdam.
Mehra, Jagdish and Helmut Rechenberg. 1982-2000. *The Historical Development of Quantum Theory.* New York, Heidelberg, Berlin: Springer.
Metz, Karlien. 2019. *De mensen achter El Pintor.* Amsterdam: Verzetsmuseum.
Minkowski, Hermann. 1915. 'Das Relativitätsprinzip.' *Annalen der Physik* 352(15): 927–938. https://doi.org/10.1002/andp.19153521505
Minkowski, Hermann. 2012. *Space and Time: Minkowski's papers on relativity*, translated by F. Lewertoff and V. Petkov. Montreal: Minkowski Institute Press.
Mittler, Elmar G. and Silke Glitsch. 2005. 'Wie der Blitz einschlägt, hat sich das Räthsel gelöst.' Carl Friedrich Gauß in Göttingen. Göttingen: Universitätsverlag Göttingen.
Molenaar, Leo. 2003. *De rok van het universum.* Amsterdam: Balans.
Moor, Ed de. 1993. 'Het 'gelijk' van Tatiana Ehrenfest-Afanassjewa.' *Nieuwe Wiskrant* 12(4): 15–24.
Moor, Ed de. 1999. *Van vormleer naar realistische meetkunde.* Doctoral dissertation, Utrecht University. Amersfoort: Wilco. https://www.fisme.science.uu.nl/publicaties/literatuur/1999_moor_de_0-693.pdf
Moor, Ed de and Adri Treffers. 1974/75. 'Het aanvankelijk meetkundeonderwijs, de jaren 1920 tot heden.' *Euclides* 50(2): 41–44.
Moore, Walther J. 1989. *Schrodinger. Life and Thought.* Cambridge: Cambridge University Press.
Morrissey, Susan K. 1998. *Heralds of Revolution. Russian Students and the Mythologies of Radicalism.* Oxford: Oxford University Press.

Moskovchenko, N. I. and Viktor Frenkel. 1990. *Ehrenfest–Ioffe– Nauchnaya Perepiska 1907–1933*. Leningrad.
Muller, Fred. 1997a. 'The equivalence myth of quantum mechanics – Part 1.' *Studies in History and Philosophy of Modern Physics* 28(1): 35–61. https://doi.org/10.1016/S1355-2198(96)00022-6
Muller, Fred. 1997b. 'The equivalence myth of quantum mechanics – Part II.' *Studies in History and Philosophy of Modern Physics* 28(2): 219–247. https://doi.org/10.1016/S1355-2198(97)00001-4
Muller, Fred. 1999. 'Addendum.' *Studies in History and Philosophy of Modern Physics*, 30: 543–545. https://doi.org/10.1016/S1355-2198(99)00016-7
Navarro, Luis and Enric Pérez. 2004. 'Paul Ehrenfest on the necessity of quanta (1911): discontinuity, quantization, corpuscularity, and adiabatic invariance.' *Archive for History of Exact Science* 58: 97–141. https://doi.org/10.1007/s00407-003-0068-z
Neuenschwander, Erwin; Burmann, Hans-Wilhelm. 1994. Die Entwicklüng der Mathematik an der Universität Göttingen. In Hans Günther Schlotter (Ed.), *Die Geschichte der Verfassung und der Fachbereiche der Georg August Universität zu Göttingen*: 141–159. Göttingen: Vandenhoeck&Ruprecht.
Neuenschwander, Dwight E. 2013–2014. 'The struggles of Paul Ehrenfest.' *The SPS Observer*, 24–29.
Neuenschwander, Dwight E. 2017. *Emmy Noether's Wonderful Theorem*. Baltimore: John Hopkins University Press.
Noether, Emmy. 1918. Invariante Variationsprobleme. *Nachrichten der Göttinger Gesellschaft der Wissenschaften S*, 235–257.
Noether, Emiliana Pasca. 2005. 'Emmy Noether.' In *Complexities: Women in Mathematics*, edited by Bettye Anne Case and Anne M. Leggett, 30–37. Princeton University Press. http://www.jstor.org/stable/j.ctt1dr35cc.15.
Neville Bonner, Thomas. 1992. *To the Ends of the Earth. Women's Search for Education in Medicine*. Harvard.
Otterspeer, Willem and J. Schuller Tot Peursum-meijer. 1997. *Wetenschap en wereldvrede. De Koninklijke akademie van wetenschappen en het herstel van de internationale wetenschap tijdens het interbellum*. Amsterdam: KNAW.
Pais, Abraham. 1982. *Subtle is the Lord. The Science and the Life of Albert Einstein*. Oxford: Oxford University Press.
Pais, Abraham. 1989. 'George Uhenbeck and the discovery of electron spin.' *Physics Today* (December), 34–40. https://doi.org/10.1063/1.881186
Pais, Abraham. 1991. *Niels Bohr's Times in Physics, Philosophy and Polity*. Oxford: Oxford University Press.
Pauleys, F. 1991. *From Prejudice to Persecution. A History of Austrian Antisemitism*. The University of North Carolina Press.
Pauli, Wolfgang. 1925. 'Über den Zusammenhang des Abschlusses der Elektronengruppen im Atom mit der Komplexstruktur der Spektren.' *Zeitschrift für Physik* 31: 765–783. https://doi.org/10.1007/BF02980631
Pauli, Wolfgang. 1945. 'Niels Bohr on His 60th Birthday.' *Reviews of Modern Physics* 17(1): 97–101. https://doi.org/10.1103/RevModPhys.17.97
Pauli, Wolfgang. 1979. *Wissenschaftlicher Briefwechsel mit Bohr, Einstein, Heisenberg u.a. Band 1*: 1919–1929, edited by A. Hermann, K. Meyenn, and V. F. Weisskopf. Berlin, Heidelberg: Springer.
Pechenkin, Alexander. 2014. *Leonid Isaakovich Mandelstam. Research, Teaching. Life*. Cham: Springer.
Pechenkin, Alexander. 2019. *L.I. Mandelstam and His School in Physics*. 2nd Edition. Cham: Springer.
Pérez, Enric. 2009. 'Ehrenfest's adiabatic theory and the old quantum theory, 1916–1918.' *Archive for History of Exact Sciences* 63(1): 81–127. http://www.jstor.org/stable/41134311

Pérez, Enric and Blai Pié i Valls. 2015. 'Ehrenfest's adiabatic hypothesis in Bohr's quantum theory.' In *One hundred years of the Bohr atom*, edited by Finn Aaserud and Helge Kragh, 272–289. Copenhagen: The Royal Danish Academy of Science and Letters.
Pietrow-Enker, Bianca. 1999. *Russlands Neue Menschen*. Frankfurt/New York: Campus Verlag.
Planck, Max. 1906. Vorlesungen Über Die Theorie der Wärmestrahlung: Mit 6 Abb. Leipzig: Barth.
Poincaré, Henri. 1912. 'Sur la théorie des quanta.' *Journal de Physique* 2(1): 5–34.
Poley, J. P. 2000. *Eroica. The Quest for Oil in Indonesia (1850–1898)*. Dordrecht: Springer.
Powell, Arthur B. and Marilyn Frankenstein. 1999. In his prime: Dirk Jan Struik reflects on 103 years of mathematical and political activities. *Harvard Educational Review* 69(4).
Presser, Jacques. 1965. *Ondergang*. Den Haag: Staatuitgeverij.
Ramsey, Norman F. 1956. 'Thermodynamics and statistical mechanics at negative absolute temperatures.' *Physical Review* 103: 20. https://doi.org/10.1103/PhysRev.103.20
Rappaport, Karen D. 1981. 'S. Kovalevsky: A mathematical lesson.' *The American Mathematical Monthly* 88: 564–573. https://doi.org/10.2307/2320506
Regt, Henk W. de. 1999. 'Ludwig Boltzmann's Bildtheorie and scientific understanding.' *Synthese* 119: 113–134. https://doi.org/10.1023/A:1005295304955
Reid, Constance. 1986. *Hilbert*. New York: Springer Verlag.
Reid, Constance. 1996a. The future of mathematics. In *Hilbert*, 1–20. New York: Springer Verlag. https://doi.org/10.1007/978-1-4612-0739-9_10
Reid, Constance. 1996b. *Courant*. New York: Springer Verlag.
Reinders, L. J. 2018. *The Life, Science and Times of Lev Shubnikov. Pioneer of Soviet Cryogenics*. Cham: Springer.
Reindersma, W. 1918/1924. *Beknopt Leerboek der Vlakke Meetkunde*. Den Haag, Groningen: J.B. Wolters.
Renn, Jurgen and Michel Janssen. 2015. Einstein was no lone genius. *Nature* 527(9), 298–300. https://doi.org/10.1038/527298a
Rentetzi, Maria. 2007. Gender science and the city. In *Trafficking Materials and Gendered Experimental Practices: Radium Research in Early 20th Century Vienna*. New York: Columbia University Press. http://www.gutenberg-e.org/rentetzi/chapter03.html
Reve, Karel van het. 2019. *Karel van het Reve voor gevorderden*. Amsterdam: Van Oorschot.
Ritz, Walther. 1908. 'Recherches critiques sur les théories électrodynamiques de Cl. Maxwell et de H.A. Lorentz.' *Archives des Sciences physiques et naturelles* 26: 209–236.
Ritz, Walther. 1911. *Gesammelte Werke – Oeuvres*. Paris: Société Suisse de Physique.
Roth, Joseph. 2018. *Spoken in Moskou*, translated by E. Snick. Amsterdam: Bas Lubberhuizen.
Rowe, David E. 1989. 'Klein, Hilbert, and the Gottingen Mathematical Tradition.' *Osiris* 5: 186–213. http://www.jstor.org/stable/301797
Rowe, David E. 2018. *A Richer Picture of Mathematics. The Göttingen Tradition and Beyond*. Cham: Springer.
Rubin, Harry. 1995. Walter Elsasser. *Biographical Memoirs* 68, 115–117.
Rutgers, Arend Joan. 1930. *Bijdrage tot de theorie der thermo-elektriciteit in kristallen*. Doctoral dissertation, Leiden University. Amsterdam: H.J. Paris. https://ilorentz.org/history/proefschriften/sources/Rutgers_1930.pdf
Ryutova-Kemoklidze, Margarita. 1995. *The Quantum Generation*. Berlin, Heidelberg, New York: Springer.
Sandner, Günther. 2019. The scientific world-conception in the making: Towards the ideological roots of logical empiricism in Berlin and in Vienna. In *Ernst Mach – Life, Work, Influence*, edited by F. Stadler, 1–20. Cham: Springer. https://doi.org/10.1007/978-3-030-04378-0_21
Schiller, Friedrich von. (1838). *Sämmtliche Werke I*. Stuttgart: Cotta.
Schilpp, Paul A. (Ed.). 1970. *Albert Einstein: Philosopher–Scientist*. La Salle: Open Court.

Schirrmacher, Arne. 2015. 'Who made quantum theory popular with physicists and beyond?' *The European Physical Journal, Special Topics* 224: 2113–2125. https://doi.org/10.1140/epjst/e2015-02526-1

Schlögel, K. 2002. *Petersburg, Das Laboratorium der Moderne, 1909–1921.* Hanser Verlag.

Smoluchowski, Marian. 1912. *Kobiety w naukach ścisłych.* https://doi.org/10.12775/18411

Smoluchowski, Marian. 2018. 'Women in exact sciences. A lecture delivered at the Scientific-Literary Association in Lwów in the year 1912.' *Acta Poloniae Historica* 117: 231–240. https://doi.org/10.12775/18411.

Schrödinger, Erwin. 1926. 'Quantisierung als Eigenwertproblem.' *Annalen der Physik* 384(4): 361–376. https://doi.org/10.1002/andp.19263840602

Seelig, Carl. 1956. *Albert Einstein. A Documentary Biography.* Translated from German by Mervyn Savill. London: Staples Press.

Seth, Suman. 2004. Quantum theory and the electromagnetic world-view. *Historical Studies in the Physical Sciences* 35(1): 67–93. https://doi.org/10.1525/hsps.2004.35.1.67

Simmons, E. J. 1968. *Introduction to Tolstoy's Writings.* Chicago: Chicago University Press.

Singer, S. 2003. *Adventures Abroad: North American Women at German-speaking Universities 1868–1915.* Westport: Praeger.

Sitter, Willem de. 1913a. 'A proof of the constancy of the velocity of light.' *KNAW Proceedings* 15 II, 1297–1298. https://dwc.knaw.nl/DL/publications/PU00013063.pdf

Sitter, Willem de. 1913b. 'Ein astronomischer Beweis für die Konstanz der Lichtgeschwindigkeit.' *Physikalische Zeitschrift* 14: 429.

Sitter, Willem de. 1913c. 'On the constancy of the velocity of light.' *KNAW Proceedings* 16: 395–396. https://dwc.knaw.nl/DL/publications/PU00012842.pdf

Sitter, Willem de. 1913d. 'Über die Genauigkeit, innerhalb welcher die Unabhängigkeit der Lichtgeschwindigkeit von der Bewegung der Quelle behauptet werden kann.' *Physikalische Zeitschrift* 14: 1267.

Sitter, Willem de. 1916–1917. 'On Einstein's theory of gravitation, and its astronomical consequences. Second paper.' *Monthly Notices of the Royal Astronomical Society* 76: 699–728. https://doi.org/10.1093/mnras/76.9.699; *MNRAS* 77: 155–184. https://doi.org/10.1093/mnras/77.2.155; *MNRAS* 78: 3–28. 10.1093/MNRAS/78.1.3

Smulders, Eelke. 2007. *The Architectural Wishes of the Viennese Bourgeoisie 1857–1873.* Master's thesis, Utrecht University.

Snelders, H.A.M. 1983. 'De natuurwetenschappen in de lokale wetenschappelijke genootschappen uit de eerste helft van de negentiende eeuw.' *De negentiende eeuw,* 7(2): 102–122.

Stark, Johannes. 1922. *Die gegenwärtige Krise der Physik.* Leipzig: J.A. Barth.

Steen, Agnes van. 2007/2008. 'Vrouwenkiesrechtstrijd in leiden.' In *Jaarboek Dirk van Eck,* 177–220. Leiden: Stichting Dirk van Eck.

Steen, Agnes van. 2011. '"Vol moed en blakende ijver". Aletta Lorentz-Kaiser en de vrouwenbeweging in Leiden (1881–1912).' In *Jaarboek Dirk van Eck,* 125–166. Leiden: Stichting Dik van Eck.

Steen, Agnes van. 2013a. 'De gehuwde vrouw moet maatschappelijk voelen.' *Jaarboek Dirk van Eck,* 135–184. Leiden: Stichting Dirk van Eck.

Steen, Agnes van. 2013b. *Vrolec. Leidsche hoogleraarsvrouwen een eeuw verenigd.* Leiden: Vrolec.

Strien, Marij van. 2013. The nineteenth century conflict between mechanism and irreversibility. *Studies in History and Philosophy of Modern Physics* 44(3): 191–205. https://doi.org/10.1016/j.shpsb.2013.05.004

Struik, Dirk. 1973. *Autobiographical Notes.* Niels Bohr Library MB 508.

Thiele, Ruediger. 2011. *Felix Klein in Leipzig.* Leipzig: EAGLE.

Tichy, Marina. 1993. 'Die geschlechtliche Un-Ordnung: Facetten des Widerstands gegen das Frauenstudium von 1870 bis zur Jahrhundertwende.' In *Durch Erkenntnis zu Freiheit und Glück ... Frauen an der Universität Wien (ab 1897),* edited by W. E. Heindl, 27–48. Wien: Universitätsverlag.

Sengers, Jan and Gijs Ooms (Eds.). 2017. Jan Burgers. Delft: JMBC. https://filelist.tudelft. nl/Websections/JM%20Burgerscentrum/Downloads/BurgersBoek.pdf
Simmons, Ernest J. 1949. *Leo Tolstoy. A Biography*. London: Lehmann.
Tinbergen, Jan. 1929. *Minimumproblemen in de natuurkunde en de ekonomie*. Doctoral dissertation, Leiden University. Amsterdam: H.J. Paris.
Tinbergen, Jan. 1979. 'Work and well-being — desires for the future.' In *Futures for Work*, edited by J. Boersma et al., 67–80. Dordrecht: Springer.
Tobies, Renate. 2019. 'Felix Klein—Mathematician, academic organizer, educational reformer.' In *The Legacy of Felix Klein*, edited by H. G. Weigand et al., 1–20. ICME-13 Monographs. Cham: Springer.
Tobies, Renate. 2021. 'Setting the course, 1892/93–1895.' In *Felix Klein. Vita Mathematica* 20, 1–20. Cham: Birkhäuser. https://doi.org/10.1007/978-3-030-75785-4_7
Tobies, Renate. 2022. 'Internationalism and women mathematicians at the University of Göttingen.' In *The Palgrave Handbook of Women and Science since 1660*, edited by C. G. Jones, A. E. Martin, and A. Wolf, 1–20. Cham: Palgrave Macmillan. https://doi.org/10. 1007/978-3-030-78973-2_11
Tolman, Richard C. 1914. 'The specific heat of solids and the principle of similitude.' *Physical Review* 7: 145–153. https://doi.org/10.1103/PhysRev.4.145
Tolman, Richard and Paul Ehrenfest. 1930. 'Temperature equilibrium in a static gravitational field.' *Physical Review* 36: 1791–1798. https://doi.org/10.1103/PhysRev.36.1791
Tolman, Richard, Paul Ehrenfest, and B. Podolsky. 1931. 'On the gravitational field produced by light.' *Physical Review* 37: 602–615. https://doi.org/10.1103/PhysRev. 37.602
Tolstoy, Leo. 1892. *The first step*. Foreword to the Russian edition of Howard Williams, *The Ethics of Diet*.
Trapeznikova, Olga N. 2019. 'Reminiscences about L. V. Shubnikov.' Translated by Y. Bazaliy. First appeared in 1990, in Russian, in 'L. V. Shubnikov. Selected works. Memoirs.' Kyiv: Naukova Dumka. https://www.lorentz.leidenuniv.nl/history/Shubnikov/ Trapeznikova_about_L.V.Shubnikov.pdf
Tropp, Eduard. A., Viktor Ya. Frenkel, and Artur D. Chernin. 1993. *A.A. Friedmann. The Man Who Made the Universe Expand*. Cambridge: Cambridge University Press.
Tsipenyuk, Yu. M. 1973. 'From the history of the journal of the Russian Physicochemical Society/JETP.' *JETP* 37(1): 1–24.
Uffink, Jos. 2000. 'Bluff your way in the Second Law of Thermodynamics.' *Studies In History and Philosophy of Science Part B Studies in History and Philosophy of Modern Physics* 32(3): 305–394. https://doi.org/10.1016/s1355-2198(01)00016-8
Uffink, Jos. 2006. 'Insuperable difficulties: Einstein's statistical road to molecular physics.' *Studies in History and Philosophy of Science, Part B* 37(1): 36–70. https://doi.org/10. 1016/j.shpsb.2005.07.004
Uffink, Jos. 2017. 'Boltzmann's work in statistical physics.' In *The Stanford Encyclopedia of Philosophy*. Retrieved from https://plato.stanford.edu/archives/spr2017/entries/ statphys-Boltzmann/
Uffink, Jos and Giovanni Valente. 2020. Afanassjewa and the foundations of thermodynamics. In *The Legacy of Tatjana Afanassjewa, Philosophical Insights from the Work of an Original Physicist and Mathematician*, edited by Jos Uffink et al., 1–20. Cham: Springer.
Uhlenbeck, George. E. 1956. 'Reminiscences of Professor Paul Ehrenfest.' *American Journal of Physics* 24: 431–433. https://doi.org/10.1119/1.1934259
Uhlenbeck, George. 1962, session I. Interview of George Uhlenbeck by Thomas S. Kuhn on March 30, 1962. Niels Bohr Library & Archives, AIP. www.aip.org/history-programs/ niels-bohr-library/oral-histories/4922-1
Uhlenbeck, George. 1962, session II. Interview of George Uhlenbeck by Thomas S. Kuhn on March 31, 1962. Niels Bohr Library & Archives, AIP. www.aip.org/history-programs/ niels-bohr-library/oral-histories/4922-2

Uhlenbeck, George. 1962, session III. Interview of George Uhlenbeck by Thomas S. Kuhn on April 5, 1962. Niels Bohr Library & Archives, AIP. www.aip.org/history-programs/niels-bohr-library/oral-histories/4922-3

Uhlenbeck, George. 1963, session V. Interview of George Uhlenbeck by Thomas S. Kuhn on December 9, 1963. Niels Bohr Library & Archives, AIP. www.aip.org/history-programs/niels-bohr-library/oral-histories/4922-5

Uhlenbeck, George, and Samuel Goudsmit. 1925. 'Ersetzung der Hypothese vom unmechanischen Zwang durch eine Forderung bezüglich des inneren Verhaltens jedes einzelnen Elektrons.' *Naturwissenschaften* 13: 953–954. https://doi.org/10.1007/BF01558878

Unna, Issachar and Tilman Sauer. 2013. 'Einstein, Ehrenfest and the quantum measurement problem.' *Annalen der Physik* 525(1–2): A15–A19. https://doi.org/10.1002/andp.201300708

Valk, S. 1918. *Sankt Peterburgskie vysshie zhenskie (Bestuzhevskie) kursy (1878–1918)*. Petersburg.

Valls Blais Pié I and Enric Pérez. 2016. 'The historical role of the Adiabatic Principle in Bohr's quantum theory.' *Annalen der Physik* 528(7–8): 530–534. https://doi.org/10.1002/andp.201600178

Vakhromeeva, Oksana B. (2018). 'Features of teaching of the scientific corporation at the Higher women's (Bestuzhev) Courses in 1878–1918' «Запад-Восток». *Научно-практический ежегодник* 11: 180–191.

Vogel-Prandtl, Johanna. 2004. *Ludwig Prandtl. A Biographical Sketch, Remembrances and Documents*. Trieste: The International Centre for Theoretical Physics.

Vooys, C. de. 1952. 'Biografie D. C. Tinbergen.' In *Jaarboek van de Maatschappij der Nederlandse Letterkunde 1950–1955*, 194–199.

Vredespaleis Carnegie Stichting. 2020. De oproep van de tsaar. Uit het archief. Website: https://www.vredespaleis.nl/vredespaleis/uit-het-archief/the-tsars-rescript/

Waals, Johannes D. van der. 1927. *Lehrbuch der Thermostatistik*, edited by Philip Kohnstamm. Leipzig: J.A. Barth.

Waals, Johannes D. van der and Philip Kohnstamm. 1912. *Lehrbuch der Thermodynamik*. Leipzig: J.A. Barth.

Watson, Derek. 2005. The making of a revolutionary, 1890–1917. In *Molotov. Studies in Russian and East European History and Society*, 4–23. London: Palgrave Macmillan.

Weinzierl, Erika. 2003. *The Jewish Middle Class in Vienna in the Late Nineteenth and Early Twentieth Centuries*. Wien: Working paper.

Westwood, J. N.1964. *A History of Russian Railways*. London: Allen & Unwin.

Weyl, Hermann. 1935. Weyl on Noether – MacTutor History of Mathematics. https://mathshistory.standrews.ac.uk/Extras/Weyl_Noether/

Weyl, Hermann. 1946. Encomium. *Science* 103, 216–218. https://doi.org/10.1126/science.103.2669.216

Winter, Robert. 1996. *Das Akademische Gymnasium in Wien. Vergangenheit und Gegenwart*. Wien: Böhlau Verlag.

Wistrich, Robert S. 1987. 'Social democracy, the Jews, and anti-semitism in Fin-de-siecle Vienna.' In *Living with Antisemitism. Modern Jewish Responses*, edited by J. Reinharz, 193–209. Hanover and London: New England University Press.

List of abbreviations

AIP	American Institute of Physics, College Park, MD USA
CPAE	Collected Papers of Albert Einstein
BA-DUT	Burgers Archive – Delft University of Technology
EA-MBL	Ehrenfest Archive – National Museum Boerhaave Leiden
EFA	Ehrenfest Family Archive
IISH	International Institute of Social History in Amsterdam
LNA	Leids Newspaper Archive
MKEAD	Martin Klein's Ehrenfest Archive Digitised (now included in EA-MBL)
NHA	Noord-Hollands Archief
HD	Hans Driessen
BM	Brendan Monaghan

Index

For the benefit of digital users, indexed terms that span two pages (e.g., 52–53) may, on occasion, appear on only one of those pages.

Aardenne, van
  Gijs (sr.), 187, 192, 194, 209, 210, 241–242, 272, 290, 292–294, 305
  Tatiana Pavlovna (Tanya, Tanitchka). *See* Ehrenfest, Tatiana Pavlovna
Abraham, Max, 35, 38, 73, 75, 79, 95, 161
adiabatic principle, 165, 176–177, 180–181, 188, 190–191, 267, 269, 323
Afanasyev
  Alexei (*father*), 14–15
  Family, 86
  Pyotr (*uncle*), 13–14, 17, 20, 56, 185, 322
Afanasyeva Ivanova, Jekaterina (baba Katya, *mother*), 14–15, 59, 88, 100, 101, 104, 146, 244–245, 272, 276–277, 280, 286–287, 306, 309–310, 314
Afanasyeva Maslova, Sofia (aunt Sonya; baba Sonya), 14, 20, 28, 30, 42–43, 45–46, 56–57, 58*f*, 59, 65–66, 70, 79, 83, 91, 100, 107–108, 129–130, 135–136, 142, 146, 162, 180, 182, 185, 187–188, 195–196, 214, 280, 283–284, 322
Afanasyeva, Tatiana Alexeyevna. *See* Ehrenfest Afanassjewa, Tatiana Alexeyevna
Aigentler, Henriette von, 25–26, 44–45
Alexander II, Aleksandr Nikolaevich Romanov, tsar, 13–14, 18–19
Alexander III, Aleksandr Aleksandrovich Romanov, tsar, 19–20
anschaulichkeit, 56, 87, 249
Asscher-Pinkhoff, Clara, 293
Atatürk, Mustafa Kemal, 291
atomic spectrum, 192, 215, 224, 253, 254. *See also* light spectrum; spectral lines
Auger, Pierre, 290, 303–304
Avogadro, Lorenzo, 24

Bach, Johann Sebastian, 174, 183, 287
Backer, Hilmar, 146, 152–153
Bad Kissingen, 100–102
Bauer, Edmond, 65, 71, 290, 303, 311
Baumgart, Karl, 88–89, 91, 200–202
Bertel, Anny, 286–287

Besso, Michele, 121, 150–151, 174–175, 341 n.1, 351 n.18
Bestuzhev courses, 19–20, 27–29, 31–32, 34, 42–43, 52, 57–58, 67–69, 79–80, 85, 87, 91, 96–97, 162, 360 n.8
Bestuzhev-Ryumin, Konstantin, 19
Beth, Hermanus J.E., 238
Bilibin, Ivan Yakovlevich, 118, 315
black body radiation, 53–54, 71–72, 111, 164
Blumenthal, Otto, 38
Boeke, Kees, 306
Bogomolov, S.A., 127–128
Bohr, Margrethe, 262, 285, 305–306, 308
Bohr, Niels, 165–166, 176–178, 179–180, 190–194, 193*f*, 197–198, 200–201, 206–208, 209–210, 215, 222, 224, 242, 253–254, 255–257, 261–262, 264–269, 272, 274–276, 285–287, 296, 305–306, 323–324, 327
Bohr's
  correspondence principle, 190–191, 208, 255–256, 267, 323
  model of the atom, 176–177, 180, 184, 188, 192, 207, 215, 218, 252–254, 265–266
  (old) quantum theory, 175, 243, 255–256, 267. *See also* quanta
Boltzmann, Ludwig, 21–22, 23–26, 33, 35, 44–45, 48–49, 53–55, 59, 62–63, 65, 68–69, 70–72, 74–75, 77–79, 84, 93, 149, 155, 164, 179, 230, 247, 326
Boltzmann's
  H-theorem, 78–79, 80–81. *See also* recurrence paradox; reversibility paradox
  Statistical mechanics, 69–70, 71–72, 77–79, 111, 131, 205, 233, 269–270, 324
Bolyai, Farkas, 38, 51
Bolyai, Janos, 38, 86–87
Bomans, Godfried, 307
Bommel, Henk van, 312, 315–316, 317*f*, 317–318, 319–320
Bommel, Tamara van, 312, 315–316
Borgman, Ivan, 84, 89, 95, 98, 133
Born, Hedwig (Hedi), 285–286, 337 n.2

Born, Max, 36, 95, 101, 218, 246, 255, 265–267, 274–276, 290–292, 298, 307–308
Borodin, Alexander, 19
Bosscha, Karel Albert Rudolf (Ru), 150. *See also* Reading Room for Mathematics and Physics in Leiden
Brahms, Johannes, 20, 114, 120–121, 150
Braun, Karl, 94. *See also* Le Chatelier-Braun principle
Breit, George, 215, 219, 274
Bridgman, Percy, 267
Brillouin, Marcel, 110
Broglie, Louis-Victor de, 110, 266
Brownian motion. *See* random motion
Bryner, Teddy, 232–234, 272
Bulgakov, Mikhail Afanasyevich, 97
Burgers, Jan, 150, 154–155, 158, 161, 169–170, 175–179, 180–181, 183–184, 188–190, 198, 201, 224, 233, 251–252, 274–275, 304, 315, 318, 323–324, 327
Burgers, Willy, 183, 327
Bursian, Victor, 91, 200–201, 220–221
Busch, Adolf, 211, 262, 305

Camp, Henri, 147
Carathéodory, Constantin, 246–247, 248–249
Carnot, Sadi, 245–246
Casimir, Hendrik, 148, 158, 161, 267–268, 274, 279f, 282, 307–308, 318, 323–324
Casimir, Rommert, 159–160, 235, 239, 276–277
Chatelier, Henri le, 94. *See also* Le Chatelier-Braun principle
Chisholm, Grace, 41, 44
Christiaan Huygens Study Association, 179, 205–206, 226
Clausius, Rudolf, 24–25, 233–235
Correspondence principle. *See* Bohr's, correspondence principle
Coster, Dirk, 158, 173, 175–176, 177–178, 180, 189, 215, 238, 252, 253, 262, 265, 274–275, 292, 305–306, 318, 323–324
Crommelin, Claude, 148
Curie, Marie. *See* Skłodowska Curie, Marie

Dalton, John, 24
Debye, Peter, 115–117, 129, 131, 149–150, 222, 240–241, 274
didactics of mathematics. *See* mathematics didactics
Dieke, Gerhard, 206, 224, 252, 278–279, 295, 323
Dijksterhuis, Eduard, 235–239, 249, 306, 316
dimensional analysis, 175
Dirac, Paul, 266, 274–275, 294

Dirichlet, Johann, 36
Dobiash, Alexander, 91
Doynikova, Valentina, 91–92
Droste, Jan, 169–170, 172, 191–192
Dutch League for Women's Suffrage (Nederlandsche Bond voor Vrouwenkiesrecht), 163–164

Eddington, Arthur, 173–174, 194–197
Educational method of Maria Montessori, 240–241, 262, 279
Ehrenfest,
    Arthur (*brother*), 3–4, 6–12, 22, 34, 36, 44, 54, 57, 105, 118, 321
    Emil (*brother*), 3–4, 7, 10, 22, 57, 118, 139, 152, 321
    Fritz (*Arthur's son*), 321
    Galinka (Anna Galinka, Galja; *daughter*), 90f, 135f, 146, 159f, 170, 171f, 181–182, 182f, 186–188, 193, 196–197, 209–211, 219, 240, 244, 262, 268f, 272, 276, 279–280, 284, 290, 293, 294–295, 304–310, 312, 316, 319–321
    Hugo (*brother*), 3–4, 6–7, 9–11, 22, 57, 112, 126, 211–212, 229, 230–231, 232–235, 242, 250, 265, 271, 280, 282, 285f, 290, 298, 303, 321, 322–323
    Johanna Jellinek (*mother*), 3, 5–6, 7–11
    Josephine Jellinek (Pepperl; *Sigmund's second wife*), 11, 26, 321
    Otto (*brother*), 3–4, 22, 54, 57–58, 321
    paradox, 95
    Pawlik (Paul jr.; Papsik; *son*), 170–171, 173f, 181–182, 182f, 187–188, 193, 195f, 196f, 209–210, 222, 227–228, 240, 261–262, 272, 275–278, 280, 286–288, 289–290, 294–295, 303–307, 314, 321, 328
    Regine Egger- (*Arthur's wife*), 11, 321
    Sigmund (*father*), 3, 5, 7–8, 10–11
    Tatiana Pavlovna (T'; Tanitchka; Tanya'; *daughter*), 65–66, 68–69, 71f, 71, 75, 76f, 81–83, 84–85, 87–89, 91, 95, 100, 101, 103, 118, 126, 134–135, 145, 155, 158, 170, 173, 180–182, 182f, 186–188, 193, 209–211, 213–214, 219, 229, 240, 257, 262, 272, 275f, 275–276, 280, 283–284, 290, 292–293, 294–296, 304, 315
    theorem, 268–270
    Wassily (Wassik; Wapsili; *son*), 186, 188–189, 196, 199, 209–210, 211–214, 222, 227–228, 242, 261, 273, 293, 294–296, 324–325, 328
Ehrenfest, Paul (Pavel) *passim*
Ehrenfest Afanassjewa, Tatiana Alexeyevna (Tanya), *passim*

Einstein,
  Hans-Albert, 61, 121, 220
  Ilse, 209, 211, 213–214, 306
  Margot, 209, 211, 213–214
  Tete, 220
Einstein, Albert *passim*
Einstein Löwenthal, Elsa, 167–168, 169–170, 174–175, 197, 299, 306
Einstein Marić, Mileva. *See* Marić, Mileva
Einthoven, Willem, 241
electron spin, 254–255
electron theory. *See* Lorentz' electron theory
Elsasser, Walter, 273
Encyclopaedia of Mathematical Sciences, 77–78, 94, 97, 107, 115, 130, 131, 149, 204, 319
Engels, Friedrich, 190–191, 284
Epstein, Minna, 114
Epstein, Paul, 84, 114, 146, 176–177, 232–233, 253–254
Ernst, Bruno, 319
Euclides, 86–87, 97–98, 127–128, 159–161, 236
Euclides, Supplement to the New Journal for Mathematics (Dutch), 237
Exner, Franz, 68, 73, 155

Fausek, Julia, 240–241
Fehr, Henri, 98
Fermi, Enrico, 224, 240–241, 267, 279*f*
Filosofova, Anna, 19
Flexner, Abraham, 188–189
Föhringer, Anna, 88, 91, 200–201, 318
Fokker, Adriaan, 146, 149, 158, 198, 206–207, 251, 264–265, 282, 292, 300
Franck, James, 139, 167, 185, 189, 240–241, 251, 262, 264, 267, 268*f*, 274, 290, 292, 295, 296
Frenkel, Victor, 243–244
Fresnel, Augustin Jean, 50, 142–143, 261
Freud, Sigmund, 65–67, 213, 233
Freudenthal, Hans, 316
Friedmann, Alexander, 91, 108, 200–201, 220–221, 322

Gauss, Carl, 35–36, 38, 86–87, 237
geometry education, 86–87, 235–239, 244–246, 277–278, 288, 306, 313–314, 319–320, 321, 324. *See also* mathematics didactics; *Übungensammlung*
Gerlach, Walter, 218. *See also* Stern-Gerlach experiment
Gernet, Nadezhda Nikolayevna, 42, 44–46, 79, 85, 97
Gibbs, Josiah, 65, 70, 179
Goeppert, Maria (Later in life: Goeppert Mayer), 267, 274, 318–319

Gordon, Walter, 292–293
Gorter, Evert, 169–170, 180, 276, 279
Goudsmit, Samuel (Sam), 204–206, 224, 252–255, 257*f*, 267, 278–279, 279*f*, 281, 286, 290–292, 295, 318, 323–324
Grossman, Marcel, 61, 151, 168

Haas, Wander de, 139, 172, 198, 267, 272, 292, 295
Haas Lorentz, Berta de, 145, 198, 202–204
Haber, Fritz, 167
hafnium (discovery of), 252, 324
Hahn, Hans, 22, 26, 39–40, 43, 47, 59, 68, 78, 100–101, 109, 122
Hahn, Otto, 203, 226, 318–319
Hasenöhrl, Fritz, 110, 118, 132, 178–179, 215
Heisenberg, Werner, 254–255, 262, 265–266, 269, 274–275
Heitler, Walter, 274, 281, 292
Herglotz, Gustav, 12, 21–22, 26, 39–40, 43, 47, 59, 75, 100–106, 109–110, 113–114, 122, 126, 136, 146, 178, 275, 286, 290, 296, 303, 312
Hertz, Gustav, 59, 80–81, 113, 139, 167, 189, 218, 241, 262
Herzen, Alexander, 17, 46, 110
Hessen, Boris, 218, 264, 318
Hevesy, Charles de, 252
Hilbert, David, 38–39, 42–44, 47, 52–53, 74–75, 78–79, 86–87, 91, 141, 150, 155, 186, 203–204, 226, 237, 266
Hitchcock, Albert, 326–327
Hoff, Jacobus Hendricus van 't, 94, 167–168
Holst, Gilles, 251, 295
Huizinga, Johan, 227, 296–297

Ioffe, Abram, 106, 107*f*, 108, 121–123, 134, 136, 146, 149–150, 151–152, 166, 178, 200–202, 209–210, 215, 222, 240–241, 243, 261–262, 272–273, 284, 289, 291, 295–296, 303, 308, 318, 322

Jacobs, Aletta, 163–164, 202–204. *See also* Dutch League for Women's Suffrage
Jeans, Sir James, 53–54, 72, 110–111, 115, 118. *See also* Rayleigh Jeans ultraviolet catastrophe
Jordan, Pascual, 266, 274–276, 281
Joule, James, 245

Kagan, Veniamin (Benjamin), 123, 128–129, 263–264, 284, 298, 318
Kamerlingh Onnes, Harm, 297f, 305–306, 309f
Kamerlingh Onnes, Heike (Onnes), 125, 133–134, 140–142, 148–149, 152–155, 173, 185, 191–192, 193f, 198–199, 202–203, 206, 215–216, 219–220, 223, 226, 241, 242, 251, 267, 272, 276–277, 279, 318–319, 324
Kapitza, Pyotr, 240–241, 272, 291
Kappers, Cor Ariëns, 300
Kaufmann, Walter, 75–76
Keesom, Willem, 173, 272
Kelvin, Lord William Thomson, 245–246, 247–248. *See also* negative absolute temperatures
Khvolson, Orest Danilovich, 27, 89, 95, 98, 108, 133, 140–141, 200–202
Kiselyov, Andrei, 127–128
Klein, Felix, 35–43, 47, 52–53, 77–78, 85–87, 97–98
Klein, Oskar, 267
Kleiner, Alfred, 116–117
Kloot, Jaap, 304, 307–308, 312
Kohnstamm, Philip, 169–170, 202, 204, 214, 228, 235, 239, 241, 246, 248–249, 251, 296, 306
Kovalevskaya, Sofia, 42, 97
Koyalovich, Boris Mikhailovich, 85–86
Kramers, Hans, 158, 177–180, 192, 203–204, 206–208, 222, 251–252, 255–256, 307–308, 318, 323–324
Krans, Roelf Luppo, 274, 282
Krutkov, Yuri, 146, 190, 191–192, 200–201, 220–221, 264–265, 318
Krylov, Aleksei, 127–128, 202
Kuenen, Johan, 141, 158, 198–199, 205, 213–214, 296

Laer Kronig, Ralph de, 254
Lampa, Anton, 119, 121–124
Landau, Lev, 303–304
Landé, Alfred, 249, 252–253
Langedijk, Sari, 309–310
Langevin, Paul, 75, 110, 115, 118, 203, 206, 208, 216, 271–272, 297
Lazarev, Petr, 242–243, 284
le Chatelier-Braun principle, 94, 115
Lebedev, Pyotr, 84, 98, 108, 114, 130–131
Leeuwen, Jo van, 162–163, 184, 204
Leeuwen, Nel van, 162–163, 184, 202–203, 228, 296–297
Leiden Jar Student Association, 205–206
Lenard, Philipp, 215–217
Lermontov, Mikhail, 288–289

light spectrum, 50, 53–54, 71–72, 88, 111–112, 164. *See also* atomic spectrum
London, Fritz, 274, 292–293
Lorentz, Hendrik, 35, 47–51, 52–53, 63–64, 73, 77–78, 84, 110, 115, 118, 129–136, 139–144, 146, 148–151, 154–155, 157–159, 162–163, 166–170, 172, 175–178, 184, 189, 191–192, 194–200, 202–204, 206–207, 208–209, 215–217, 219–220, 226, 229–230, 242, 244–245, 254–257, 264–265, 271–272, 274, 282, 291, 324, 327
Lorentz' electron theory, 35, 47, 51
Lorentz Kaiser, Aletta, 49, 145, 163–164, 271
Loschmidt, Johann, 25, 78. *See also* reversibility paradox

Mach, Ernst, 24–26, 65
Mahler, Gustav, 20, 66, 106
Maksheyev, Zakhar, 91, 103–104
Mandelstam, Leonid Isaakovich, 263–264, 272, 298, 318
Mandelstam Solomonovna, Lydia, 263–264, 284
Mannoury, Gerrit, 235, 239
Marić, Mileva, 61, 65, 121–122, 167–168, 170, 174–175
Marx, Karl, 84, 182
Maxwell, James Clerk, 23, 34, 50, 63, 64, 155, 257–258, 271
mathematics didactics, 52, 123, 235–239, 258, 261, 270, 277, 315–316, 322, 324–326. *See also* geometry education; mathematics education reform
mathematics education reform, 103, 151, 159–161, 238, 315–316, 324. *See also* mathematics didactics
Mathematics Study Group (WVO Wiskunde Werkgroep), 306–307, 315, 319–320
Meitner, Lise, 23, 68–69, 73–74, 113, 203, 226, 233, 285, 318–319, 323, 326
Mendeleev, Dmitri, 18–19, 24, 83, 102
Michelson-Morley experiment, 63–64, 143–144
Millikan, Robert, 229, 240–241, 253
Miloradovich, Kitty, 27, 146
Minkowski, Hermann, 52, 75, 77
Minnaert, Marcel, 227, 251–252, 306, 323

negative absolute temperatures, 247–248
Nernst, Walther, 41, 109–110, 161, 166–168
Newton, Isaac, 25, 50, 73, 142, 144, 232, 270
Newtonian physics, 34, 63, 73, 270
Nicholas II, Nikolai Aleksandrovich Romanov, tsar, 20, 28–29, 32–33, 58–59, 80, 84, 182, 198–199

Nieuwenhuis von Uexküll Güldenband, Margarethe, 139, 146, 190
Noether, Emmy, 203–204, 226, 318–319
Nordström, Gunnar, 158, 169–170, 172–173, 184, 198, 202–203, 206–207, 228
Nordström van Leeuwen, Nel. *See* Leeuwen, Nel van

Obreimov, Ivan, 133, 272
Oppenheimer, Robert, 274–275, 279
Ostwald, Wilhelm, 25–26, 74, 78

Pannekoek, Anton, 181–182
Paschen, Friedrich, 214, 224
Pauli, Wolfgang, 190–191, 222, 250, 252–253, 254, 265–266, 267–268, 273–275, 282
Pavlov, Ivan, 18–19, 88–89, 102, 202
Pedagogic Museum St. Petersburg, 90–91, 97–98, 103–104, 123, 127–128, 281
Peter, Pyotr Alekseyevich Romanov, tsar, 15, 18–19, 48, 59, 89
Philips
 Anton and Gerard, 49, 251
 Natlab (Natuurkundig Laboratorium), 318, 324
Pinkevich, Albert Petrovich, 261
Pirogov, Nikolay, 17
Planck, Max, 25, 53–54, 71–73, 95, 109–110, 111–114, 118, 132–134, 139, 150, 166–167, 176–177, 198, 206–207, 209, 226, 269, 274
Poincaré, Henri, 110–112, 113–116
Popov, Alexander, 34
Posthumus Meyjes, Nelly, 286–287, 288–290, 295, 299–300, 327–328
Prandtl, Ludwig, 42–44
probability theory, 69, 88–89, 96, 102, 104, 107, 123, 145, 162–163, 227, 319, 322, 324

quanta, 63, 72–73, 109–113, 118, 131, 149, 157–158, 164–165, 201, 233, 265, 288, 323
quantum mechanics, 63, 187–279, 282, 285, 315, 323, 325
quantum theory. *See* Bohr's, (old) quantum theory

Ramsey, Gordon, 247–248. *See also* negative absolute temperatures
random motion, 62, 131, 145, 162–163
Rayleigh Jeans ultraviolet catastrophe, 72–73, 111
Rayleigh, John William Strutt, Baron Lord, 53–54, 72, 96, 115. *See also* Rayleigh Jeans ultraviolet catastrophe

Reading Room for Mathematics and Physics in Leiden (Bosscha Library), 148–150, 153, 158, 184, 318
recurrence paradox, 78–79, 325–326
relativity theory
 general relativity, 36, 155, 159–160, 168–169, 172, 173–174, 176–177, 194–195, 201, 203–204, 206–207, 216–217, 223, 256–257, 288, 323
 special relativity, 63–64, 75–77, 95, 98, 101, 119, 121, 133, 143–144, 150–151, 205, 323
reversibility paradox, 78, 325–326
Ritz, Walter, 35, 39–40, 45f, 47–49, 51–54, 59, 65–66, 73, 100–101, 115–116, 129–130, 142, 144, 155, 178, 181
Röntgen, Wilhelm, 121–122, 166
Roosenschoon, Nettie, 188–189

Schaap, Mau en Bep, 41
Schiff, Vera Iosifovna, 27, 30, 32–33, 41–42, 45–46, 79, 85, 97
Schmoluchowski, Marian, 122, 162–163
Schrödinger, Erwin, 178–179, 266–269, 274, 286–287, 291–292, 298, 323
Schwab, Arthur, 230, 234, 280
Schwab, Teddy. *See* Bryner, Teddy
Selivanov, Dmitry Fyodorovich, 27, 78, 96–97, 98–99
Shubnikov, Lev, 267, 272, 279, 283, 318
Shubnikov, Olga Trapeznikova. *See* Trapeznikova, Olga
Sitter, Willem de, 144, 151, 152–153, 169–170, 172, 173–174, 194–195, 198, 206–207, 255, 297–298
Skłodowska Curie, Marie, 110, 203, 206, 216, 217, 255, 256–257, 267, 272, 318–319
Solvay
 Conferences, 113, 118, 129–132, 150, 157–158, 166–167, 203, 206–209, 215, 229, 256–257, 268, 297, 303
 Foundation and Institute, 128, 130, 157–158
Solvay, Ernest, 109–110, 112, 115, 157–158, 166–167, 215–216, 246–247
Sommerfeld, Arnold, 73, 77–78, 109–110, 112, 114–115, 116–118, 129–130, 131–134, 141, 165, 176–177, 185, 191, 201, 206–207, 222, 224, 253–255, 265–266
spectral lines, 51, 192, 201, 224, 241, 254, 291. *See also* atomic spectrum
Spin. *See* electron spin
Sponer, Hertha, 240–241
Stalin, Joseph, 281, 289–290, 304, 318
Stark effect, 291

Stark, Johannes, 35, 42–43, 216–217, 291–292. *See also* Stark effect
statistical mechanics. *See* Boltzmann's, statistical mechanics
Stern-Gerlach experiment, 218, 229
Stern, Otto, 218. *See also* Stern-Gerlach experiment
Stokes, George, 142–143, 150
Stolypin, Pyotr, 80, 89, 104–105
Struik, Dirk, 158, 161, 172, 175–176, 177–178, 180, 225, 252, 306, 318, 323–324

Tamarkin, Yakov Davidovich, 88, 91, 200–201
Tamm, Igor, 272, 283, 318
theory of probability. *See* probability theory
thermodynamics
  equilibrium states in, 25, 69–71, 78–79, 141, 226, 247, 314, 325–326
  first law of, 246
  second law of, 24–25, 78–79, 111, 228, 246–248. *See also* recurrence paradox; reversibility paradox
Thomson, J.J., 35, 51, 150
Thorbecke, Johan, 48–49
Tietze, Heinrich, 22, 59, 68, 87, 100–101, 112, 118–119
Timoshenko, Stephan, 107
Tinbergen, Jan, 224–226, 251–253, 262, 273, 276–277, 281, 295, 314, 315, 319, 323–324, 325–326
Tolman, Richard, 175–176, 178, 296
Tolstoy, count Ivan (minister of education in Russia, 1905–1906), 80, 83
Tolstoy, Dmitry (minister of education in Russia, 1866–1880), 18
Tolstoy, Leo, 30, 51, 57, 84, 91, 107–108, 126–127, 155, 186, 242, 326. *See also* Tolstoy family estate

Tolstoy family estate, 123–125
Trapeznikova, Olga, 267, 272, 283, 318
Trüper Institute in Jena, 211, 212–213, 227–228, 276–277, 293, 325

Übungensammlung, 277–278, 288, 306, 319–320
Uhlenbeck, George, 204–207, 224, 251–255, 257*f*, 265, 267, 278–279

Vegetarianism, 121–122, 132, 176, 186, 326, 328
Voigt, Waldemar, 35, 38
Vollenhoven, Cornelis van, 198–199, 216
Vosmaer, Carel, 106

Waals, Johannes van der, 49, 54, 131, 141, 144, 248
Waterink Institute, 293–294
Weierstrass, Carl, 27, 42, 97
Weiss, Pierre, 116, 122, 215–216, 251
Westerweel, Joop, 311
Wien's displacement law, 111
Wien, Wilhelm, 53–54, 110–111, 115, 118, 166. *See also* Wien's displacement law
Wiersma, Eli, 241, 279, 295, 318
Wolfskehl Lectures, 150, 186
Woude, Adriaan van der, 275–276, 290

Zapolskaya, Lyubov, 42, 44–46
Zeeman effect, 50–51, 254, 291
Zeeman, Pieter, 50–51, 54, 191–192, 198, 226, 249, 252, 254, 292, 294, 305–306. *See also* Zeeman effect
Zeeman, Pieter (mathematician, *nephew* of Pieter Zeeman), 148, 170
Zermelo, Ernst, 25, 38, 78–79. *See also* recurrence paradox
Zwicky, Fritz, 280, 312

The manufacturer's authorised representative in the EU for product safety is
Oxford University Press España S.A. of el Parque Empresarial San Fernando de
Henares, Avenida de Castilla, 2 – 28830 Madrid (www.oup.es/en or product.
safety@oup.com). OUP España S.A. also acts as importer into Spain of products
made by the manufacturer.

www.ingramcontent.com/pod-product-compliance
Ingram Content Group UK Ltd.
Pitfield, Milton Keynes, MK11 3LW, UK
UKHW021142260426
470302UK00029B/73/J